The Structure of Matter:

A Survey of Modern Physics

The Structure of Matter:

A Survey of Modern Physics

STEPHEN GASIOROWICZ

University of Minnesota

ADDISON-WESLEY PUBLISHING COMPANY

Reading, Massachusetts · Menlo Park, California
London · Amsterdam · Don Mills, Ontario · Sydney

This book is in the
ADDISON-WESLEY SERIES IN PHYSICS

Library of Congress Cataloging in Publication Data

Gasiorowicz, Stephen.
    The structure of matter: a survey of modern physics

    Bibliography: p.503
    Includes index.
    1. Matter--Constitution. 2. Physics. I. Title.
QC173.G348   530.1   78-18645
ISBN 0-201-02511-6

ISBN 0–201–02511–6
BCDEFGHIJK–MA–89876543210

For Hilde

Preface

The Structure of Matter grew from lecture notes prepared for a one-year course on modern physics, taken by university juniors majoring in one of the science or engineering fields. My aim in designing the course was to achieve the following goals:

a) To expose the students to the new ideas ("new" to the students with their one- to two-year background in classical physics) of relativity, statistical physics, and quantum physics.

b) To keep the mathematical level of the technical treatment as low as possible so as to avoid putting a double burden on the student, who has to assimilate a large number of ideas in the course.

c) To stress the importance of quantitative aspects of physics, and to teach the students the importance of, and the possibility of, getting a roughly correct answer to simple questions without the full use of the machinery of quantum mechanics.

d) To discuss a large number of phenomena whose explanation lies in quantum physics, but whose manifestation looms very large in present research areas.

Let me elaborate: the students for whom this book is written are expected to have, as their physics background, a one-year course on calculus-based general physics, and, preferably, a course on wave phenomena. Their mathematical background should be a year of calculus, some acquaintance with simple differential equations, and a quarter course on algebra, that is, vectors, linear transformations, and matrices. In general, such students will not have had any exposure to the theory of relativity, and the first eight chapters deal with that subject.

The mathematics of the theory of relativity is very simple, and I have kept the development at a simple level. Much attention is paid to the basic ideas of relativity, and the main phenomena are discussed on three levels: first, using clocks and rods, second, using the formulas of the Lorentz transformation, and third, using Minkowski diagrams. Relativity finds a very important application in the kinematics of elementary particle physics scattering and production, and a chapter is devoted to that subject. The equivalence principle can also be discussed at the mathematical level reached by the students, and Chapter 8 deals with some aspects of Einstein's theory of gravitation. The chapter on Relativity and Electromagnetism was sometimes omitted from the course, the material being left to the student for reading on the "cultural" rather than the "essential" level. This separation is not in any way stressed in the text: only the *Notes and Comments* at the end of each chapter are distinctly cultural—they are a substitute for conversations with the interested student.

Statistical physics is included in the book because it is important to teach students something about probability theory and about statistical reasoning. The main accomplishment, using the ideas developed in Chapter 9, is the derivation of the ubiquitous Boltzmann factor $e^{-E/kT}$ whose importance in physics, chemistry, and biology cannot be overestimated. A brief discussion of kinetic theory is essential to an understanding of the structure of matter, and a large number of applications, ranging from the barometer formula to Brownian motion and equipartition, are discussed in the last chapter of Part II. There is usually no time to discuss all of the applications, and the division into essential and supplemental is left up to the individual instructor.

Quantum mechanics is the method that must be used for the discussion of the structure of matter. The theory grew out of radiation theory and out of classical electron theory. These roots of quantum physics are discussed in Part III on the Old Quantum Theory. A special feature of this section is a more detailed than usual discussion of the Lorentz electron theory, which still provides an excellent qualitative understanding of many optical phenomena.

Our introduction to quantum mechanics deals primarily with one-dimensional problems, so that the mathematical complications are not allowed to obscure the important new physical ideas. A great deal of attention is paid to the role of the uncertainty relations in the interpretation of quantum mechanical results. New phenomena, such as barrier penetration, are discussed in detail, with attention paid to physical applications and to numerical estimates. The harmonic-oscillator problem is solved in detail, providing instruction in the handling of ordinary differential equations. The discussion of the two-particle Schrödinger equation brings with it a totally new quantum phenomenon, the Pauli Exclusion Principle, which is discussed in great detail and illustrated by a discussion of the Fermi sea and neutron stars.

With the introduction of the quantum mechanical angular momentum, the mathematical level goes somewhat above what has been required heretofore: in my experience, angular momentum and spin are perhaps the hardest subjects to develop a feeling for at this level. Given the facts about angular momentum, the discussion of the hydrogen atom is relatively straightforward, in that it follows the steps that appeared in the solution of the harmonic oscillator. The section on the more advanced topics in quantum mechanics ends with a series of topics in which the manifestations of spin in a variety of physical systems is discussed. This material is treated more qualitatively, and falls into the cultural category, although some of the material appears later in the discussion of atoms and molecules.

In Part VI of the book a variety of physical systems is discussed semiquantitatively. Here I have tried to build on the exact results obtained for simple one-dimensional systems and to get a feeling for the magnitudes involved. The structure of atoms can be discussed with the help of what is known about the hydrogen atom and the exclusion principle, and that approach easily extends to a discussion of simple molecules. The new aspect of rotational and vibrational motion, and associated spectra, is discussed, as is the Raman effect. In Chapter 33 the mathematical treatment becomes a little more intense again, with a brief discussion of time-dependent perturbation theory. Much of this material can be omitted, but the notion of matrix element that appears here is important, since that is used to study selection rules. Radioactivity and applications are also discussed in this chapter. The development naturally leads to a discussion of stimulated emission, and a more thorough than usual discussion of lasers, and their uses.

The remaining chapters of the book deal with (i) solid state physics, (ii) nuclear physics, and (iii) elementary partical physics. It is obviously impossible to cover more than the basic phenomena in a survey course, and even then, the material is more than can be covered in a reasonable allotment of time. My own interests have dictated a light coverage of the first two topics, and a detailed discussion of the third, but other instructors may make different choices and assign the additional material as cultural reading. The solid state section deals with crystal lattices, specific heats of solids (the Einstein and Debye theories), and the Mössbauer effect; with metals and the Fermi distribution function, electronic specific heat; and with semiconductors and superconductors. The material discussed in the nuclear physics section includes the semi-empirical mass formula, the liquid drop model, and the shell model, with a brief discussion of collective effects. The section on elementary particle physics is, because of my own interests, somewhat more extensive. It covers antiparticles, neutrinos, and beta decay, the strong interactions and their symmetries, and the quark model, including the notions of "flavor" and "color."

The transition from lecture notes to book manuscript was much aided by the useful comments of Professors A. Goldman and R. N. Dexter; by some general suggestions of Professor Paul Stoler, and above all, by the very detailed criticism and the countless suggestions for improvements that were provided by Professor Herbert Kabat. I am very grateful to all of them, and to the individuals and institutions who have kindly consented to the reproduction of figures. My special thanks go to the students of the modern physics course that I taught for four years: their excitement about the new ideas of relativity and quantum physics, and their healthy skepticism strongly motivated me to write and rewrite the notes on which this book is based.

Minneapolis, Minnesota   S.G.
February 1979

Contents

PART I
Relativity

The bulk of this book deals with the structure of matter and the understanding that quantum mechanics provides of it. It is nevertheless impossible, even though we live in a world in which most atomic constituents move with velocities very much smaller than that of light, to ignore the theory of relativity, as the view of space-time ushered in by its discovery has become an integral part of every physicist's thinking about the world of matter. In particular, relativity plays a crucial part in our thinking about the fundamental particles. I have therefore chosen to begin this book with a section on relativity. In keeping with the assumed level of mathematical training of the potential reader, I have kept the mathematics as simple as I could. No tensor notation is used, and the four-dimensional treatment of electromagnetism is only briefly discussed. I have not discussed the energy-momentum tensor, and the Lorentz transformation laws for quantities more complicated than four-vectors. The same mathematical inhibitions, together with a lack of space (and time in lecturing) have kept me from a more thorough discussion of the elements of the Einstein theory of gravitation. Nevertheless, I feel that the topics covered do provide an adequate undergraduate education in this fascinating subject. The bibliography at the end of the book attempts to point the student and the instructor in directions of deeper coverage.

1
The Relativity Postulates

Galilean relativity Observers find it convenient to use reference frames to describe physical phenomena. An important question is how different observers describe the same phenomena, and the theory of relativity is concerned with this question. An *event*, such as a collision between two infinitesimally small bodies, the absorption of a flash of light by a small body, an explosion, and so on, can be described by its location, and the time at which it is observed to occur. The location may be described by the coordinates (x, y, z) relative to a set of coordinate axes, called the reference frame. Newton's first law of motion singles out a set of frames. In these frames, called *inertial frames*, the first law, according to which an absence of forces implies uniform motion, is true. There are frames in which the first law does not hold: a body at rest relative to a rotating frame, for example on a turntable, will experience an acceleration. It is conventional, to preserve the equation

$$\text{Force} = \text{mass} \times \text{acceleration}$$

to introduce fictitious centrifugal forces, but these forces are not due to any external agency; they reflect noninertial effects.

The laws of mechanics are described by an equation of the type

$$M_i \frac{d^2 \mathbf{r}_i}{dt^2} = \mathbf{F}_i \tag{1.1}$$

with the force, if it is due to another particle, for example, of the form (Fig. 1.1)

$$\mathbf{F}_i = \mathbf{F}_i(\mathbf{r}_i - \mathbf{r}_j). \tag{1.2}$$

It should be stressed that all known forces, gravitational, electromagnetic, nuclear, and weak have the characteristic that they only depend on the *relative* location of the source of the force and the body acted on, and not on the location of the body relative to some "center of the universe."

An observer in a frame moving at a uniform velocity relative to the original frame in which the coordinates of the two particles in interaction are $\mathbf{r}_1$ and $\mathbf{r}_2$, say, will assign coordinates (Fig. 1.2)

$$\begin{aligned} \mathbf{r}_1' &= \mathbf{r}_1 - \mathbf{u}t \\ \mathbf{r}_2' &= \mathbf{r}_2 - \mathbf{u}t \end{aligned} \tag{1.3}$$

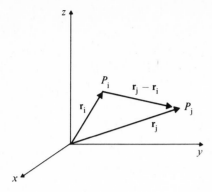

Figure 1.1. Vectors describing the positions and the separation between two interacting particles.

to the particles, provided the two frames coincided at $t = 0$ and the observer is moving with velocity **u** relative to the initial frame. The new observer will see different velocities for the particles

$$\frac{d\mathbf{r}'}{dt} = \frac{d\mathbf{r}}{dt} - \mathbf{u} \qquad (1.4)$$

but the accelerations are the same for constant **u**, since

$$\frac{d^2\mathbf{r}'}{dt^2} = \frac{d^2\mathbf{r}}{dt^2}. \qquad (1.5)$$

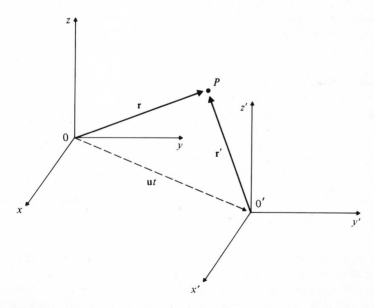

Figure 1.2. The primed frame moves with velocity **u** relative to the unprimed frame, and they coincide at $t = 0$. The point P is described differently relative to the two coordinate frames. The figure shows how $\mathbf{r}' = \mathbf{r} - \mathbf{u}t$.

Also, since
$$\mathbf{r}'_1 - \mathbf{r}'_2 = \mathbf{r}_1 - \mathbf{r}_2, \tag{1.6}$$
the force, which only depends on the difference, is unchanged under the transformation. Thus the observer would write down exactly the same equation of motion. What we have shown is that

the laws of mechanics are form-invariant

under the Galilean transformation

$$\mathbf{r} \rightarrow \mathbf{r}' = \mathbf{r} - \mathbf{u}t$$

or, equivalently,

the laws of mechanics are the same in

all inertial frames.

We do not assert that the solutions of the equations of motion are identical. To determine the motion we need initial conditions, and Eq.(1.4) shows that a given initial velocity in the unprimed frame differs from the initial velocity in the primed frame. What is important is that upon examination of the trajectories of the particles in interaction, an observer should not be able to distinguish which inertial frame is singled out. There is no singling out if the differences in trajectories can be ascribed to different initial conditions.

We do not know at this time what determines an inertial frame. Newton worried about this problem and satisfied himself with the assertion that there must exist an "absolute space," and inertial frames are those that are at rest or in a state of uniform motion relative to it. The Austrian physicist and philosopher Ernest Mach had another point of view, namely that inertial frames are somehow determined by the distribution of matter in the universe. There is a simple experiment that brings out this point: the surface of water in a bucket is flat when the bucket is at rest relative to the stars in the sky, and it is curved when the stars are rotating relative to the bucket. Is the acceleration of the distant stars the source of the noninertial forces? Mach believed that it is, and his point of view is one that is very attractive to many cosmologists. It is, however, difficult to incorporate Mach's Principle into a model of the universe.

The aether With the development of the laws of electrodynamics by James Clerk Maxwell, a crisis of sorts arose. The laws of electrodynamics were not invariant under the Galilean transformation
$$\mathbf{r} \rightarrow \mathbf{r}' = \mathbf{r} - \mathbf{u}t \tag{1.7}$$
according to which the velocity undergoes the transformation
$$\mathbf{v} \rightarrow \mathbf{v}' = \mathbf{v} - \mathbf{u}. \tag{1.8}$$
The most direct way to see this without actually examining Maxwell's equations is to recall that these equations predict that electromagnetic radiation propagates with the velocity $c = 3 \times 10^{10}$ cm/sec. We may ask: velocity relative to what? It is not correct to say that this is the velocity relative to the source of the radiation; that, at least, is *not* what Maxwell's equations say. An answer that had won a certain amount of acceptance in the latter part of the nineteenth century was that c was the velocity relative to the *aether*, a substance that filled the universe and made wave propagation possible. The aether was favored by old models of wave motion, according to which waves

could only propagate in a medium, in which there was something that could be induced to oscillate by the action of the source. The medium was recognized to be very peculiar, since it only sustained transverse oscillations, but the assumption of its existence removed the possibility of invariance of the laws of electromagnetism under the galilean transformations. Since the aether provided a preferred frame, it removed the problem.

It seemed reasonable to assume that the aether was not tied in its motion to the earth, and it then became an interesting question of what the velocity of the earth relative to the aether was. The velocity of the earth relative to the sun is approximately 30 km/sec (the distance to the sun is approximately 1.5×10^8 km and it takes a year to get around a nearly circular orbit), and one might expect that an "aether wind" of such speed might be detectable. In 1887 A.A. Michelson and E.W. Morley carried out the first of a series of brilliant experiments designed to measure the speed of this aether wind, and they found the remarkable result that the velocity of light is the same for light traveling along the earth's orbital motion as for light moving transversely to that direction. This equality has now been established to within 1 km/sec.

Michelson-Morley experiment The idea of the experiment designed by Michelson and Morley is the following: consider light moving from a point P to a mirror M_1 (Fig. 1.3) along the east-west axis, and assume the aether wind is blowing from the east with speed v. The propagation of light against the wind proceeds with speed $c - v$; the speed is $c + v$ when the light goes with the wind. Thus the time of flight from P to M_1 and back is

$$t_1 = \frac{l_1}{c+v} + \frac{l_1}{c-v} = \frac{2l_1 c}{c^2 - v^2} \approx \frac{2l_1}{c}\left(1 + \frac{v^2}{c^2}\right) \qquad (1.9)$$

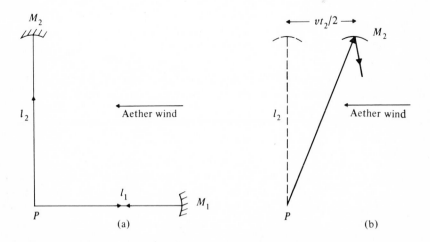

Figure 1.3. The optical paths in the Michelson Morley interferometer. (a) The location of the mirrors and the effective source P relative to the direction of the aether wind. (b) The actual direction that the light ray must travel in the aether wind to travel from P to M_2 and back.

where l_1 is the distance from P to M_1. When the light is moving transversely to the wind, a calculation of the time of flight from the point P to the mirror M_2 and back to P must take into account the delay due to the fact that the light is effectively being "blown off course." If the time of flight is t_2, that is, $t_2/2$ for the one-way trip, the distance covered is really the hypotenuse of a triangle whose sides are l_2 and $vt_2/2$, respectively. Thus

$$ct_2/2 = [l_2^2 + (vt_2/2)^2]^{1/2}. \tag{1.10}$$

Hence

$$c^2 t_2^2 = 4l_2^2 + v^2 t_2^2,$$

and therefore the time t_2 is given by the formula

$$t_2 = \frac{2l_2}{c} \frac{1}{(1-v^2/c^2)^{1/2}} \approx \frac{2l_2}{c}\left(1 + \frac{1}{2}\frac{v^2}{c^2}\right). \tag{1.11}$$

The time difference is given by

$$\Delta_{12} = t_1 - t_2 = \frac{2}{c}(l_1 - l_2) + \frac{v^2}{c^2}\left(\frac{2l_1}{c} - \frac{l_2}{c}\right). \tag{1.12}$$

Suppose we now rotate the apparatus through $90°$ so that PM_2 now points against the aether wind, and PM_1 points southward. We are now effectively interchanging l_1 and l_2, and, taking into account that we always want the time difference (parallel to wind) $-$ (transverse to wind), we get

$$\Delta_{12}^{\text{rot}} = \frac{2}{c}(l_1 - l_2) + \frac{v^2}{c^2}\left(\frac{l_1}{c} - \frac{2l_2}{c}\right). \tag{1.13}$$

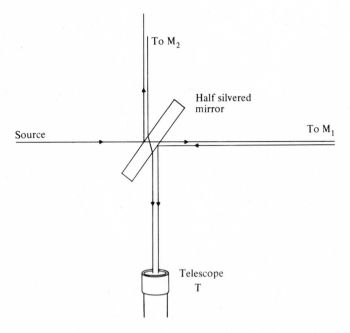

Figure 1.4. The schematics of the Michelson-Morley experiment.

The change in the time difference as a result of the rotation is

$$\Delta_{12} - \Delta_{12}^{\text{rot}} = \frac{v^2}{c^2} \frac{l_1 + l_2}{c} \equiv \Delta t \tag{1.14}$$

which is very tiny indeed, being proportional to $(v/c)^2$. If we make a reasonable guess that $v \approx 3 \times 10^6$ cm/sec, then $(v/c)^2 \approx 10^{-8}$. Such an accurate measurement became possible through the use of an optical device, the interferometer, designed by Michelson. A beam of light is split in two by a half-silvered mirror (Fig. 1.4); part of the beam is reflected by M_2 and part is reflected by M_1. The two parts of the beam recombine and are observed with the telescope T. Because the wavefronts are not exactly parallel (it is impossible to make the mirrors exactly perpendicular to each other, and so on) interference fringes will appear in the field of vision of the telescope. If care is taken that the mirrors and the source are well stabilized, the interference fringes will not change. Upon rotation of the apparatus, a change in the difference of the two path lengths of the beams will be introduced. That change is $c\Delta t$, and if the light has wavelength λ, there will be a shift in the pattern of interference fringes (this is equivalent to moving one mirror a little). The number of fringes shifted is

$$\Delta n = \frac{c\Delta t}{\lambda} = \frac{l_1 + l_2}{\lambda} \left(\frac{v}{c}\right)^2. \tag{1.15}$$

Michelson and Morley used light with wavelength 5.9×10^{-5} cm ($= 5900$ Å) and their apparatus was such that $l_1 \approx l_2 \approx 11$ meters. Thus the expected shift was

$$\Delta n \simeq \frac{2.2 \times 10^3}{5.9 \times 10^{-5}} (10^{-4})^2 \simeq 0.37.$$

A shift of as few as 0.04 fringes could have been detected, and *the absence of any observable shift* was a remarkable result.

Stellar aberration A possible way to evade the conclusion that there was no aether could be the assertion that perhaps the aether is locally "dragged" by the earth, and thus more or less at rest relative to the terrestrial laboratory. This explanation is in conflict with the observation of *stellar aberration*. Consider the observation of a distant star with a telescope (Fig. 1.5). If the earth were at rest relative to the distant star, and the star were directly overhead, the telescope would have to be pointed straight up. With the earth moving with speed v, it is evident that the telescope must be tilted at a slight angle relative to the original vertical direction to allow the light from the star to pass through the telescope without hitting the sides. (An everyday analogy is the tilt of an umbrella carried by a running pedestrian in the rain on a windless day). The light, as seen by the observer, is emitted by a source moving with horizontal velocity v, so that the components of the light velocity are $c \sin \theta = v$ and $c \cos \theta$. For $v/c \ll 1$ the tilt angle is $\theta \simeq v/c$. One cannot, of course, be sure that a given star is exactly overhead, but, whatever its azimuth, the earth will be moving toward it at some time of the year, and away from it six months later. Thus a maximum change in the angle, given by

$$\Delta\theta \simeq 2\frac{v}{c} \simeq 2 \times 10^{-4} \text{ radians} = 44''$$

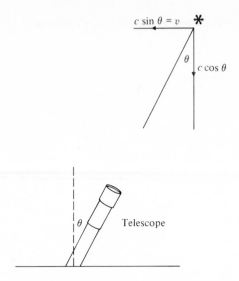

Figure 1.5. Aberration of light. The angle θ measures the tilt of the telescope necessary to avoid having the light strike the walls of the telescope.

is expected, and is in fact observed. These observations were first made by James Bradley in 1728, and provided clear evidence for the motion of the earth around the sun. Were the aether to be dragged by the earth out to some distance, the maximum angle would be different in some model-dependent way (in terms of our analogy, the angle of the umbrella would change if the running pedestrian were accompanied by a wind in his or her vicinity).

Lorentz-Fitzgerald contraction Faced with the need to find another explanation for the results of the Michelson-Morley experiment, G.F. Fitzgerald, and independently, H.A. Lorentz (perhaps the leading theoretical physicist of the latter part of the nineteenth century) proposed an ingenious, though *ad hoc* solution. This was, that as a result of some mechanism intrinsic to the molecular forces that bind the atomic constituents of matter, materials in motion with velocity v through the aether should be contracted in length in the direction of motion by a factor of $(1 - v^2/c^2)^{1/2}$. It is clear from our discussion of the experiment of Michelson and Morley that a replacement of l_1 by $l_1(1 - v^2/c^2)^{1/2}$ and a replacement of l_2 by $l_2(1 - v^2/c^2)^{1/2}$ in the two orientations, respectively, will lead to the observed null result. This explanation was not satisfying in that, without an explanation of the contraction of moving bodies, it merely replaced one mystery by another. We shall soon see that in a sense it was technically the correct explanation, but it did leave the aether untouched, though unobservable.

The Einstein postulates Apparently uninfluenced by the surprising outcome of the Michelson-Morley experiment, Albert Einstein in 1905 culminated his researches into the nature of time and motion by a most remarkable paper, in which he postulated that

I. *All laws of physics are the same in all inertial frames,* and

II. *The speed of light, taken to be the maximum speed of propagation of a signal, is the same in all inertial frames.*

The first postulate maintains the integrity of the belief in the relativity of motion; the second postulate, in effect, restates in generalized form the results of Michelson and Morley. This postulate, by contradicting the general validity of invariance under Galilean transformations, that is, invariance under

$$\mathbf{v} \rightarrow \mathbf{v}' = \mathbf{v} - \mathbf{u},$$

forces us to modify the laws of mechanics. The modification must, of course, be consistent with Newtonian mechanics in the domain where $(v/c)^2$ effects can be neglected. We shall soon see that this modification can indeed be carried out.

NOTES AND COMMENTS

1. In the interest of brevity only the Michelson-Morley experiment was described in this chapter. There were earlier attempts to detect motion through the aether, and these are discussed in most textbooks on Special Relativity. I found the discussion in A.P. French, *Special Relativity*, W.W. Norton, New York, 1968, very clear.

2. Einstein's paper which appeared in *Annalen der Physik*, *17*, p. 891 (1905) appears in translation in *The Principles of Relativity*, Dover Publications, 1951, together with some of the other classic papers on special as well as general relativity. It is a remarkably readable paper.

3. Some historical background on the theory of relativity, and in particular on the question of the significance of the Michelson-Morley experiment for the actual development of the theory may be found in an extremely stimulating series of essays in *Thematic Origins of Scientific Thought, Kepler to Einstein*, by Gerald Holton, Harvard University Press, Cambridge, Massachusetts, 1973. Biographical material may be found in many books. I particularly enjoyed *Albert Einstein, Creator and Rebel*, by Banesh Hoffmann, New American Library, New York, 1972 and *Einstein* by Jeremy Bernstein, Viking Press, New York, 1973.

4. One might be tempted to ask the question: "Why is the maximum speed of propagation also the speed of light?" I do not know whether I have a really good answer to that. When the notion of rest mass is discussed in Chapter 5 we shall see that it is objects with zero rest mass that move with the maximum speed. It is generally believed that both electromagnetic radiation and neutrinos move in this way, and there are some deep reasons for believing that this *must* be so for radiation. The question "why is the speed of light 3×10^{10} cm/sec?" has no deep meaning. The particular value happens to be a consequence of our choice of unit of length in terms of the circumference of the earth, and the unit of time in terms of how long it takes the earth to go around the sun. A sensible set of units would be one in which $c = 1$. Thus if we insist on the meter as the unit of length, a

sensible set of relativistic units would have the time unit, the "tick," say, be approximately 3.3×10^{-9} sec. Astronomers keep the year as a unit of time for certain matters, and define the unit of distance as the light-year, which is roughly 9.5×10^{15} meters.

5.   A useful mnemonic is
$$1 \text{ year} \approx \pi \times 10^7 \text{ sec.}$$

6.   A modern version of the Michelson-Morley experiment was performed with lasers. In this experiment one actually carried out a measurement of a "classical" Doppler shift (see Chapter 2) of light moving with the aether wind relative to light moving across the aether wind. The absence of a shift yields the same conclusions as the Michelson-Morley experiment, but the accuracy was such that the effect was less than 10^{-3} of the classically expected effect, compared with the less than 10% effect in the original experiment. The laser experiment was performed by T.S. Jaseja, A. Javan, J. Murray, and C.H. Townes, *Physical Review*, 133 A, 1221 (1964).

7.   It is important to remember that it is only the speed of light in a vacuum that is to be identified with the maximum speed. Light in a medium travels with speed c/n, where n is the refractive index, and in such a medium an electron, for example, can travel faster than that. When its speed exceeds c/n, it emits radiation, called Čerenkov radiation after its discoverer. The radiation is emitted in a forward cone of half angle given by $\cos \theta = c/nv$, where v is the electron speed.

8.   It is very difficult to test Mach's Principle. A possible consequence might be that different cosmological conditions aeons ago implied different values for physical constants such as the electric charge and the gravitational constant, G. Such differences would have observable effects on planetary history. The case for such changes in G is made in F. Hoyle *From Stonehenge to Modern Cosmology*, W.H. Freeman, San Francisco, 1969, but I believe that there is fairly universal scepticism about the conclusions among astronomers. A recent experiment on the measurement of two spectral lines from a quasar with a red shift of the order of 0.524 gives us a measure of how the two spectral lines were related thousands of millions of years ago. The upper limit on the change in the electric charge is of the order of 10^{-12} per year.

9.   Recently observed pulses from neutron stars (pulsars) in double star formations emitting X-rays (binary x-ray sources) have been used to show that the speed of light is independent of the velocity of the source to an accuracy of better than one part in 10^9. The most recent summary of tests of Special Relativity may be found in D. Newman, G.W. Ford, A. Rich, and E. Sweetman, *Physical Review Letters*, **40**, 1355 (1978). This reference is probably more useful for instructors than for students.

PROBLEMS

1.1 Consider a collision between two objects, in which both energy and momentum are conserved. Show that an observer in a frame moving with velocity v relative to the original frame also sees energy and momentum conserved, provided there is conservation of mass in the process.

1.2 A deuteron is a bound state of a proton and a neutron, with binding energy of 2.2 MeV. What is the minimum kinetic energy a proton must have when it is incident on a deuteron at rest in order that the deuteron can break up in the collision? What is it when the deuteron is incident on a proton at rest?

1.3 Consider light propagating from a point A to a mirror M and back, given that there is an aether wind of velocity v making an angle θ with the line MA. What is the time of travel $(A \rightarrow M \rightarrow A)$ given that the distance AM is L? Show that your results go over into the longitudinal and transverse results of the Michelson-Morley experiment discussion when $\theta = 0°$ and $90°$, respectively.

1.4 Discuss stellar aberration when the star is not vertically overhead. Suppose a star in the sky is observed from earth to describe an elliptical path. If the star makes an angle of $75°$ with the plane of the earth's motion, what is the angular variation of the altitudes of the star?

1.5 The Lorentz force experienced by a particle of charge q moving with velocity $\mathbf{v}$ in a uniform electric and magnetic field $(\mathbf{E}, \mathbf{B})$ is
$$\mathbf{F} = q(\mathbf{E} + \mathbf{v} \times \mathbf{B}/c).$$
If the expression of the force is invariant under Galilean transformations, what are the transformation laws for $\mathbf{E}$ and $\mathbf{B}$, that is, given $\mathbf{E}$ and $\mathbf{B}$ in one frame, what are $\mathbf{E}'$ and $\mathbf{B}'$ in another frame?

1.6 Two identical particles, each of mass M are moving toward each other with velocity v and $-v$, respectively. After the collision they move away from each other along a line making an angle θ^* with the original line of motion. Assuming Galilean invariance, compute the angle of deflection θ of one of the particles, as seen by an observer at rest relative to the other particle during the initial motion.

2
Kinematical Consequences
of the Einstein Postulates

Time in relativity Implicit in the Galilean transformation law between inertial frames is the notion that there is no transformation for the time t, that is, one should add the rule

$$t' = t \tag{2.1}$$

to

$$\mathbf{r}' = \mathbf{r} - \mathbf{u}t. \tag{2.2}$$

We immediately observe that the principle of the constancy of the speed of light is incompatible with the above. Consider two frames S and S' which coincide at $t = 0$, at which time a pulse of light is emitted at the origin. In the frame S the spherical front of the pulse is described by the equation

$$x^2 + y^2 + z^2 = \mathbf{r}^2 = c^2 t^2. \tag{2.3}$$

In the primed system it should also be described by

$$\mathbf{r}'^2 = c^2 t'^2, \tag{2.4}$$

and if we set $t' = t$ we cannot use the transformation (2.2). Thus the relation $t' = t$ must be modified. Any such modification will have consequences for the notion of simultaneity of events — for the notion "these things happened at the same time" — as seen by different observers. This, and other consequences of the Einstein postulates can be discussed without knowing the modified transformation laws. Such a discussion was carried out in Einstein's original paper, and is a fascinating aspect of that work.

Simultaneity To speak of simultaneity, we must be able to synchronize clocks. Even though signals have a maximum velocity, this is easily done. Thus, if the distance between the source of the signal A and the receiver B is known, then the light is received by B at a time d_{AB}/c later, and B's clock can be set accordingly. Consider

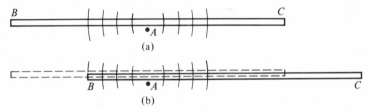

Figure 2.1. The relativity of simultaneity (a) Signals leaving A reach B and C simultaneously. (b) With train moving signals reach B before they do C.

now the situation shown in Fig. 2.1. If B and C are at rest relative to A, then pulses emitted by A will reach B and C at the same time, that is, when both clock B and clock C read the same, provided $d_{AB} = d_{AC}$. Suppose B and C are put at the rear and front end of a train, respectively, and that the train is now moving with velocity v to the right, relative to A. The time at which the signal reaches B will be earlier, since B effectively comes forward to meet the signal. The distance that the light travels in time t_B is $L/2$ (where L is the length of the train) less the distance that B came forward, that is, $L/2 - vt_B$. Since light moves with the speed c, this equals ct_B, so that

$$t_B = \frac{L/2}{c + v}. \tag{2.5}$$

On the other hand, the distance that the light must cover to get to C is $L/2 + vt_C$ and this must equal ct_C, so that

$$t_C = \frac{L/2}{c - v}. \tag{2.6}$$

Thus $t_B \neq t_C$ and events that are simultaneous in the frame in which the train is at rest relative to A, are no longer so in other frames. Note that

$$t_C - t_B = \frac{L}{2}\left(\frac{1}{c - v} - \frac{1}{c + v}\right) = \frac{Lv}{c^2 - v^2} \tag{2.7}$$

so that these effects are not observable unless v is close to c.

More dramatic are the strange consequences that a clock moving with velocity v relative to an observer is measured to tick more slowly by a factor $(1 - v^2/c^2)^{1/2}$, and a body of length L seen by an observer moving with velocity v relative to the body (along the dimension L) is contracted to $L(1 - v^2/c^2)^{1/2}$. Let us discuss these effects:

The possibility of contractions of lengths, suggested by Fitzgerald and Lorentz implies that we must be careful about lengths, just as we were about simultaneity. Consider two rods parallel to each other, in relative motion with respect to each other, with the relative velocity perpendicular to the rods. When the rods are at rest relative to each other, they are made to be of equal length. We now assert that when they are in the above motion relative to each other, they will still be equal in length. (Fig. 2.2.) Suppose that in the frame in which B is at rest (the *rest-frame* of B) rod A appears longer. Then, as A passed B, a piece of it could be cut off, as proof of its greater

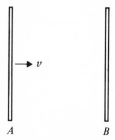

Figure 2.2. The equality of two rods moving transversely to their extension.

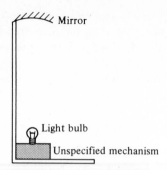

Figure 2.3. An idealized clock useful for thought-experiments.

extension. It is important to note that the two ends of A would be passing B *simultaneously* in B's rest frame. By the relativity principle, however, the physical situation should be the same in the frame in which A is at rest, and B moving. By symmetry, however, it would now have to be B that is longer, and it would be the rod B that would lose its tips. These conclusions are incompatible, unless the lengths are indeed the same. Since we will obtain a different answer when the rods are lined up in the direction of relative motion, we should stress again that in the present configuration, the ends of the rods pass each other simultaneously, in either frame.

Time dilation Given this fact, we can now construct a clock. The clock consists of a lightbulb at one end of a rod, a mirror at the other, and some mechanism that makes the lightbulb flash whenever the reflection of the previous flash reaches it. This is not a technologically realizable clock, but it is simple, periodic, and well suited to thought experiments. The clock, in its rest frame, will flash at intervals $2L/c$, where L is the length of the rod. Any other type of clock (for example, a Caesium atomic clock) if synchronized with our clock in one frame, will remain synchronized with it in another frame in which the two clocks are moving: if this were not so, we could single out inertial frames in violation of the first postulate. In a moving frame, the light still travels with speed c, but it has farther to go (see Figs. 2.3 and 2.4). In the time t'

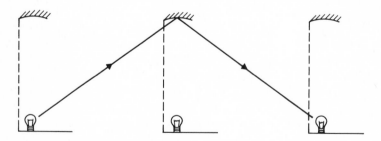

Figure 2.4. Path taken by light ray in clock in order to trigger periodic response.

between flashes, where t' is the time measured by an observer, relative to whom the clock is in motion, the light has to travel a distance

$$2\left(L^2 + \left(\frac{vt'}{2}\right)^2\right)^{1/2} \tag{2.8}$$

and this must equal ct'. Thus, solving, we get

$$t' = \frac{2L}{c} \frac{1}{(1 - v^2/c^2)^{1/2}}. \tag{2.9}$$

Thus we see that the interval between flashes is longer, that is, the clock is slowed down by a factor of $(1 - v^2/c^2)^{1/2}$.

Experimental evidence for this effect is provided by the measurement of the mean lives of radioactive nuclei or unstable particles in motion. An unstable particle decays with some mean lifetime that can be measured both in a situation when it is at rest, and when it is moving. For example, the *pion* $(\pi^\pm)$ has a mean life of 2.56×10^{-8} sec in its rest frame. Without the time dilation effect, the distance that it could travel, when moving with velocity v is, when v is very close to c,

$$c\tau = 3 \times 10^{10} \times 2.56 \times 10^{-8} \simeq 7.7 \times 10^2 \text{ cm} = 7.7 \text{ m}.$$

The time dilation effect extends this to

$$\frac{c\tau}{(1 - v^2/c^2)^{1/2}} \approx \frac{7.7}{\sqrt{\epsilon}} \text{ m} \tag{2.10}$$

where $v/c = 1 - \epsilon/2$ and $\epsilon \ll 1$. For $\epsilon = 10^{-6}$, say, this distance is quite large. The prediction of the theory of relativity is in complete accord with experience. Advantage is taken of this time dilation effect in high-energy laboratories, where long beam lines of particles that are very short-lived, are constructed. In this way experiments involving pions produced at the machine can be carried out hundreds of meters from the place of production. Similarly, unstable particles produced at the top of the atmosphere when cosmic ray primary particles strike nuclei there, manage to reach ground level if they have a large enough velocity (or more specifically, if they have a large enough $\gamma = (1 - v^2/c^2)^{-1/2}$ even if the product of c and the *proper*, that is, rest-frame, lifetime is much smaller than the height of the atmosphere.

The prediction of time dilation, and its uncontestable confirmation in everyday life (at high-energy laboratories!) leads to a seeming paradox, the so-called *twin paradox*. Imagine identical twins on earth. Twin A takes off on a long journey at high v. After traveling for a long time, decelerating gently and returning, twin A will be younger than the stay-at-home twin B. This is certainly B's observation: twin A is moving relative to B and thus A's clock is slowed down. We assume that the deceleration is gentle enough that nothing drastic happens, and we can still make the straight part of the journey long enough to dwarf to any degree of accuracy our uncertainties of what happens during the deceleration. The seeming paradox arises when one argues that from the point of view of A it is B (and the earth) that are on the journey, and it is therefore B that ought to be younger at the time of reunion. This is false. Twin A is not in an inertial frame (it is twin A that experiences the deceleration, no matter how gentle), and cannot therefore make the same assertion as twin B does about the clock

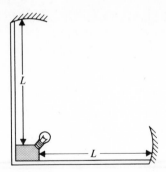

Figure 2.5. Idealized double-arm clock.

of the other. We shall return to this subject twice: once, to show graphically how the twins really agree about the age if they send signals on their birthdays to each other, and once again when we discuss the Equivalence Principle. Although there is no paradox in the theory of relativity, there is a way for the traveling twin to take into account his or her acceleration in the working out of the twins' relative age, and this must await the developments to be discussed in Chapter 8[†].

Length contraction If the clock is tilted in the direction of motion, it still ticks at the rate

$$t' = \frac{2L}{c} \frac{1}{(1 - v^2/c^2)^{1/2}}.$$

To convince oneself of that, it is useful to consider a somewhat more complex clock (Fig. 2.5) consisting of two rods that are equal in the rest frame, fixed at right angles to each other, with mirrors at both ends, and a bulb mechanism designed so that it flashes only when both reflected rays reach it simultaneously. With rods of equal length, the clock will flash. Such a flash is an *event*, and if it occurs in one frame, it must occur in all frames, as otherwise we could distinguish between inertial frames in violation of the relativity principle. Thus the horizontal clock ticks at the same rate as the vertical one. We will now argue that to an observer, relative to whom the clock is moving with velocity v, the horizontal rod will have length $L' = L(1 - v^2/c^2)^{1/2}$. To see this, observe that since the mirror at the end of the horizontal rod is moving to the right, the time that it takes the light to get to it is t_1, where

$$ct_1 = L' + vt_1. \tag{2.11}$$

The time that it takes for the light to return to the bulb is t_2 where

$$ct_2 = L' - vt_2. \tag{2.12}$$

Thus the total time is

$$t_1 + t_2 = \frac{L'}{c - v} + \frac{L'}{c + v} = \frac{2L'}{c} \frac{1}{1 - v^2/c^2}. \tag{2.13}$$

---

† See Problem 8.1.

This, however, we argued, must equal

$$\frac{2L}{c}\frac{1}{(1-v^2/c^2)^{1/2}}.$$

From this it follows that

$$L' = L(1-v^2/c^2)^{1/2}. \tag{2.14}$$

Thus a rod moving with velocity v relative to an observer, is contracted. The Fitzgerald-Lorentz contraction is thus real, but it is not a consequence of some special properties of interactions between the atoms; it is a consequence of the properties of space and time. Incidentally, it will not escape the reader that our two-armed clock is just an idealized Michelson-Morley experiment, and the null-result implies the length contraction.

A more direct way of seeing the necessity for the contraction is to consider an unstable particle traveling from the top of the atmosphere to the ground. If the proper lifetime of the particle is τ, then it must travel with velocity v, where

$$d = \frac{v\tau}{(1-v^2/c^2)^{1/2}} \tag{2.15}$$

and d is the distance traveled. The particle gets as far as it does, from the point of view of the terrestrial observer, because of the dilation of its proper lifetime. An observer traveling with the particle will see it live for a time τ. Yet that observer will also see the particle reach ground level. This is only possible if the depth of the atmosphere, seen by the moving observer, is foreshortened to $d(1-v^2/c^2)^{1/2}$.

Doppler shift Another interesting effect is the *Doppler shift* in the frequency of light as seen by a moving observer. Let us first review the nonrelativistic (for example, acoustical) Doppler effect. Consider a source, such as a siren, moving with velocity v_s and emitting waves with frequency ν. When the source is at rest, the distance between peaks in the wave train (the wavelength for harmonic waves) is $\lambda = c/\nu$, where c is the velocity of sound. When the source is moving, the second peak is emitted after a time $1/\nu$, from a point v_s/ν closer to the first peak. Thus the effective wavelength is

$$\lambda' = \lambda - \frac{v_s}{\nu} = \frac{c-v_s}{\nu}. \tag{2.16}$$

In other words, a source moving towards the observer "crams" more waves into a given distance. If the observer is moving with velocity v_0 in the same direction as the source, that observer experiences a velocity of sound equal to $c-v_0$, and thus counts waves with a frequency

$$\nu' = \frac{c-v_0}{\lambda'} = \nu\,\frac{c-v_0}{c-v_s}. \tag{2.17}$$

Note that this expression distinguishes between source and observer velocities. The result for ν'/ν with $v_0 = 0$, $v_s = u$ is not the same as that with $v_0 = -u$ and $v_s = 0$, albeit the difference is of order u^2/c^2.

The relativistic Doppler shift cannot, according to the relativity principle, distinguish between these two cases. Consider a periodic emitter moving with velocity v away from an observer. In a time t', measured in the emitter's frame, N pulses are emitted, so that the frequency of emission is $\nu_0 = N/t'$. The observer sees the N waves fitted into a distance $(c + v)t$ where t is the time of emission measured in the observer's frame. Note that we are *not* using the old (incorrect) nonrelativistic law of addition of velocities. The factor $(c + v)t$ comes about because the distance into which the waves can be fitted, ct, is augmented by extra space created by the motion of the source, that is vt, in the time t. Hence the observer sees a frequency $v = c/\lambda$, where the wavelength is

$$\lambda = \frac{(c + v)t}{N}.$$ (2.18)

Thus

$$v = \frac{cN}{(c + v)t} = \frac{N}{t} \frac{1}{1 + v/c} = \frac{N}{t'}\left(\frac{t'}{t}\right)\frac{1}{1 + v/c}$$

$$= \frac{t'}{t} \frac{v_0}{1 + v/c}.$$ (2.19)

However the relation between t' and t is

$$t' = t(1 - v^2/c^2)^{1/2},$$

that is, the emitter clock runs slowly, as seen by the observer. Hence

$$v = v_0 \frac{(1 - v^2/c^2)^{1/2}}{1 + v/c} = v_0\left(\frac{1 - v/c}{1 + v/c}\right)^{1/2}$$ (2.20)

If the emitter is emitting in a direction making an angle θ with the direction of motion (Fig. 2.6), then the only change is that the N waves are now fitted into a distance $t(c + v\cos\theta)$ so that the final formula reads

$$v = v_0 \frac{(1 - v^2/c^2)^{1/2}}{1 + v\cos\theta/c}.$$ (2.21)

For $\theta = 90°$ we have the *transverse Doppler shift*, which is purely relativistic, and entirely due to the time dilation effect.

Astronomical applications The Doppler shift is an important tool in the measurement of velocities of radiating bodies. This is because radiation emitted by atoms is characterized by well-defined *spectral* lines, whose wavelengths (and therefore frequencies) can be accurately measured. Thus a radiating body, moving away from the observer will emit radiation of lower frequency (*red shift*), and the shift of the spectral line

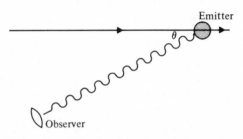

Figure 2.6. Doppler shift for emitter moving at angle θ relative to light path from emitter to absorber.

relative to the same line as measured in the laboratory measures the velocity of recession. One of the most fascinating uses of the Doppler shift was made by the astronomer Edwin Hubble, who analyzed the spectra of a large number of stars, whose distance from the solar system was known. He computed their velocities of recession from their red shifts, and was led to the discovery of the law, that *the velocity of recession is proportional to the distance from the solar system*. The fact that the distance and the velocity are measured relative to the earth should not be taken as implying a privileged role for the earth; in an expanding universe, any point could equally well serve as an observation center. (The analogy of the stars with raisins in an expanding raisin cake is sometimes made; this might be helpful to some readers). The law may be written as

$$D = \frac{v}{H} \tag{2.22}$$

where $H = 75$ km/sec/Megaparsec $= 2.5 \times 10^{-18}$ sec^{-1} (1 parsec is the distance at which the earth's orbit around the sun subtends one second of arc). Astronomers have used Hubble's law to estimate distances of very distant stars, from measurements of their red shifts. In recent years some powerful radiators, called *quasistellar objects*, have been observed with radio telescopes. Their red shifts are very large, with $\Delta\lambda/\lambda$ in excess of 2.0 having been observed. An application of Hubble's law would place these quasistellar objects several billion light years away; given their luminosity, that is, the amount of energy arriving at the telescope, astrophysicists have had a hard time inventing mechanisms that would lead to the prodigious energy production rate necessary to maintain this luminosity. The red shift could, however, be of gravitational origin. Gravitational red shifts will be discussed in Chapter 8.

The Doppler shift can also be a nuisance. To study spectra of atoms it is often necessary to heat up a gas of the atoms under consideration, and thus the atoms move about with quite large velocities. For atoms approaching the observer, the frequency is shifted upward; for atoms moving away, the frequency is shifted downwards. The random motion thus broadens the spectral lines, making accurate measurements much more difficult. Overcoming the Doppler broadening is one of the arts of making spectral measurements, and only when this is done does spectroscopy become the superb tool that it can be.

NOTES AND COMMENTS

1. A very beautiful discussion of the kinematical consequences of the Einstein postulates may be found in the book *Space and Time in Relativity* by N.D. Mermin, McGraw-Hill, New York, 1968. This introductory textbook uses a minimum of mathematics. For other books, with treatments on a level comparable with ours, see the bibliography.
2. Experiments with short-lived elementary particles continue to provide tests of the predictions of special relativity. An experiment measuring the magnetic dipole moment of a *muon*, a heavy electron with a proper lifetime of $2.200 \pm 0.0015 \times 10^{-6}$ sec, gave as a by-product a test of the theory: muons were injected into a

large "storage ring," in a circular orbit in an almost uniform magnetic field. The detection of the decay products of the muons provided a way of timing the decay. This turned out to be within 2% of the result predicted by the time-dilation factor. In another experiment, particles (π^0) that decay into gamma rays were used. The π^0 particles were moving with a velocity extremely close to c, and the γ rays emitted by them were timed with very sophisticated electronic equipment as they crossed the laboratory. Their speed was found equal to c within 1 part in 10^4, showing that the speed of the source of the radiation did not affect the speed of the radiation. These and other tests were reported by F.J.M. Farley, J. Bailey, and E. Picasso in *Nature*, **217** (1968): 17.

3. A "kinematical" consequence of the special relativity theory is the impossibility of maintaining the idealized concept of a *rigid body*. A rigid body would allow us to transmit a signal instantaneously, by pushing on one end, and having the other end move right away. In reality a rigid body transmits a push on one end with the speed of sound (a compressional wave) in the material that the rod is made of.

4. The notion of a "contraction of length" has captured the imagination of illustrators of popularizations of relativity theory, and many books (and films) exhibit the visual distortion expected from a reduction of size along the motion without the accompanying distortion in the transverse direction. It is amusing that it was not until 1960 that this expectation was properly analyzed. J.L. Terrell, *Physical Review* **116**, (1959): 1041, pointed out that what the eye registers is the arrival of light rays that leave different parts of a moving object at different times, and that in fact the distortions are not present when an object is seen broadside. This fascinating subject is also discussed by V. Weisskopf in *Lectures in Theoretical Physics*, vol. III, Summer Institute for Theoretical Physics, University of Colorado, Boulder, Interscience Publishers, New York, 1961, and in V.F. Weisskopf, *Physics in the Twentieth Century*, MIT Press, Cambridge, Mass., 1972.

PROBLEMS

2.1 A meter stick moves with speed $0.96c$ relative to you. Find the length of the stick as measured by you. How long does it take for the stick to pass you? How long a time would it take from the point of view of an observer traveling with the stick?

2.2 An unstable particle has a lifetime of 0.8×10^{-10} sec in its rest frame. An experiment with such a particle requires it to travel 30 m before decaying. Calculate what the velocity of the particle must be for this to become feasible. Express your answer by giving x where $x = 1 - v/c$.

2.3 The radius of the galaxy is 3×10^{20} m. How fast would a spaceship have to travel to cross it in 300 yr? Express your result in terms of $\gamma = (1 - v^2/c^2)^{-1/2}$.

2.4 Quasistellar objects exhibit a Doppler shift of magnitude $z = (\lambda_{obs} - \lambda)/\lambda = 1.95$, where λ is the wavelength that would have been measured by an observer at rest relative to the emitter of radiation. Use this information to calculate the velocity with which these objects move relative to us. How far away are they (in light years) if they obey Hubble's law?

2.5 A beam of muons, elementary particles with a proper lifetime of 2.2×10^{-6} sec are injected into a storage ring of circular shape and diameter 60 m. They are injected with velocity such that $\gamma = 15$. How many times will an "average" muon go around in the storage ring?

2.6 An observer sees a spaceship coming from the west at a speed $0.6c$ and a spaceship coming from the east at a speed $0.8c$. The western spaceship sends a signal with a frequency of 10^4 Hz in its rest frame. What is the frequency of the signal as perceived by the observer? If the observer sends on the signal immediately upon reception, what is the frequency with which the eastern spaceship receives the signal?

2.7 If the eastern spaceship in Problem 2.6 were to interpret the signal as one that is Doppler shifted because of the relative velocity between the western and eastern spaceships, what would the eastern spaceship conclude about the relative velocity?

2.8 Repeat the above calculation with velocities of the western and eastern spaceships v_1 and v_2, respectively. Show that the relative velocity must be $(v_1 + v_2)/(1 + v_1 v_2/c^2)$.

2.9 Suppose that you wanted to test the time dilation by taking a clock on a commercial airliner on a trip around the world. Assuming some reasonable speed for the airliner (e.g., 1000 km/hr) and some reasonable route, how accurate would the clock have to be to check the time dilation formula to an accuracy of 5%?

2.10 Rocket A moves with a velocity $v = 0.8c$ in the northerly direction relative to an observer O. Rocket B moves in a westerly direction with velocity $0.6c$ relative to the observer. Rocket B emits radiation with a wavelength $\lambda = 1000$ Å normal to its line of motion, in the northerly direction. What is the wavelength of the radiation seen by rocket A?

2.11 A source flashes with a frequency of 10^{15} Hz. The signal is reflected by a mirror moving away from the source with speed 10 km/sec. What is the frequency of the reflected radiation as observed at the source?

2.12 Two spaceships pass each other with relative velocity $0.8c$. The spaceships are 100 m long in their respective rest frames. How long will it take for spaceship B to pass the pilot at the front end of spaceship A, as reported by that pilot? How long will it take for all of spaceship B to pass all of spaceship A, as reported by the crew of spaceship A?

3
The Lorentz Transformation

The Lorentz transformation The Galilean transformation law

$$x' = x - vt$$
$$y' = y$$
$$z' = z$$
$$t' = t$$

(3.1)

that relates the space-time coordinate description of an event in one frame (x, y, z, t) to the space-time description of that same event, used by an observer that moves with speed v in the x direction relative to the unprimed frame, is *not* the correct transformation law, except in the nonrelativistic limit, when $v/c \to 0$. What is the proper transformation law? Consider the frames (Fig. 3.1) S and S' which coincide when $t = t' = 0$. The frame S' moves with velocity v in the positive x direction relative to S. Our arguments in Chapter 2 about the length of rods perpendicular to their direction of motion indicates that the y and z coordinates are not affected by the transformation so that we may, without hesitation, continue to use

$$y' = y$$
$$z' = z.$$

(3.2)

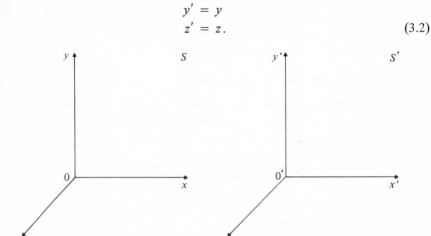

Figure 3.1.

The transformation for x' and t' must be *linear*, that is, of the form

$$x' = a_1(v)x + a_2(v)t$$
$$t' = a_3(v)x + a_4(v)t. \tag{3.3}$$

This must be so, since by the relativity principle, the *inverse transformation*, express-
ing (x, t) in terms of (x', t') must be the same as Eq. (3.3) with the replacement
$v \to -v$, as the unprimed observer is moving with velocity $-v$ along the positive x axis
relative to the primed observer. It is easy to convince oneself by playing around with
some examples, that only the linear transformation law will work. Since Eq. (3.3)
implies

$$x = \frac{a_4 x' - a_2 t'}{a_1 a_4 - a_2 a_3} \qquad t' = \frac{a_1 t' - a_3 x'}{a_1 a_4 - a_2 a_3},$$

we have

$$a_1(-v) = \frac{a_4(v)}{a_1(v)a_4(v) - a_2(v)a_3(v)}; \qquad a_2(-v) = -\frac{a_2(v)}{a_1(v)a_4(v) - a_2(v)a_3(v)} \tag{3.4}$$

and so on.

Suppose the "event" is the presence of a rocket that is sent out from the origin at
$t = t' = 0$ in the positive x direction with velocity v. For such an event

$$x = vt$$

and, for the primed observer, whose coordinate frame is fastened to the rocket, that
rocket never leaves the origin, so that

$$x' = 0.$$

When these are substituted into the first of the relations (3.3) we get

$$va_1(v) = -a_2(v). \tag{3.5}$$

We determine the unknown functions by requiring that the equation describing the
propagation of a light pulse, emitted at the origin at $t = 0$, that is,

$$x^2 + y^2 + z^2 = c^2 t^2, \tag{3.6}$$

be described by the same equation in the primed frame, that is,

$$x'^2 + y'^2 + z'^2 = c^2 t'^2. \tag{3.7}$$

Thus, using Eq. (3.2) we have

$$x^2 - c^2 t^2 = x'^2 - c^2 t'^2 = a_1^2(x - vt)^2 - c^2(a_3 x + a_4 t)^2$$
$$= x^2(a_1^2 - c^2 a_3^2) + t^2(a_1^2 v^2 - c^2 a_4^2)$$
$$- 2xt(va_1^2 + c^2 a_3 a_4)$$

which leads to three equations

$$va_1^2 + c^2 a_3 a_4 = 0$$
$$a_1^2 - c^2 a_3^2 = 1$$
$$c^2 a_4^2 - v^2 a_1^2 = c^2. \tag{3.8}$$

Some algebra leads to the solution

$$a_1 = a_4 = \frac{1}{\sqrt{1 - v^2/c^2}}; \qquad a_3 = -\frac{v}{c^2} \frac{1}{\sqrt{1 - v^2/c^2}}. \tag{3.9}$$

Thus the correct transformation law reads

$$x' = \gamma(x - vt)$$

$$t' = \gamma\left(t - \frac{v}{c^2}x\right),$$ (3.10)

where

$$\gamma = (1 - v^2/c^2)^{-1/2}.$$ (3.11)

A neater form, which takes into account that in a more sensible description distance and time ought to have the same dimensions uses

$$\beta = v/c$$ (3.12)

to write the transformation law in the form

$$x' = \gamma(x - \beta ct)$$
$$ct' = \gamma(ct - \beta x)$$ (3.13)

together with (3.2). Note that in the limit that $\beta \to 0$,

$$\gamma = (1 - \beta^2)^{-1/2} \approx 1 + \tfrac{1}{2}\beta^2 \to 1,$$

and we get the Galilean transformation law (3.1).

Addition of velocities We may use the transformation law to understand how the velocity of light is the same in all frames, while in the nonrelativistic domain the law of addition of velocities

$$u'_x = u_x - v$$ (3.14)

continues to hold. To do so we write

$$u'_x = \frac{dx'}{dt'} = \frac{dx'/dt}{dt'/dt}.$$

Now

$$\frac{dx'}{dt} = \gamma\left(\frac{dx}{dt} - v\right); \qquad \frac{dt'}{dt} = \gamma\left(1 - \frac{v}{c^2}\frac{dx}{dt}\right).$$

Hence

$$u'_x = \frac{dx'}{dt'} = \frac{\gamma(u_x - v)}{\gamma(1 - vu_x/c^2)},$$

that is[†]

$$u'_x = \frac{u_x - v}{1 - vu_x/c^2}.$$ (3.15)

The velocities in the transverse direction are also altered:

$$u'_y = \frac{dy'}{dt'} = \frac{dy/dt}{dt'/dt} = \frac{u_y}{\gamma(1 - vu_x/c^2)},$$ (3.16)

and similarly

$$u'_z = \frac{u_z}{\gamma(1 - vu_x/c^2)}.$$ (3.17)

[†] This result was already derived in Problem 2.8 with the help of the Doppler shift formula.

We can show that no matter how large v is (it must be less than c), $u'_x \leq c$. Let us write

$$\beta_x = u_x/c, \qquad \beta = v/c.$$

Then

$$\beta'^2_x = \frac{(\beta_x - \beta)^2}{(1 - \beta\beta_x)^2} = \frac{\beta^2_x + \beta^2 - 2\beta\beta_x}{1 + \beta^2\beta^2_x - 2\beta\beta_x}$$

$$= \frac{1 - 2\beta\beta_x + \beta^2\beta^2_x - (1 - \beta^2 - \beta^2_x + \beta^2\beta^2_x)}{(1 - \beta\beta_x)^2}$$

$$= 1 - \frac{(1 - \beta^2)(1 - \beta^2_x)}{(1 - \beta\beta_x)^2} \leq 1. \tag{3.18}$$

The equality sign holds only when $u_x = c$, $u_y = u_z = 0$. This remarkable law for the addition of velocities shows that c is transformed into c independent of the value of v, and it has the correct limiting form when $\beta\beta_x \to 0$.

We quote, without derivation, the Lorentz transformation law for the case that the primed frame moves relative to the unprimed one with velocity $\mathbf{v}$ in an arbitrary direction:

$$\mathbf{r'} = \mathbf{r} + \mathbf{v}\left[\mathbf{r}\cdot\mathbf{v}\frac{\gamma - 1}{v^2} - \gamma t\right]$$

$$t' = \gamma(t - \mathbf{r}\cdot\mathbf{v}/c^2). \tag{3.19}$$

Time dilation and length contraction The Lorentz transformation is the proper mathematical way to describe the correct relativistic relationship between descriptions by different inertial observers. We may use it to discuss the time dilation and length contraction phenomena. First, consider a clock that in the unprimed frame remains at the origin. Let the first tick of the clock occur at $t = t' = 0$ when the two frames S and S' coincide. The first event is thus described by $x = x' = 0$, $t = t' = 0$. The second tick, seen by the observer in the unprimed frame is described by $x = 0$ (the clock is at rest) and $t = T$, the period of the clock. The second tick, that event, is described by the primed observer by the primed coordinates x', t', which according to Eq. (3.10) are given by

$$x' = -\gamma v T$$

$$t' = \gamma T. \tag{3.20}$$

The first expression just tells us that the clock has receded according to the second observer. The second expression shows that in the primed frame the time interval between ticks is longer by $(1 - v^2/c^2)^{-1/2}$, so that the clock runs slower.

To discuss the length contraction, consider a rod that is at rest in the unprimed frame. The rod has length L, and we may choose its coordinates as

$$x_1 = 0$$

$$x_2 = L.$$

If the primed observer wishes to measure the length of the rod, he or she must measure the position of the ends of the rod "simultaneously" in his or her frame. Let us take the time when the measurement is carried out to be $t'_1 = t'_2 = 0$. Then

$$t'_1 = 0 = \gamma(t_1 - vx_1/c^2) = \gamma t_1,$$

$$t'_2 = 0 = \gamma(t_2 - vx_2/c^2) = \gamma(t_2 - vL/c^2),$$

that is, $t_1 = 0$ and $t_2 = vL/c^2$. Now

$$x'_1 = \gamma(x_1 - vt_1) = 0$$
$$x'_2 = \gamma(x_2 - vt_2) = \gamma(L - v^2L/c^2)$$
$$= (1 - v^2/c^2)^{1/2}L. \tag{3.21}$$

Thus the rod is seen to have length $x'_2 - x'_1 = L/\gamma$, that is, it is contracted. The fact that the measurement of the length in the primed system determines the front and back end positions of the rod at different times in the unprimed system is essential to the argument, and allows us to resolve some apparent paradoxes. We illustrate this by the example that follows.

Example. A pole vaulter holding a pole that is 16 ft long in his rest frame, and running with relativistic speed, such that $\gamma = 2$, approaches a shed that, in its rest frame, is 8 ft long. An observer, at rest relative to the shed, sees the pole as being only 8 ft long, and arranges for gates at the two ends of the shed to shut as soon as the front of the pole reaches the far end of the shed. From the observer's point of view, no harm will be done. The runner sees the shed as only 4 ft long, and anticipates losing 12 ft of his pole as the gates fall. Should he worry?

Let us use the Lorentz transformation equations to describe the situation. The unprimed frame will be the rest frame of the shed, and in that frame, the entrance and exit of the shed will be described by $x = 0$ and $x = L \; (= 8\text{ ft})$, respectively (Fig. 3.2). The length of the pole in the unprimed frame is 8 ft, that is, L. In the primed frame, the coordinates of the pole are conveniently chosen as $x' = -\gamma L$ ($- 16$ ft for the rear) and $x' = 0$ for the front. This choice of coordinates is consistent with the choice of times $t = t' = 0$ for the first event, the entry of the front pole into the shed. The front end of the pole reaches the exit of the shed when the event characterized by $x = L$, $x' = 0$ occurs. This occurs at a time such that

$$x' = \gamma(x - vt) = \gamma(L - vt) = 0,$$

that is, the time is $t = L/v$. The time measured by the primed observer, the runner is

$$t' = \gamma\left(t - \frac{v}{c^2}x\right) = \gamma(L/v - vL/c^2)$$
$$= L/\gamma v.$$

The third event, when the rear end of the pole coincides with the entrance of the shed, is described by $x = 0, x' = -\gamma L$. This occurs at a time when

$$x' = \gamma(x - vt) = -\gamma vt = -\gamma L$$

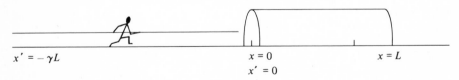

$x' = -\gamma L$ $x = 0$ $x = L$
                $x' = 0$

Figure 3.2. Schematic layout for paradox of the pole vaulter. The figure is drawn for the nonrelativistic tryouts.

that is, simultaneously with the exit time of the front of the pole,

$$t = L/v$$

as per design. The time for this event, as seen by the runner, is given by

$$t' = \gamma\left(t - \frac{v}{c^2}x\right) = \gamma L/v,$$

which is much later than $L/\gamma v$, the time of entry into the shed. Thus the closing of the gate will not lead to a disaster. If there is no disaster in one frame, there cannot be a disaster in another, as the principle of relativity tells us!

NOTES AND COMMENTS

1. The transformation law (3.10) was actually discovered by H.A. Lorentz (1903) as the one that leaves Maxwell's equations unaltered, and that is why the relativistic transformations are called Lorentz transformations.

PROBLEMS

3.1 Check that formula (3.19) approaches the correct limits when the velocity is taken along one of the axes, say the x axis. Check explicitly that the inverse transformation, that is, $\mathbf{r} = f(\mathbf{r}', t')$ and $t = g(\mathbf{r}', t')$ obtained by solving (3.19) for $\mathbf{r}$ and t coincides with what is obtained from (3.19) by changing $\mathbf{v}$ into $-\mathbf{v}$.

3.2 Use the transformation laws (3.10) to calculate the transformation laws for the acceleration d^2x/dt^2.

3.3 Prove that the differential operator

$$\frac{1}{c^2}\frac{\partial^2}{\partial t^2} - \frac{\partial^2}{\partial x^2} - \frac{\partial^2}{\partial y^2} - \frac{\partial^2}{\partial z^2}$$

is invariant under Lorentz transformations. Use the chain rule

$$\frac{\partial}{\partial t'} = \frac{\partial t}{\partial t'}\frac{\partial}{\partial t} + \frac{\partial x}{\partial t'}\frac{\partial}{\partial x} + \cdots$$

and so on. This differential operator appears commonly in wave equations. Thus if $\mathbf{B}$, for example, is eliminated from Maxwell's equations in free space, the equation for the electric field is

$$\frac{1}{c^2}\frac{\partial^2 \mathbf{E}}{\partial t^2} - \nabla^2\mathbf{E} = 0.$$

3.4 The velocity of light in a medium of refractive index n is c/n. Consider light propagating in a liquid with refractive index n when the liquid is moving with velocity w. What is the velocity of light relative to the moving liquid? Write your expression in a form that ignores terms of order $(w/c)^2$.

3.5 An atomic clock emits signals with a frequency ν. Describe the events of the emission in the frame of the clock. An observer, moving away from the clock with velocity v intercepts the signals. What are the x and t coordinates of these events? What are the coordinates in the frame of the observer? Can you calculate the Doppler shift formula from this?

3.6 This problem is for students who are familiar with matrix multiplication:

The formula (3.19) for the general Lorentz transformation may be written in the form

$$\begin{pmatrix} ct' \\ x' \\ y' \\ z' \end{pmatrix} = M \begin{pmatrix} ct \\ x \\ y \\ z \end{pmatrix}$$

where M is some 4×4 matrix. What is that matrix? Show that $M(\mathbf{v})M(-\mathbf{v}) = 1$. What is the meaning of this matrix relation?

3.7 Use the general Lorentz transformation in Eq. (3.19) to prove that if $(ct_1, \mathbf{r}_1)$ and $(ct_2, \mathbf{r}_2)$ both transform as given by that formula, then the product

$$(ct_1)(ct_2) - \mathbf{r}_1 \cdot \mathbf{r}_2$$

is invariant, that is, unchanged under Lorentz transformations. This is a generalization of the statement of invariance of $(ct)^2 - \mathbf{r}^2$, used in Eqs. (3.6) and (3.7).

4
Minkowski Diagrams

Events in one space and one time dimension are visually well described on an $x - t$ graph. This continues to be true for relativistic kinematics, but here, it is more convenient to use the relativistic time scale and use an $x - ct$ graph. The Lorentz transformation law in the form (3.13) uses such coordinates. An event will be described by a point in the (x, ct) plane. The motion of a particle may be viewed as a sequence of events, and thus is represented by a curve in the $x - ct$ plane. Such a curve is called the *world line* of the particle, and the only constraint on it is that its slope must always be larger than unity (cf. Fig. 4.1) since its instantaneous velocity can never exceed that of light. The line of unit slope is called the *light cone* (the description as *cone* refers to the more general situation of motion in more than one-space dimensions), and it is given by

$$x = \pm ct. \tag{4.1}$$

A particle moving with uniform velocity traces a straight line, inclined to the vertical by an angle θ, with

$$\tan \theta = \frac{x}{ct} = \frac{v}{c} = \beta. \tag{4.2}$$

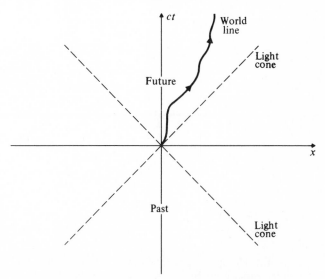

Figure 4.1. The Minkowski space-time diagram.

An observer moving with this particle may set up a different, primed coordinate system. The world line of the particle may serve as the ct' axis, since the location of the particle, as seen by the primed observer, will always be the same, say $x' = 0$. To determine the location of the x' axis, we note that all observers agree that the light cone is the same for all of them. Thus the line $x = ct$ and the line $x' = ct'$ must coincide. This suggests that the x' axis be given by the line that makes an angle θ with the x axis, with

$$\tan \theta = \beta. \tag{4.3}$$

We may check the correctness of this using (3.13) with $ct' = 0$, the definition of the x' axis (Fig. 4.2). The next question concerns the size of the grid that determines the units in the two frames. To determine this size, we use the fact that

$$x^2 - (ct)^2 = x'^2 - (ct')^2, \tag{4.4}$$

that is, $x^2 - (ct)^2 \equiv -\tau^2$ *is an invariant*, as can easily be checked with the help of Eq. (3.13). The curves of constant $x^2 - (ct)^2$ are hyperbolas. Inside the light-cone (forward and aft) $\tau^2 > 0$, on the light cone $\tau^2 = 0$, and outside the light cone $\tau^2 < 0$. For $x = 0$, $\tau = ct$. We call τ the *proper time* (when $\tau^2 > 0$) since it is the time ticked off by a clock in its own rest frame. Thus a unit of ct' is obtained by finding the intersection of the $\tau^2 = 1$ curve with the ct' axis, that is, the intersection of

$$x^2 - (ct)^2 = -1 \tag{4.5}$$

and

$$x = \beta ct. \tag{4.6}$$

Similar considerations allow us to determine all of the grid points in the primed, skewed, coordinate frame. The diagrams were invented by H. Minkowski, and are named after him.

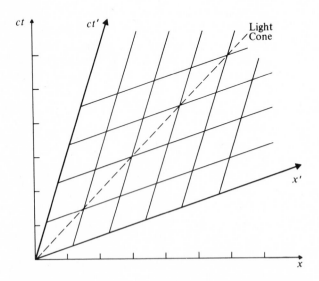

Figure 4.2. The space-time diagram of a moving (primed) observer superimposed on the space-time diagram of a stationary observer.

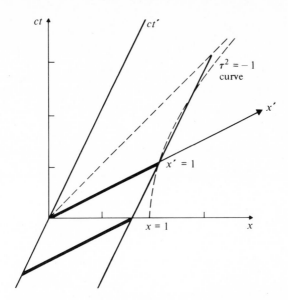

Figure 4.3. Picture of Lorentz contraction. The rod is at rest in the primed frame. Unprimed observer measures front end of rod at $ct = 0$ and finds it at a value $x < 1$. The rear end is measured "later" in the primed frame.

Lorentz contraction Let us discuss a few aspects of relativistic kinematics using the Minkowski diagram. First, how does the Lorentz contraction appear in the space-time diagram? (Fig. 4.3.) In the figure we draw a rod that is always at rest in the S' frame. One end is always on the $x' = 0$ line (that is, the ct' axis), and the other on the $x' = 1$ line, and it is always oriented parallel to the x' axis, so that the times at the two ends are always the same in the S' frame, as required for a rod at rest in that frame. To see how long the rod looks in the unprimed frame, we must determine the positions on the x axis for the rear and front of the rod, since a length measurement in any frame (now the unprimed frame) must involve locating the two ends at the same time; locating the front of a moving train at noon and the rear at 10 a.m. does not tell us anything about the length of the train directly! Note that these measurements are not simultaneous from the point of view of the S' frame, since one end is measured on the upper dark line, and the other on the lower dark line which represents an "earlier" primed time. The length of the rod is thus from the origin to where the $x' = 1$ line intersects the $ct = 0$ line. How big is that interval? A unit length occurs when the hyperbola through $x' = 1$ and $ct' = 0$ intersects the $ct = 0$ line. It is clear that the rod is shortened, and the value of that shortening is easily calculated by noting that the point $x' = 1$ and $ct' = 0$ has coordinates

$$x = \gamma(x' + \beta ct') = \gamma \qquad (4.7)$$

and

$$ct = \gamma(ct' + \beta x') = \gamma\beta \qquad (4.8)$$

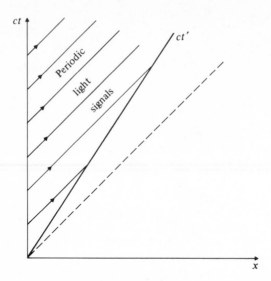

Figure 4.4. Stationary observer in unprimed frame sends periodic light signals to traveler, whose world line is the ct' axis.

in the S frame. A line through this point with slope $ct/x = 1/\beta$ is given by

$$ct - \gamma\beta = \frac{1}{\beta}(x - \gamma). \tag{4.9}$$

This intersects the $ct = 0$ line at the point

$$x = \gamma(1 - \beta^2) = (1 - \beta^2)^{1/2} = \frac{1}{\gamma}. \tag{4.10}$$

Thus the length of a unit rod is seen in a moving frame as shortened by a factor γ.

Doppler shift As another illustration of the utility of the Minkowski diagram, let us work out the Doppler shift formula (Fig. 4.4). The clock, stationary in the S frame, that is, on the ct axis, sends out a signal every unit of time. This signal is represented by a line making a $45°$ angle with the axes. The signal is received by the observer moving away from the clock with velocity v. That observer is always at $x' = 0$, so that the reception time that is measured, is measured on the ct' axis. The light lines are described by

$$ct - N = x \qquad N = 0, 1, 2, \ldots \tag{4.11}$$

and they intersect the line

$$x = \beta ct \tag{4.12}$$

at

$$x = \beta\frac{N}{1 - \beta}, \qquad ct = \frac{N}{1 - \beta} \tag{4.13}$$

This corresponds to

$$ct' = \gamma(x - \beta ct) \doteq \gamma N \frac{1 - \beta^2}{1 - \beta} = \gamma N(1 + \beta)$$

$$= N\sqrt{\frac{1 + \beta}{1 - \beta}} .$$
(4.14)

Thus the frequency of reception is

$$\frac{N}{ct'} = \left(\frac{1 - \beta}{1 + \beta}\right)^{1/2} ,$$
(4.15)

which is just the red shift calculated in Chapter 2.

Twin paradox As a final illustration, let us consider the twin paradox. Suppose the traveling twin moves with some high velocity, say $v/c = 24/25$ (so chosen because then $\gamma = 25/7$ is simple). The twin travels for 25 years, as seen by the earthbound twin, and then turns around, so that moving with the same speed he will be gone a total of 50 years, earth time (Fig. 4.5). The earthbound twin will calculate that the traveler's clock is slowed down to 7/25 of its rate, so that the traveler will have aged only 14 years. We can see how this works in the diagram. The earthbound twin sends a message every $2\frac{1}{2}$ years, and it is clear that every message will get to the traveler. The traveler will, however, get the messages quite infrequently on the way out, and very frequently on the way back. During his journey away for Y years (in his frame) he will receive the messages that are sent out at a rate of 0.4 per year, with the Doppler-shifted frequency

$$v' = v\sqrt{\frac{1 - \beta}{1 + \beta}} = v\sqrt{\frac{1 - 24/25}{1 + 24/25}} = \frac{0.4}{7}$$

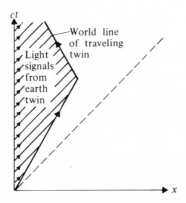

Figure 4.5. Minkowski diagram for twin paradox. The bent line is the world line of the traveling twin, and the lines at 45° indicate the signals sent from the earthbound twin.

per year. On the way back he will receive them at the rate of

$$v'' = v\sqrt{\frac{1+\beta}{1-\beta}} = 0.4 \times 7$$

per year. The total number of messages is 20, so that

$$20 = 0.4\,Y(1/7 + 7)$$

from which it follows that $2\,Y = 14$ years. Suppose the traveler decides to send a message on every one of *his* birthdays. The first message will be sent when $ct' = 1$, $x' = 0$, and this is described by the coordinates in (x, ct) obtained by the intersection of the hyperbola

$$(ct)^2 - x^2 = 1$$

with

$$\frac{x}{ct} = \frac{24}{25}.$$

Thus this occurs when

$$ct = \frac{25}{7}$$

and continues at equal intervals until the traveler returns. The intervals between receptions again change depending on whether the traveler is still outward bound or is returning, and some straightforward algebra shows that the receiving earthbound twin sees himself as having aged fifty years in the time that the traveler was away.

Superluminal signals The space-time diagram is useful in that it shows clearly that for signals, particles, travelers, and so on, there is a clear distinction between future and past. No skewing of axes will move a point in the forward light cone into the backward light cone. The meaning of future and past does not exist outside of the light cone. By

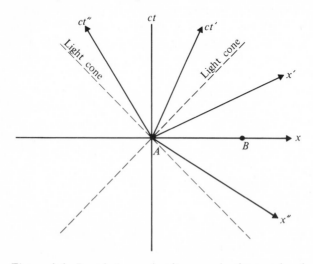

Figure 4.6. B and A are simultaneous in the unprimed frame. B is earlier than A in the primed frame and later than A in the double primed frame.

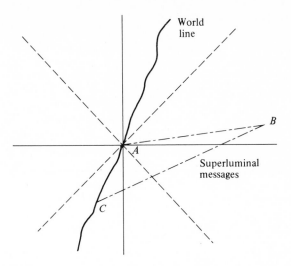

Figure 4.7. The difficulties with superluminal $(v > c)$ messages. A sends a superluminal message to B which is later than A in some frame. B sends it on to C, which is later than B in some other frame. But C is in the past of A. Can A arrange not to have been born?

a transformation to the primed frame with $\beta > 0$ or to the double-primed frame with $\beta < 0$, two points "simultaneous" in one frame, may change that relationship to A being later or earlier than B (Fig. 4.6). If it were possible to send signals faster than light (established by experiment to be the maximum speed signal), it would be possible to affect the past by going through the region outside of the light-cone (Fig. 4.7). This would imply a violation of causality, the requirement that the cause must precede the effect, and would lead to all kinds of logical difficulties (could you shoot all your ancestors 100 years ago?!). The theory of relativity is a consistent theory only if there is a maximum speed of propagation. It is possible to observe effects that appear to move faster than light, but such effects do not contradict the theory of relativity as they do not permit the transmission of a signal. Imagine a person rotating a flashlight with an angular velocity ω. A sequence of light detectors on a circle of radius R designed to flash upon the arrival of the light, would, if $R\omega$ were larger than c, present an effect propagating faster than light. This does not contradict the theory of relativity, since it is not possible to use this mechanism to send a signal from one light detector to another. Such apparent violations of the limit on the speed of a signal are actually observed in nature. Radio telescope observations of a quasistellar object (NC 345) during 1974 and 1975 showed that certain peaks in brightness increased their separation at the rate of 0.2 milliseconds of arc per year. The red shift of $z = \Delta\lambda/\lambda = 0.595$ combined with Hubble's Law indicates that the separation is increasing with $v \approx 8\,c$! A possible proposed explanation is the excitation of existing material by a relativistic shock wave propagating in the material. The expanding surface of relativistic electrons reaches different regions of material in a perfectly causal way, but we see radiation emitted from different places in an apparently acausal way.

NOTES AND COMMENTS

1. Minkowski's original paper is quite readable, and appears in translation in *The Principles of Relativity*, a collection of original papers, Dover Publications, 1951. The space-time diagram method is extensively used in *Spacetime Physics* by E.F. Taylor and J.A. Wheeler, W.H. Freeman, San Francisco, 1963. If I had to limit my recommendation to a single introductory book, my choice would be this one. It is a very stimulating book, with a marvellous choice of problems.

PROBLEMS

4.1 A stick 1 m long slides (along its extension) with a speed of 0.96 c over a manhole that is 1 m wide in its rest frame. In the manhole frame the stick is shortened and should fall in the hole; in the frame of the stick, the hole is shortened and the stick should not be able to fall in. Analyze what happens by carefully drawing a space-time diagram for this problem.

4.2 Work out the time dilation effect using a Minkowski diagram.

4.3 Work out the law of addition of velocities using a Minkowski diagram. The formula

$$\tan(\theta_1 + \theta_2) = \frac{\tan\theta_1 + \tan\theta_2}{1 - \tan\theta_1 \tan\theta_2}$$

will prove useful.

5
Relativistic Mechanics

Relativistic momentum Newtonian mechanics is not consistent with the Einstein postulates, and we must therefore look for a modification of the equation

$$m \frac{d^2 \mathbf{r}}{dt^2} = \mathbf{F}. \tag{5.1}$$

To do so, we will postulate that the third law of Newton, according to which *momentum is conserved* for a mechanical system, still holds in relativistic mechanics. We shall also assume that the momentum of a particle is parallel to its velocity, so that we shall write

$$\mathbf{p} = m(v)\mathbf{v}. \tag{5.2}$$

The function $m(v)$ must have the following properties: it should not depend on the direction in which the particle is traveling, that is, it depends only on the magnitude of v, and for $v/c \to 0$ it should approach what we usually mean by the mass, that is, the *rest mass of the particle*. It must have the dimensions of a mass, and since there is no other parameter with mass dimensions in the problem, we expect that

$$m(v) = mf((v/c)^2) \tag{5.3}$$

where m is the rest mass. Let us now consider an elastic collision between identical particles, that is, particles that have the same rest mass. First consider the collision to

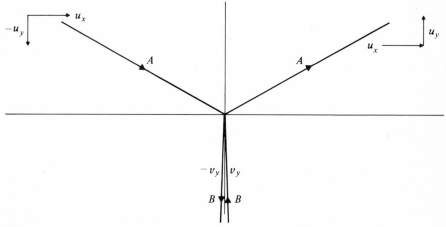

Figure 5.1. Collision in "brick-wall" frame.

take place in the "brick-wall" frame (Fig. 5.1) so that particle A is deflected and particle B sees the collision as a head-on collision. Although the particles are identical, their masses are different in the collision, since their velocities are different. Conservation of momentum along the y direction implies that

$$m_B v_y = m_A u_y. \tag{5.4}$$

Momentum is automatically conserved in the x direction in the frame shown in the figure. Let us now examine the same collision from the point of view of an observer moving in the positive x direction with velocity w. The observer sees the velocities transformed, according to Eqs. (3.15–3.17) as follows:

$$u_x \to u_x' = \frac{u_x - w}{1 - u_x w/c^2}$$

$$v_x \to v_x' = \frac{v_x - w}{1 - v_x w/c^2} = -w$$

$$u_y \to u_y' = \frac{1}{\gamma(w)} \frac{u_y}{1 - u_x w/c^2}$$

$$v_y \to v_y' = \frac{1}{\gamma(w)} v_y. \tag{5.5}$$

The masses will also change to m_A' and m_B' in a manner as yet unknown. Let us choose w so that the collision is completely symmetric, as in Fig. 5.2. The symmetric situation will be obtained when

$$u_x' = -v_x', \tag{5.6}$$

that is, when

$$\frac{u_x - w}{1 - u_x w/c^2} = w.$$

This implies that

$$u_x \frac{w^2}{c^2} - 2w + u_x = 0.$$

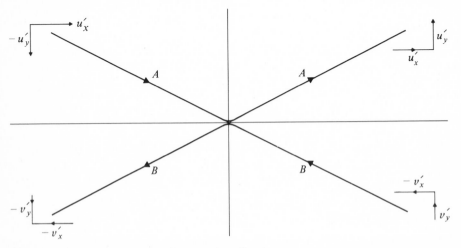

Figure 5.2. Collision in center of momentum frame.

The quadratic equation has solutions

$$\frac{w}{c} = (1 \pm \sqrt{1 - u_x^2/c^2})/(u_x/c).$$

In the nonrelativistic limit we obviously want

$$w = u_x/2$$

so that the solution with the minus sign is the correct one. With the choice

$$\frac{w}{c} = \frac{1 - \sqrt{1 - u_x^2/c^2}}{u_x/c} \tag{5.7}$$

we will obtain the symmetric configuration, for which we will also have

$$m_A' = m_B' \tag{5.8}$$

by symmetry. Momentum conservation in the y direction implies that

$$m_A' \frac{1}{\gamma(w)} \frac{u_y}{1 - u_x w/c^2} = m_B' \frac{1}{\gamma(w)} v_y \tag{5.9}$$

so that

$$\frac{u_y}{v_y} = 1 - \frac{u_x}{c} \frac{w}{c} = 1 - (1 - \sqrt{1 - u_x^2/c^2})$$
$$= \sqrt{1 - u_x^2/c^2}. \tag{5.10}$$

The momentum conservation equation in the y direction in the brick-wall frame may thus be written as

$$\frac{m_A}{m_B} = \frac{v_y}{u_y} = \frac{1}{\sqrt{1 - u_x^2/c^2}}. \tag{5.11}$$

If we now take the limit $v_y \to 0$, the particle B comes to rest, and m_B becomes the rest mass of the particle. In that limit particle A has velocity u_x and the above formula takes the form

$$m(u) = \frac{m}{\sqrt{1 - u^2/c^2}} = \gamma(u)m. \tag{5.12}$$

(Note that the rest mass of A is the same as the rest mass of B and we have replaced u_x by the symbol u). It is easily checked that the original momentum conservation law holds with this velocity dependence of the mass even when v_y does not vanish. This means showing that

$$\frac{mu_y}{\sqrt{1 - (u_x^2 + u_y^2)/c^2}} = \frac{mv_y}{\sqrt{1 - v_y^2/c^2}}$$

which is easily done with the help of Eq. (5.10). Thus the relativistic momentum formula is

$$\mathbf{p} = m\gamma\mathbf{v}. \tag{5.13}$$

Newton's second law may now be written in the form

$$\mathbf{F} = \frac{d}{dt}\mathbf{p} = \frac{d}{dt}(m\gamma\mathbf{v}). \tag{5.14}$$

Note that $\mathbf{F}$ and $d\mathbf{v}/dt$ no longer point in the same direction.

Kinetic energy We may use this expression to calculate the work done to bring a particle with rest mass m and velocity v to rest. This is defined to be the *kinetic energy*, and is given by

$$K = \int \mathbf{F} \cdot d\mathbf{r}$$

$$= \int \mathbf{F} \cdot \frac{d\mathbf{r}}{dt} \, dt. \tag{5.15}$$

The integrand has the form

$$\frac{d}{dt}(m\gamma\mathbf{v}) \cdot \mathbf{v} = \frac{d}{dt}(m\gamma)v^2 + m\gamma\mathbf{v} \cdot \frac{d\mathbf{v}}{dt}.$$

We may easily work out the time derivative in the first term

$$\frac{d}{dt}m\gamma = \frac{d}{dt}m(1 - v^2/c^2)^{-1/2} = \frac{1}{2}\frac{m}{c^2}(1 - v^2/c^2)^{-3/2}\frac{d}{dt}v^2$$

$$= \frac{m}{c^2}\gamma^3\mathbf{v} \cdot \frac{d\mathbf{v}}{dt},$$

so that

$$\frac{d\gamma}{dt} = \frac{1}{c^2}\gamma^3\mathbf{v} \cdot \frac{d\mathbf{v}}{dt}. \tag{5.16}$$

The integrand becomes

$$\left(\frac{m}{c^2}\gamma^3 v^2 + m\gamma\right)\mathbf{v} \cdot \frac{d\mathbf{v}}{dt} = m\gamma\left(1 + \frac{v^2/c^2}{1 - v^2/c^2}\right)\mathbf{v} \cdot \frac{d\mathbf{v}}{dt}$$

$$= m\gamma^3\mathbf{v} \cdot \frac{d\mathbf{v}}{dt}$$

$$= mc^2\frac{d\gamma}{dt}. \tag{5.17}$$

Hence

$$K = \int dt \, mc^2\frac{d\gamma}{dt} = mc^2\gamma(v)\bigg|_0^v$$

$$= mc^2(\gamma(v) - 1). \tag{5.18}$$

Note that in the nonrelativistic limit

$$\gamma = (1 - v^2/c^2)^{-1/2} \approx 1 + \frac{1}{2}\frac{v^2}{c^2} \tag{5.19}$$

so that

$$K = \tfrac{1}{2}mv^2 \tag{5.20}$$

in agreement with the form familiar from Newtonian mechanics.

Rest energy It is possible to associate an energy with a particle at rest, and if we conjecture that

$$E_{\text{rest}} = mc^2, \tag{5.21}$$

then the total energy of a particle is given by

$$E = E_{rest} + K$$
$$= mc^2\gamma, \tag{5.22}$$

a very elegant formula. It should be stressed, however, that as long as mass is conserved, which the relatively inaccurate chemical experiments of the nineteenth century seemed to imply, there is no physical significance to the rest energy, as it is never measurable. It is a sign of Einstein's deep intuition that he foresaw the possibility that mass might not be conserved in radioactive decays, and attached great significance to the physical reality of rest energy. That reality has been thoroughly tested in innumerable nuclear and elementary particle reactions. Thus, in a fission reaction, for example, the sum of the rest masses of the fission products is less than the rest mass of the fissioning nucleus, and the net mass difference Δm, multiplied by c^2, is then equal to the sum of the kinetic energies of the fission products, as demanded by energy conservation. When masses can change, *energy conservation means the conservation of kinetic, potential, and rest energy.*

Energy-momentum transformation laws Our next task is to study the transformation laws of energy and momentum. In the frame S, a particle will be taken to have energy and momentum given by

$$E = mc^2\gamma(u),$$
$$\mathbf{p} = m\mathbf{u}\gamma(u). \tag{5.23}$$

We will take u in the direction of the x axis, for simplicity, where that is the axis along which the observer in the frame S' is moving with velocity v. The velocity of the particle in the new frame will be

$$u' = \frac{u-v}{1-uv/c^2}. \tag{5.24}$$

We need to calculate $\gamma(u')$. This is given by

$$\gamma(u') = \left[1 - \frac{1}{c^2}\frac{(u-v)^2}{(1-uv/c^2)^2}\right]^{-1/2}$$

$$= \left(1 - \frac{uv}{c^2}\right)\left[\left(1 - \frac{uv}{c^2}\right)^2 - \frac{1}{c^2}(u-v)^2\right]^{-1/2}$$

$$= \left(1 - \frac{uv}{c^2}\right)\left[1 + \frac{u^2v^2}{c^4} - \frac{u^2}{c^2} - \frac{v^2}{c^2}\right]^{-1/2}$$

$$= \left(1 - \frac{uv}{c^2}\right)\left(1 - \frac{u^2}{c^2}\right)^{-1/2}\left(1 - \frac{v^2}{c^2}\right)^{-1/2}$$

$$= \left(1 - \frac{uv}{c^2}\right)\gamma(u)\gamma(v). \tag{5.25}$$

The momentum in the primed frame may now be expressed in terms of u and v:

$$p' = mu'\gamma(u') = m\frac{u-v}{1-uv/c^2}(1-uv/c^2)\gamma(u)\gamma(v)$$

$$= \gamma(v)(mu\gamma(u)-mv\gamma(u))$$

$$= \gamma(v)\left(p-mc^2\gamma(u)\frac{v}{c^2}\right)$$

$$= \gamma(v)\left(p-\frac{v}{c^2}E\right). \tag{5.26}$$

Similarly,

$$E' = mc^2\gamma(u') = mc^2\left(1-\frac{uv}{c^2}\right)\gamma(u)\gamma(v)$$

$$= \gamma(v)(mc^2\gamma(u)-vmu\gamma(u))$$

$$= \gamma(v)(E-vp). \tag{5.27}$$

These transformation laws, when rewritten in the form

$$E'/c = \gamma(E/c-\beta p)$$
$$p' = \gamma(p-\beta E/c) \tag{5.28}$$

are identical with the transformation laws for the space-time coordinates

$$ct' = \gamma(ct-\beta x)$$
$$x' = \gamma(x-\beta ct). \tag{5.29}$$

Just as

$$(ct)^2 - \mathbf{r}^2 \equiv \tau^2 \tag{5.30}$$

is an invariant, that is, the same in all frames, so is

$$\left(\frac{E}{c}\right)^2 - \mathbf{p}^2 = (mc)^2. \tag{5.31}$$

Because of these transformation laws, one frequently calls both $(ct, \mathbf{r})$ and $(E/c, \mathbf{p})$ Lorentz four-vectors. Invariants under Lorentz transformations are called Lorentz scalars.

It follows from Eq. (5.31) that

$$2\frac{E}{c^2}\frac{dE}{dt} - 2\mathbf{p}\cdot\frac{d\mathbf{p}}{dt} = 0,$$

that is,

$$\frac{dE}{dt} = \frac{c^2}{E}\mathbf{p}\cdot\frac{d\mathbf{p}}{dt} = \mathbf{u}\cdot\mathbf{F}. \tag{5.32}$$

This will be useful in discussing the transformation laws for the force. We have

$$F'_x = \frac{dp'_x}{dt'} = \frac{dp'_x/dt}{dt'/dt} = \frac{\gamma(dp_x/dt-(v/c^2)dE/dt)}{\gamma(1-vu_x/c^2)}$$

$$= \frac{F_x - v(\mathbf{u}\cdot\mathbf{F})/c^2}{1-vu_x/c^2}, \tag{5.33}$$

while for the components of the force perpendicular to the motion

$$F'_y = \frac{dp'_y}{dt'} = \frac{dp'_y/dt}{dt'/dt} = \frac{dp_y/dt}{\gamma(1 - vu_x/c^2)}$$

$$= \frac{F_y}{\gamma(1 - vu_x/c^2)}. \tag{5.34}$$

If the particle is at rest in the unprimed frame (it can only be so instantaneously, as it is being accelerated), the law is simply

$$F'_x = F_x, \qquad F'_y = \frac{1}{\gamma}F_y, \qquad F'_z = \frac{1}{\gamma}F_z. \tag{5.35}$$

Massless particles Before concluding this chapter, let us return to the relations (5.23) and consider what happens when we have a particle with zero rest mass. The formulas appear to imply that such particles would not be observable: they carry no energy or momentum, *unless they move with the speed of light*, in which case $m \to 0$ is accompanied by $\gamma(u) \to \infty$. The ratio of the two equations in (5.23) shows that for massless particles propagating with the speed of light

$$E = |\mathbf{p}|c. \tag{5.36}$$

There exist particles, such as the neutrino (and the quantum of electromagnetic radiation, the photon) that have rest mass zero within experimental errors. Such particles can never be brought to rest, though in a collision the momentum and the energy may change in a way that is consistent with Eq. (5.36). They also have inertia whose magnitude is E/c^2.

NOTES AND COMMENTS

---

1. In nonrelativistic mechanics force is a vector, just like momentum. In relativistic mechanics we see that the transformation laws of forces are quite different from those of energy and momentum. We can define a relativistic force by differentiating the four components of the four-momentum $(E/c, \mathbf{p})$ with respect to the proper time τ. We shall see in Chapter 7 that electric and magnetic fields also do not transform like the momentum.

2. Taylor and Wheeler in *Spacetime Physics* (W.H. Freeman, San Francisco, 1963) discuss a high precision test of the equivalence of energy and rest mass in the reaction $^2H + {}^2H \to {}^1H + {}^3H$. In this reaction the deuteron mass, the hydrogen mass and the tritium mass are known to high accuracy, and the initial kinetic energy and the final kinetic energies were measured with sufficient precision to check the equivalence to one part in 10^5.

3. There is actually a close relationship between the energy-momentum transformation laws and the Doppler shift formulas, because, as we shall learn in Chapter 14, radiation of a frequency ν may be associated with *quanta* of energy $h\nu$ (where h is a universal constant, Planck's constant) and momentum $h\nu/c$. Thus the transformation laws with zero rest mass also apply to the frequency of radiation. By

the same token, a quantum of radiation with frequency ν, has inertia as if it had mass $h\nu/c^2$.

4. For any two four-vectors such as $(ct, \mathbf{r})$ and $(E/c, \mathbf{p})$ the "scalar product" such as $(ct)^2 - \mathbf{r}^2$, $(E/c)^2 - \mathbf{p}^2$ and also $ct(E/c) - \mathbf{r} \cdot \mathbf{p} = (Et - \mathbf{r} \cdot \mathbf{p})$ are invariant.

5. A very pretty demonstration that radiation has inertia given by E/c^2 is given by Einstein in a paper reprinted in *The Principles of Relativity*, Dover Publications.

PROBLEMS

5.1 The earth receives energy in the form of radiation from the sun at the rate of 1.34 $\times 10^3$ w/m^2. What is the rate at which the sun is losing energy?

5.2 A beam of pions consisting of 10^{10} pions/cm^2 sec moving with velocity such that the total energy of each pion is 10^5 MeV (the pion rest mass is such that $m_\pi c^2 = 140$ MeV) is absorbed by a surface. What is the pressure that the beam exerts on the surface?

5.3 Show that a γ ray cannot create an $e^+ - e^-$ pair without the presence of a third body, as such a process would violate energy or momentum conservation.

5.4 Radiation behaves like a stream of particles with zero rest mass, so that the propagation is always with velocity c. Suppose these particles had a tiny rest mass. Show that γ-rays with different energies would have different velocities. Suppose a hydrogen bomb is exploded in outer space, 10^6 km from earth. If γ-rays with energies ranging from 1 MeV down to 1 eV arrive at a detector at the same time within 10^{-9} sec, what upper bound can you set on $m_\gamma c^2$?

5.5 An atom of rest mass M decays from an excited state to the ground state, with a change of rest mass ΔM, with $\Delta M \ll M$. In the decay process it emits a γ ray. Use energy and momentum conservation to determine the energy of the γ-ray, when the atom is initially at rest.

5.6 The reaction $p + d \rightarrow {}^3\text{He} + \gamma$ is exothermic, that is, it takes place even when the initial particles have zero kinetic energy. If the energy of the γ-ray is 5.5 MeV with the initial particles at rest, what is the mass of the ${}^3\text{He}$ nucleus? The proton mass is 1.6724 $\times 10^{-24}$ gm, and that of the deuteron is 3.3432×10^{-24} gm.

5.7 Use the transformation law for energy and momentum to show that given two particles with energies and momenta (E_1, p_1) and (E_2, p_2), respectively, the quantity

$$E_1 E_2 - \mathbf{p_1} \cdot \mathbf{p_2} c^2$$

is invariant.

5.8 A radio station broadcasts with power 500 kilowatts. How much "mass" does this use up in a year?

5.9 An electron of energy E collides head on with a γ-ray of energy ϵ. What is the electron energy loss?

5.10 A twenty megaton hydrogen bomb explodes. Given that 1 ton of TNT equivalent $= 4.18 \times 10^9$ j, calculate the equivalent amount of "mass" used up in the reaction.

5.11 Since E/c and $\mathbf{p}$ transform just like ct and $\mathbf{r}$, it is possible to draw a Minkowski type of diagram relating energy to momentum. Plot such an $E/c - p$ diagram, and draw on it the lines of constant rest mass. What does the region analogous to "outside the light cone" represent?

5.12 Consider the angular momentum $\mathbf{L} = \mathbf{r} \times \mathbf{p}$. Use the transformation properties of $\mathbf{r}$ and $\mathbf{p}$ to calculate $\mathbf{L}'$ the angular momentum in the frame of an observer moving along the x axis with velocity v relative to the unprimed frame. The transformation laws will bring in the combination of the form $(\mathbf{p}t - \mathbf{r}E/c^2)$. Obtain the transformation properties of these combinations as well.

6
Relativistic Kinematics
of Particles

The most immediate applications of the relativistic transformation laws for energy and momentum occur in the kinematics of collisions and decays of nuclear and subnuclear particles. In high-energy accelerators, charged particles are accelerated to energies that are very large compared with their rest masses, and allowed to collide with targets, consisting of other (or the same) particles effectively at rest. The energies, momenta, scattering angles, and so on, are all measured in the laboratory. It turns out that the theoretical predictions of the outcomes of such high-energy experiments are generally simplest in the center-of-momentum frame, and one thus faces the technical problem of translating measurements in the *laboratory frame* to the *center-of-momentum frame*.

Center-of-momentum frame Our first task is to find the velocity that an observer must move with so that the center-of-momentum frame is at rest relative to that observer. To do this, let us consider a two-body collision (for practical reasons associated with the smallness of the elementary particles, there are no three-body collisions) between particle A, at rest, and particle B, with momentum p_L (lab-momentum) chosen in the x direction for convenience. In the laboratory

$$E_A = m_A c^2, \qquad\qquad p_A = 0$$
$$E_B = (p_L^2 c^2 + m_B^2 c^4)^{1/2}, \qquad p_B = p_L. \qquad (6.1)$$

In a frame moving with velocity v in the positive x direction (the direction of motion of particle B) these become, using Eqs. (5.26 and 5.27)

$$E_A' = \gamma m_A c^2 \qquad\qquad p_A' = -\beta\gamma m_A c$$

$$E_B' = \gamma(E_B - p_L \beta c) \qquad p_B' = \gamma\left(p_L - \frac{1}{c}\beta E_B\right). \qquad (6.2)$$

The *center-of-momentum* frame is that in which

$$\mathbf{p}_A' + \mathbf{p}_B' = 0, \qquad (6.3)$$

that is,

$$\gamma\left(p_L - \frac{1}{c}\beta E_B\right) - \beta\gamma m_A c = 0.$$

The resulting expression for v is

$$v = \beta c = \frac{p_L c}{E_B/c + m_A c}. \qquad (6.4)$$

In the nonrelativistic limit, $E_B/c \approx m_B c$ and $p_L \approx m_B u_L$ so that

$$v = \frac{m_B u_L}{m_A + m_B} \tag{6.5}$$

the classical result. In the ultrarelativistic limit $E_B/c \approx p_L$ so that

$$v \approx \frac{p_L c}{p_L + m_A c} \approx \frac{c}{1 + m_A c/p_L}. \tag{6.6}$$

Transformation of scattering angles Suppose in the collision

$$A + B \rightarrow C_1 + C_2 + \cdots$$

a particle emerges that in the center-of-momentum frame has momentum q^* making an angle θ with the collision axis, that is

$$q_x = q^* \cos \theta,$$
$$q_y = q^* \sin \theta. \tag{6.7}$$

We denote the energy of the particle in the center-of-momentum frame by E^*. We can now transform back to the laboratory frame:

$$q_{xL} = \gamma(v)\left(q^* \cos \theta + \frac{v}{c^2} E^*\right)$$

$$q_{yL} = q^* \sin \theta$$
$$E_L = \gamma(v)(E^* + vq^* \cos \theta). \tag{6.8}$$

The angle that the particle makes with the x axis in the laboratory frame is given by

$$\tan \theta_L = \frac{q_{yL}}{q_{xL}} = \frac{1}{\gamma(v)} \frac{q^* \sin \theta}{q^* \cos \theta + \frac{v}{c^2} E^*}. \tag{6.9}$$

In the nonrelativistic limit $E^* \approx m_n c^2$ and $q^* = m_n u^*$, so that

$$\tan \theta_L \approx \frac{u^* \sin \theta}{u^* \cos \theta + v} \tag{6.10}$$

In the ultrarelativistic limit we use Eq. (6.6) to calculate

$$1 - \beta^2 \approx 1 - (1 - m_A c/p_L)^2 \approx 2 m_A c/p_L,$$

so that

$$\frac{1}{\gamma(v)} \approx \left(\frac{2 m_A c}{p_L}\right)^{1/2}.$$

Hence

$$\tan \theta_L \approx \left(\frac{2 m_A c}{p_L}\right)^{1/2} \frac{u^* \sin \theta}{u^* \cos \theta + v}. \tag{6.11}$$

This is the nonrelativistic result, except that the angle is "squeezed down" into the forward direction by the factor $1/\gamma$. Thus all particles, except those that come off in a very small backward cone in the center-of-momentum frame, will be projected into a very narrow cone in the laboratory. This has important consequences for experimentalists who undertake to detect and study the properties of particles produced in such high-energy reactions. All equipment must be "downstream" at a very small angle to the incident beam direction. This is not the disaster that it could be, because the time-dilation factor for unstable particles gives experimentalists the freedom to go a long way from the target and gives them room to place their detectors. At the Fermi National Accelerator Laboratory, protons are accelerated to a velocity so close to c that $\gamma \approx 400$, and it is a measure of the forward collimation (aided by the fact that θ tends to be small in the center-of-momentum frame for high-energy collisions), that most of the particles that are produced in a high-energy proton-proton collision can be found within an area of 10^3 cm^2 15 m downstream (that is the typical aperture of the analyzing magnets used for the detection of charged particles).

The center-of-momentum energy in a reaction is much smaller than the laboratory energy, and it is the former that is of theoretical significance in reaction theory; much of the laboratory energy brought in by the incident particle is expended in pushing the center of momentum downstream, and only a small fraction is left for close penetration and/or production of new particles. An efficient way of converting between the laboratory frame and the center-of-momentum frame is to use *invariants*.

Invariants It will be recalled that the quantity

$$E^2 - (pc)^2 = m^2 c^4 \tag{6.12}$$

is an invariant, that is, it takes on the same value in every frame. If we have two different particles, then the transformation laws for the two, for example,

$$E'_A = \gamma(E_A - vp_A)$$
$$E'_B = \gamma(E_B - vp_B) \tag{6.13}$$

imply that

$$(E_A + E_B)' = \gamma((E_A + E_B) - v(p_A + p_B)) \tag{6.14}$$

and similarly

$$(p_A + p_B)' = \gamma\left(p_A + p_B - \frac{v}{c^2}(E_A + E_B)\right). \tag{6.15}$$

Consequently sums of energies and sums of momenta transform just like single energies and momenta, and an immediate consequence of Eqs. (6.14) and (6.15) is that

$$(E_A + E_B)^2 - c^2(\mathbf{p}_A + \mathbf{p}_B)^2 \equiv s \tag{6.16}$$

is an invariant. In analogy with the nomenclature used in Eq. (6.12) we call s an invariant (mass)^2. Thus to find the laboratory energy, given the center of momentum energies or vice versa, all we need to do is compute s in the frame in which the data are given. For example, in the center-of-momentum frame, in which $\mathbf{p}'_A + \mathbf{p}'_B = 0$, we have

$$s = (E'_A + E'_B)^2. \tag{6.17}$$

In the laboratory frame it is given by

$$\begin{aligned} s &= (m_A c^2 + E_B)^2 - c^2 \mathbf{p}_B^2 \\ &= m_A^2 c^4 + 2 m_A c^2 E_B + E_B^2 - c^2 \mathbf{p}_B^2 \\ &= m_A^2 c^4 + m_B^2 c^4 + 2 m_A c^2 E_B \end{aligned} \tag{6.18}$$

so that

$$E_B = \frac{s - (m_A c^2)^2 - (m_B c^2)^2}{2 m_A c^2}. \tag{6.19}$$

In this way the laboratory energy required to produce a certain center-of-momentum energy s is easily calculated. In this formula it is particularly clear that laboratory energies grow as the square of the desired center-of-momentum energy, and this has had profound implications for the art of accelerator building. In order to go up by an order of magnitude in useful energy, it is necessary to go up by two orders of magnitude in laboratory energy, and that means an enormous cost in materials, primarily iron for the magnets that confine the beam, and the cost of maintaining a vacuum system. The Fermi National Acclerator Laboratory proton accelerator has a diameter in excess of 2 km. In order to achieve much higher center-of-momentum energies, primarily for the production of high-mass particles, several *colliding-beam* machines have been designed. In each case an existing accelerator produces particles that are deflected by means of magnets into a storage ring in which no acceleration takes place, and the vacuum is so good that particles can circulate in it for a long time. That same accelerator then produces more particles (of the same or opposite charge) that are injected into another storage ring. The two storage rings intersect at several points, and there collisions at enormous center-of-momentum energies can take place. There exist electron-positron colliding-beam facilities in several laboratories around the world, the highest energy being achieved at DESY in Hamburg and PEP at Stanford, and there exists a proton-proton colliding-beam facility at CERN in which 28 GeV protons collide in the center of momentum with 28 GeV protons (28 GeV energy corresponds to a γ of approximately 30). Colliding-beam machines do not provide all the answers: clearly the number of collisions is much smaller, since a solid target is replaced by a beam of particles. Also the collision region of overlap of the rings is right in the middle of the accelerating apparatus, and shielding materials render the experimental area small and awkward to work in, so that a fraction of the total solid angle of 4π into which particles can be produced cannot be covered with detectors. Nevertheless, these new accelerators have helped physicists uncover totally new phenomena that would have otherwise remained inaccessible to experimental study, and plans are being made to build several very high-energy colliding-beam facilities.

Production thresholds The use of invariants is very convenient in the calculation of energies required to produce new particles of a given mass. For example, the only way to produce antiprotons is in a collision reaction. For reasons that will be discussed in Chapter 41 the reaction must conserve the number of protons minus the number of antiprotons, so that a typical reaction is of the form

$$p + p \rightarrow p + p + p + \bar{p}.$$

What is the minimum energy of the incident proton in the laboratory to produce an antiproton in such a reaction? The question is easily answered in the center of momentum frame. There the initial momentum is zero, and by *momentum conservation* the sum of all the final momenta is also zero. The lowest energy in the final state is that for which all the particles in the final state have zero momentum. That is the *threshold* for the reaction because all of the energy goes into the production of the particles and none into their kinetic energy. At threshold

$$s = (4M_p c^2)^2, \tag{6.20}$$

and hence

$$E_L = \frac{s - 2(M_p c^2)^2}{2M_p c^2} = 7M_p c^2. \tag{6.21}$$

Quite generally, for the production of a final state of particles with masses $M_1, M_2,$ $\ldots M_k$, the threshold value of s is

$$s = \left(\sum_{n=1}^{k} M_n c^2 \right)^2. \tag{6.22}$$

Relativistic kinematics prove particularly useful in the study of decays of very short-lived unstable particles, when only the decay products live long enough to have their energies and momenta measured. To see how the masses of such particles can be measured, consider a particle A at rest, and suppose that it decays into a number of decay products

$$A \rightarrow B_1 + B_2 + \cdots + B_n.$$

The emerging particles are measured to have energies and momenta $(E_i, \mathbf{p}_i)$. The *conservation of energy* implies that

$$\sum_{i=1}^{n} E_i = M_A c^2 \tag{6.23}$$

and the conservation of momentum implies that

$$\sum \mathbf{p}_i = 0. \tag{6.24}$$

Hence

$$s = \left(\sum_{i=1}^{n} E_i \right)^2 - \left(\sum_{i=1}^{n} \mathbf{p}_i c \right)^2 = (M_A c^2)^2. \tag{6.25}$$

Because s is an invariant, it will take on the same value in any other frame. For example, in the reaction

$$\pi^- + p \rightarrow \omega^0 + \pi^- + p$$

in which a short-lived particle ω^0, with the decay

$$\omega^0 \rightarrow \pi^+ + \pi^- + \pi^0$$

was first produced, it was found by the following technique. The reaction was studied in a bubble chamber, a detection device in which charged particles leave well-defined tracks that can be photographed and measured. The above reaction may give rise to a picture in the bubble chamber that looks as follows (Fig. 6.1): only four prongs are visible, and all such "four-prong" events in a given collection of photographs are measured. Since the liquid in the bubble chamber is usually hydrogen, one knows that the target particle was a proton, and the experiment is designed in such a way that one knows the momentum, and hence the energy, of the incoming particle with sufficient accuracy. The four tracks are measured, and thus the sum of the visible energies $\sum_{i=1}^{4} E_i$ and the sum of the visible momenta $\sum_{i=1}^{4} \mathbf{p}_i$ can be determined. The question next is: are there any missing neutral particles? Suppose a single neutral particle were produced in the reaction. In that case, the *missing energy*, that is, $E_L - \sum_{i=1}^{4} E_i$ and the *missing momentum* should be those of a single particle, and

$$s_{\text{missing}} = \left(E_L - \sum_{i=1}^{4} E_i \right)^2 - c^2 \left(\mathbf{p}_L - \sum_{i=1}^{4} \mathbf{p}_i \right)^2$$

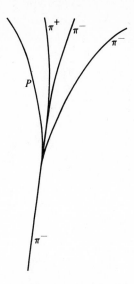

Figure 6.1. Schematic drawing of reaction $\pi^- p \to \pi^- p \omega^0$ as it might manifest itself in a bubble chamber. The magnetic field curves positive particle and negative particle tracks in opposite directions. The unobserved π^0 in the decay $\omega^0 \to \pi^+\pi^-\pi^0$ is deduced from kinematical measurements made on the remaining tracks.

should be fixed. Those pictures, for which $s_{\text{missing}} = (m_{\pi^0} c^2)^2$ within the errors, then describe the reaction

$$\pi^- + p \to \pi^- + p + \pi^+ + \pi^- + \pi^0.$$

We now know the energies of all the particles (including the unseen one), and their momenta. If one then suspects that a triplet of particles, π^+, π^-, π^0, are the decay product of a single state, one instructs the computer to calculate

$$s_{\text{triplet}} = (E_{\pi^+} + E_{\pi^-} + E_{\pi^0})^2 - c^2(\mathbf{p}_{\pi^+} + \mathbf{p}_{\pi^-} + \mathbf{p}_{\pi^0})^2$$

for all triplets, that is, all combinations of π^+, π^- and π^0 in each picture. Not all such triplets will be the decay products of a single state, but if there is such a state, a picture of the number of triplets, as a function of s_{triplet} such as shown in Fig. 6.2 will emerge. The well-defined peak shows that the postulated ω^0 really exists. It is in this way, with various refinements, that a large number of the presently known subnuclear particles were discovered. (Figure 6.3 shows what a real bubble chamber picture, with its large number of noninteracting particles, looks like).

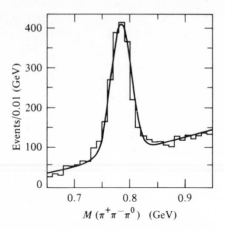

Figure 6.2. Plot of the number of $(\pi^+\pi^-\pi^0)$ events as a function of the invariant mass $M(\pi^+\pi^-\pi^0)$. (Courtesy of the Lawrence Berkeley Laboratory, University of California).

NOTES AND COMMENTS

1. The effect discussed in connection with Eq. (6.11) is also present when a system radiates; the radiation is strongly concentrated in the forward direction. Thus when an electron is circulating in a synchrotron, a circular accelerator, the loss in energy due to radiation associated with the acceleration is concentrated in a direction tangential to the orbit.

2. For reasons that are not yet completely understood, particles produced in high-energy collisions in the center-of-momentum-frame tend to have small transverse momenta, that is, they tend to be produced with angles small relative to the collision axis. Under these circumstances it is a good approximation to treat the problem as if it involved only one space and one time dimension. If we introduce the variable α by writing $E = mc^2 \cosh \alpha$, $pc = mc^2 \sinh \alpha$, then the transformation law after the factor mc^2 has been divided out, reads

$$\cosh \alpha' = \gamma(\cosh \alpha - \beta \sinh \alpha)$$
$$\sinh \alpha' = \gamma(\sinh \alpha - \beta \cosh \alpha).$$

If we now write $\gamma = \cosh \chi$, $\beta\gamma = \sinh \chi$

$$\cosh \alpha' = \cosh \chi \cosh \alpha - \sinh \alpha \sinh \chi = \cosh(\alpha - \chi)$$

which means that a Lorentz transformation just shifts the variable α by χ. Thus for the variables

$$e^\alpha = (E + pc), \qquad e^{-\alpha} = (E - pc)$$

a Lorentz transformation is just a multiplicative transformation. Such variables, called *rapidity* variables are frequently used in high-energy physics.

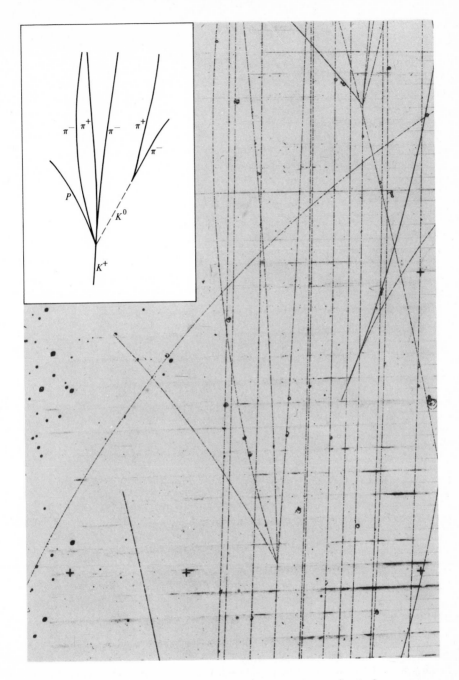

Figure 6.3. Bubble chamber picture of the reaction $K^+p \rightarrow K^0p\pi^+\omega^0$. (Courtesy of the Lawrence Berkeley Laboratory, University of California).

PROBLEMS

6.1 A gram of antimatter falls on earth and annihilates with a gram of matter, giving off only radiant energy. How much energy is there? How much matter can be lifted to a height of 1 km with the energy liberated?

6.2 Recently two new particles were discovered, the ψ' with rest mass such that $M_1 c^2 = 3700\,\text{MeV}$, and the other, the ψ/J as it is called, with $M_2 c^2 = 3100\,\text{MeV}$. Suppose the process

$$\psi' \to \psi + \gamma$$

could occur. What would be the energy of the emerging γ ray, if the ψ' decayed from rest?

6.3 Given that $m_\pi c^2 = 140\,\text{MeV}$ and $M_p c^2 = 940\,\text{MeV}$, calculate the threshold for antiproton production ($M_{\bar{p}} = M_p$) in the reaction

$$\pi + p \to p + p + \bar{p}.$$

6.4 Suppose the pion (π) above is incident upon a nucleus in which the target proton can be in motion. If the maximum momentum of the proton is $pc = 200\,\text{MeV}$, what is the laboratory threshold energy for antiproton production according to the reaction in Problem 6.3?

6.5 In the reactions below:

$$\text{i)} \quad \pi + p \to \Lambda + K,$$
$$\text{ii)} \quad \pi + p \to p + K + K,$$

the masses of the particles involved are $m_\pi c^2 = 140\,\text{MeV}, m_p c^2 = 940\,\text{MeV}, m_K c^2 = 495\,\text{MeV}, m_\Lambda c^2 = 1115\,\text{MeV}$. Calculate the laboratory thresholds for the two reactions.

6.6 What is the threshold for the production of pions in the reaction

$$\gamma + p \to p + \pi?$$

(Use the rest masses given in Problem 6.5.)

6.7 Theorists suspect that there exists a fundamental particle whose rest mass is such that $m_W c^2 = 8 \times 10^4\,\text{MeV}$. What is the threshold for the production of such a particle in a reaction like

$$p + p \to p + p + W?$$

6.8 Suppose the search for such a particle is carried out with a colliding beam set up. If the energy of the protons in one of the beams is $4 \times 10^5\,\text{MeV}$, what is the minimum energy of protons in the other storage ring, so that in a head-on collision the reaction in Problem 6.7 takes place?

6.9 In a colliding beam experiment, in which the beams of identical particles each have energy E, the total available energy for reactions is $2E$. Suppose the beams are not exactly adjusted for a head-on collision, but make a very small angle θ. What is the available energy?

7
Relativity and Electromagnetism

The laws of electrodynamics are form invariant under the Lorentz transformation, that is, they look the same in all inertial frames, as required by the first postulate of special relativity. Rather than just list the transformation laws for the electric and magnetic fields, we shall find it useful to discuss a particular electromagnetic phenomenon in detail and see how different observers interpret the same physical events. Consider a wire, carrying a current i, flowing to the right (Fig. 7.1). The current gives rise to a magnetic field circling the wire, as indicated in the figure. The magnitude of the field at a distance r from the center of the wire is[†]

$$B = \frac{2i}{cr}. \tag{7.1}$$

If a particle of charge q is moving in the negative x direction with velocity v, it will experience a force in a direction perpendicular to the wire, with magnitude

$$F_y = q(v/c)B. \tag{7.2}$$

Thus the particle will be accelerated in the y direction. Consider now an observer at rest relative to the particle. That observer will still see the particle experience an

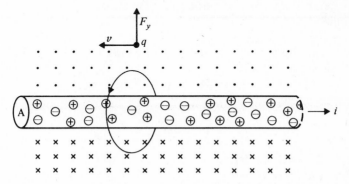

Figure 7.1. Electric current i flows to the right, giving rise to a magnetic field circling the wire in the manner shown. A particle of charge q moving to the left with velocity v experiences the force shown in the figure.

[†] We use Gaussian units for electromagnetic quantities. See Appendix A.

acceleration in the y direction. Since the particle has no x-component of velocity in this frame, the accelerating force cannot be due to the magnetic field, since the force has the form

$$\mathbf{F} = q \, \frac{\mathbf{v} \times \mathbf{B}}{c}. \tag{7.3}$$

How does the second observer explain the motion of the charged particle?

Forces in moving frame To analyze this, let us consider the motion of the charges that give rise to the current. The current is generated by the motion of negatively charged electrons moving with the drift velocity u in the negative x direction. If the charge density of electrons in the laboratory frame is ρ_-, then the current is

$$i = \rho_- A u \tag{7.4}$$

where A is the cross-sectional area of the wire. Since the wire is electrically neutral, the charge of the positive ions ρ_+ is given by

$$\rho_+ = \rho_-. \tag{7.5}$$

If the density of electrons in the frame in which they are at rest is ρ_0, then

$$\rho_- = (1 - u^2/c^2)^{-1/2} \rho_0. \tag{7.6}$$

This increase of density comes about because of the Lorentz contraction of the volume in the direction of motion, and because the *total charge is an invariant*. This statement can be established experimentally, and also is part and parcel of the enormously successful Maxwell equations. A simple way to convince oneself of that is to heat a piece of metal. As we shall learn, heating speeds up the electrons and the ions in the metal differently, and if the charge depended on the velocity, neutrality of the metal would be disturbed. This does not happen.

Consider the special case in which the charged particle that is deflected by the field happens to be moving with velocity v equal to the drift velocity u. This will simplify the algebra in what follows.

In the frame of the observer moving with the charged particle, the electrons are at rest, and therefore

$$\rho_-' = \rho_0. \tag{7.7}$$

On the other hand, the ions are now moving to the right with velocity of magnitude u, so that their charge density is

$$\rho_+' = (1 - u^2/c^2)^{-1/2} \rho_+ = (1 - u^2/c^2)^{-1} \rho_0 = \gamma^2 \rho_0. \tag{7.8}$$

Thus for the primed observer, the wire acquires a net positive charge density given by

$$\rho_+' - \rho_-' = (\gamma^2 - 1) \rho_0 = \frac{u^2}{c^2} \gamma^2 \rho_0. \tag{7.9}$$

This corresponds to a charge density λ per unit length, with

$$\lambda = \frac{u^2}{c^2} \gamma^2 \rho_0 A. \tag{7.10}$$

Such a line charge gives rise to an electric field in the direction perpendicular to the wire, with magnitude

$$E_\perp = \frac{2\lambda}{r} = \frac{1}{r} \frac{2u^2}{c^2} \gamma^2 \rho_0 A. \tag{7.11}$$

This exerts a force on the charged particle, whose magnitude is

$$F'_y = qE_\perp = \frac{2q}{r}\frac{u^2}{c^2}\gamma^2\rho_0 A. \tag{7.12}$$

Let us compare this with the magnetic force in the unprimed frame. There

$$F_y = q(u/c)B = q(u/c)(2i/cr)$$
$$= \frac{2q}{r}\frac{u}{c^2}\rho_- Au$$
$$= \frac{2q}{r}\frac{u^2}{c^2}\gamma\rho_0 A. \tag{7.13}$$

We thus find

$$F'_y = \gamma F_y \tag{7.14}$$

as expected from the transformation law of forces (5.35). Our conclusion is that the two observers see the same effect, a perpendicular force on the charged particle, but the two observers will give different explanations for the acceleration. The observable effect will be the same, when the γ coming from the time dilation effect is taken into account.

Transformation Laws for Electromagnetic Fields The power and beauty of relativity is that one does not need to know anything about the actual mechanism responsible for the current. All one needs to know are the transformation properties of the fields. These are given by

$$E'_\parallel = E_\parallel$$

$$E'_\perp = \gamma\left(E_\perp + \frac{1}{c}\mathbf{v}\times\mathbf{B}\right)$$

$$B'_\parallel = B_\parallel$$

$$B'_\perp = \gamma\left(B_\perp - \frac{1}{c}\mathbf{v}\times\mathbf{E}_\perp\right). \tag{7.15}$$

We shall discuss these transformation laws after indicating one way in which they can be derived. This is to require that the Lorentz force expression

$$\mathbf{F} = q\left(\mathbf{E} + \frac{1}{c}\mathbf{u}\times\mathbf{B}\right) \tag{7.16}$$

read

$$\mathbf{F}' = q\left(\mathbf{E}' + \frac{1}{c}\mathbf{u}'\times\mathbf{B}'\right) \tag{7.17}$$

in a different frame, with

$$u'_\parallel = (u_\parallel - v)/(1 - u_\parallel v/c^2)$$

$$\mathbf{u}'_\perp = \frac{1}{\gamma}\mathbf{u}_\perp/(1 - u_\parallel v/c^2) \tag{7.18}$$

derived in (3.15, 3.16) and

$$F'_\| = \left(F_\| - \frac{v}{c^2} \mathbf{u} \cdot \mathbf{F}\right)\bigg/(1 - u_\| v/c^2)$$

$$F'_\perp = \frac{1}{\gamma} F_\perp/(1 - u_\| v/c^2) \tag{7.19}$$

derived in (5.33 and 5.34). Let us choose the x-direction to be the longitudinal ($\|$) direction.

$$F'_x = q\left(E'_x + \frac{1}{c} u'_y B'_z - \frac{1}{c} u'_z B'_y\right)$$

for example, reads

$$\frac{1}{1 - u_x v/c^2}\left(F_x - \frac{v}{c^2}(u_x F_x + u_y F_y + u_z F_z)\right)$$

$$= q\left(E'_x + \frac{1}{\gamma c}\frac{1}{1 - u_x v/c^2}(u_y B'_z - u_z B'_y)\right). \tag{7.20}$$

If we now substitute

$$F_x = q\left(E_x + \frac{1}{c}(u_y B_z - u_z B_y)\right)$$

and so on, on the left side, and compare coefficients of $u_x, u_y,$ and u_z we find that

$$q\left(E_x + \frac{1}{c} u_y B_z - \frac{1}{c} u_z B_y\right) - \frac{v u_y/c^2}{1 - u_x v/c^3} q\left(E_y + \frac{1}{c} u_z B_x - \frac{1}{c} u_x B_y\right)$$

$$- \frac{v u_z/c^2}{1 - u_x v/c^2} q\left(E_z + \frac{1}{c} u_x B_y - \frac{1}{c} u_y B_x\right)$$

$$= q\left(E'_x + \frac{1}{\gamma}\frac{1}{1 - u_x v/c^2}\left(\frac{u_y}{c} B'_z - \frac{u_z}{c} B'_y\right)\right).$$

The velocity-independent term gives us

$$E_x = E'_x. \tag{7.21}$$

The coefficient of u_y gives, after a little algebra,

$$\frac{1}{\gamma c} B'_z = \frac{1}{c} B_z\left(1 - \frac{u_x v}{c^2}\right) - \frac{v}{c^2}\left(E_y + \frac{1}{c} u_z B_x - \frac{1}{c} u_x B_y\right)$$

$$+ \frac{1}{c}\frac{v u_z}{c^2} B_x$$

which simplifies to

$$B'_z = \gamma\left(B_z - \frac{v}{c} E_y\right). \tag{7.22}$$

The remaining terms give

$$\frac{1}{c\gamma} B'_y = \frac{1}{c} B_y\left(1 - \frac{u_x v}{c^2}\right) + \frac{v}{c^2}\left(E_z + \frac{1}{c} u_x B_y\right)$$

from which follows

$$B'_y = \gamma\left(B_y + \frac{v}{c} E_z\right). \tag{7.23}$$

Thus we have shown that

$$E'_\parallel = E_\parallel$$

$$B'_\perp = \gamma\left(B_\perp - \frac{1}{c} v \times E\right). \tag{7.24}$$

The other equations in (7.15) follow from a similar treatment of F'_y or F'_z.

The implications of (7.15) are varied. One interesting application that we will have occasion to refer to is the fact that an isolated charge at rest in one frame gives rise to a magnetic field of magnitude

$$B'_\parallel = 0 \qquad B'_\perp = -\frac{1}{c}\gamma v \times E_\perp \tag{7.25}$$

in a moving frame. The electric field is also changed to

$$E'_\parallel = E_\parallel \qquad E'_\perp = \gamma E_\perp. \tag{7.26}$$

We notice that

$$E'_\perp \cdot B'_\perp = 0 \tag{7.27}$$

and if γ is very large, we can neglect $E'_\parallel$ in comparison with $E'_\perp$, so that the field looks like the field of an electromagnetic wave! This realization is of great utility in computing the interaction of radiation with matter at high energies. For example, an electron passing a nucleus at very high energies will experience the Coulomb field of the nucleus in its own rest frame as if it were an electromagnetic wave, with amplitude and spatial dependence determined by the transformation laws. An electron, struck by radiation, will scatter it: the process known as Thomson scattering will be discussed in Chapter 14. In the laboratory frame the reaction will appear as the emission of radiation by the electron as it passes the nucleus. In the laboratory frame we explain this by noting that the electron is deflected by the Coulomb field, and whenever a charge is accelerated, it radiates. The calculations are just much simpler (albeit only approximate) in the electron rest frame.

The transformation laws may be used to show that

$$E \cdot B = E' \cdot B' \tag{7.28}$$

and

$$E^2 - B^2 = E'^2 - B'^2, \tag{7.29}$$

thus establishing the existence of two invariants.

What the transformation laws show is that electric and magnetic fields are not separate entities, since they transform into each other. They are different manifestations of the same fundamental entity, the electromagnetic field, and must be treated on the same footing. This unity is not affected by the fact that there are isolated charges present in the world, while there do not seem to exist isolated magnetic monopoles: Nature manifests its symmetries in very subtle ways.

PROBLEMS

7.1 A charge Q at rest gives rise to an electric field given by

$$\mathbf{E} = \frac{Q}{r^2} \, \hat{i}_r.$$

An observer approaches the charge with relativistic speed v. What is the electric field observed by the moving observer?

7.2 Consider the discussion of the description of an electric current from the point of view of two observers given in this chapter. Make a Minkowski diagram analysis of the discussion, and show, in particular, how the wire can appear to have a net charge in the electron rest frame.

7.3 Maxwell's equations are constructed so as to be consistent with the law of conservation of charge, which reads

$$\boldsymbol{\nabla} \cdot \mathbf{J} + \frac{\partial}{\partial t} \rho = 0$$

where $\mathbf{J}$ is the current density and ρ is the charge density. If this law is to be form invariant, what must be the transformation properties of $\mathbf{J}$ and ρ?

8
Beyond Special Relativity

Equivalence principle Special relativity theory deals with the equivalence of all inertial frames. This equivalence clearly cannot extend to frames that are accelerated relative to each other. The physics is manifestly different in the two frames, even when the speeds involved are nonrelativistic. Thus an observer at rest can easily determine whether he or she is in an inertial frame or standing on a rapidly turning turntable. Similarly, an observer standing in an elevator that is accelerated upwards will experience a downward pressure on the feet not shared by an observer in an inertial frame. Our experience with gravity does, however, suggest an equivalence between a downward force of gravity and an upward acceleration. Common experience would suggest that if an observer were to drop a tennis ball in a closed elevator, the behavior of that tennis ball could not be used to distinguish between the following two situations: (i) the elevator is in free space, far from any massive body, but is pulled upwards, and (ii) the elevator is at rest relative to a massive body that exerts a gravitational pull on it and everything in it. Actually this is not quite true: in a large enough elevator, two tennis balls dropped at opposite ends would move slightly towards each other in case (ii), since they would be attracted to the same center of the massive attracting body. On a small enough scale, that is, *locally*, everybody would be willing to accept that it should not be possible to distinguish by mechanical observation whether a system is undergoing uniform acceleration in a gravity-free region or whether it is in a uniform gravitational potential. As a prelude to his formulation of the General Theory of Relativity (which is really a General Theory of Gravitation), Albert Einstein in 1911 postulated the *Equivalence Principle*, which is a generalization of the above observation. Einstein postulated that

> *It is not possible to distinguish by any experiment whatsoever between an accelerated frame and an inertial frame with an appropriate gravitational field, provided the observations are limited to a small enough region of space-time.*

The above principle has some extraordinary consequences, which are in complete agreement with experiment.

Consider a mass on a spring hanging in our elevator. If the elevator is accelerated upwards with acceleration g, the spring extends, and the extension is that due to an equivalent force

$$F = m_i g. \tag{8.1}$$

The coefficient m_i is the *inertial mass* of the body. The same coefficient m_i appears in the expression for the centrifugal force. If the elevator is unaccelerated, but a distance R from the center of a very massive body of mass M, the downward force on the mass on the spring will be

$$F = G \frac{m_g M}{R^2} \qquad (8.2)$$

where G is the gravitational constant ($G = 6.67 \times 10^{-8}$ cm^3 gm^{-1} sec^{-2}) and m_g *is the gravitational mass*. Normally we would replace both m_i and m_g by the "mass," but that is making an assumption. There is no *a priori* reason why the gravitational law of interaction, which in the nonrelativistic limit is represented by the potential energy of interaction

$$V(\mathbf{r}_1 - \mathbf{r}_2) = G \frac{m_1 m_2}{|\mathbf{r}_1 - \mathbf{r}_2|}, \qquad (8.3)$$

should in any way involve the inertial mass. Coulomb's law follows from a similar kind of potential, but it involves some other property (the net electrical charge) of the bodies in interaction. The gravitational mass is a measure of the strength with which the gravitational field couples to a body, and one could easily imagine a theory in which the potential is

$$V(\mathbf{r}_1 - \mathbf{r}_2) = f \frac{N_1 N_2}{|\mathbf{r}_1 - \mathbf{r}_2|} \qquad (8.4)$$

where N_i is the total number of protons and neutrons in each body. Such a theory would, to an accuracy of a few parts in a thousand, look like the theory of Newton (because the electron mass is 2000 times smaller than the proton or neutron mass, and binding effects are even smaller), but it would really be a different theory, and it would not satisfy the equivalence principle. That principle requires that

$$m_i = m_g. \qquad (8.5)$$

That this is indeed the case was determined to an accuracy of one part in 10^9 by Baron Eötvös in a series of experiments beginning in 1889. More recently, the equality was confirmed by R.H. Dicke and collaborators (1967) to an accuracy of one part in 10^{11}. The principle of the experiment is the following: when two weights, made of different materials, are balanced at a given time, the equilibrium results from the equality of the product (force) × (lever arm) for the two weights. The force acting on a given weight is

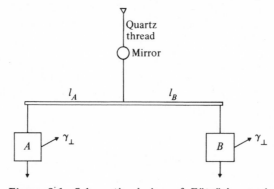

Figure 8.1. Schematic design of Eötvös' experiment.

a composite of the gravitational force of the earth, the gravitational force of the sun, and the centripetal forces due to the earth's rotation, due to the motion of the earth around the sun, and so on. Let us denote the acceleration due to gravity by g and the acceleration due to the earth's rotation and other noninertial effects by $(\gamma_{\parallel}, \gamma_{\perp})$, where the first component is parallel to the direction of g and the second is in the horizontal direction (Fig. 8.1). The equilibrium condition is

$$l_A(m_g(A)g + m_i(A)\gamma_{\parallel}) = l_B(m_g(B)g + m_i(B)\gamma_{\parallel}) \qquad (8.6)$$

where l_A and l_B are the lengths of the lever arms on the two sides of the point of suspension. There is a torque on the bar, whose magnitude is

$$T = \gamma_{\perp}(m_i(A)l_A - m_i(B)l_B). \qquad (8.7)$$

Using Eq. (8.6) we can rewrite this in the form

$$\frac{1}{\gamma_{\perp}}T = l_A\left(m_i(A) - m_i(B)\frac{m_g(A)g + m_i(A)\gamma_{\parallel}}{m_g(B)g + m_i(B)\gamma_{\parallel}}\right) \qquad (8.8)$$

and since $\gamma_{\parallel} \ll g$, we neglect $\gamma_{\parallel}$ to get the simpler form

$$T = m_g(A)l_A\gamma_{\perp}\left(\frac{m_i(A)}{m_g(A)} - \frac{m_i(B)}{m_g(B)}\right). \qquad (8.9)$$

Thus if the two ratios are not equal, the rod will start to rotate. The rotation can be determined by studying the reflection of a beam of light from a mirror attached to the quartz thread that supports the balance. Eötvös' experiment was sufficiently accurate to determine the equality of the ratios to one part in 10^9. The experiment of Dicke, Roll, and Krotkov was more sophisticated, and allowed for an improvement by two and one-half orders of magnitude.

Gravitational red shift Another consequence of the Equivalence Principle is the gravitational red shift of radiation. Consider an emitter of radiation of frequency ν at a height x above an absorber. The emitter is at a higher potential gx. The Equivalence Principle states that the physics should be the same if, instead of being in a potential field due to a large mass below the absorber, the system were accelerated in the upward direction with acceleration g (Fig. 8.2). If at the time of emission, $t = 0$, both

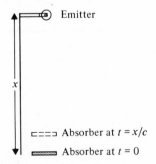

Figure 8.2. Demonstrating why a "falling" light ray changes frequency.

the emitter and the absorber are at rest, then in the time that the radiation has reached the absorber, $t = x/c$ to good approximation, the absorber will have acquired a velocity $gt = gx/c$. Thus the absorber detects the radiation with a frequency that is Doppler shifted upwards (blue shift) to ν', where

$$\frac{\nu'}{\nu} = \sqrt{\frac{1 + v/c}{1 - v/c}} \approx 1 + \frac{v}{c} = 1 + \frac{gx}{c^2}$$

$$= 1 + \phi/c^2. \tag{8.10}$$

In the last step we have introduced ϕ, the potential difference between the emitter and the detector. In the case of observation of light from a star, we are "up" relative to the emitter. The earth is effectively at zero potential, and that on the surface of the star is $- GM/R$, where M is the mass of the star, and R is its radius. Thus for the red shift

$$\frac{\Delta\nu}{\nu} = \frac{\nu' - \nu}{\nu} = \frac{\phi}{c^2} = -\frac{GM}{Rc^2}. \tag{8.11}$$

For the sun, $M = 2 \times 10^{33}$ gm, $R = 7 \times 10^7$ cm, so that with $G = 6.67 \times 10^{-8}$ cm^3 gm^{-1} sec^{-2} and $c = 3 \times 10^{10}$ cm sec^{-1}, the red shift is

$$\left|\frac{\Delta\nu}{\nu}\right| = 2.12 \times 10^{-6}. \tag{8.12}$$

Recent measurements carried out on the sodium D lines in the solar spectrum have confirmed this to an accuracy of 5%. Terrestrial measurements, using extremely accurate clocks that utilize the Mössbauer effect (see Chapter 36), have also confirmed the gravitational red shift to a 10% accuracy. To see how accurate a clock is needed, we note that the light "fell" from a height of 30 m so that with $g = 9.8$ m sec^{-2},

$$\frac{\Delta\nu}{\nu} = \frac{gx}{c^2} \simeq 3.3 \times 10^{-15}. $$

Deflection of light The Equivalence Principle also predicts that light is deflected by a gravitational field. Consider again our elevator that is being accelerated upwards. If there is a pinhole in the side of the elevator, a ray of light entering it will, after traversing the elevator, hit the other side somewhat below where it would have hit had the elevator not been accelerating (Fig. 8.3). In the time of traversal of the width w, $t = w/c$, the elevator moves upwards a distance $gt^2/2 = \frac{1}{2}g(w/c)^2$ where g is the acceleration. This corresponds to a deflection by an angle

$$\delta\theta \approx \frac{1}{2}g\frac{w}{c^2} \approx \frac{1}{2}\frac{GM}{R^2c^2}w. \tag{8.13}$$

An observer in the elevator, viewed as being at rest in a gravitational field, would interpret the effect as due to the light "falling" in the gravitational field. Thus light possesses *inertia*. The deflection of light by a massive body, the sun, has been observed during solar eclipses. This is done by comparing the angle subtended by double stars when the light rays pass on the two sides of the sun with the angle when they do not go near the sun. The first measurement can clearly be carried out only when the over-

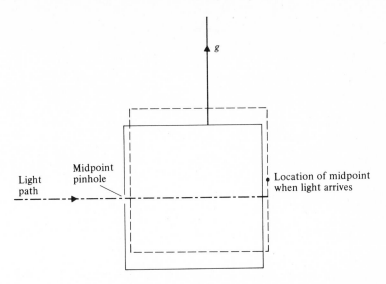

Figure 8.3. Demonstration that a light ray is bent by a gravitational field.

whelming brightness of the sun is blocked out in a total eclipse. The prediction of the Equivalence Principle is a deflection of

$$\delta\theta = \frac{GM}{Rc^2}.$$ (8.14)

This result is not in agreement with experiment. The Einstein theory of gravitation predicts double that amount because of the distortion of space in the vicinity of a massive body. The prediction of $1.66''$ of arc deflection is in good agreement with experiment, and its confirmation was one of the earliest triumphs of General Relativity theory.

Distortion of space-time Any discussion of Einstein's theory of gravitation involves mathematical tools, such as tensor analysis, that are beyond the reach of the intended reader of this book. Thus we can give only a qualitative discussion of how it comes about that space-time is distorted. Such a distortion is difficult to visualize, except in the much simpler case of a two-dimensional space with time ignored. Imagine a bug living in a two-dimensional space, that is, on a surface. If the surface were a table, an intelligent bug might, through experiences with measurements, be led to the invention of Euclidean geometry. The bug would derive certain theorems. One of them would be that the distance between two points nearby can always be written in the form

$$(ds)^2 = (dx)^2 + (dy)^2.$$ (8.15)

More precisely, the bug would find it possible to always find a local orthogonal coordinate system in which the above form holds. In polar coordinates the bug would find

$$(ds)^2 = (dr)^2 + r^2(d\theta)^2$$ (8.16)

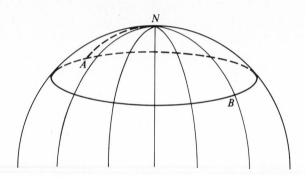

Figure 8.4. The diameter of the circle, ANB, is larger in spherical geometry than it is in Euclidean geometry for the same circle, so that (circumference)/(diameter) $< \pi$.

from which it would be possible to predict that the circumference of a circle is related to the radius by

$$s = \int_{r=\text{const.}} ds = r\int_0^{2\pi} d\theta = 2\pi r. \tag{8.17}$$

Suppose now that the bug begins to wonder whether its space is really a table, or whether it might be living on the surface of a large sphere, which is only locally Euclidean. A way of checking this is to measure whether for a circle, the ratio (circumference)/(diameter) is indeed π. A circle is defined as the locus of points equidistant from the center, and the diameter is the longest "straight line" across the circle, where "straight line" must be defined as the curve taken by the most tautly tied string between two points (Fig. 8.4). Such a "shortest distance" is called a *geodesic*. If a bug were living on a sphere, a measurement on a large circle would show that the ratio is actually smaller than π, as the inspection of a globe easily shows. In fact, for the extreme case of an equator, the ratio is 2. The deviation of the ratio from π is characterized by a parameter called the *curvature*. The definition of curvature is so arranged that it is $1/R$ for a sphere of radius R, that is, it is positive for a sphere, and zero for a plane (the limit $R \to \infty$). An example of a surface with negative curvature is the surface of a trumpet. On a general surface, the distance between two adjacent points in an arbitrary coordinate system (still labeling the coordinates by x, y) is

$$(ds)^2 = g_{11}(x,y)(dx)^2 + 2g_{12}(x,y)\,dx\,dy + g_{22}(x,y)(dy)^2 \tag{8.18}$$

which looks different from Eq. (8.15). The point where the more advanced subject of Riemannian geometry comes is in determining whether the form (8.18) is different because we have chosen a novel set of coordinates (cf. Eq. (8.16) which still holds for a plane) or whether it is different because of an intrinsic curvature in the space.

Distortion of space in rotating frame We will now illustrate how a gravitational field may "distort" space. Consider, for example, a rotating disc. Any object on the disc, not at the center, will experience an acceleration, whose magnitude is $\omega^2 R$ where ω is the angular velocity of the disc and R the distance from the center. Locally the object would react as if it were in a gravitational field. The gravitational potential may be found by noting that the force at a point r from the center is (for a unit mass)

$$F = \omega^2 r. \tag{8.19}$$

Hence

$$\frac{d\phi}{dr} = -\omega^2 r. \tag{8.20}$$

Thus, with $\phi(0) = 0$ we have

$$\phi(r) = -\frac{\omega^2 r^2}{2} \tag{8.21}$$

Thus clocks will be distorted, since a periodic source will have its frequency changed by

$$\frac{\Delta\nu}{\nu} = 1 + \frac{\phi}{c^2} = 1 - \frac{\omega^2 r^2}{2c^2} \tag{8.22}$$

as seen at the zero potential point, that is, the center. Note that this is identical to the transverse Doppler shift

$$\frac{\Delta\nu}{\nu} = (1 - v^2/c^2)^{1/2} \cong 1 - \frac{v^2}{2c^2}$$

$$= 1 - \frac{(\omega r)^2}{2c^2}. \tag{8.23}$$

Because clocks get distorted in a way that depends on their location we must be careful about the definition of time and also space measurements. We will agree to define lengths, on a small scale, as if they were determined in an inertial frame whose velocity coincides with their instantaneous velocity when the latter is not constant. Thus a small cord on the circumference of a circular turntable moves instantaneously in a tangential direction with velocity $v = \omega r$, and the length of that cord, measured by an observer at rest relative to the center of the circle, will be that of a similar length moving with uniform velocity $v = \omega r$. In this spirit, let us consider the ratio of circumference to radius for a circle centered about the axis of rotation. To the inertial observer, measuring rods laid out on the disc in a radial direction from the center to the circumference move transversely to their extension. An inertial observer may count R unit rods. When such unit rods are placed in a way to form a polygon that approximates the circumference of the circle, they move in the direction of their extension. Thus the inertial observer will see them shortened by a fraction $(1 - v^2/c^2)^{1/2} = (1 - \omega^2 r^2/c^2)^{1/2}$. The inertial observer will therefore conclude that the number of rods necessary to go round the circle is larger than $2\pi R$, by a factor of $(1 - \omega^2 r^2/c^2)^{-1/2}$. Hence in the gravitational field for which the local gravitational potential is

$$\phi = -\tfrac{1}{2}\omega^2 r^2, \tag{8.24}$$

the ratio

$$\frac{\text{circumference}}{\text{diameter}} = \frac{\pi}{(1 - \omega^2 r^2/c^2)^{1/2}}$$

$$= \frac{\pi}{(1 + 2\phi/c^2)^{1/2}}$$

$$> \pi. \tag{8.25}$$

Thus the geometry is noneuclidean. The expression for the separation between two neighboring points is

$$(ds)^2 = (dr)^2 + \frac{r^2(d\theta)^2}{1 + 2\phi/c^2} \tag{8.26}$$

to take the above into account. We note that merely the taking account of special relativity effects already introduces the noneuclidean features. In the general theory of relativity the space-time separation (with $ct = x_4$),

$$(d\tau)^2 = \sum_i \sum_j g_{ij}(dx_i)(dx_j) \tag{8.27}$$

(with $g_{ij} = g_{ji}$) involves ten quantities, $g_{11}, g_{12}, \ldots g_{44}$, that is, a ten-component gravitational field. Einstein wrote down the simplest equation governing the changes in the g_{ij} consistent with special relativity and the Equivalence Principle. These equations involve the distribution of energy and momentum of matter, and are quite complicated.

Some of the consequences were worked out by Einstein. These are (i) the deflection of light additional to that caused by the inertia of radiation, (Fig. 8.5) and (ii) the precession of the perihelion of mercury. For a pure $1/r$ potential, the orbits are, as is well-known, Keplerian ellipses. These elliptical orbits of the planets are stationary, that is, the major axis always points in the same direction. If the potential is altered by a small perturbing potential, the shape of the orbit will not change, but each time the planet comes around it misses the previous year's point by just a little, and the effect is that the ellipse appears to precess. Thus planetary orbits are precessing because of the presence of other planets, particularly Jupiter and Saturn with their large masses. After the calculable effect of all the planets was taken into account, it was found that a small precession appeared to be left over. In the most pronounced case, that of the orbit of Mercury a precession of $42''$ of arc per century was unaccounted for. We should add, parenthetically, that it is a tribute to the incredible care with which astronomers have made their observations, that a trustworthy number for such a tiny effect was established. The deformation of space gives a small $1/r^2$ contribution to the potential and the coefficient that emerged from the calculation was of just the right magnitude to explain the $42''$ precession.

Black holes In recent years much attention has been paid to large gravitational field effects. The most dramatic one is the prediction of *black holes*. In an ordinary star, the mass of the material will tend to collapse because of the mutual gravitational

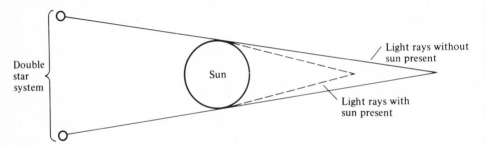

Figure 8.5. Measuring the deflection of light by the sun.

attraction of the various parts of the mass with each other. This collapse is resisted by short-range repulsive forces between the particles that make up the atoms in the stars, and by the outward pressure caused by the nuclear reactions going on inside the star. When the burning ceases, a star will collapse until the repulsive forces stabilize it. A small star will turn into a "white dwarf," something the size of a planet. If the mass of the material is larger (say a couple of solar masses), then the first line of defense, the electronic repulsive forces, become inadequate, and a neutron star (see Chapter 26) with diameter of the order of 10 km is formed. If the mass is larger still, the nuclear repulsive forces are inadequate, and a gravitational collapse occurs. When the radius of the mass distribution becomes small enough, that is, when

$$2\frac{GM}{Rc^2} > 1, \tag{8.28}$$

which occurs when

$$R < R_S \equiv \frac{2GM}{c^2}, \tag{8.29}$$

where R_S is the so-called *Schwartzschild radius*, then radiation cannot escape the gravitational potential $(\Delta\nu/\nu = 1)$ and the star becomes invisible. It still has its gravitational field, but nothing can be emitted from it. This rather extraordinary object should, in principle, be quite common, and in recent years it appears that astronomers have discovered a candidate for it. There exists a double-star system known as Cygnus X, in which one of the stars is visible and very massive (of the order of 20 solar masses). The partner is not visible, but its presence and mass can be inferred from the motion of the visible member. This is possible because the period appears to be very short, of the order of days. The mass of the invisible partner is of the order of ten solar masses, and the binary system is a source of a strongly fluctuating flux of X-rays. The interpretation of the X-ray flux is that it is due to matter leaving the visible star and falling into the black hole, the invisible partner. The identification is somewhat speculative, but many astronomers believe that they have found a candidate for a black hole, only about 6000 light years away.

Finally, the Einstein theory of the gravitational field predicts that any sharp changes in the source of the field will give rise to gravitational radiation, just as a sudden change (acceleration) in the charge will give rise to electromagnetic radiation. Gravitational radiation is very difficult to detect, because the signal is so weak, and recent reports of the detection of gravitational radiation have not been confirmed by subsequent experiments. Nevertheless, detecting gravitational radiation is a technological problem, and it is quite possible that within a decade or two, some of the problems can be solved. Such a signal would give us direct information about some of the mysterious things going on in the center of galaxies, invisible to detectors of electromagnetic signals.

NOTES AND COMMENTS

1. The equality of the inertial and gravitational masses was already noted by Newton, who carried out some experiments with pendulums of equal length but

different materials. The experiments of Roland von Eötvös and the modern refinements of Roll, Krotkov, and Dicke are discussed in detail in R.H. Dicke, *The Theoretical Significance of Experimental Relativity*, Gordon and Breach, Science Publishers, New York, 1964.

2.   The time dilation factor $(1 + \phi/c^2)$ is an approximation for $(1 + 2\phi/c^2)^{1/2}$. When the general form for a space-time interval is written down, then at the same point

$$(d\tau)^2 = c^2(dt)^2(1 + 2\phi/c^2),$$

that is, the gravitational field also affects the time.

3.   The most elementary book on general relativity that I have seen is W. Rindler, *Essential Relativity*, Van Nostrand Reinhold Company, New York, 1969. A non-mathematical account may be found in P.G. Bergmann, *The Riddle of Gravitation*, Charles Scribner's Sons, New York, 1968. A brief account of black-hole physics readable to someone who has gone through this book may be found in the lectures of R.U. Sexl, *Acta Physica Austriaca*, vol. 42 (1975): 303. A number of fascinating articles from *Scientific American* are reprinted in *Cosmology + One*, O. Gingerich (Ed.), W.H. Freeman, San Francisco, 1977. See also D.W. Sciama, *Modern Cosmology*, Cambridge University Press, 1971.

PROBLEMS

8.1 Discuss the twin paradox as explained by the traveling twin. The traveling twin experiences a deceleration at the turning around time. He interprets it as due to a strong gravitational field ahead of him. Thus from his point of view the earthbound twin is at a much higher potential. During the deceleration (and return acceleration) period, the earthbound twin's clock is blue-shifted. Work out the details and show that the traveling twin will get the same result as the earthbound twin who can use special relativity theory.

8.2 A neutron star has a mass of about 4×10^{30} kg and a radius of 10 km. What is the gravitational red shift? If the angular velocity of the pulsar is 100 radians/sec, what is the Doppler shift due to the motion of the source of radiation? Calculate the spread between the effect when the source is coming toward us and when the source is moving away from us.

8.3 The gravitational red shift for the sun is 2.12×10^{-6}. With what velocities would the hot gases on the surface of the sun have to move to mask this shift by an ordinary Doppler shift? Can you think of how it might still be possible to measure the gravitational red shift in spite of the masking?

PART II
Statistical Physics

One of the major achievements of nineteenth-century physics was the recognition of the atomic nature of matter. The work of Dalton and Gay-Lussac led to the conclusion of Avogadro, that at the same temperature and pressure, a given volume of any dilute gas contains the same number of molecules. *Avogadro's number, N_0, defined as the number of molecules in one gram-mole of substance*, has been experimentally determined to be

$$N_0 = 6.02252 \pm 0.00009 \times 10^{23} \text{ molecules/gm-mole}$$

and is a very large number. Nowadays its magnitude is easily understood as follows: molecules consist of electrons and nuclei, and it is the latter that determine the mass of the molecule, since the ratio of the electron mass to nuclear mass is approximately $1/2000A$ where A is the number of protons and neutrons (of nearly equal mass) in the nucleus. A molecule of molecular weight M will thus contain M nucleons, and have an approximate mass of $1.6M \times 10^{-24}$ gm. Thus the number of molecules in one gram-mole is

$$N_0 \cong \frac{M \text{ (gm)}}{1.6 \times M \times 10^{-24} \text{ (gm)}} \cong 6 \times 10^{23}.$$

The value of N_0 tells us something about the size of atoms. A mole of water, for example, occupies 18 cm^3 since the density is 1 gm/cm^3, and thus the number of molecules per cm^3 is $6.02 \times 10^{23}/18 = 3.3 \times 10^{22}$. If close packing is assumed for the molecules, their size is

$$d = (3.3 \times 10^{22})^{-1/3} = 3.1 \times 10^{-8} \text{ cm}$$
$$= 3.1 \text{ Å}.$$

Since molecules are so small, any macroscopic volume of material will contain a huge number of them. Statistical physics rests on the notion that the behavior of a macroscopic amount of material depends only on the average behavior of this large number of molecules. One argues that the behavior of gas in a container does not depend on the individual histories of 10^{23} molecules but only on some parameters that characterize the gas macroscopically. It seems very reasonable that if some gas is let into a box, the particular details of how the gas flowed in become irrelevant after a short time, since collisions among the gas molecules, and collisions with the walls of the container will very quickly randomize any initial organization. This expectation holds only for average properties: clearly, if a single molecule is somehow "tagged," its behavior, assuming it is of interest, has to be followed in detail. Even for average properties one must make some assumptions about the environment of the container, to dispose of the following criticism:

We know that the laws of motion of individual molecules, expressed by

$$m_i \frac{d^2 \mathbf{r}_i}{dt^2} = \mathbf{F}_i = -\nabla_i \left(\sum_j V(\mathbf{r}_i - \mathbf{r}_j) \right),$$

are invariant under the interchange $t \to -t$, since the potential energy does not explicitly depend on the time. Thus if we were to reverse the velocity of every one of the 10^{23} molecules at the same instant, these molecules would revert to the initial state as could be seen if we made a film of the filling of the box and ran it backwards. According to the above equation, the molecules would end up by running out of the hole that they were let in through. This would imply that the molecules "remember" how they got where they are, at any time. Common sense tells us that the criticism is nonsensical (visualize a film of an egg falling on the floor run backwards!), and there is a reason why Newton's equations do not tell the whole story. The molecules are not isolated. The container is in contact with its environment, so that its molecules are bombarded by molecules from outside; if the container is kept at a fixed temperature, in a "heat bath," our statistical considerations are meant to apply only to the container, and not to the heat bath. The newtonian equations of motion, or the Schrödinger equation, are not by themselves enough to describe the system. The time reversal argument could invalidate statistical arguments about a closed system (the universe as a whole), but for physical systems that are not closed, statistical physics is applicable.

To prepare ourselves for this, we begin this part of the book with a study of some elementary probability theory, including the notion of fluctuations and a normal distribution. After a discussion of the statistical meaning of pressure and temperature, the Boltzmann distribution is derived. This is then applied to a variety of systems, including Brownian motion and the Equipartition Law which plays an important role in the classical treatment of blackbody radiation. Statistical notions appear in many other applications that are discussed in the book, and that is why this material is viewed as such an important part of a course dealing with the structure of matter.

NOTES AND COMMENTS

1. A very useful reference for the early history of the understanding of the atomic nature of matter is F.L. Friedman and L. Sartori, *The Classical Atom*, Addison-Wesley, Reading, Mass., 1965.

2. The question of reversibility in statistical physics is discussed very thoroughly in Chapter 1 of F. Reif, *Statistical Physics* (Berkeley Physics Course Vol. 5), McGraw-Hill, New York, 1964.

3. A superb treatment of the subject matter of this part of the book may be found in Chapters 39 to 46 of R.P. Feynman, R.B. Leighton, and M. Sands, *The Feynman Lectures on Physics*, Vol. 1, Addison-Wesley, Reading, Mass., 1963. The coverage of statistical physics is particularly good.

9
An Introduction to Probability Theory

The Binomial Distribution The definition of the probability of an event E is given by

$$P(E) = \lim_{N \to \infty} \left(\frac{n}{N} \right) \tag{9.1}$$

where N is the total number of equally probable outcomes and n is the number of outcomes that constitute the event E. In everyday parlance we might leave out the words "equally probable" because we always take care to make sure that this criterion is satisfied. For example, in estimating the survival probability during an operation, say an appendectomy, one would look up statistics for appendectomies and not statistics for the use of an operating room. In the latter case one may be unduly frightened by chancing on a place where 20% of the operations are heart transplants! The reason for taking the limit of an infinite number of trials is to wipe out the fluctuations. In an operational way one does not, of course, go to the limit. If a coin is tossed 100,000 times, and 48,000 heads turn up, then, for this biased coin the probability of getting heads is 0.48. One could get a good reading of the probability with many fewer tosses, but, say, ten tosses would not be enough, since fluctuations about the mean can easily occur. One of our tasks in this brief introduction is to obtain a quantitative understanding of what are reasonable fluctuations. To do this for the coin problem, we will work out in detail the general problem of tossing a biased coin. We shall find the remarkable result that our conclusions have a very wide range of applicability to statistical physics.

Consider a coin for which the probability of getting heads is p and the probability of getting tails is q. Evidently

$$p + q = 1, \tag{9.2}$$

but for technical reasons we will not replace q by $1 - p$ in our expressions till the very end of the calculation. Suppose we have N coins (identical) and toss the whole batch repeatedly. What is the probability of getting n heads and $N - n$ tails? By our definition, we are asking for the number of heads in a large number of tosses. The number of ways of getting all heads, that is, N heads and no tails, is one: all the coins must turn up heads. Hence the probability of getting N heads is

$$\underbrace{p \times p \times p \times \ldots \ldots \times p}_{N \text{ times}} = p^N. \tag{9.3}$$

We multiply probabilities, since the tossings of different coins constitute independent events. It is obvious that the outcome of one toss will not affect the outcome of the next one. (This obvious fact does not deter gamblers who insist that a run of bad luck must be followed by a good luck run — today!). Hence with the notation $P_N(n)$ for the probability of getting n heads in N tosses, we have

$$P_N(N) = p^N. \tag{9.4}$$

Next, consider the probability of getting $N-1$ heads in N tosses. The probability of getting that outcome is $p^{N-1}q$, multiplied by the number of ways in which this event can happen. If we picture the N tosses in some orderly array, then tails may turn up on the first toss, *or* the second toss, *or* the third toss, ... *or* the Nth toss. Thus there are N ways of getting tails once, so that

$$P_N(N-1) = Np^{N-1}q. \tag{9.5}$$

What about two tails and $N-2$ heads? The probability of this outcome is obviously $p^{N-2}q^2$ but the number of ways of this happening is a little harder to calculate. If we again order the tosses by formally making little boxes for them and numbering the boxes from 1 to N

1	2	3	4	5	....	$N-2$	$N-1$	N

then we find that the first tail (T) can go into any one of the N boxes, so that there are N ways of doing *that*. The next T has only $N-1$ boxes to go into, so that there are $(N-1)$ ways of placing the second tail. This leaves the number of ways of getting TT is $N(N-1)$. This, however, is overcounting, since we do not distinguish between the two T's. Having the first tail go into box 5 and the second into box 3 is the same as having the first tail go into box 3 and the second into box 5. We must therefore divide by the number of permutations of the two T's, which is 2. Thus the number is $N(N-1)/2$. Equivalently, we could count directly: with the first T in box 1, there are $(N-1)$ boxes for the second T. With the first T in box 2, there are only $(N-2)$ boxes for the second T — box 1 is excluded to avoid duplication. With the first T in the third box, the second T can only be in $(N-3)$ boxes, and so on. Thus the number of ways is

$$N + (N-1) + (N-2) + \cdots + 1 = \frac{N(N-1)}{2}.$$

We can write this as

$$\frac{N(N-1)}{1.2} \frac{(N-2)(N-3)\cdots 1}{(N-2)(N-3)\cdots 1} = \frac{N!}{2!(N-2)!} \equiv \binom{N}{2}. \tag{9.6}$$

Thus we find that

$$P_N(N-2) = \binom{N}{2} p^{N-2}q^2. \tag{9.7}$$

Let us now jump ahead and ask for the probability of getting $N-r$ heads and r tails. The probability of the event is $p^{N-r}q^r$ but the number of ways of getting that situation must be examined. The first T can be found in any one of N boxes; the second in any one of $(N-1)$ boxes, the third in any one of $(N-2)$ boxes, ... the r-th T in any one of $(N-r+1)$ boxes. The product $N(N-1)(N-2)\ldots(N-r+1)$ overcounts considerably, since it treats all outcomes as different. If we recognize that all permutations of the r boxes with T's are equivalent, we see that we must divide by the number of permutations of r objects, which is $r!$. Thus the number of ways of filling the boxes is

$$\frac{N(N-1)\ldots(N-r+1)}{r!} = \frac{N!}{r!(N-r)!} = \binom{N}{r}. \tag{9.8}$$

Hence

$$P_N(N-r) = \frac{N!}{r!(N-r)!} \, p^{N-r}q^r. \tag{9.9}$$

Thus the answer that we were seeking is

$$P_N(n) = \binom{N}{n}p^n q^{N-n}. \tag{9.10}$$

This probability function is known as the *binomial distribution* since the coefficients

$$\binom{N}{n} = \frac{N!}{n!(N-n)!} = \binom{N}{N-n}$$

are just the binomial coefficients.

First of all, let us check that the probabilities do, indeed, add up to unity, which must be the case, since *something* must happen. We have

$$\sum_{n=0}^{N} P_N(n) = \sum_{n=0}^{N} \binom{N}{n}p^n q^{N-n}$$
$$= (p+q)^N \tag{9.11}$$

since the series is just the binomial expansion of the final form. At this stage we may set $p+q=1$ and get the result that

$$\sum_{n=0}^{N} P_N(n) = 1. \tag{9.12}$$

Suppose we use the above coin game to run a casino. A head earns \$$a$ and a tails loses \$$b$. What is the expectation of a gain per toss, if the game is played for a long time? In N tosses, the gain is $an - b(N-n)$ dollars, with a probability $P_N(n)$. The *expectation value* of some function $f(n)$ if n occurs with probability $P(n)$ is just

$$\langle f \rangle = \sum_{n=0}^{N} f(n)P_N(n). \tag{9.13}$$

We thus need to calculate $\langle (a+b)n - bN \rangle$, that is

$$(a+b)\sum_{n=0}^{N} nP_N(n) - bN\sum_{n=0}^{N} P_N(n).$$

It is here that our decision not to set $q = 1-p$ pays off. We observe that in

$$\sum_{n=0}^{N} nP_N(n) = \sum_{n=0}^{N} \binom{N}{n} np^n q^{N-n} \tag{9.14}$$

we may use the fact that

$$np^n = p\frac{d}{dp}p^n.$$ (9.15)

Thus we write

$$\sum_{n=0}^{N} nP_N(n) = p\frac{d}{dp}\sum_{n=0}^{N}\binom{N}{n}p^n q^{N-n}$$

$$= p\frac{d}{dp}(p+q)^N$$

$$= Np(p+q)^{N-1}.$$ (9.16)

Hence

$$(a+b)\langle n\rangle - bN(p+q)^N =$$

$$(a+b)Np(p+q)^{N-1} - Nb(p+q)^N.$$ (9.17)

At this stage we may set $p + q = 1$ to get

$$(a+b)\langle n\rangle - bN\langle 1\rangle = Np(a+b) - bN$$

$$= N(pa - qb).$$ (9.18)

The expected gain per toss is obtained by dividing the above by N. If there is to be a net gain, $pa - bq$ must be positive.

Random walks Exactly the same approach works for a problem that is of great interest in physics, albeit formulated somewhat differently, the *random-walk problem*. Consider a one-dimensional space and a grain of dust in it. The grain of dust is struck randomly from the left and from the right. If the space is not homogeneous, so that the probability of getting struck from the left is p and that of getting struck from the right is q, and if upon being struck the grain of dust moves a distance d in a direction opposite to that from which it is struck, we may be interested in asking how far the grain of dust will travel after being struck N times. This is *completely identical* to the biased coin-tossing problem. To see the equivalence of the two problems it is only necessary to recognize that the assumption that a given collision is a new independent event is equivalent to the assumption that each new coin toss is an independent event. Thus after N collisions, the mean distance traveled by the grain of dust is

$$\langle D\rangle = Nd(p - q),$$ (9.19)

the result (9.18) with the "gain" replaced by the distance traveled, independent of whether the collision came from the right or from the left.

If $p = q = \frac{1}{2}$, then on the average the particle will have gone nowhere. A quantity of interest is: how far from the original position could we expect the particle to be? In principle an average of zero could be obtained in a large number of ways: for example, if the average on a test is 50, this could happen because half the class gets 10 and half the class gets 90, but it is more likely that the grades will be clustered around 50. What happens in the random walk problem? It will be useful to calculate the expectation of the square of the distance, that is,

$$\langle D^2\rangle = \langle (nd - (N-n)d)^2\rangle = d^2\langle(2n - N)^2\rangle.$$ (9.20)

We therefore need to calculate

$$4d^2 \langle n^2 \rangle - 4d^2 N \langle n \rangle + d^2 N^2$$

$$= 4d^2 \langle n^2 \rangle - 4d^2 N^2 p + d^2 N^2 .$$

Now

$$\langle n^2 \rangle = \sum_{n=0}^{N} \binom{N}{n} n^2 p^n q^{N-n}$$

$$= \left(p \frac{d}{dp} \right) \left(p \frac{d}{dp} \right) \sum_{n=0}^{N} \binom{N}{n} p^n q^{N-n}$$

$$= \left(p \frac{d}{dp} \right) \left(p \frac{d}{dp} \right) (p+q)^N$$

$$= p \frac{d}{dp} (Np(p+q)^{N-1})$$

$$= pN(p+q)^{N-1} + p^2 N(N-1)(p+q)^{N-2}$$

$$= pN + p^2 N(N-1), \tag{9.21}$$

so that

$$\langle D^2 \rangle = d^2 (4pN + 4p^2 N(N-1) - 4pN^2 + N^2)$$

$$= d^2 (N^2 (4p^2 - 4p + 1)$$

$$+ N(4p - 4p^2)). \tag{9.22}$$

The mean square distance is not yet a measure of the deviation from the mean, since a difference between p and q implies a drift to the right or to the left, on the average. The quantity of interest is the dispersion ΔD, defined by

$$\langle \Delta D \rangle^2 \equiv \langle D^2 \rangle - \langle D \rangle^2 \tag{9.23}$$

since it is equal to

$$\langle (D - \langle D \rangle)^2 \rangle = \langle D^2 \rangle - 2 \langle D \rangle \langle D \rangle + \langle D \rangle^2 . \tag{9.24}$$

In our case we get

$$\sqrt{\langle D^2 \rangle - \langle D \rangle^2} = (d^2 N^2 (2p-1)^2 + 4p(1-p)N$$

$$- d^2 N^2 (2p-1)^2)^{1/2}$$

$$= (4pqN)^{1/2} . \tag{9.25}$$

Thus the root mean square deviation from the mean, divided by the number of steps N goes to zero with increasing N as $1/N^{1/2}$. The particle thus will not get very far from its mean value.

If the probability of getting hit from the left is the same as getting hit from the right, then $p = q = \frac{1}{2}$, and

$$\Delta D = \sqrt{\langle D^2 \rangle - \langle D \rangle^2} = \sqrt{N} d. \tag{9.26}$$

The result is readily generalized to a three-dimensional random walk. There we have, for $p = q = \frac{1}{2}$, or more generally, for equal probability of getting hit from anywhere,

$$\langle D_x \rangle = \langle D_y \rangle = \langle D_z \rangle = 0$$

$$\langle D_x^2 \rangle = \langle D_y^2 \rangle = \langle D_z^2 \rangle = d^2 N \tag{9.27}$$

so that

$$\sqrt{\langle D^2 \rangle - \langle D \rangle \cdot \langle D \rangle} = d\sqrt{3N}. \tag{9.28}$$

Stirling formula We can get more precise information about the probability of deviating from the mean by a more detailed inspection of $P_N(n)$. A difficulty is that factorials are very difficult to calculate when the numbers are large (imagine 10^{23}!), but fortunately there is an approximate expression for factorials due to *Stirling* that is much easier to handle. The approximate formula states that for N large enough so that $N \gg \log N$

$$\log N! = N \log N - N + \tfrac{1}{2} \log N + \tfrac{1}{2} \log 2\pi + \cdots \tag{9.29}$$

(here, as everywhere else in the book, log refers to the natural logarithm, to base e). To prove this formula requires techniques that are slightly beyond the scope of the book. We may, however, establish the correctness of the first two terms without difficulty. We have

$$\log N! = \log N + \log (N-1) + \log (N-2) + \cdots + \log 2 + \log 1$$

$$= N \log N + \log \frac{N-1}{N} + \log \frac{N-2}{N} + \cdots + \log \frac{2}{N} + \log \frac{1}{N}$$

$$= N \log N + \sum_{r=1}^{N-1} \log \frac{N-r}{N}$$

$$= N \log N + \sum_{r=1}^{N-1} \log \left(1 - \frac{r}{N}\right).$$

To calculate the sum, we write

$$\sum_{r=1}^{N-1} \log \left(1 - \frac{r}{N}\right) = \sum_{r=1}^{N-1} \Delta r \log \left(1 - \frac{r}{N}\right)$$

where we have $\Delta r = (r - (r-1)) = 1$. Now

$$\sum_{r=1}^{N-1} \Delta r \log \left(1 - \frac{r}{N}\right) = N \sum_{r=1}^{N-1} \frac{\Delta r}{N} \log \left(1 - \frac{r}{N}\right).$$

When N is large, the steps between adjacent terms are small compared with N. We can convert the sum to an integral by introducing the variable $x = r/N$, which for large N approaches a continuous variable. Thus $\Delta r/N = dx$ and

$$N \sum_{r=1}^{N-1} \frac{\Delta r}{N} \log \left(1 - \frac{r}{N}\right) \to N \int_0^1 dx \log (1-x) = -N.$$

We thus get

$$\log N! \approx N \log N - N \tag{9.30}$$

which is good enough for all practical purposes.

Normal distribution Let us use Stirling's formula to plot the shape of $P_N(n)$ when N is very large:

$$\log P_N(n) = \log N! - \log n! - \log (N-n)! + n \log p + (N-n) \log q$$

$$\approx N \log N - N + \tfrac{1}{2} \log 2\pi N - n \log n + n$$

$$- \tfrac{1}{2} \log 2\pi n - (N-n) \log (N-n) + (N-n)$$

$$- \tfrac{1}{2} \log 2\pi (N-n) + n \log p + (N-n) \log q. \tag{9.31}$$

To find where $P_N(n)$ peaks as a function of n we need to know where $dP/dn = 0$. Now, treating n as a continuous variable,

$$\frac{d}{dn} \log P_N(n) = -\log n - \frac{1}{2n} + \log (N-n)$$

$$+ \frac{1}{2(N-n)} + \log p - \log q$$

$$= \log \frac{(N-n)p}{nq} + \frac{2n-N}{n(N-n)}.$$

The peak occurs where the right-hand side vanishes. We guess that if p does not differ too much from q, then the second term can be neglected. This is because the result of the calculation is expected to be that $\langle n \rangle = pN$ and then the second term goes like $1/N$. Let us ignore the second term. We then find that the peak occurs where

$$\frac{(N-n)p}{nq} = 1$$

that is, for

$$n = pN. \tag{9.32}$$

Note that near this value the second term is $(1/q - 1/p)/2N$ which is negligible for large N unless p or q are pathologically small. In any case, for fixed p and q we can always go to large enough N.

The fact that the peaking in $P_N(n)$ occurs at the average value of n tells us that *the average value of n is also the most probable value of n*. How sharp is the peak in $P_N(n)$? To investigate this, let us expand $\log P_N(n)$ about the mean value. We will write

$$n = pN + v. \tag{9.33}$$

We then find that

$$\log P_N(n) = N \log N - N + \tfrac{1}{2} \log 2\pi N - (pN + v) \log (pN + v)$$

$$- \tfrac{1}{2} \log 2\pi (pN + v) - (qN - v) \log (qN - v) + N$$

$$- \tfrac{1}{2} \log 2\pi (qN - v) + (pN + v) \log p + (qN - v) \log q$$

where we have used $q = 1 - p$ in some places. We will expand this to second order in v. A little algebra, in which we use

$$\log (pN + v) = \log pN + \log \left(1 + \frac{v}{pN}\right)$$

$$= \log pN + \frac{v}{pN} - \frac{1}{2}\left(\frac{v}{pN}\right)^2$$

and so on, gives

$$\log P_N(n) \approx -\frac{1}{2}\log 2\pi pqN + \frac{v}{2N}\left(\frac{1}{q} - \frac{1}{p}\right)$$

$$- \frac{v^2}{2Npq}\left(1 + \frac{1}{2N}\left(\frac{p}{q} + \frac{q}{p}\right)\right). \tag{9.34}$$

For $v \ll N$, the second term vanishes, and so does the second term in the coefficient of v^2. Thus to first approximation we may write

$$P_N(n) \approx \frac{1}{(2\pi pqN)^{1/2}} e^{-v^2/2Npq} \tag{9.35}$$

where

$$v = n - \langle n \rangle = n - pN. \tag{9.36}$$

This formula is called the *normal distribution*. It is immeasurably easier to calculate with than is the formula involving the factorials. It also shows at a glance (Fig. 9.1) that all of the probability is distributed very close to the mean. For an example, consider an unbiased coin ($p = q = \frac{1}{2}$) with 10^6 tosses. If we wanted to calculate the probability of getting 510,000 heads, we would have to calculate

$$\frac{(1,000,000)!}{510,000! \ 490,000!}\left(\frac{1}{2}\right)^{1000,000}$$

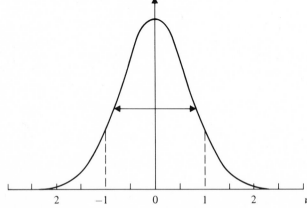

Figure 9.1. The Gaussian distribution of Eq. (9.35). The width at half maximum is drawn in. The dashed lines represent the value of v for which the exponent is unity.

which just cannot be done. Here the result is

$$P \approx \sqrt{\frac{2}{\pi}} \frac{1}{10^3} e^{-10^8/0.5 \times 10^6} \approx e^{-200}$$

which is infinitesimal. The normal distribution has a *width* of $(8Npq)^{1/2}$ and deviations from the mean that are two or three widths away from the mean are increasingly improbable. For large N, the average value of n is overwhelmingly the most probable one, and this lies at the heart of statistical physics.

Poisson distribution For completeness we consider a case that was explicitly excluded in our derivation of the normal distribution, namely the case in which p is very tiny. Let

$$s = pN$$

be fixed as N becomes very large. We then have

$$P_N(n) = \frac{N!}{n!(N-n)!} p^n (1-p)^{N-n}. \tag{9.37}$$

Let us consider this for values of n that are of the same order of magnitude as s, that is $n \ll N$. We then have

$$P_N(n) = \frac{p^n}{n!} N(N-1) \dots (N-n+1) \left(1 - \frac{s}{N}\right)^N \left(1 - \frac{s}{N}\right)^{-n}$$

$$\approx \frac{p^n}{n!} N^n e^{-s}$$

$$\approx \frac{s^n}{n!} e^{-s}. \tag{9.38}$$

We made use of the formula

$$\lim_{N \to \infty} \left(1 + \frac{x}{N}\right)^N = e^x \tag{9.39}$$

in deriving this. The distribution is known as the *Poisson distribution* and it describes rare, independent events. Note that

$$s = \langle n \rangle \tag{9.40}$$

and that

$$\sum_{n=0}^{\infty} P_N(n) = \sum_{n=0}^{\infty} \frac{s^n}{n!} e^{-s} = 1 \tag{9.41}$$

as it should be.

NOTES AND COMMENTS

There exist many books on probability theory, and the subject is treated in a number of physics textbooks, such as F. Reif, *Statistical Physics*, McGraw-Hill,

New York, 1964. There exists a delightful introduction to the subject, that is nontechnical and very stimulating: W. Weaver, *Lady Luck*, Doubleday and Company, New York, 1963. The standard, somewhat more advanced book, is W. Feller, *An Introduction to Probability Theory and Its Applications*, 3rd Edition, John Wiley, New York. 1967.

PROBLEMS

9.1 What is the probability of getting thirteen spades in a bridge deal?

9.2 What is the probability of getting the A, K, Q, J, 10 of clubs in a game of poker?

9.3 What is the probability of getting a total of nine in a throw of three dice?

9.4 What is the probability that at a party attended by twenty-five people, at least two have the same birthday? [*Hint*: Instead of calculating the probability of 2, 3, 4, . . . 25 having the same birthday, calculate the probability that none have the same birthday.]

9.5 A coin is tossed 10,000 times. What is the probability of getting (a) 5,000 heads, (b) 5400 heads, (c) 6,000 heads, if the coin is unbiased?

9.6 Given the probability distribution

$$P_n = \frac{N!}{n!(N-n)!} p^n q^{N-n}$$

$$p + q = 1,$$

calculate (a) $\langle n \rangle$, (b) $\langle n^2 \rangle$ (c) $\langle n^3 \rangle$ (d) $\langle n^4 \rangle$.

9.7 What is the probability of getting at least one six in four throws of the dice?

9.8 Consider a gas of N_0 noninteracting molecules in a volume V_0. What is the average number of molecules in a cell of volume V in the larger volume? If N is the number of molecules in the volume V, calculate $\langle N^2 \rangle$ and $\sqrt{\langle N^2 \rangle - \langle N \rangle^2} / \sqrt{\langle N^2 \rangle}$.

9.9 In an expression of the transcendental number $e = 2.7182. . .$ to two thousand digits, the various digits appear with the following frequencies

0	1	2	3	4	5	6	7	8	9
196	190	208	202	201	197	204	198	202	202

Use the normal distribution curve to check whether this is consistent with a random distribution of integers in the expression.

9.10 The Poisson distribution for the probability of getting n events in a sampling is given by

$$P_n = \frac{\lambda^n}{n!} e^{-\lambda}.$$

a) Show that
$$\langle n \rangle = \lambda.$$

b) Calculate
$$\langle n^2 \rangle, \langle n^3 \rangle, \langle n^2 \rangle - \langle n \rangle^2, \quad \left(\frac{\langle n^2 \rangle - \langle n \rangle^2}{\langle n^2 \rangle} \right)^{1/2}.$$

[*Hint*: Use the fact that
$$\sum_{n=0}^{\infty} n \frac{\lambda^n}{n!} = \lambda \frac{d}{d\lambda} \sum_{n=0}^{\infty} \frac{\lambda^n}{n!} \Bigg].$$

9.11 A bank sends out statements to 5000 customers. There are 1200 errors in a month's batch. What is the probability of finding an error-free statement? What is the probability that a statement has one error? that it has two errors? Assume that the distribution of errors is a Poisson distribution.

10
Pressure and Temperature

Molecular pressure If we picture a gas as consisting of a very large number of molecules that collide with each other and the walls of the container, we expect the gas to exert a pressure on the walls, because collisions with the walls involve momentum transfer to the walls. *Pressure* is defined as the average force exerted on a unit area of the wall, and the average force is the average momentum transferred to the wall per unit time, since

$$\mathbf{F} = d\mathbf{p}/dt. \tag{10.1}$$

If a molecule moving with velocity $\mathbf{v}$ strikes an area dA of the wall, it will transfer twice its normal momentum to the wall, if the collision is elastic, because the tangential component of the velocity is unaltered, but the normal component reverses sign. Thus if the molecule has mass m the momentum transfer per molecule is $2m\mathbf{n} \cdot \mathbf{v}$, where $\mathbf{n}$ is the unit vector representing the outward normal to the wall of the container at the area dA (Fig. 10.1). The number of molecules that strike the wall in a time dt will be the number contained in a cylinder of length $|\mathbf{v}|dt$ and base area dA. The volume of this cylinder is dA $(\mathbf{n} \cdot \mathbf{v})dt$, and this must be multiplied by the number of molecules per cm^3 that have the velocity v, or, more precisely, the number that have velocities

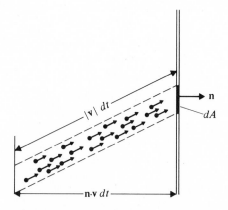

Figure 10.1. The contribution to the pressure on dA from molecules moving with velocity v.

in the range (v_x, v_y, v_z) to $(v_x + dv_x, v_y + dv_y, v_z + dv_z)$, which we denote by $n(\mathbf{v})d^3v$. Thus the momentum transfer for a given velocity range $(\mathbf{v}, \mathbf{v} + d\mathbf{v})$ is

$$(2m\mathbf{n} \cdot \mathbf{v})(dAdt\mathbf{n} \cdot \mathbf{v})n(\mathbf{v})d^3v. \tag{10.2}$$

The total momentum transfer is obtained by integrating over all velocities, subject to the condition that $\mathbf{n} \cdot \mathbf{v} > 0$, since only the molecules going toward the wall will transfer momentum to it in the next short-time interval. The pressure is obtained by dividing the momentum transfer by the time interval dt and by the area dA. From the momentum transfer

$$\Delta P = dAdt2m \int_{\mathbf{n} \cdot \mathbf{v} > 0} d^3v(\mathbf{n} \cdot \mathbf{v})^2 n(\mathbf{v}), \tag{10.3}$$

the pressure p is calculated to be

$$p = 2m \int_{\mathbf{n} \cdot \mathbf{v} > 0} d^3v(\mathbf{n} \cdot \mathbf{v})^2 n(\mathbf{v}). \tag{10.4}$$

We may arbitrarily define the outward normal direction to be the z direction, and then we have

$$p = 2m \int_{v_z > 0} d^3v\, v\, v_z^2 n(\mathbf{v}). \tag{10.5}$$

If the velocity distribution is uniform, that is, if $n(\mathbf{v})$ does not depend on the sign of v_z, then we may integrate over all v_z and divide by 2. Thus

$$p = m \int d^3v\, v\, v_z^2 n(\mathbf{v}). \tag{10.6}$$

If the distribution $n(\mathbf{v})$ does not depend on the direction of $\mathbf{v}$, then

$$\int d^3v\, n(\mathbf{v})v_z^2 = \int d^3v\, n(\mathbf{v})v_x^2 = \int d^3v\, n(\mathbf{v})v_y^2$$

$$= \tfrac{1}{3} \int d^3v\, n(\mathbf{v})(v_x^2 + v_y^2 + v_z^2)$$

$$= \tfrac{1}{3} \int d^3v\, n(\mathbf{v})\mathbf{v}^2, \tag{10.7}$$

and hence

$$p = \tfrac{1}{3}m \int d^3v\, n(\mathbf{v})\mathbf{v}^2. \tag{10.8}$$

The total number density of molecules per cm^3 is given by

$$n = \int d^3v\, n(\mathbf{v}). \tag{10.9}$$

Hence $n(\mathbf{v})/n$ may be considered to be the *probability* of finding a molecule with velocity $\mathbf{v}$. Hence, rewriting (10.8) in the form

$$p = \frac{1}{3}mn \int d^3v\, \frac{n(\mathbf{v})}{n}\, \mathbf{v}^2,$$

we see that it involves

$$\int d^3v \frac{n(\mathbf{v})}{n} \mathbf{v}^2 = \langle \mathbf{v}^2 \rangle, \tag{10.10}$$

the expectation value of the square of the velocity. Thus finally

$$p = \tfrac{1}{3} mn \langle \mathbf{v}^2 \rangle$$
$$= \tfrac{2}{3} n (\tfrac{1}{2} m \langle \mathbf{v}^2 \rangle). \tag{10.11}$$

The pressure is therefore directly proportional to the average kinetic energy of the molecules. If we consider a volume V and a gram mole of molecules, then the number of molecules is Avogadro's number N_0 and

$$n = N_0/V. \tag{10.12}$$

Thus for a gram mole of gas we have

$$pV = \tfrac{2}{3} N_0 (\tfrac{1}{2} m \langle \mathbf{v}^2 \rangle). \tag{10.13}$$

If we consider a monatomic gas, then the kinetic energy of a molecule is all the energy that it can have, that is, the internal energy U is just the kinetic energy. This is not true for diatomic molecules, since the latter have not only kinetic energy, but also rotational energy and even vibrational energy. For monatomic gases, the total internal energy U is the number of molecules multiplied by the average kinetic energy, so that

$$pV = \tfrac{2}{3} U. \tag{10.14}$$

If we want to consider the pressure exerted on the walls of a cavity containing radiation, we can go through the same series of steps, except when considering the "gas" of radiation we replace $m\mathbf{n} \cdot \mathbf{v}$, the momentum in the direction of the outward normal by the expression $\mathbf{n} \cdot \mathbf{p}$, where $\mathbf{p}$ is the momentum of the ray of radiation. Then in the expression for the pressure

$$p = \tfrac{1}{3} n \langle m\mathbf{v} \cdot \mathbf{v} \rangle$$

we replace $m\mathbf{v} \cdot \mathbf{v}$ by

$$\mathbf{p} \cdot \frac{\mathbf{p}c^2}{E} = \frac{\mathbf{p}^2 c^2}{E} = |\mathbf{p}|c = E \tag{10.15}$$

where E is the energy of a "particle" with zero rest mass. In the same way, $n \langle E \rangle$ becomes the "energy per particle" $\times$ "particles per cm^3" that is, the energy density U. Hence

$$p = \tfrac{1}{3} U \quad ergs/cm^3. \tag{10.16}$$

We will return to the problem of radiation pressure and radiation energy density later in the book.

Temperature Comparison of the relation (10.13) with the perfect gas law is suggestive. That law has the form

$$pV = RT \tag{10.17}$$

where R is the gas constant. For a gram mole of gas, $R = 8.315$ J/degree K = 8.315×10^7 erg/degree K. This suggests that

$$\tfrac{1}{3} m N_0 \langle v^2 \rangle = RT. \tag{10.18}$$

If we introduce *Boltzmann's constant*, k, by

$$k = \frac{R}{N_0} = 1.381 \times 10^{-16} \text{ erg/degree K}, \tag{10.19}$$

we get

$$\tfrac{1}{2} m \langle v^2 \rangle = \tfrac{3}{2} kT. \tag{10.20}$$

The identification suggests that the temperature is proportional to the mean kinetic energy of the molecules. This relation is not restricted to ideal gases.

One can give a qualitative, and not very rigorous argument which suggests a connection between the mean kinetic energy and temperature. Consider a collision between molecules in a gas. If the molecules approach each other with velocities v_1 and v_2 respectively, then the center of mass velocity is

$$\mathbf{v}_{cm} = \frac{m_1 \mathbf{v}_1 + m_2 \mathbf{v}_2}{m_1 + m_2} \tag{10.21}$$

and the relative velocity is

$$\mathbf{v}_r = \mathbf{v}_1 - \mathbf{v}_2 \quad \text{(Fig. 10.2)}. \tag{10.22}$$

For a randomized distribution, which is what we expect for a gas *in equilibrium*, when all memory of the initial state is lost, it is reasonable to expect that

$$\langle \mathbf{v}_1 \cdot \mathbf{v}_2 \rangle = 0 \tag{10.23}$$

and

$$\langle \mathbf{v}_r \cdot \mathbf{v}_{cm} \rangle = 0. \tag{10.24}$$

The first of these follows from the expectation that different molecules have uncorrelated velocities, so that

$$\langle \mathbf{v}_1 \cdot \mathbf{v}_2 \rangle = \langle \mathbf{v}_1 \rangle \cdot \langle \mathbf{v}_2 \rangle = 0 \tag{10.25}$$

since $\langle \mathbf{v}_1 \rangle = \langle \mathbf{v}_2 \rangle = 0$. The second is really a statement about the independence of successive collisions. This implies that

$$\langle (\mathbf{v}_1 - \mathbf{v}_2) \cdot (m_1 \mathbf{v}_1 + m_2 \mathbf{v}_2) \rangle = 0, \tag{10.26}$$

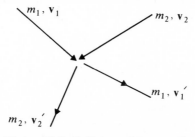

Figure 10.2. Collision of two particles leading to randomization.

that is,

$$m_1 \langle v_1^2 \rangle + (m_2 - m_1)\langle v_1 \cdot v_2 \rangle - m_2 \langle v_2^2 \rangle$$

$$= m_1 \langle v_1^2 \rangle - m_2 \langle v_2^2 \rangle = 0, \qquad (10.27)$$

from which it follows that

$$\tfrac{1}{2} m_1 \langle v_1^2 \rangle = \tfrac{1}{2} m_2 \langle v_2^2 \rangle. \qquad (10.28)$$

Since in equilibrium a mixture of gases is at the same temperature, the above indicates that the mean kinetic energy should be a function of the temperature. The connection with the ideal gas law suggests the linear relation.

To get some feel for the magnitudes that we are considering, let us work out some numbers. For air at $20°C$, that is, a gas of molecular weight approximately 29 and a temperature of $293°K$, we have

$$\tfrac{1}{2} m \langle v^2 \rangle = \tfrac{3}{2}(1.381 \times 10^{-16} \times 293) = 6.1 \times 10^{-14} \text{ ergs.}$$

Since

$$m = 29 \times 1.6 \times 10^{-24} \text{ gm,}$$

we have

$$\langle v^2 \rangle \cong 0.25 \times 10^{10} \text{ cm}^2/\text{sec}^2,$$

and therefore

$$\sqrt{\langle v^2 \rangle} \simeq 0.5 \times 10^5 \text{ cm/sec.}$$

At normal temperature and pressure, that is, $p = 1$ atmosphere, or $p = 10^6$ dynes/cm^2 and $T = 0°C = 273°K$, the volume occupied by one mole of gas is

$$V = \frac{8.315 \times 10^7 \times 273}{10^6}$$

$$= 2.27 \times 10^4 \text{ cm}^3$$

$$= 22.7 \text{ liters.} \qquad (10.29)$$

If we have interstellar clouds of atomic hydrogen with a density of 1 atom/cm^3 at a temperature of $100°K$, we have

$$\tfrac{1}{2} m \langle v^2 \rangle = \tfrac{3}{2} \times 1.381 \times 10^{-16} \times 100 = 2.07 \times 10^{-14} \text{ ergs}$$

and since $m \doteq 1.6 \times 10^{-24}$ gm, we get

$$\langle v^2 \rangle = \frac{4.14 \times 10^{-14}}{1.6 \times 10^{-24}} = 2.58 \times 10^{10} \text{ cm}^2/\text{sec}^2.$$

Hence

$$\sqrt{\langle v^2 \rangle} \simeq 1.6 \times 10^5 \text{ cm/sec} \simeq 1.6 \text{ km/sec.}$$

The pressure exerted by this gas is

$$p = \frac{2}{3}\left(\frac{1}{2} m \langle v^2 \rangle\right) n = \frac{2}{3} \times 2.07 \times 10^{-14} \times 1 = 1.38 \times 10^{-14} \frac{\text{dynes}}{\text{cm}^2}$$

It is sometimes necessary to calculate expectation values of powers of v other than the second, for example $\langle v^4 \rangle$. We expect, on dimensional grounds that

$$\langle v^4 \rangle = \text{const.} \left(\frac{kT}{m} \right)^2, \tag{10.30}$$

since for a noninteracting gas of molecules this is the only quantity with the right dimensions. To find out what the constant is, we must know the distribution $n(\mathbf{v})$ in velocity-space. We shall calculate this function using probability theory, and will do this in great generality, since the resulting distribution, the Boltzmann distribution, is of great interest:

NOTES AND COMMENTS

1. The relation $pV = 2U/3$ is correct only for monatomic gases. The more general relation is $pV = (\gamma - 1)U$. A way of determining γ is to consider the slow compression of a gas by a piston. The compression is done under *adiabatic conditions*, that is, in a way that no energy is lost due to heat losses during the compression. The work done is $dW = -pdV$ and this is all transferred to the molecules, increasing their internal energy. Thus we have $dU = pdV$. We also have $(\gamma - 1)$ $dU = pdV + Vdp$. From the two relations it follows that during an adiabatic compression $pV^\gamma = \text{constant}$. Thus for an ideal gas we expect $pV^{5/3} = \text{const.}$, and for a "gas" of radiation we expect $pV^{4/3} = \text{const.}$

2. If the distribution function for velocity is nonuniform, the calculations are more complicated. If we have a simple case of the whole collection of gas molecules drifting with a uniform velocity $\mathbf{u}$, then $n(\mathbf{v}') = n(\mathbf{v} - \mathbf{u})$ is still uniform. Then

$$n\langle \mathbf{v} \rangle = \int d^3v\, \mathbf{v}\, n(\mathbf{v} - \mathbf{u}) = \int d^3v'(\mathbf{v}' + \mathbf{u})n(\mathbf{v}') = \mathbf{u}n$$

and

$$n\langle v^2 \rangle = \int d^3v'(\mathbf{v}' + \mathbf{u})^2 n(\mathbf{v}') = n\langle v'^2 \rangle + n\mathbf{u}^2.$$

The mean velocity is just the drift velocity u, and the mean kinetic energy is the sum of the kinetic energy of the gas at rest plus the kinetic energy of the drift motion.

PROBLEMS

10.1 Use the gas law

$$pV = RT$$

with $R = 8.3 \times 10^7$ erg/degree K to calculate the volume occupied by one mole of ideal gas at 1 atmosphere pressure (10^6 dynes/cm^2) at $0°$C. Given that the average molecular weight of air is 28.9, what is the mass density of air at the above conditions, in gm/cm^3?

10.2 Use the above, and Avogadro's number $N_0 = 6.02 \times 10^{23}$ molecules/gm mole to estimate

  a) the number of molecules in the lecture hall;

  b) the mass density of Helium gas ($A = 4$) at 1 atmosphere pressure and $0°C$;

  c) the amount of Helium needed to lift a payload (including the mass of the balloon) of $100\,$kg.

10.3 A mixture of nitrogen and hydrogen gas is in equilibrium at $-40°C$. What is the root mean square velocity of the hydrogen molecules (H_2)? What is the root mean square velocity of the nitrogen (N_2) molecules? (The molecular weight of H_2 is 2, and that of N_2 is 28.)

10.4 The random velocity of galaxies is roughly $100\,$km/sec, and their number density is 2.88×10^{-3} per cubic light year. If the average mass of a galaxy is 3×10^{44} gm, what is the pressure of a gas of galaxies? What is the temperature?

10.5 The *solar constant* is the radiation energy incident from the sun on earth per unit area per unit time, 1.37×10^6 erg/cm^2 sec.

  a) Given that the radiation moves with velocity c, what energy density U does such a flow of energy/cm^2 sec correspond to?

  b) If the radiation is reflected by a thin metallic foil, what pressure does it exert on that foil? If the foil were totally absorbing instead of perfectly reflecting, what would the pressure be?

10.6 The radiation energy density in the universe is estimated to be 6×10^{-13} erg/cm^3. What is the pressure due to the radiation in the universe, and how does it compare with the pressure calculated in Problem 10.4?

10.7 A water droplet 1 micron (10^{-6} m) in diameter is in equilibrium with air molecules at $300°K$. What is the root mean square velocity of the droplet?

11
The Boltzmann Distribution

Distribution in Velocity Space Let us consider a container filled with gas in equilibrium at a temperature T. If we were to focus on a single molecule and follow its history in detail, we would see it collide with other molecules (including those of the walls of the container) with great frequency, and after each collision its velocity vector would change. The assumed random character of the collision assures us that the molecule will, in the course of its history, have just about any velocity at some time or other. To bypass the time reversal argument of the introduction, we invoke the random fluctuations in the condition of the wall molecules caused by the maintenance of a fixed temperature (with perfectly elastic walls we could imagine very unrandom patterns such as one might see in the motion of balls on a billiards table). Some velocity vectors will, of course, have to be excluded, for example, velocities so large that the molecule "carries" the entire energy of the container, but within a large range, say many times the root-mean square velocity ($\sqrt{\langle v^2 \rangle}$) we would find that no region in *velocity space*, the space whose coordinates are the points (v_x, v_y, v_z) is preferred. This expectation is extremely plausible and it has almost, but not quite, been proved as following directly from the basic laws of mechanics. We need not be concerned with the mathematical problem of proving this, since we now believe that it is quantum mechanics that determines the basic laws of behavior of systems, and instead take it to be a very physical and fundamental assumption that

> *As a result of randomizing collisions among*
> *the molecules, the probability that a given*
> *molecule is in some cell in velocity space*
> *of size $\Delta \mathbf{v}$ is independent of the location*
> *of the cell.*

We picture velocity space as divided up into cells, which are small enough so that within each cell v may be taken as constant, yet large enough so that each cell contains many molecules on the average. A reasonable upper limit on the magnitude of the velocity might be, say, 10^9 cm/sec. If we then take the cells with $\Delta v_x = \Delta v_y = \Delta v_z = 10^3$ cm/sec, then the variation of the velocity inside each cell will be very tiny for any reasonably smooth distribution. On the other hand, there are $(10^6)^3 = 10^{18}$ cells, and with a sample of molecules containing 10^{23} molecules, we still have an average of 10^5

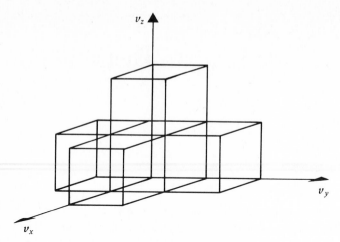

Figure 11.1. Cells in velocity space.

molecules per cell. Thus our picture of cells in velocity space is a reasonable one (Fig. 11.1).

We now ask the following question: given N molecules, what is the probability of finding n_1 molecules in cell $1, n_2$ molecules in cell 2, n_3 molecules in cell 3, and so on. This is a generalization of the coin-tossing problem where the question was: given N coins, and two cells (H and T), what is the probability of finding n_1 in the H cell and n_2 in the T cell. There we found that the number of ways of distributing the N coins among the two boxes was

$$W_N(n_1, n_2) = \frac{N!}{n_1! n_2!} \qquad (n_1 + n_2 = N). \tag{11.1}$$

From this, the probability can be calculated by dividing the above by the total number of events, that is, by the quantity

$$\sum_{\substack{n_1 \\ (n_1 + n_2 = N)}} \sum_{n_2} W_N(n_1, n_2) = \sum_{\substack{n_1 \\ (n_1 + n_2 = N)}} \sum_{n_2} \frac{N!}{n_1! n_2!}. \tag{11.2}$$

Now

$$\sum_{\substack{n_1 \\ (n_1 + n_2 = N)}} \sum_{n_2} \frac{N!}{n_1! n_2!} = \sum_{n_1} \frac{N!}{n_1!(N - n_1)!} (1)^{n_1} (1)^{N - n_1}$$

$$= (1 + 1)^N = 2^N \tag{11.3}$$

and hence

$$P_N(n_1, n_2) = \frac{W_N(n_1, n_2)}{2^N} = \binom{N}{n_1} \left(\frac{1}{2}\right)^N \tag{11.4}$$

as already seen in Chapter 9. The generalization to k cells is

$$W_N(n_1, n_2, \ldots, n_k) = \frac{N!}{n_1! n_2! \cdots n_k!}. \tag{11.5}$$

To see this, we first note that the number of ways of putting n_1 molecules in cell number 1 and the rest outside is

$$\frac{N!}{n_1!(N-n_1)!}.$$

Out of the remaining $N-n_1$, the number of ways of putting n_2 in cell 2 and the rest elsewhere is

$$\frac{(N-n_1)!}{n_2!(N-n_1-n_2)!}$$

Again, out of the remaining $N-n_1-n_2$ the number of ways of putting n_3 in cell 3 and the rest elsewhere is

$$\frac{(N-n_1-n_2)!}{n_3!(N-n_1-n_2-n_3)!},$$

and so on. The total number of ways of dividing the N molecules among the various boxes is the product of all of the above

$$\frac{N!}{n_1!(N-n_1)!}\frac{(N-n_1)!}{n_2!(N-n_1-n_2)!}\frac{(N-n_1-n_2)!}{n_3!(N-n_1-n_2-n_3)!}\cdots$$

$$=\frac{N!}{n_1!n_2!\ldots n_k!}$$

which is the result (11.5). The corresponding probability is

$$P_N(n_1\cdots n_k)=\frac{W_N(n_1,n_2,\ldots,n_k)}{\displaystyle\sum_{n_1}\cdots\sum_{n_k}W_N(n_1,n_2,\cdots,n_k)}.\qquad(11.6)$$
$$\scriptstyle(n_1+\cdots+n_k=N)$$

In our discussion of the normal distribution in Chapter 10, we saw that for large numbers, the average distribution is also the most probable distribution, since the distribution peaks so strongly about the mean. In this, more general case, we expect the same to be true, and we will find the most probable distribution by maximizing $\log W_N(n_1,n_2,\ldots n_k)$. We must, however, impose the condition that

$$\sum_{i=1}^{k}n_i=N.\qquad(11.7)$$

Furthermore, we want to consider a situation in which the total energy is fixed. Thus if the cell "i" has energy E_i associated with it, we must also impose the condition that

$$\sum_{i=1}^{k}n_iE_i=E.\qquad(11.8)$$

To minimize or maximize a function subject to constraints it is most convenient to use the method of *Lagrange Multipliers*. Since this may not be familiar to the reader, we digress briefly to discuss this.

Lagrange multipliers Consider a function $f(x,y)$ that may represent the altitude at the location (x,y). If we want to find the highest point in the region, we find all the peaks, which are those points at which a displacement in any direction does not change the altitude to first order in the small displacement. It is enough to consider a displacement in the x direction and a displacement in the y direction independently. Thus we require that

$$df(x, y) = \frac{\partial f}{\partial x} dx + \frac{\partial f}{\partial y} dy = 0 \tag{11.9}$$

and this must hold for two infinitesimal displacements independently. Thus we get the extremum condition

$$\frac{\partial f}{\partial x} = 0, \qquad \frac{\partial f}{\partial y} = 0. \tag{11.10}$$

If we ask for the maximum of $f(x, y)$ subject to a constraint condition $\phi(x, y) = 0$, we are asking not for the peaks, but, for example, for the highest points on a road. Thus we are no longer dealing with arbitrary infinitesimal displacements dx and dy, but must constrain these so that we are still along the road. The constraint equation

$$\phi(x, y) = 0 \tag{11.11}$$

implies that

$$\frac{\partial \phi}{\partial x} dx + \frac{\partial \phi}{\partial y} dy = 0 \tag{11.12}$$

so that the displacements are related by

$$dy = - \frac{\partial \phi / \partial x}{\partial \phi / \partial y} dx. \tag{11.13}$$

Now a change in altitude along the road is given by

$$df = \left[\left(\frac{\partial f}{\partial x} \right) - \left(\frac{\partial f}{\partial y} \right) \frac{\partial \phi / \partial x}{\partial \phi / \partial y} \right] dx \tag{11.14}$$

and this will vanish when the quantity in the brackets vanishes.

The method of Lagrange Multipliers instructs us to do the following:

a) Write out df in differential form

$$df = \frac{\partial f}{\partial x} dx + \frac{\partial f}{\partial y} dy = 0. \tag{11.15}$$

b) Write out the constraining equations in differential form

$$d\phi = \frac{\partial \phi}{\partial x} dx + \frac{\partial \phi}{\partial y} dy = 0. \tag{11.16}$$

c) Multiply the constraint differential forms by arbitrary constants, the Lagrange Multipliers, and then add the equations. Here we have

$$df + \lambda d\phi = 0. \tag{11.17}$$

d) The displacements may now be treated as independent and the extremum conditions now become

$$\left(\frac{\partial f}{\partial x} + \lambda \frac{\partial \phi}{\partial x}\right) dx + \left(\frac{\partial f}{\partial y} + \lambda \frac{\partial \phi}{\partial y}\right) dy = 0,$$

that is,

$$\frac{\partial f}{\partial x} + \lambda \frac{\partial \phi}{\partial x} = 0$$

$$\frac{\partial f}{\partial y} + \lambda \frac{\partial \phi}{\partial y} = 0. \tag{11.18}$$

Elimination of λ from one of these equations leads to

$$\frac{\partial f}{\partial x} - \frac{\partial f}{\partial y} \frac{\partial \phi/\partial x}{\partial \phi/\partial y} = 0, \tag{11.19}$$

which is the same result obtained in Eq. (11.14).

More generally, to maximize $f(x_1, x_2, \ldots x_n)$ subject to several constraints

$$\phi_1(x_1, x_2, \ldots, x_n) = 0$$

$$\phi_2(x_1, x_2, \ldots, x_n) = 0$$

$$\vdots$$

$$\phi_r(x_1, x_2, \ldots, x_n) = 0 \tag{11.20}$$

we write

$$df = \frac{\partial f}{\partial x_1} dx_1 + \frac{\partial f}{\partial x_2} dx_2 + \cdots = 0$$

$$\lambda_1 d\phi_1 = \lambda_1 \frac{\partial \phi_1}{\partial x_1} dx_1 + \lambda_1 \frac{\partial \phi_1}{\partial x_2} dx_2 + \ldots = 0$$

$$\vdots$$

$$\lambda_r d\phi_r = \lambda_r \frac{\partial \phi_r}{\partial x_1} dx_1 + \lambda_r \frac{\partial \phi_r}{\partial x_2} dx_2 + \ldots = 0,$$

add these and treat the displacements $dx_1, dx_2, \ldots, dx_n$ as independent, obtaining the extremum conditions as

$$\frac{\partial f}{\partial x_1} + \lambda_1 \frac{\partial \phi_1}{\partial x_1} + \lambda_2 \frac{\partial \phi_2}{\partial x_1} + \cdots + \lambda_r \frac{\partial \phi_r}{\partial x_1} = 0$$

$$\frac{\partial f}{\partial x_2} + \lambda_1 \frac{\partial \phi_1}{\partial x_2} + \lambda_2 \frac{\partial \phi_2}{\partial x_2} + \cdots + \lambda_r \frac{\partial \phi_r}{\partial x_2} = 0$$

$$\vdots$$

$$\frac{\partial f}{\partial x_n} + \lambda_1 \frac{\partial \phi_1}{\partial x_n} + \lambda_2 \frac{\partial \phi_2}{\partial x_n} + \cdots + \lambda_r \frac{\partial \phi_r}{\partial x_n} = 0 \tag{11.21}$$

Most probable distribution For our problem we take

$$d(\log W_N(n_1, \ldots, n_k)) - \lambda_1 dN(n_1, \ldots, n_k)$$
$$- \lambda_2 dE(n_1, \ldots, n_k) = 0. \tag{11.22}$$

This takes the form

$$\sum_{i=1}^{k} \left(\frac{\partial}{\partial n_i} \log W_N - \lambda_1 \frac{\partial N}{\partial n_i} - \lambda_2 \frac{\partial E}{\partial n_i} \right) dn_i = 0$$

and treating the dn_i as independent we get the extremum conditions

$$\frac{\partial}{\partial n_i} \log W_N - \lambda_1 \frac{\partial N}{\partial n_i} - \lambda_2 \frac{\partial E}{\partial n_i} = 0; \qquad i = 1, 2, \ldots, k. \tag{11.23}$$

If we use the leading terms in Stirling's formula, we have

$$\log W_N \cong N \log N - N - \sum_{i=1}^{k} (n_i \log n_i - n_i) \tag{11.24}$$

so that

$$\frac{\partial}{\partial n_i} \log W_N = -\log n_i. \tag{11.25}$$

Using Eqs. (11.7) and (11.8), we get

$$-\log n_i - \lambda_1 - \lambda_2 E_i = 0, \tag{11.26}$$

that is,

$$n_i = e^{-\lambda_1} e^{-\lambda_2 E_i}. \tag{11.27}$$

The still unknown multipliers can be determined by the subsidiary conditions

$$N = \sum_i n_i = e^{-\lambda_1} \sum_i e^{-\lambda_2 E_i}$$

$$E = \sum_i n_i E_i = e^{-\lambda_1} \sum_i E_i e^{-\lambda_2 E_i}. \tag{11.28}$$

We see that the average energy is

$$\langle E \rangle = \frac{E}{N} = \frac{\sum_i E_i e^{-\lambda_2 E_i}}{\sum_i e^{-\lambda_2 E_i}}. \tag{11.29}$$

Since

$$-\frac{\partial}{\partial \lambda_2} \sum_i e^{-\lambda_2 E_i} = \sum_i E_i e^{-\lambda_2 E_i},$$

we may rewrite the average energy in the form

$$\langle E \rangle = \left(-\frac{\partial}{\partial \lambda_2} \sum_i e^{-\lambda_2 E_i} \right) \Big/ \sum_i e^{-\lambda_2 E_i}$$

$$= -\frac{\partial}{\partial \lambda_2} \log \sum_i e^{-\lambda_2 E_i} \tag{11.30}$$

We may relate the multiplier λ_2 to the temperature by considering a gas of molecules that is noninteracting except for the collisions that randomize the motion. For the energies of the molecules we therefore take just the kinetic energy

$$E_i = \tfrac{1}{2}mv_i^2, \tag{11.31}$$

where m is the mass of the molecule and v_i is the velocity in the cell denoted by "i".

In our counting argument we used the fundamental assumption that the probability of finding a given molecule in some cell is independent of the *location* of the cell in velocity space. It is clearly not independent of the size of the cell, about which we only specified that it not be too large or too small. Thus we should really write

$$n_i = Cg_i(\Delta v_i)^3 e^{-\lambda_2 E_i} \tag{11.32}$$

where $(\Delta v_i)^3$ is the volume of the i-th cell and g_i is a *degeneracy parameter* that takes into account the possibility that there may be several states with the same velocity that have the same energy. This will become relevant when we consider quantum mechanical problems involving angular momentum. C is a constant which also absorbs the quantity previously written as $e^{-\lambda_1}$. Let us go on with simple molecules for which $g_i = 1$. Then

$$N = C \sum_i (\Delta v_i)^3 e^{-\lambda_2 E_i} \tag{11.33}$$

and

$$\langle E \rangle = -\frac{\partial}{\partial \lambda_2} \log \sum_i (\Delta v_i)^3 e^{-\lambda_2 E_i}. \tag{11.34}$$

In the limit that the cells become small and there are many of them, the sum becomes an integral, since the energy and the velocity change in an almost continuous manner from cell to cell. Thus we write

$$dn(\mathbf{v}) = Cd^3v e^{-\lambda_2 mv^2/2}$$

$$N = C \int d^3v e^{-\lambda_2 mv^2/2}$$

$$\langle E \rangle = -\frac{\partial}{\partial \lambda_2} \log \int d^3v e^{-\lambda_2 mv^2/2}. \tag{11.35}$$

Although we at one point argued that we would pick some large v_0 as a cut-off on the size of the velocity, we see that the integrand falls very rapidly when v gets to be large, and that the contribution from the region $|\mathbf{v}| > v_0$ is negligible when the velocity integration is carried out over all velocities. We thus need to evaluate

$$Z = \int d^3v e^{-\lambda_2 mv^2/2}. \tag{11.36}$$

To do this, we proceed as follows: on one hand

$$Z = \int_{-\infty}^{\infty} dv_x \int_{-\infty}^{\infty} dv_y \int_{-\infty}^{\infty} dv_z e^{-\lambda_2 m(v_x^2 + v_y^2 + v_z^2)/2} = \left(\int_{-\infty}^{\infty} dv e^{-\lambda_2 mv^2/2} \right)^3. \tag{11.37}$$

With the change of variables $\lambda_2 mv^2/2 = u^2$, we get

$$\int_{-\infty}^{\infty} dv e^{-\lambda_2 mv^2/2} = \left(\frac{2}{\lambda_2 m}\right)^{1/2} \int_{-\infty}^{\infty} du e^{-u^2}. \tag{11.38}$$

The integral is just a number, and as will be shown

$$\int_{-\infty}^{\infty} du e^{-u^2} = \sqrt{\pi}. \tag{11.39}$$

Consequently,

$$Z = \int d^3 v e^{-\lambda_2 mv^2/2} = \left(\frac{2\pi}{\lambda_2 m}\right)^{3/2}. \tag{11.40}$$

Thus (11.35) implies that

$$N = C \left(\frac{2\pi}{\lambda_2 m}\right)^{3/2}. \tag{11.41}$$

Let us now consider one mole of gas. We then have

$$C = N_0 \left(\frac{\lambda_2 m}{2\pi}\right)^{3/2} \tag{11.42}$$

and

$$\langle E \rangle = \frac{E}{N_0} = -\frac{\partial}{\partial \lambda_2} \log Z = -\frac{\partial}{\partial \lambda_2} \log \left(\frac{2\pi}{\lambda_2 m}\right)^{3/2}$$

$$= -\frac{\partial}{\partial \lambda_2} \left[\frac{3}{2} \log \frac{2\pi}{m} - \frac{3}{2} \log \lambda_2\right] = \frac{3}{2\lambda_2}. \tag{11.43}$$

If we now make the identification

$$\langle E \rangle = \tfrac{1}{2} m \langle \mathbf{v}^2 \rangle = \tfrac{3}{2} kT, \tag{11.44}$$

we see that

$$\lambda_2 = \frac{1}{kT}. \tag{11.45}$$

For noninteracting molecules we therefore have

$$dn(\mathbf{v}) = N_0 \left(\frac{m}{2\pi kT}\right)^{3/2} e^{-mv^2/2kT} d^3 v. \tag{11.46}$$

We also list

$$Z = \sum_i e^{-E_i/kT} = \left(\frac{2\pi kT}{m}\right)^{3/2} \tag{11.47}$$

and

$$\langle E \rangle = -\frac{\partial}{\partial \lambda_2} \log \left(\frac{2\pi}{\lambda_2 m}\right)^{3/2} = \frac{3}{2} \frac{1}{\lambda_2} = \frac{3}{2} kT. \tag{11.48}$$

The identification of λ_2 with $1/kT$ holds even for interacting molecules, so that quite generally

$$Z = \sum_i g_i e^{-E_i/kT} \tag{11.49}$$

and

$$\langle E \rangle = -\frac{\partial}{\partial(1/kT)} Z. \tag{11.50}$$

If we look over the derivation of the Boltzmann distribution formula, we observe that we did not use the fact that we were dealing with cells in *velocity* space until the very end. The cells could equally well have been cells in velocity + ordinary space (the so-called phase space) and then n_i would have ended up being the number of molecules with velocities in the range $(\mathbf{v}, \mathbf{v} + d\mathbf{v})$ *and* with location in the spatial region $(\mathbf{r}, \mathbf{r} + d\mathbf{r})$. The result would have been, then,

$$dn(\mathbf{r}, \mathbf{v}) = Ce^{-E(\mathbf{r}, \mathbf{v})/kT} d^3\mathbf{r} d^3\mathbf{v} \tag{11.51}$$

provided that the randomization makes the probability of being in any spatial region independent of where that spatial region is situated, exactly as for the case of velocity-space cells. It is reasonable to assume that collisions provided that kind of randomization as well.

Since the fraction of molecules in a given cell is equal to the probability of finding a molecule in that cell, we may summarize the results of this chapter by stating that

> *the probability of finding a molecule in a state characterized by the energy E_i is given by*
>
> $$P_i = \frac{1}{Z} g_i e^{-E_i/kT}$$
>
> *where $Z = \sum_i g_i e^{-E_i/kT}$ and g_i is the number of states with that energy.*

We shall soon have occasion to use this result in a number of applications.

NOTES AND COMMENTS

1. To evaluate $I = \int_{-\infty}^{\infty} du\, e^{-u^2}$ we note that

$$I^2 = \int_{-\infty}^{\infty} du \int_{-\infty}^{\infty} dv\, e^{-(u^2 + v^2)}$$

which is the integral over the area of the whole plane. If we transform to polar coordinates, we have $u = \rho \cos\theta$, $v = \rho \sin\theta$,

$$du\, dv = \rho\, d\rho\, d\theta, \quad u^2 + v^2 = \rho^2$$

$$I^2 = \int_0^{2\pi} d\theta \int_0^{\infty} \rho\, d\rho\, e^{-\rho^2} = 2\pi \times \tfrac{1}{2} \int_0^{\infty} dz\, e^{-z} = \pi$$

so that

$$I = \sqrt{\pi}.$$

2. Quite generally we may evaluate integrals of the form

$$\int_{-\infty}^{\infty} du\, u^{2k} e^{-\alpha u^2}$$

$$\int_{-\infty}^{\infty} du\, u^{2k} e^{-\alpha u^2} = \underbrace{\left(-\frac{\partial}{\partial \alpha}\right)\left(-\frac{\partial}{\partial \alpha}\right)\cdots\left(-\frac{\partial}{\partial \alpha}\right)}_{k \text{ times}} \int_{-\infty}^{\infty} du\, e^{-\alpha u^2}$$

$$= \left(-\frac{\partial}{\partial \alpha}\right)^k \left(\frac{\pi}{\alpha}\right)^{1/2}.$$

For odd powers we have

$$\int_0^{\infty} du\, u^{2k+1} e^{-\alpha u^2} = \frac{1}{2}\int_0^{\infty} (2u\,du)(u^2)^k e^{-\alpha u^2}$$

$$= \frac{1}{2}\int_0^{\infty} dw\, w^k e^{-\alpha w}$$

$$= \frac{1}{2\alpha^{k+1}}\int_0^{\infty} dy\, y^k e^{-y}$$

$$= \frac{k!}{2\alpha^{k+1}}.$$

3. Although it is not our intent to get into the field of thermodynamics, it is not out of place to point out that the free energy is given by

$$F = \log Z,$$

the entropy by

$$S = k\log Z + \langle E\rangle/T.$$

If the volume of the system is V, then the pressure may be written in the form

$$p = kT\frac{\partial}{\partial V}\log Z.$$

For a treatment of these matters based on the approach used in this chapter, see the beautiful little book of lectures: E. Schrödinger, *Statistical Thermodynamics*, Cambridge University Press, 1948. Some of the material is clearly beyond the scope of the readers of this book, but the first five chapters are at a level comparable to ours. See also F. Reif, *Statistical Physics*, McGraw-Hill, New York, 1964.

4. There might be some confusion about the sudden appearance of the degeneracy factor g_i. After all, Eq. (11.27) has no reference to it, and suddenly there it is in Eq. (11.32)! The reason is that to simplify the argument and the algebra, we started out with Eq. (11.5) without mentioning degeneracy. If there is degeneracy, then Eq. (11.5) really must be changed. Suppose the molecule can be in two states when it is in cell 1. Then the number of ways in which n_1 molecules can be accommodated in cell 1 has an additional factor 2^{n_1}. More

generally, if the degeneracy in cell i is g_i, then the result (11.5) must be changed to

$$W_N = \frac{N!}{n_1! n_2! \cdots} (g_1)^{n_1} (g_2)^{n_2} \cdots$$

Now

$$dW_N = d(N \log N - \sum_i (n_i \log n_i - n_i) + \sum n_i \log g_i - N) = - \sum dn_i \log n_i / g_i$$

which shows that $n_i \to n_i/g_i$ in Eq. (11.27).

PROBLEMS

11.1 Consider a gas whose velocity distribution is of the Maxwell Boltzmann form. If the gas is being blown along with a velocity **u**, then the function representing the distribution is

$$(m/2\pi kT)^{3/2} e^{-m(v-u)^2/2kT}.$$

Use this to calculate $\langle v \rangle$ and $\langle v^2 \rangle$. How will the pressure calculation of Chapter 11 be affected?

11.2 Ten dice are tossed. What is the number of ways of getting 4 ones, 2 twos, and 4 threes? What is the probability of that event happening?

11.3 A gas of $N_2(A = 28)$ molecules is at a temperature of $300°K$. What is the probability of finding molecules with velocity larger in magnitude than 1.2×10^6 cm/sec? What is it for 1.2×10^5 cm/sec?

11.4 Find the extremum of

$$4x^2 - 8xy + 14y^2 - 12x + 5y$$

subject to the condition

$$3x - 6y = 10$$

(a) by eliminating one of the variables, and (b) using Lagrange multipliers.

11.5 A sphere in N dimensions is the surface defined by

$$x_1^2 + x_2^2 + \cdots + x_N^2 = R^2.$$

Calculate the surface area of this sphere and the volume of the sphere. The procedure to be used is the following:

a) the volume of a sphere in N dimensions is given by

$$\int_0^R r^{N-1} dr \int d\Omega$$

where $\int d\Omega$ is the integral over the solid angle. That integral can be obtained by observing that on one hand

$$\int d^N r e^{-\alpha r^2} = \int d\Omega \int_0^\infty r^{N-1} dr e^{-\alpha r^2}$$

and on the other

$$\int d^N r e^{-\alpha r^2} = \int_{-\infty}^\infty dx_1 e^{-\alpha x_1^2} \int_{-\infty}^\infty dx_2 e^{-\alpha x_2^2} \cdots \int_{-\infty}^\infty dx_N e^{-\alpha x_N^2}.$$

All of these integrals can be done using the material in the notes and comments to this chapter.

b) The surface is given by dV/dR.

11.6 Write down an expression for $\langle E^2 \rangle$ in terms of

$$Z = \sum e^{-\beta E} \qquad \beta = 1/kT.$$

Use this to calculate the mean square deviation of the energy from the average, that is

$$\langle (E - \langle E \rangle)^2 \rangle.$$

Show that in terms of the temperature,

$$\langle (E - \langle E \rangle)^2 \rangle = kT^2 C_v,$$

where C_v is the heat capacity of the system, defined by

$$C_v = \partial \langle E \rangle / \partial T.$$

11.7 Use the distribution (11.46) to calculate the number of molecules that have velocity in the range $(v, v + dv)$ independent of direction. Use the distribution to calculate the number that have energy in the range $(E, E + dE)$. To do this, use $E = mv^2/2$ and thus $dE = mv\,dv$.

11.8 The energy of a molecule in a potential is given by

$$E = p^2/2m + V_0(r/a)^n.$$

Use Eq. (11.51) to calculate the average energy. Proceed by first calculating

$$Z = \int d^3v e^{-mv^2/2kT} \int d^3r e^{-V_0(r/a)^n/kT}$$

and then using Eq. (11.34). You may evaluate the T dependence of the second integral using dimensional analysis.

12
Some Applications of Statistical Physics

In this chapter we discuss some physical applications of the Boltzmann distribution and our identification of the average kinetic energy with the temperature. The first topic we discuss is the use of the Maxwell-Boltzmann distribution for noninteracting molecules Eq. (11.46), for the calculation of certain averages of powers of velocity.

Test of Maxwell-Boltzmann Distribution Let us calculate the velocity distribution of molecules escaping through a small hole in a container holding a gas of molecules at temperature T. The number of molecules that emerge from the hole is given by the sum of molecules coming from the various cylinders as shown in Fig. 12.1. The number that emerge from a given cylinder in time Δt is a product of the volume of the cylinder, which is $nv\Delta t\Delta A$ (ΔA is the size of the hole) and the number of molecules in that cylinder. Thus the number emerging is

$$dn_{\text{emerg}} = \Delta t\Delta A\, \mathbf{n} \cdot \mathbf{v}dn(\mathbf{v}). \tag{12.1}$$

If we define the z axis to coincide with the direction $\mathbf{n}$, the normal to the hole, then the number of molecules emerging per unit time, per cm^2 is obtained by dividing by

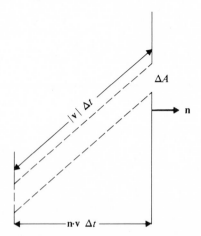

Figure 12.1. Cylinder contribution to effusion of molecules.

$\Delta t \Delta A$, and integrating over the x and y components of the velocity. Thus the number emerging with velocity v_z per cm^2 per second is

$$dn_{\text{emerg.}}(v_z) = \left(\int_{-\infty}^{\infty} dv_x \int_{-\infty}^{\infty} dv_y \, N_0 \left(\frac{m}{2\pi kT} \right)^{3/2} e^{-mv^2/2kT} \right) \times v_z dv_z$$

$$= N_0 \left(\frac{m}{2\pi kT} \right)^{3/2} \left(\frac{2\pi kT}{m} \right)^{1/2} \left(\frac{2\pi kT}{m} \right)^{1/2} e^{-mv_z^2/2kT} v_z dv_z$$

$$= N_0 \left(\frac{m}{2\pi kT} \right)^{1/2} e^{-mv_z^2/2kT} v_z dv_z. \tag{12.2}$$

In the above, we have made use of the fact that

$$e^{-mv^2/2kT} = e^{-mv_x^2/2kT} e^{-mv_y^2/2kT} e^{-mv_z^2/2kT}$$

and

$$\int_{-\infty}^{\infty} dv_x e^{-mv_x^2/2kT} = \left(\frac{2\pi kT}{m} \right)^{1/2}.$$

This prediction was tested experimentally by Stern in 1920 and by Zartman and by Ko in the early 1930s. The experiment utilizes an oven with a small hole, from which molecules with the distribution (12.2) emerge (Fig. 12.2). A rapidly rotating drum with a slit on one side was used to collect the molecules on its inside walls. The passage of the slit in front of the oven effectively opens up a rapid shutter letting in a bunch of molecules. The fast ones cross the drum quickly, the slower ones lag, hitting a different portion of the inside wall of the rapidly rotating drum. The distribution of molecules on the drum can be translated into a velocity distribution of the incoming molecules, and that turned out to be in perfect agreement with Eq. (12.2). If we want the total number of molecules emitted, we must integrate (12.2), taking care to integrate only over positive values of v_z, since molecules moving away from the hole do not pass through it.

The Barometric Formula In the last chapter we pointed out that the Boltzmann factor could also be used to describe a joint distribution of molecules in space as well as in velocity space. If the energy of a molecule is given by

$$E = \tfrac{1}{2} m v^2 + U(\mathbf{r}), \tag{12.3}$$

then

$$dn(\mathbf{r}, \mathbf{v}) = C e^{-mv^2/2kT} e^{-U(\mathbf{r})/kT} d^3 v d^3 r. \tag{12.4}$$

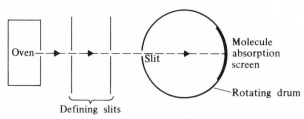

Figure 12.2. Schematic diagram of experiment testing Maxwell-Boltzmann distribution.

If we are not interested in the velocity distribution, then we integrate over velocities and get

$$dn(\mathbf{r}) = C'e^{-U(\mathbf{r})/kT}d^3r. \tag{12.5}$$

Consider molecules in a uniform gravitational field. In that case the potential energy is given by

$$U(\mathbf{r}) = mgz \tag{12.6}$$

where z is the height above some reference level. We then have

$$\frac{n(z)}{n(0)} = e^{-mgz/kT}. \tag{12.7}$$

The density falls to half its ground level value when

$$z = z_{1/2} = 0.69\,kT/mg. \tag{12.8}$$

If we consider molecules of molecular weight M, then $m = 1.6 \times 10^{-24}M$ gm, and then

$$z_{1/2} \simeq \frac{0.69 \times 1.38 \times 10^{-16}\,T}{1.6 \times 10^{-24} \times 980\,M} \cong 0.6\,\frac{T}{M}\,\text{km}. \tag{12.9}$$

For O_2 molecules ($M = 32$) and $T = 300°$K, this gives approximately 5.6 km. Actually the temperature in the atmosphere is not constant. It falls with altitude, and the density therefore falls off more rapidly. The above calculation does, however, give us a rough approximation to the right number.

Doppler Broadening Consider molecules moving with a Maxwellian distribution. If a molecule moving with velocity $\mathbf{v}$ radiates, the frequency of the radiation is shifted from the frequency characterizing the molecule at rest, ν_0. The shifted frequency seen along the z direction is

$$\nu = \nu_0(1 + v_z/c). \tag{12.10}$$

The distribution of velocities v_z of the molecules implies that the distribution of frequencies will be given by replacing v_z in

$$dn(v_z) = Ce^{-mv_z^2/2kT}dv_z \tag{12.11}$$

by

$$v_z = c\,\frac{\nu - \nu_0}{\nu_0} \tag{12.12}$$

$$dn(\nu) = C'e^{-mc^2(\nu-\nu_0)^2/2kT\nu_0^2}d\nu. \tag{12.13}$$

This is a gaussian shape, and the line has half its maximum intensity at a frequency $\bar{\nu}$ such that

$$\frac{mc^2}{2kT}\left(\frac{\bar{\nu} - \nu_0}{\nu_0}\right)^2 = 0.69, \tag{12.14}$$

that is,

$$\frac{\Delta\nu}{\nu_0} = \left(\frac{1.38\,kT}{mc^2}\right)^{1/2}. \tag{12.15}$$

A spectral line is therefore broadened by thermal effects, and this leads to a loss of precision in the determination of frequencies of radiated light. To get a rough idea of the magnitude of $\Delta\nu/\nu_0$, we consider the above for hydrogen at 1000°K. For hydrogen, $m = 1.6 \times 10^{-24}$ gm and we have

$$\frac{\Delta\nu}{\nu_0} = \left(\frac{1.38 \times 1.38 \times 10^{-16} \times 10^3}{1.6 \times 10^{-24} \times 9 \times 10^{20}}\right)^{1/2}$$

$$= 1.15 \times 10^{-5}.$$

This corresponds to quite a sizeable loss of precision. In a clock, for example, this would mean a discrepancy of about six minutes per year. Recall that a precision of 10^{-15} is needed to measure the gravitational red shift of light "falling" from a 100 ft tower.

Paramagnetic Gases Consider a gas of molecules that are characterized by a magnetic dipole moment μ. Let us put a sample of the gas in equilibrium at temperature T in an external magnetic field B (Fig. 12.3). The potential energy of a molecule is given by

$$U = -\mathbf{\mu} \cdot \mathbf{B}. \tag{12.16}$$

The probability of finding a molecule pointing in a direction is proportional to the solid angle $d\Omega$, and the Boltzmann factor $e^{+\mathbf{\mu}\cdot\mathbf{B}/kT}$. Thus with $\mathbf{B}$ defining the z direction,

$$P(\theta,\phi)d\theta d\phi = \frac{e^{\mu B \cos\theta/kT}\sin\theta d\theta d\phi}{\int_0^\pi \sin\theta d\theta \int_0^{2\pi} d\phi e^{\mu B \cos\theta/kT}}. \tag{12.17}$$

The average energy of a molecule in the external field is

$$\langle -\mathbf{\mu}\cdot\mathbf{B}\rangle = \frac{\int_0^\pi 2\pi \sin\theta d\theta\, e^{\mu B \cos\theta/kT}(-\mu B \cos\theta)}{\int_0^\pi 2\pi \sin\theta d\theta\, e^{\mu B \cos\theta/kT}}. \tag{12.18}$$

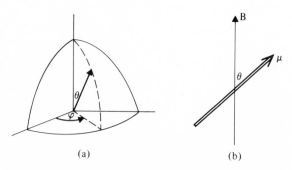

(a)   (b)

Figure 12.3. (a) Insert showing definition of spherical angles. (b) Orientation of magnetic dipole relative to magnetic field.

To evaluate the integrals, make the change of variables $\cos \theta = u$, then

$$\langle E \rangle = \frac{-\mu B \int_{-1}^{1} du\, u\, e^{\mu B u/kT}}{\int_{-1}^{1} du\, e^{\mu B u/kT}}$$

$$= -\mu B \left. \frac{\dfrac{\partial}{\partial \alpha} \int_{-1}^{1} du\, e^{u\alpha}}{\int_{-1}^{1} du\, e^{u\alpha}} \right|_{\alpha = \mu B/kT}$$

$$= -\mu B \frac{\dfrac{\partial}{\partial \alpha} \dfrac{e^{\alpha} - e^{-\alpha}}{\alpha}}{(e^{\alpha} - e^{-\alpha})/\alpha}$$

$$= -\mu B \left(\coth \frac{\mu B}{kT} - \frac{kT}{\mu B} \right)$$

$$= kT \left(1 - \frac{\mu B}{kT} \coth \frac{\mu B}{kT} \right). \qquad (12.19)$$

In the high temperature limit, when $\alpha = \mu B/kT \ll 1$, we have

$$\alpha \coth \alpha = \alpha \frac{\cosh \alpha}{\sinh \alpha} = \alpha \frac{1 + \alpha^2/2! + \cdots}{\alpha + \alpha^3/3! + \cdots}$$

$$= \frac{1 + \alpha^2/2 + \cdots}{1 + \alpha^2/6 + \cdots}$$

$$\simeq \left(1 + \frac{\alpha^2}{2} \right) \left(1 - \frac{\alpha^2}{6} \right)$$

$$\simeq 1 + \frac{\alpha^2}{3} \qquad (12.20)$$

so that

$$\langle E \rangle \simeq \tfrac{1}{3}(\mu B)^2 / kT. \qquad (12.21)$$

This is to be contrasted with the low temperature limit in which $\alpha \to \infty$. There

$$\alpha \coth \alpha = \alpha \frac{e^{\alpha} + e^{-\alpha}}{e^{\alpha} - e^{-\alpha}} = \alpha(1 + e^{-2\alpha})(1 - e^{-2\alpha})^{-1}$$

$$\simeq \alpha(1 + 2e^{-2\alpha}),$$

so that

$$\langle E \rangle \cong kT \left(1 - \frac{\mu B}{kT} \right)$$

$$\cong -\mu B \left(1 - \frac{kT}{\mu B} \right). \qquad (12.22)$$

We see that in the low temperature limit the average energy approaches the value it would have if all the molecules lined up in the external field so as to minimize the energy. This tendency is only slightly disturbed by the correction term: at low temperatures thermal randomizing effects are not important. At high temperatures all orientations become equally likely since the energy differences become negligible compared with kT. Thus the average energy tends to zero.

Evaporation Consider a liquid in contact with vapor of the same substance, and let these be in equilibrium at temperature T. The equilibrium is not static (as when the tension of a rope balances the force of gravity, for example), but dynamic. As many molecules enter the liquid from the vapor as leave the liquid. Molecules, in order to leave the liquid, must break loose from the molecular forces due to adjacent molecules. The simplest model for the molecular forces is that in which each molecule experiences an attraction W when it is in the liquid, and no potential when it is in the gas. This is realized by the picture of a short-range attractive potential energy (Fig. 12.4). When the molecules are in the gaseous phase, they are so far apart that no potential energy contributes to their total energy. We may now use the Boltzmann formula to relate the number of molecules in the vapor to the number of molecules in the liquid. If we consider a fixed volume, we obtain the relation between the number densities

$$\frac{n_{\text{vapor}}}{n_{\text{liquid}}} = e^{-W/kT}. \tag{12.23}$$

In writing this, we have canceled the kinetic energy terms in the numerator and in the denominator, on the assumption that the velocity distribution is the same in the liquid as in the vapor phase. We may estimate W in two ways. First, consider the latent heat of evaporation, the energy required to convert one mole of liquid to one mole of gas, L. Work must be done to evaporate liquid at its boiling point. Some of it goes into expanding the gas at a constant temperature, and the rest, by far the larger part, goes into breaking the molecular bonds that keep the molecules at the surface of the liquid.

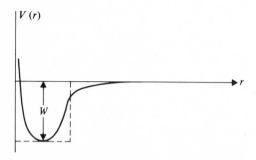

Figure 12.4. Sketch of assumed potential energy for molecule which, because of its short range ($\sim 1 - 2\,\text{Å}$) is only effective when the molecule is in the liquid. Dashed line represents approximation used in our model.

For water, $L = 540 \times 18$ Cal/mole. (The molecular weight of water is 18.) This corresponds to an energy of

$$L = 540 \times 18 \times 4.2 \times 10^7 = 4.08 \times 10^{11} \text{ ergs}$$

since 1 Cal $= 4.2$ J $= 4.2 \times 10^7$ ergs. Thus the energy per molecule $= L/N_0 = 6.8 \times 10^{-13}$ ergs. The amount of energy required to expand the gas from the liquid phase density to the gaseous phase density is approximately pV where p is the pressure at the boiling point and V is the volume of one mole of gas. By the ideal gas law,

$$pV = RT = N_0kT$$

and therefore the energy of expansion per molecule is

$$pV/N_0 = kT.$$

For water this is $1.38 \times 10^{-16} \times 373°\text{K} = 0.51 \times 10^{-13}$ erg. Hence

$$W = (6.8 - 0.5) \times 10^{-13} = 6.3 \times 10^{-13} \text{ erg.}$$

A convenient unit of energy is the *electron volt*, where

$$1 \text{ eV} = 1.60210 \pm 0.00002 \times 10^{-12} \text{ erg.} \tag{12.24}$$

Thus our estimate is that

$$W = 0.4 \text{ eV} \tag{12.25}$$

for water. For the other estimate we use the formula (12.23). We note that at the boiling point of water, the vapor pressure is $p = 1$ atmosphere $= 10^6$ dynes/cm^2. (1 dyne $= 1$ gm cm/sec^2 is the c.g.s. unit of force), and the temperature is $373°\text{K}$. If we consider one mole of water vapor, then

$$n_{\text{vapor}} = N_0/18V = pN_0/18pV = pN_0/18RT = p/18kT$$
$$= 1.1 \times 10^{18} \text{ cm}^{-3}.$$

For the liquid the density of water is 1 gm/cm^3. Thus the density in terms of molecules /cm^3 is

$$n_{\text{liquid}} = 1 \frac{\text{gm}}{\text{cm}^3} \times \frac{6.02 \times 10^{23}}{18} \frac{\text{molec.}}{\text{gm}} = 3.3 \times 10^{22} \text{ cm}^{-3}.$$

Thus using Eq. (12.23) with $T = 373°\text{K}$ we find

$$W = kT \log \frac{n_{\text{liquid}}}{n_{\text{vapor}}}$$
$$= 1.38 \times 10^{-16} \times 373 \times \log(3 \times 10^4)$$
$$= 5.31 \times 10^{-13} \text{ ergs}$$
$$= 0.33 \text{ eV.} \tag{12.26}$$

The 25% discrepancy between the two estimates is not surprising for such a crude model. The formula may be used to calculate (very approximately) the variation in vapor pressure with temperature. Since the perfect gas law yields

$$n \propto p/T, \tag{12.27}$$

we predict that

$$\frac{p_1}{p_2} = \frac{T_1}{T_2} \frac{e^{-W/kT_1}}{e^{-W/kT_2}}. \tag{12.28}$$

With the larger estimate of W (12.25), we obtain the following results by adjusting a proportionality constant at $373°\text{K}$.

Temperature	Experimental vapor pressure (atmosphere)	$562T e^{-W/kT}$
353	0.474	0.479
333	0.199	0.208
313	0.074	0.081
293	0.0234	0.0281

Estimation of evaporation rates is more difficult. One argues that the number of molecules that get captured from the vapor by the liquid is proportional to the rate at which the molecules hit the liquid ($\propto \sqrt{\langle v^2 \rangle}$), the density of the vapor, n_{vapor}, and some unknown "sticking probability" that the molecule from the vapor will be captured, rather than bounce off. One would like to assume that this is of order $0.3 - 1.0$ rather than some tiny number like 10^{-3}, say. Thus the rate of molecules coming in per second per cm^2 is

$$N_{\text{in}} = \xi(2kT/m)^{1/2} n_{\text{vapor}}. \tag{12.29}$$

In equilibrium this is also the rate of molecules leaving, so that

$$N_{\text{out}} = \xi(2kT/m)^{1/2} n_{\text{liquid}} e^{-W/kT}. \tag{12.30}$$

One now argues that if the vapor is being blown away, there will be no incoming molecules, but there will still be molecules leaving at the old rate. Thus the evaporation rate is

$$N_{\text{evap}} = \xi(2kT/m)^{1/2} n_{\text{liquid}} e^{-W/kT} \tag{12.31}$$

for evaporation into the vacuum. This does not work for evaporation in the atmosphere, since there are air molecules above the water, so that water molecules tend to be reflected back. There are also water molecules that reflect other water molecules and the formula (12.31) is pretty useless. In fact, with $\xi = 1$ it predicts an evaporation rate of 40 gm/hr/cm^2 at 300°K!

Brownian Motion A British botanist named Brown was the first to observe that tiny particles immersed in a liquid move randomly. Their motion, called *Brownian motion*, was explained by Albert Einstein and M. Smoluchowski in 1905. The motion is caused by random collisions with the molecules that make up the liquid (Fig. 12.5). The

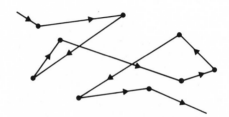

Figure 12.5. Schematic drawing of observation of path of particle suspended in liquid.

particles, typically 10^{-4} cm in size are much more massive than the molecules, and they do not move as far or as fast as the molecules. In fact, if they are in equilibrium with the liquid at temperature T, then the mean kinetic energy is

$$\tfrac{1}{2} M \langle \mathbf{v}^2 \rangle = \tfrac{3}{2} kT \tag{12.32}$$

where M is the mass of the particle, so that

$$\langle \mathbf{v}^2 \rangle = \frac{m_{\text{molec.}}}{M} \langle \mathbf{v}^2 \rangle_{\text{molec.}} \cdot \tag{12.33}$$

The particles can be observed under a microscope, and their mean square displacement can be measured. The theoretical predictions follow from writing down the equation of motion for the particles. The forces consist of a viscous force, which is a frictional force of the form $- \alpha v$, where, by Stokes law

$$\alpha = 6\pi a \eta, \tag{12.34}$$

with a the radius of the particle and η the viscosity of the liquid, and of a random force $\mathbf{F}(t)$ due to the impulsive interaction with the molecules of the liquid. Thus we have

$$M \frac{d^2 \mathbf{r}}{dt^2} = -\alpha \frac{d\mathbf{r}}{dt} + \mathbf{F}(t). \tag{12.35}$$

This equation, known as the *Langevin equation*, describes the motion approximately. Now upon multiplying (12.35) by $\mathbf{r}$, we can rewrite the left side of the equation as follows:

$$M \mathbf{r} \cdot \frac{d^2 \mathbf{r}}{dt^2} = M \frac{d}{dt} \left(\mathbf{r} \cdot \frac{d\mathbf{r}}{dt} \right) - M \left(\frac{d\mathbf{r}}{dt} \right)^2. \tag{12.36}$$

Since the force is random, and uncorrelated with the position of the particle, on the right side of the equation

$$\langle \mathbf{r} \cdot \mathbf{F}(t) \rangle = 0. \tag{12.37}$$

We thus get the equation

$$M \frac{d}{dt} \left\langle \mathbf{r} \cdot \frac{d\mathbf{r}}{dt} \right\rangle = -\alpha \left\langle \mathbf{r} \cdot \frac{d\mathbf{r}}{dt} \right\rangle + M \left\langle \left(\frac{d\mathbf{r}}{dt} \right)^2 \right\rangle$$

and since in equilibrium at temperature T we have

$$\frac{1}{2} M \left\langle \left(\frac{d\mathbf{r}}{dt} \right)^2 \right\rangle = \frac{3}{2} kT, \tag{12.38}$$

we get the equation

$$\frac{d}{dt} \left\langle \mathbf{r} \cdot \frac{d\mathbf{r}}{dt} \right\rangle = -\frac{\alpha}{M} \left\langle \mathbf{r} \cdot \frac{d\mathbf{r}}{dt} \right\rangle + \frac{3kT}{M}. \tag{12.39}$$

This is a simple differential equation of the form

$$\frac{df}{dt} = -\gamma f + C. \tag{12.40}$$

To solve this we introduce another function $g(t)$ through

$$f(t) = \frac{C}{\gamma} + g(t) \tag{12.41}$$

with the constant so chosen that the equation for $g(t)$ is simply

$$\frac{dg}{dt} = -\gamma g. \tag{12.42}$$

We know the solution for this equation: it is

$$g(t) = Ae^{-\gamma t}. \tag{12.43}$$

Thus we get

$$\left\langle \mathbf{r} \cdot \frac{d\mathbf{r}}{dt} \right\rangle = Ae^{-\alpha t/M} + \frac{3kT}{\alpha} \tag{12.44}$$

with the constant A to be determined by initial conditions. Now

$$\left\langle \mathbf{r} \cdot \frac{d\mathbf{r}}{dt} \right\rangle = \frac{1}{2} \frac{d}{dt} \langle r^2 \rangle \tag{12.45}$$

so that

$$\frac{d}{dt} \langle \mathbf{r}^2 \rangle = 2Ae^{-\alpha t/M} + \frac{6kT}{\alpha}. \tag{12.46}$$

This can be integrated, and we get

$$\langle \mathbf{r}^2 \rangle = B - \frac{2MA}{\alpha} e^{-\alpha t/M} + \frac{6kT}{\alpha} t \tag{12.47}$$

where B is an integration constant. After a long time

$$\langle \mathbf{r}^2 \rangle \to \frac{6kT}{\alpha} t. \tag{12.48}$$

This result is strongly reminiscent of the random walk problem discussed in Chapter 10. The root mean square distance traveled is proportional to $\sqrt{t}$. The measurement of $\langle \mathbf{r}^2 \rangle$ thus yields a knowledge of $k = R/N_0$ and hence Avogadro's number. The first good determination of N_0 was made by Perrin in 1911 using this method.

Equipartition and Heat Capacities of Molecules Consider a gas of point particles. With the assumption that they are noninteracting except for the short-range force that is effective during the randomizing collisions, we found that

$$\langle E \rangle = \tfrac{3}{2} kT \tag{12.49}$$

so that the energy per mole of gas is

$$U = N_0 \langle E \rangle = \tfrac{3}{2} RT. \tag{12.50}$$

The *heat capacity*, or specific heat at constant volume, is defined by

$$C_v = \left(\frac{\partial U}{\partial T} \right)_v. \tag{12.51}$$

Thus for the gas under consideration

$$C_v = \tfrac{3}{2} R. \tag{12.52}$$

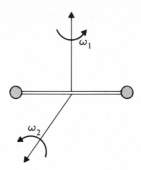

Figure 12.6. Rotational motion of diatomic molecule.

Had we considered motion in one dimension only, we would have found

$$\langle E \rangle = \tfrac{1}{2}kT. \tag{12.53}$$

The result $3kT/2$ is the sum of three contributions of $kT/2$ each from the three independent degrees of freedom of a point particle: the particle can move in a three-dimensional space, and the motion along the three axes is a superposition of three independent motions.

Consider next a gas of diatomic molecules, such as O_2, N_2, H_2, and so on. One may view these molecules classically as two point masses connected by an almost rigid bond. The molecules, in addition to moving through space with velocity v (this is really the motion of the center of mass) can also have an *internal motion*: they can rotate about two axes (not about the line joining the masses), and the two rotations are independent (Fig. 12.6). An expression for the rotational contributions to the energy is

$$E_{\text{rot}} = \tfrac{1}{2}I\omega_1^2 + \tfrac{1}{2}I\omega_2^2, \tag{12.54}$$

where I is the moment of inertia and ω_1, ω_2 are the angular velocities about the two axes. If we now consider the motion of such molecules, we see that each one has *two* additional degrees of freedom. In terms of coordinates needed to specify the state of the molecule, we would need to describe the orientation of the axes of the molecule in three-dimensional space, which requires two spherical angles (latitude and longitude, for example). During collisions among molecules in a gas, there is again randomization. Angular momentum is exchanged in collisions between molecules, and the rotation angles and the angular velocities vary. We may, following the discussion in Chapter 11, take our *cells* to be not only in velocity space, but also in the "space" of the possible values of ω_1 and ω_2 for each molecule. With the assumption that all such values are equally probable, the derivation of the Boltzmann distribution goes through unchanged and we end up with the probability for a given value of $\mathbf{v}$, ω_1 and ω_2 as

$$Ce^{-mv^2/2kT}\, e^{-I(\omega_1^2 + \omega_2^2)/2kT}. \tag{12.55}$$

The average rotational energy is given by

$$\langle E_{\text{rot}} \rangle = \frac{\int_{-\infty}^{\infty} d\omega_1 \int_{-\infty}^{\infty} d\omega_2\, e^{-I(\omega_1^2 + \omega_2^2)/2kT}\, \tfrac{1}{2}(I\omega_1^2 + I\omega_2^2)}{\int_{-\infty}^{\infty} d\omega_1 \int_{-\infty}^{\infty} d\omega_2\, e^{-I(\omega_1^2 + \omega_2^2)/2kT}}. \tag{12.56}$$

This is just the sum of two terms, each of which is of the form

$$\frac{\int_{-\infty}^{\infty} d\omega \, e^{-I\omega^2/2kT} \, I\omega^2/2}{\int_{-\infty}^{\infty} d\omega \, e^{-I\omega^2/2kT}}. \qquad (12.57)$$

The integral is identical to the one that appears for the mean kinetic energy in a given direction

$$\frac{\int_{-\infty}^{\infty} dv_z \, e^{-mv_z^2/2kT} \, mv_z^2/2}{\int_{-\infty}^{\infty} dv_z \, e^{-mv_z^2/2kT}} \qquad (12.58)$$

so that the result is again $kT/2$ for each term. We therefore get

$$\langle E_{\text{rot}} \rangle = \tfrac{1}{2}kT + \tfrac{1}{2}kT = kT \qquad (12.59)$$

and thus

$$U = N_0 \langle E_{\text{translat}} \rangle + N_0 \langle E_{\text{rot}} \rangle = \tfrac{5}{2}RT, \qquad (12.60)$$

so that

$$C_v = \left(\frac{\partial U}{\partial T}\right)_v = \frac{5}{2}R. \qquad (12.61)$$

The heat capacity is larger, since the gas molecules can store heat not only in their transitional motion, but also in their rotational motion. For more complex molecules, there are three independent axes about which rotation can take place. It is irrelevant that the three moments of inertia are generally different: the contribution to the average energy per molecule will now be $3kT/2$!

Let us now return to our diatomic molecule and observe that the "bond" that ties the two atoms together can be stretched or compressed (Fig. 12.7). To first approximation the two atoms undergo simple harmonic motion relative to each other. If the separation between the atoms is written as $d + \eta$, where d is the equilibrium separation, then the vibrational energy is, to first approximation

$$E_{\text{vib}} = \tfrac{1}{2}\mu\dot{\eta}^2 + \tfrac{1}{2}K\eta^2. \qquad (12.62)$$

How many degrees of freedom does this represent? The motion is described by the solution

$$\eta(t) = A \cos(\Omega t + \varphi) \qquad (12.63)$$

where A is the amplitude, $\Omega = \sqrt{K/\mu}$ (K is the spring constant) is the frequency of the vibration, and φ is a phase angle. Now in randomizing collisions, both A and φ change, and presumably have equal probability of taking on any value in the range of possible values. Thus there are two degrees of freedom. Instead of dealing with A and φ, we could equally well deal with η and $\dot{\eta}$, since the knowledge of η and $\dot{\eta}$ determine A and φ, and vice versa. Thus our cells could include the specification of $(\eta, \dot{\eta})$ and the Boltzmann factor would include a factor

$$e^{-\frac{1}{2}\mu\dot{\eta}^2/kT} \, e^{-\frac{1}{2}K\eta^2/kT}. \qquad (12.64)$$

K

Figure 12.7. Vibrational motion of diatomic molecule, with spring constant K.

Thus

$$\langle E_{\text{vib}} \rangle = \frac{\int_{-\infty}^{\infty} d\eta \int_{-\infty}^{\infty} d\dot{\eta}\, e^{-\frac{1}{2}\mu\dot{\eta}^2/kT}\, e^{-\frac{1}{2}K\eta^2/kT} \left(\frac{1}{2}\mu\dot{\eta}^2 + \frac{1}{2}K\eta^2\right)}{\int_{-\infty}^{\infty} d\eta \int_{-\infty}^{\infty} d\dot{\eta}\, e^{-\frac{1}{2}\mu\dot{\eta}^2/kT}\, e^{-\frac{1}{2}K\eta^2/kT}}$$

and the average vibrational energy is

$$\langle E_{\text{vib}} \rangle = \frac{\int_{-\infty}^{\infty} d\eta\, e^{-\frac{1}{2}K\eta^2/kT}\, \frac{1}{2}K\eta^2}{\int_{-\infty}^{\infty} d\eta\, e^{-\frac{1}{2}K\eta^2/kT}} + \frac{\int_{-\infty}^{\infty} d\dot{\eta}\, e^{-\mu\dot{\eta}^2/2kT}\, \frac{1}{2}\mu\dot{\eta}^2}{\int_{-\infty}^{\infty} d\dot{\eta}\, e^{-\mu\dot{\eta}^2/2kT}}. \qquad (12.65)$$

Again, comparison with previous arguments shows that this is kT, with $kT/2$ coming from the η contribution and $kT/2$ coming from the $\dot{\eta}$ contribution. Thus

$$U = \langle E_{\text{translat}} \rangle + \langle E_{\text{rot}} \rangle + \langle E_{\text{vib}} \rangle$$
$$= \tfrac{7}{2}RT \qquad (12.66)$$

when all the degrees of freedom are taken into account, and

$$C_v = \tfrac{7}{2}R \qquad (12.67)$$

for a gas of diatomic molecules.

Our procedure shows that whenever the energy of a system may be written as a sum of independent terms, each of which is quadratic in the variable representing the associated degree of freedom, then in equilibrium at temperature T, each of the terms, that is, each degree of freedom contributes $kT/2$ to the average energy of each molecule. This is known as the law of *equipartition of energy*.

The general rule may now be applied to more complex systems. Consider a one-dimensional crystal (Fig. 12.8). In equilibrium we have atoms or ions, spaced a distance d apart. If the ith atom is displaced by η_i, then the energy is given by

$$E = \tfrac{1}{2}m \sum_{i=1}^{N} \dot{\eta}_i^2 + \tfrac{1}{2}K \sum_{i=1}^{N} (\eta_i - \eta_{i+1})^2. \qquad (12.68)$$

To make the crystal finite, we make it periodic in that we choose

$$\eta_{i+N} = \eta_i. \qquad (12.69)$$

The energy is a quadratic form, and one may show that by an appropriate transformation of variables, to

$$u_l = \sum_{i=1}^{N} A_{li}\eta_i \qquad (12.70)$$

where the A_{li} are appropriately determined coefficients (see Chapter 36), the energy becomes

$$E = \sum_{l=1}^{N} (\tfrac{1}{2}m\dot{u}_l^2 + \tfrac{1}{2}m\omega_l^2 u_l^2) \qquad (12.71)$$

that is, a sum of N independent oscillators, each with its own frequency. This looks like N independent diatomic molecules (with vibrational energy only) but how does randomization occur? Actually the potential is somewhat more complicated than the

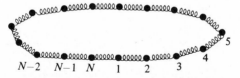

Figure 12.8. One-dimensional crystal.

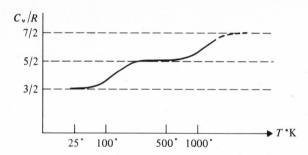

Figure 12.9. Specific heat at constant volume for diatomic molecule. The temperature scale is on a logarithmic scale (schematic only).

simple "spring" model indicates, and the anharmonic forces do randomize the individual normal modes, as they are called. If we generalize to three dimensions, we find that the average energy for the whole crystal is

$$\langle E \rangle = 3NkT. \tag{12.72}$$

One mole of crystal contains N_0 ions, and thus the specific heat per mole is

$$C_v = \frac{\partial}{\partial T} 3N_0 kT = 3R. \tag{12.73}$$

How well do these predictions work? It turns out that for monatomic gases, such as Helium, Argon, Neon, the prediction $C_v = 3R/2$ agrees with experiment, except for quite low temperatures. For diatomic gases, say, H_2, one finds the behavior shown in Fig. 12.9. Thus at moderately low temperatures H_2 acts as if it were a monatomic gas, and the shape of the curve suggests that different degrees of freedom make their contributions at different temperatures. For crystals such as copper, there are several difficulties. First, the value of $3R$ is reached only at high temperatures (Fig. 12.10); second, the calculation did not take into account the fact that in copper, there is approximately one free electron for each ion, and one would expect a contribution of $3R/2$ to the specific heat from the translational motion of the electrons. This is not seen. What this suggests is that electrons do not store heat for the crystal. This violation of equipartition cannot be understood in the framework of classical physics. We need quantum theory to explain these effects, and we shall soon be able to understand them.

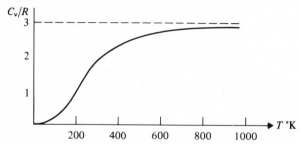

Figure 12.10. Sketch of specific heat of a crystal such as copper.

NOTES AND COMMENTS

1. Virtually every book that deals with statistical physics will illustrate the subject matter with some applications. The major reference on this point is R.P. Feynman, R.B. Leighton, and M. Sands, *The Feynman Lectures on Physics*, vol. 1, Addison-Wesley, Reading, Mass., 1963. Chapters 39–46 give a very good and immensely interesting account of the field, with many applications.
2. The Maxwell-Boltzmann distribution is discussed in detail in F.L. Friedman and L. Sartori, *The Classical Atom*, Addison-Wesley, Reading, Mass., 1965, where Brownian motion is also discussed.

PROBLEMS

12.1 Use the barometric formula to calculate the total number of molecules in the atmosphere of the earth in terms of the number density at the surface of the earth. The mean molecular weight is $A = 28.8$, $g = 980 \, \text{gm/cm}^2$ and $T = 270°\text{K}$ in the calculation. Take the radius of the earth to be $6370 \, \text{km}$, the number density from the ideal gas law (at 10^6 dynes pressure and $293°\text{K}$) to get an actual number. This will, of course, be very rough.

12.2 A system can be in one of two energy states, with energies E_1 and E_2, respectively (for example, if the magnetic dipole in the section on paramagnetic gases were restricted to pointing along or opposite to the direction of the magnetic field). Write down an expression for the probability that the system in equilibrium at temperature T is in the state of energy E_1. Write down an expression for the average energy, and plot a curve showing the temperature dependence. Write down an expression for the heat capacity $C_v = \partial \langle E \rangle / \partial T$ and plot it on the same graph as the mean energy.

12.3 A system can be in one of three energy states, with energies $-E$, 0 and $+E$. Repeat the calculations of Problem 12.2 for this system.

12.4 A model of rubber is that it consists of links of length a, and these links may either be extended or folded back on themselves. Consider a chain of N such links hanging freely in a gravitational field g. What is the length of the chain as a function of temperature, given that the mass of each link is m?

12.5 Hydrogen can be found in many possible atomic states. The lowest state is the ground state, and the energy difference between it and the next lowest state is $10.2 \, \text{eV}$. At what temperature would you expect a fraction 10^{-6} of the atoms in the ground state to be excited to the first excited state? Ignore all the other possible excited states of hydrogen.

12.6 A line in the spectrum of hydrogen has frequency $\nu = 2.5 \times 10^{15}$ Hz (cycles per second). If this radiation is emitted by hydrogen on the surface of a star, where the temperature is $6000°\text{K}$, what is the Doppler broadening?

12.7 Molecules consisting of two atoms have rotational energy that, according to the laws of quantum mechanics, has the form

$$E = AL(L + 1) \qquad L = 0, 1, 2, 3, \ldots$$

and the degeneracy of a state with a given L is

$$g_L = 2L + 1.$$

Suppose that at a certain temperature only the states with $L = 0, 1, 2, 3$ are appreciably occupied, and the fraction of molecules in the state with $L = 3$ is 7×10^{-12}. What is the fraction in the state with $L = 1$? Is the assumption that the fraction in $L = 4$ and higher is negligible justified?

12.8 Consider a gas at high temperature that may be in one of two states: either it is a gas of neutral atoms, or it consists of electrons and ions. Let the number of ions be n_i cm^{-3}, the number of electrons n_e cm^{-3} and the number of neutral atoms n_o cm^{-3}. The Saha formula takes the form

$$\frac{n_i n_e}{n_o} = 2.4 \times 10^{15} \, T^{3/2} \, \frac{g_i}{g_0} \, e^{-I/kT}$$

where the g's are the number of states the ions/atoms can be in, and I is the ionization potential. For hydrogen $g_i = g_0 = 1$ and $I = 13.6$ eV. Calculate the temperature for which half the atoms are ionized ($n_i = n_e = n_0$) for $n_e = 0.3, 3 \times 10^7$ and 3×10^{21}.

12.9 Calculate the root mean square distance of travel of a suspended particle of colloid in water, as a function of time, given that the particle has a radius of 10^{-4} cm and $\eta = 10^{-2}$ gm/cm sec for water.

12.10 Consider the "Brownian motion" of a small mirror suspended from a quartz strand. The equation of motion of the mirror is an equation for the angle θ that it makes with some equilibrium position, and it reads

$$I \frac{d^2\theta}{dt^2} + A \frac{d\theta}{dt} + B\theta = F(t)$$

where I is the moment of inertia of the mirror, A is a frictional force term due to the viscosity of the air, B represents a torsional restoring force constant (this can be measured by timing the frequency of oscillations of the mirror), and $F(t)$ represents the fluctuating force due to the impact of air molecules on the mirror. Follow the procedure used in the discussion of Brownian motion to show that

$$\frac{1}{2} B \langle \theta^2 \rangle = \frac{1}{2} I \left\langle \left(\frac{d\theta}{dt} \right)^2 \right\rangle$$

where the average is an average over many periods, and may be defined as

$$\langle f \rangle = \frac{1}{t} \int_0^t dt' f(t'), \quad t \to \infty.$$

Note that each of these equals $kT/2$ by equipartition, and thus a measurement of $\langle \theta^2 \rangle$ and B can be used to determine k. Do that, using the data $\langle \theta^2 \rangle = 4.18 \times 10^{-6}$, $B = 9.43 \times 10^{-9}$ dyne cm and $T = 287°$K.

PART III
The Origins of
the Quantum Theory

Scientific revolutions occur in different ways: sometimes, as in the case of gravitation, a genial mind (Newton and Einstein) illuminates a whole science. In other cases, the revolution is started through an insight that resolves a very specific problem, with the full consequences of that insight leading far from the original application. The Quantum Theory was born in this way. At the end of the nineteenth century there were many areas of physics in which the absence of an underlying theory was very evident: the structure of atoms and molecules, the peculiar behavior of specific heats discussed in the last chapter, the values of a variety of parameters such as the conductivity of various metals, the newly discovered phenomenon of radioactivity, all needed explanation. The breakthrough came in an area in which there was a clearcut conflict between classical theory and experiment, in the field of blackbody radiation. This is frequently the way in which scientific progress is made — an imaginative insight into the solution of a concrete puzzle. Chapter 13 deals primarily with the problem of blackbody radiation and the solution discovered by Max Planck in 1900. Out of this first suggestion of quantization of physical quantities grew a partial theory, which took its most complete form through the work of Niels Bohr, which completes this section. Although these early attempts were completely superseded by the formulation of quantum mechanics, there are good reasons to spend time on the old quantum theory: the machinery is generally simpler than that of quantum mechanics; the formulation is still within the framework of the more familiar classical physics, and above all, the old theory allows us to make order of magnitude estimates that hold for quantum mechanics, and is therefore a useful tool.

There are some topics that historically belong in this part of the book: the Stern-Gerlach experiment, the DeBroglie theory, and the theory of the Einstein A and B coefficients could have been treated here, but they make their appearance more naturally later in the book. I have, on the other hand, taken up some space with a discussion of the prequantum Lorentz theory of electrons, because so much of our understanding of the interaction of light with matter finds its simplest form in that approach. This rearrangement of history is pedagogically useful, but one does lose something of the sense of wonder at the ingenuity with which the giants of modern physics saw their way to the truth. Even a beginner can get some sense of the journey by reading at least the introduction to *The Sources of Quantum Mechanics*, B.L. van der Waerden (Ed.), Dover Publications, New York, 1967.

13
The Particle Nature of Radiation

There were many areas in physics in 1900 in which there existed a conflict between theory and observation, such as in the specific heat measurements for gases and solids. It turned out that the breakthrough to the new *Quantum Theory* came through the study of radiation, and it is this subject that we turn to now.

Blackbody Radiation When a solid is heated, it is seen to radiate. In equilibrium the radiation emitted ranges over the whole spectrum of frequencies v, with a spectral distribution that depends on the temperature T. One may define the *emissive power* $E(\lambda, T)$, where $\lambda = c/v$ as the energy emitted at wavelength λ and temperature T per unit time per unit area (W/cm^2). Theoretical research in this area began with the work of Kirchhoff in 1859. Kirchhoff showed that for a given λ, the ratio of the emissive power to the *absorptivity* A, defined as the fraction of incident radiation of wavelength λ that is absorbed by the body, is the same for all bodies. Kirchhoff considered two emitting and absorbing parallel plates, and showed from the equilibrium condition that the energy absorbed by each plate had to be equal to the energy emitted, and hence the ratio E/A had to be the same for the two plates. He soon thereafter observed that for a *black body*, defined to be one which totally absorbs all radiation that falls on it so that $A = 1$, the function $E(\lambda, T)$ is a universal function. To study this function experimentally, it is necessary to obtain as good a source of blackbody radiation as is possible. A practical solution is to consider the radiation emitted from a small hole in an enclosure heated to a temperature T. Given the imperfections in the surface on the inside of the cavity, it is clear that any radiation falling on the hole will have no chance of emerging again. Thus the surface presented by the hole is for all practical purposes "totally absorbing" and consequently the radiation coming from it is indeed *blackbody radiation*. If the hole is small enough, this radiation will be the same as that which falls on the walls of the cavity. Thus the emissivity of a blackbody is related to the distribution of radiation inside a cavity, whose walls are at a temperature T.

It was also shown by Kirchhoff that the second law of thermodynamics would be violated unless the radiation in the cavity was *isotropic*, that is, the flux were independent of the direction, that it was *homogeneous*, that is, the same at all points inside the cavity, and that it was the same in all cavities for a given temperature T, all of this. for each wavelength λ. The connection between the emissive power and the *energy density* of the radiation inside the cavity, $u(\lambda, T)$ follows from simple geometrical considerations, and is given by

$$u(\lambda, T) = 4E(\lambda, T)/c \qquad (13.1)$$

where c is the velocity of light.

It is the energy density that is the quantity of theoretical interest. Further understanding of this problem came from the work of Wien, who in 1894 showed, using very general thermodynamic arguments, that $u(\lambda, T)$ had to be of the form

$$u(\lambda, T) = \lambda^{-5} f(\lambda T) \qquad (13.2)$$

with $f(\lambda T)$ a still unknown function of its variable.

We can also write this in terms of the frequency. If $u(\nu, T)$ is the energy density for a frequency $\nu = c/\lambda$, then

$$u(\nu, T) = u(\lambda, T)\,|d\lambda/d\nu|$$

$$= \frac{c}{\nu^2} u\left(\frac{c}{\lambda}, T\right), \qquad (13.3)$$

and Wien's law reads

$$u(\nu, T) = \nu^3 g(\nu/T). \qquad (13.4)$$

The implications of Eq. (13.2), say, are the following:

a) Given the spectral distribution of blackbody radiation at one temperature, the distribution at any other temperature can be found from Eq. (13.2);

b) If the function $f(x)$ has a maximum for some value of $x > 0$, then the wavelength at which the energy density (and hence the emissive power) peaks is of the form

$$\lambda_{\max} = \frac{b}{T} \qquad (13.5)$$

where b is a *universal constant*.

The blackbody spectrum as measured by Lummer and Pringsheim 1897, conformed with these predictions. The shape of $u(\lambda, T)$ is shown in Fig. 13.1 and from it the value of the constant b can be determined. The measured value is

$$b = 0.2898\,\mathrm{cm}^\circ\,\mathrm{K}. \qquad (13.6)$$

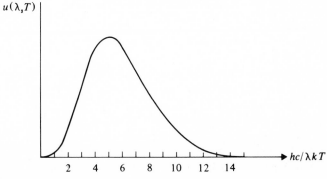

Figure 13.1. The blackbody spectrum as a function of the variable $hc/kT\lambda$. The peak occurs at $hc/kT\lambda = 4.965$.

Wien had obtained an empirical formula for the blackbody distribution

$$u(\nu, T) = C\nu^3 e^{-\beta\nu/T} \tag{13.7}$$

with two adjustable constants C and β, and that formula fit earlier experiments. The more detailed experiments of Lummer and Pringsheim showed that at low frequencies the Wien formula did not work. In that domain the data agreed with the classical prediction of Rayleigh. According to Rayleigh (with a minor correction by Jeans), radiation in a cavity is described by electric and magnetic fields in a standing wave pattern. These fields may be decomposed into normal modes. The number of independent "harmonic oscillators" in the frequency range ν to $\nu + d\nu$ can be determined to be

$$2\frac{4\pi\nu^2 \, d\nu}{c^3}, \tag{13.8}$$

the factor 2 in front being due to the two possible states of polarization of the electromagnetic field. We do not go through the derivation of this result here, since a similar counting of states will be worked out in Chapter 26. According to the *equipartition* rule each oscillator has an energy kT. Thus the energy density was predicted to be

$$u(\nu, T) = \frac{8\pi\nu^2}{c^3} kT. \tag{13.9}$$

The Planck Formula In 1900 Max Planck found the correct formula by an extremely ingenious interpolation between the low (Rayleigh) and high (Wien) frequency regimes. To appreciate Planck's procedure requires a deeper knowledge of thermodynamics than the reader is likely to possess, so that we merely quote the formula

$$u(\nu, T) = \frac{8\pi h}{c^3} \frac{\nu^3}{e^{h\nu/kT} - 1}. \tag{13.10}$$

Here h, called *Planck's constant*, is found by adjusting the curve to the data. Its value is

$$h = 6.626916 \pm 0.000024 \times 10^{-27} \text{ erg sec}. \tag{13.11}$$

The formula shows the following features:

a) When $h\nu \ll kT$, which is what we mean by "low" frequencies, we have

$$u(\nu, T) = \frac{8\pi h}{c^3} \frac{\nu^3}{(h\nu/kT)}$$

$$= \frac{8\pi\nu^2}{c^3} kT \tag{13.12}$$

in agreement with the Rayleigh-Jeans classical prediction. At high frequencies, $h\nu \gg kT$, we get the Wien empirical fit

$$u(\nu, T) = \frac{8\pi h}{c^3} \nu^3 e^{-h\nu/kT}. \tag{13.13}$$

b) We may rewrite the Planck formula in terms of a product of the number of modes (13.8) and the average energy per mode. We have

$$u(\nu, T) = \frac{8\pi\nu^2}{c^3} \frac{h\nu}{e^{h\nu/kT} - 1}. \tag{13.14}$$

Thus the average energy per oscillator depends on the frequency, and only for $hv \ll kT$ does it yield the classical value kT. High frequency modes have a very small average energy.

c) The total energy per unit volume in the cavity is obtained by integrating $u(v, T)$ over all frequencies. We have

$$U(T) = \int_0^\infty dv\, u(v, T) = \frac{8\pi h}{c^3} \int_0^\infty dv\, \frac{v^3}{e^{hv/kT} - 1}$$

$$= \frac{8\pi h}{c^3} \left(\frac{kT}{h}\right)^4 \int_0^\infty dx\, \frac{x^3}{e^x - 1} = aT^4 \tag{13.15}$$

which is the *Stefan-Boltzmann law*. This law was derived from thermodynamic considerations long before Planck's work, but the constant was not determined by this classical derivation. It can now be evaluated

$$a = \frac{8\pi k^4}{h^3 c^3} \int_0^\infty dx\, \frac{x^3}{e^x - 1} = \frac{8\pi k^4}{h^3 c^3} I \tag{13.16}$$

$$I = \int_0^\infty dx\, x^3 e^{-x}(1 - e^{-x})^{-1}$$

$$= \int_0^\infty dx\, x^3 e^{-x} \sum_{n=0}^\infty e^{-nx}$$

$$= \sum_{n=0}^\infty \int_0^\infty dx\, x^3 e^{-(n+1)x} = \sum_{n=0}^\infty (n+1)^{-4} \int_0^\infty dy\, y^3 e^{-y}$$

$$= 6 \sum_{n=0}^\infty (n+1)^{-4}$$

$$= \frac{\pi^4}{15} \tag{13.17}$$

so that

$$a = \frac{8\pi^5 k^4}{15 h^3 c^3} \cong 7.56 \times 10^{-15}\ \text{erg/cm}^3\ \text{deg}^4 \tag{13.18}$$

in excellent agreement with experiment. Similarly, the total emissive power of a black body is given by

$$E(T) = cU(T)/4 = \sigma T^4 \tag{13.19}$$

with $\sigma = 5.42 \times 10^{-5}\ \text{erg/cm}^2\ \text{sec deg}^4$.

Radiation from stars is approximately blackbody radiation. One may thus use the Stefan-Boltzmann law to estimate the surface temperature of star, either by measuring the amount of radiant energy incident on one cm^2 of detector normal to the path of the radiation, or by measuring the wavelength at which the emission spectrum peaks, and using

$$\lambda_{\max} T \simeq 0.29\ \text{cm}\ {}^\circ\text{K}. \tag{13.20}$$

Consider, for example, the sun. The radius of the sun is 0.7×10^{11} cm and therefore the total radiation emitted is

$$4\pi (0.7 \times 10^{11})^2 (5.42 \times 10^{-5}\, T^4) = 3.34 \times 10^{18}\, T^4. \tag{13.21}$$

The amount falling on 1 cm^2 on earth, which is 1.5×10^{13} cm away from the sun is

$$\frac{3.34 \times 10^{18} T^4}{4\pi(1.5 \times 10^{13})^2} = 0.96 \times 10^{-9} T^4. \tag{13.22}$$

This quantity, known as the *solar constant*, is measured to be 1.94 Cal/cm^2 min, or equivalently,

$$\frac{1.94 \times 4.2 \times 10^7}{60} = 1.36 \times 10^6 \text{ erg/cm}^2 \text{ sec}. \tag{13.23}$$

We thus get

$$T^4 = 1.52 \times 10^{15}$$

that is,

$$T \simeq 6000\,^\circ\text{K}. \tag{13.24}$$

When this is inserted into the formula for λ_{max} we find that

$$\lambda_{max} = \frac{0.29}{6000} = 4.8 \times 10^{-5} \text{ cm}$$

$$= 4800\,\text{Å} \tag{13.25}$$

(where $1\,\text{Å} = 10^{-8}$ cm is a convenient unit) which is not far off from the wavelength of light in the yellow part of the visible spectrum. These estimates are within 20% of the correct values.

Derivation of Planck formula The unqualified success of his formula drove Planck to an intensive search for an explanation of its form. On December 14, 1900, the "Birthday of Quantum Theory," Planck presented his derivation of the formula. What emerged from his work, with subsequent clarification by Einstein was the following:

The average energy is expected to be determined with the help of the formula

$$\langle E \rangle = \sum_{\substack{\text{all } E \\ \text{values}}} E P(E) \tag{13.26}$$

where

$$P(E) = \frac{e^{-E/kT}}{\sum\limits_{\substack{\text{all } E \\ \text{values}}} e^{-E/kT}}. \tag{13.27}$$

If the energy at a given frequency is given by

$$E = nh\nu \qquad n = 0, 1, 2, 3, \ldots \tag{13.28}$$

then, with $x = h\nu/kT$,

$$P(E) = \frac{e^{-nx}}{\sum\limits_{n=0}^{\infty} e^{-nx}} = \frac{e^{-nx}}{(1 - e^{-x})^{-1}}$$

$$= (1 - e^{-x})e^{-nx} \tag{13.29}$$

and the average energy is given by

$$\langle E \rangle = h\nu(1 - e^{-x}) \sum_{n=0}^{\infty} n e^{-nx}$$

$$= h\nu(1 - e^{-x})\left(-\frac{d}{dx}\right) \sum_{n=0}^{\infty} e^{-nx}$$

$$= h\nu(1 - e^{-x})\left(-\frac{d}{dx}\right)(1 - e^{-x})^{-1}$$

$$= h\nu\frac{e^{-x}}{1 - e^{-x}} = \frac{h\nu}{e^{h\nu/kT} - 1}. \tag{13.30}$$

This is just what the Planck formula requires. The postulate (13.28) is quite radical. The interpretation of this form, given by Einstein, is that *radiation consists of quanta*, whose energy is given by

$$E = h\nu \qquad (13.31)$$

where ν is the frequency associated with the radiation. There may be $0, 1, 2, 3, \ldots$ quanta of a given frequency present, and hence the energy may take on the values $nh\nu$.

How do we reconcile the discreteness proposed by Planck with our experience that electromagnetic radiation is continuous and consists of waves? Quantum theory explains this. For the time being, we just observe that the energy per quantum is very tiny indeed. For light in the optical region, say with $\lambda = 6000$ Å, we have

$$E = h\nu = \frac{hc}{\lambda} = \frac{6.63 \times 10^{-27} \times 3 \times 10^{10}}{6000 \times 10^{-8}} = 3.3 \times 10^{-12} \text{ ergs}$$

so that the number of light quanta of this wavelength, emitted by a 100 watt bulb is

$$\frac{100 \times 10^7}{3.3 \times 10^{-12}} \simeq 3 \times 10^{20} \text{ quanta/sec.} \qquad (13.32)$$

With so many quanta present, it is not surprising that we do not experience the particle nature of light directly. In general on a macroscopic scale, no deviations from classical Maxwell theory are to be expected. What sets the "macroscopic scale" is whether physical quantities, with the dimensions of (erg) $\times$ (sec), are very much larger than h. This quantity, called the *quantum of action* (action in mechanics has the dimensions of energy $\times$ time) sets the scale at which quantum theory must be used.

Cosmological blackbody radiation Blackbody radiation has again become of central interest in the past decade or so. In the late 1940s George Gamow and subsequently R. Alpher, H. Bethe, and G. Gamow studied some consequences of a "big bang" model of the creation of the universe. One of the consequences was that a residue of the intense radiation field created in the initial moments of the creation should appear in the form of a background blackbody radiation field. The calculations that predicted such a field at a temperature of $25°$K were shown to be unreliable, and no attempt was made to measure this until 1964, when A.A. Penzias and R.W. Wilson discovered excess thermal noise in their radio-astronomical detector. A group, led by R.H. Dicke, and consisting of P.J. Peebles, P.G. Roll, and D.T. Wilkinson were about to make measurements of the cosmic blackbody background, and immediately understood the significance of the thermal noise. It corresponded to blackbody radiation at a temperature that is now believed to be $2.65 \pm 0.09\,°$K (Fig. 13.2). The measurements are not easy, since an antenna is generally swamped by other signals from the ground, the atmosphere, and various cosmic point sources, as well as noise from the circuits used in the measurement. Dicke had invented a radiometer in 1945 that could be used for the measurement: the idea is to have a radio receiver that switches back and forth, a hundred times a second, between the sky and a bath of liquid helium. The receiver output is so filtered that only the signal that varies with a frequency of 100 Hz is measured, and this gives the difference between the radiation from the sky and the liquid helium. Pointing in different directions allows the separation of the atmospheric component. The detection of the radiation, at a temperature corresponding to the calculations of Dicke and collaborators, is one of the strongest arguments in support of

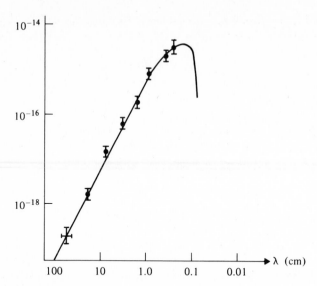

Figure 13.2. A rough sketch of the measurements of the background radiation in erg/cm^2 sec, sterad Hz as a function of the wavelength in cm. The solid curve is the predicted spectrum for $T = 2.7°$ K.

the "big bang" model of the origin of the universe. Detailed measurements may also establish how fast we (the earth, the solar system, the local group of galaxies) are moving with respect to this background of radiation. At this time our speed relative to this radiation field is less than 300 km/sec, which is roughly the speed of the solar system relative to the local group of galaxies due to the rotation of our own galaxy.

The Photoelectric Effect Support for the quantum nature of radiation came from the work of Albert Einstein, who in 1905 used the concepts of Planck to explain the photoelectric effect (Fig. 13.3). The effect was discovered by Hertz in 1887. His experiments, and those of successors (especially R. Millikan) established that:

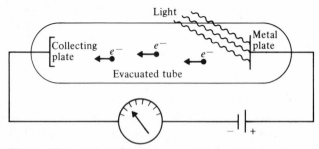

Figure 13.3. Schematic arrangement for observation of photoelectric effect. A retarding potential is maintained between the plate and the collector plate. Only the electrons which have kinetic energy large enough to overcome the potential difference reach the collector plate and give rise to a photoelectric current. Varying V allows the determination of the threshold for the effect.

a) when polished metal plates are irradiated they may emit electrons. They do not emit positive ions.

b) Whether electrons are emitted or not depends on the wavelength of the light. In general there will be emission for $\nu > \nu_0$ where ν_0 depends on the metal.

c) The magnitude of the current of electrons emitted from the plate is proportional to the intensity of the light source, but it does not depend on the frequency of the radiation, provided that the latter exceeds the threshold.

d) The energy of the photoelectrons is independent of the intensity of the light source, but varies linearly with the frequency of the incident radiation.

Classical electromagnetic theory cannot explain these observations. It was known that metals contained electrons. One might expect that some of them could absorb incident radiation and be accelerated by it enough to break loose from the metal. The energy carried by an electromagnetic wave is proportional to the intensity of the source, however, so that classically one does not expect the observed frequency dependence. Furthermore, the classical explanation of the effect carries with it the notion of concentration of the deposited energy upon single photoelectrons, and this implies that for low intensity irradiation, there should be a time delay between the arrival of the radiation and the appearance of the photoelectrons. As a matter of fact, no such time delay has ever been observed (even though looked for). Einstein explained the effect by describing the incident radiation in terms of individual quanta, which are now called *photons*. Each photon carries energy $h\nu$, where ν is the frequency of the radiation. An intense source emits many such photons per unit time, a weak source few. If there is a fixed energy W, called the *work function*, that is required to liberate an electron from the surface of the metal, then there will be no photoelectric current for $h\nu < W$, provided the mechanism is that *individual electrons absorb individual photons* and gain the energy $h\nu$. Under these circumstances, if the frequency exceeds W/h, then there will be energy available for electron kinetic energy, with

$$\tfrac{1}{2}mv^2 = h\nu - W. \tag{13.33}$$

This explains the linear relation (Fig. 13.4) between electron energy and also the frequency threshold. The mechanism of absorption of photons also explains why no

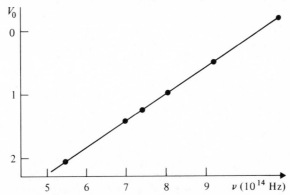

Figure 13.4. Schematic drawing of experimental results of Millikan showing linear relation between minimum retarding voltage and frequency.

time delay has been observed: a photoelectron only appears if it absorbed a photon, and this has associated with it a time short relative to the times that could be measured. Incidentally, the work function is a characteristic of the metals, and its magnitude is generally several electron volts. For example, for potassium, the threshold wavelength is 5640 Å. This implies that

$$W = \frac{hc}{\lambda_{thr}} = \frac{6.63 \times 10^{-27} \times 3 \times 10^{10}}{5.64 \times 10^{-5}} \cdot \frac{1}{1.6 \times 10^{-12}} = 2.2\,\text{eV}.$$

The Compton Effect The experiment that provided the most compelling evidence for the particle nature of radiation was the so-called Compton Effect. A.H. Compton discovered in 1922 that X-rays sent through thin metallic foils scattered in a manner not consistent with classical theory. That theory interprets the scattering of radiation as the re-radiation of light by electrons set in motion by the absorption of the incident radiation. Classical theory predicted that the intensity of the scattered radiation at an angle θ with respect to the direction of the incident radiation should vary as $(1 + \cos^2 \theta)$ independent of the wavelength of the incident radiation. Compton found that the scattered radiation at a given angle consisted of two components: one that had the same wavelength as the incident radiation; another, whose wavelength was shifted by an amount that varied with the angle θ. Compton treated the radiation as consisting of photons that collided with the free electrons in the metal foil as if they were billiard balls, that is, elastically, with energy and momentum being conserved. In this way he was able to explain the shift in the wavelength of the second component.

The derivation is quite simple. Consider an incident photon with energy E and momentum $\mathbf{p}$ colliding with an electron at rest (Fig. 13.5). The total energy is conserved, as is the momentum, so that if the recoil momentum of the electron is $\mathbf{q}$ and that of the scattered photon is $\mathbf{p}'$, then

$$\mathbf{p} = \mathbf{p}' + \mathbf{q} \tag{13.34}$$

and

$$E + mc^2 = E' + ((qc)^2 + m^2 c^4)^{1/2}. \tag{13.35}$$

Now for a photon,

$$E = h\nu = hc/\lambda,$$
$$E' = h\nu' = hc/\lambda'. \tag{13.36}$$

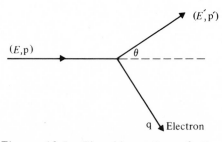

Figure 13.5. The kinematics of the "billiard ball" collision between a photon and an electron at rest, showing the momenta of the scattered photon and the recoil electron.

Also,

$$|\mathbf{p}| = h\nu/c = E/c$$
$$|\mathbf{p'}| = h\nu'/c = E'/c \tag{13.37}$$

since the photon moves with the velocity of light and thus has zero rest mass. Now

$$q^2 = (\mathbf{p} - \mathbf{p'})^2 = \mathbf{p}^2 + \mathbf{p'}^2 - 2 |\mathbf{p}| |\mathbf{p'}| \cos \theta,$$

so that

$$q^2 c^2 = \mathbf{p}^2 c^2 + \mathbf{p'}^2 c - 2pp'c^2 \cos \theta$$
$$= E^2 + E'^2 - 2EE' \cos \theta. \tag{13.38}$$

On the other hand, Eq. (13.35) implies that

$$q^2 c^2 = (E + mc^2 - E')^2 - m^2 c^4$$
$$= E^2 + E'^2 - 2EE' + 2mc^2 (E - E'). \tag{13.39}$$

Consequently,

$$2EE'(1 - \cos \theta) = 2mc^2 (E - E'),$$

that is,

$$1 - \cos \theta = mc^2 \left(\frac{1}{E'} - \frac{1}{E} \right)$$

$$= mc^2 \left(\frac{\lambda'}{hc} - \frac{\lambda}{hc} \right). \tag{13.40}$$

Therefore

$$\lambda' - \lambda = \frac{h}{mc}(1 - \cos \theta) \tag{13.41}$$

which is in excellent agreement with experiment. The unmodified component is due to the scattering by the ions, which are of the order of 10^4 more massive than the electrons.

The quantity

$$\frac{h}{mc} = \frac{6.63 \times 10^{-27}}{0.9 \times 10^{-27} \times 3 \times 10^{10}} \cong 2.4 \times 10^{-10} \text{ cm} \tag{13.42}$$

has the dimensions of a length, and is called the Compton wavelength of the electron.

The Compton effect showed that one cannot merely treat photons as being quantized entities within an electromagnetic wave, but that one must deal with them individually. Subsequent experiments by Compton showed that one could not split photons in the sense that part of a photon, say with some fraction of the energy $h\nu$, scattered. This raises conceptual difficulties: how can individual photons ever give rise to interference effects, known to be associated with electromagnetic radiation? A progressive reduction of the intensity in a source means that photons are being sent out at a reduced rate. In the limiting case where, say, one photon a second is being radiated, it is incomprehensible how an interference pattern should arise, since it would imply that a single photon that presumably goes through one slit in a diffraction grating knows about all the others. These matters are resolved by quantum theory, and will be discussed later, but it is important to bring them up just to indicate that the new discoveries and their interpretation ran totally counter to established theory and to established ways of understanding phenomena. It should not surprise us that quantum theory was difficult to accept. We should rather admire the courage that the pioneers of the quantum theory had to propagate their heresies.

NOTES AND COMMENTS

1. The pre-Planck theory of blackbody radiation is discussed in F.K. Richtmeyer, E.H. Kennard, and J.N. Cooper, *Introduction to Modern Physics*, McGraw-Hill, New York, 1969, and in A. Sommerfeld, *Thermodynamics and Statistical Mechanics*, Academic Press, New York, 1956.

2. A very interesting account of the birth of quantum theory and the work of Planck, as well as the contributions of Einstein, may be found in Armin Hermann, *The Genesis of Quantum Theory 1899–1913*, MIT Press, Cambridge, Mass., 1971.

3. The classical prediction of the blackbody energy density of Eq. 13.9 is manifestly wrong, because it implies that the energy density integrated over all frequencies is infinite! It is interesting to read in Hermann (see above) that as late as 1908, H.A. Lorentz, the leading theoretical physicist of the turn of the century, could not quite give up his belief in the validity of Eq. 13.9, until Lummer and Pringsheim pointed out that if it were correct, then the fact that melting steel at $1700°C$ emits light of blinding intensity would imply that a black body at room temperature would emit light with one sixth of that intensity.

4. The history of the discovery of the photoelectric effect may be found in M. Jammer, *The Conceptual Development of Quantum Mechanics*, McGraw-Hill, New York, 1966.

PROBLEMS

13.1 Using the distribution function Eq. (13.10) in terms of frequency, calculate $u(\lambda, T)$. Find the value of λ for which $u(\lambda, T)$ is a maximum. You will need to solve the equation

$$5 - x = 5e^{-x}$$

which is easily done by successive approximations, that is, by choosing $x = 5 - \epsilon$ and keeping only terms linear in ϵ, etc.

13.2 A spherical satellite of radius r painted black, travels around the sun at a distance D from the center. The sun radiates as a blackbody at a temperature of $6000°K$. If the sun subtends an angle of θ radians, as seen by the satellite (with $\theta \ll 1$), find an expression for the equilibrium temperature of the satellite in terms of θ. To proceed, calculate the energy absorbed by the satellite, and the energy radiated per unit time.

13.3 A hydrogen bomb upon explosion develops a temperature of the order of 10^8 degrees K. If the fireball emits radiation during a short time as if it were a blackbody, what is the value of λ_{max}? What is the energy of the corresponding photon?

13.4 Given that the sun radiates as a blackbody at a temperature $T = 6000°K$, how much energy is emitted per cm^2 per second in the range of wavelengths 5790 Å — 5810 Å?

13.5 The cosmic blackbody radiation background has a temperature of approximately $3°K$. What is the value of λ_{max} for this radiation? What is the energy of a photon with this wavelength?

13.6 The night-adapted eye can detect a lit cigarette at 500 m, say. Assuming that the pupil of the eye is 0.6 cm in diameter, and that the tip of the cigarette is a hemi-sphere, 1 cm in diameter, radiating as a black body at temperature 800°C, make some guesses to estimate how many photons hit the retina per second.

13.7 Plot the Planck distribution as a function of frequency for a temperature of 1000°K and estimate the frequency which splits the density curve in the sense that half the energy is radiated with frequencies below it, and half above it.

13.8 Light of frequency 0.85×10^{15} Hz falls on a metal surface. If the maximum energy of the photoelectrons is 1.7 eV, what is the work function of the metal?

13.9 A sodium surface emits 6.25×10^7 electrons per cm^2 per sec. Assuming that the sodium atoms are regularly spaced, and given that for sodium $A = 23$ and the mass density is $0.97 \, gm/cm^3$, and furthermore, assuming that the photoelectrons are sup-plied by the top ten layers of sodium atoms, how many atoms, on the average, furnish one photoelectron per second?

13.10 A metal has work function 4.7 eV. What is the maximum kinetic energy of a photoelectron if radiation of 2000 Å falls on the surface?

13.11 A 200 MeV photon collides with a proton at rest. What is the maximum energy loss of the photon?

13.12 In a Compton scattering experiment, the initial wavelength of the X-rays is 0.7078 Å and the final wavelength is 0.7314 Å. At what angle was the scattered radiation measured?

13.13 An electron with energy 10^3 MeV collides with a photon at λ_{max} of the three-degree blackbody background. What is the maximum energy loss that the electron can suffer? What will be the wavelength of the photon after this maximal collision?

13.14 A photon of frequency ν collides with a photon of wavelength λ_{max} of the three degree blackbody background, and an electron-positron pair is produced. The frequency is such that in the most favorable collision (head on) the threshold for the production of an electron-positron pair is reached ($m_e c^2 = 0.51$ MeV). What is ν?

14
Early Atomic Models

The Discovery of the Electron The notion that electricity is carried by particles arose quite early in the connection with the studies of the conduction of electricity in electrolytes by Faraday in 1833. Faraday himself did not view the atomic theory of matter as more than a convenient way of describing phenomena, but he did postulate that the electric current arose through the motion of ions, which carried the basic units of electric charge. Further progress was made through the study of electrical discharges through gases at low pressures. These investigations became possible when vacuum technology permitted the reduction of pressures to 10^{-6} atmosphere. In 1869 *cathode rays* were discovered, and they were found to be particles carrying negative charge. This was confirmed by the discovery of Roentgen that when cathode rays struck a metallic anticathode placed in the tube, they produced penetrating radiation named *X-rays* by him. These were interpreted by Schuster to be electromagnetic radiation of high frequency, emitted because of the deceleration of charged particles, in accordance with Maxwell's radiation theory. The negative particles in a discharge tube could be deflected by electric and magnetic fields, and in this way J.J. Thomson first measured e/m for the negatively charged particles.

If the particles pass between plates of length l with an electric field between them, then the particles are accelerated in the y direction (Fig. 14.1) with magnitude

$$a_y = \frac{eE_y}{m}. \tag{14.1}$$

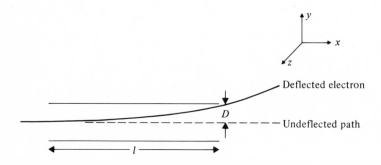

Figure 14.1. Schematic sketch of Thomson experiment.

This leads to a deflection of magnitude

$$D = \frac{1}{2} a_y t^2 = \frac{1}{2} a_y (l/v)^2 = \frac{eE_y}{2m} \left(\frac{l}{v}\right)^2 \tag{14.2}$$

where v is the velocity of the particles. To determine that velocity, Thomson imposed a magnetic field at right angles (in the z direction) and adjusted its magnitude until there was no net deflection of the beam. It follows from the Lorentz force expression

$$\mathbf{F} = e(\mathbf{E} + \mathbf{v} \times \mathbf{B}/c) = e(\hat{i}_y E_y + \hat{i}_x \times \hat{i}_z v B_z/c)$$
$$= e\hat{i}_y(E_y - vB_z/c) \tag{14.3}$$

that an absence of acceleration in any direction implies that

$$v = cE_y/B_z. \tag{14.4}$$

With the velocity measured, a measurement of D in the absence of a magnetic field yielded a value for e/m

$$e/m = 0.53 \times 10^{18} \frac{\text{e.s.u}}{\text{gm}} = 1.76 \times 10^{11} \frac{\text{C}}{\text{kg}} \tag{14.5}$$

for the negatively charged particles. When this is combined with a measurement of the charge carried by the particles, first attempted by Townsend, and then carried out with great accuracy by Millikan, who found that

$$e = 4.8 \times 10^{-10} \text{ e.s.u}$$
$$= 1.6 \times 10^{-19} \text{ C} \tag{14.6}$$

(the precise number is $4.803250 \pm 0.000021 \times 10^{-10}$ e.s.u) one finds that

$$m_e = 0.9 \times 10^{-27} \text{ gm} \tag{14.7}$$

with the precise number currently $9.109558 \pm 0.000054 \times 10^{-27}$ gm. This number is approximately 2000 times smaller than the mass of a hydrogen atom.

The newly discovered particles were called *electrons*. Their existence clearly had an important bearing on theories of the structure of atoms. The latter are electrically neutral, so that they must be made up of equal numbers of electrons and positively charged particles. The question of how many electrons there were in an atom was addressed by Thomson, who studied the absorption of radiation by matter. If one considers a slab of matter, the intensity of the radiation at a depth x denoted by $I(x)$ will be reduced in the region $(x, x + dx)$ by an amount $dI(x)$, which is proportional to $I(x)$, and also to dx (Fig. 14.2). Writing

$$dI(x) = -\frac{1}{l} I(x)dx \tag{14.8}$$

$$x \qquad x+dx$$

Figure 14.2. Intensity reduction in a slab of material.

where l is a proportionality constant with the dimension of a length, we obtain a differential equation

$$\frac{dI(x)}{dx} = -\frac{I(x)}{l} \tag{14.9}$$

whose solution is

$$I(x) = I(0)e^{-x/l}. \tag{14.10}$$

Note that the minus sign in Eq. (14.8) is there because there is a reduction in the intensity, so that $dI(x)$ must be negative. The parameter l is called the *mean free path* of radiation in the material. This quantity can be related to a more fundamental quantity, the *cross section σ for absorption of radiation*. Clearly, the mean free path depends on the density of target particles in the slab of material. If that density is given by n particles/cm^3, then there will be on the average one collision between the incident ray (or any incident particle, for that matter) and a target particle, if the volume of the cylinder whose height is l and whose base is the effective area for a collision presented by the target particle, σ, contains one target particle, that is, if

$$n\sigma l = 1 \quad \text{(Fig. 14.3)}. \tag{14.11}$$

Thus a measurement of the mean free path through an absorption measurement yields

$$\sigma = \frac{1}{nl}. \tag{14.12}$$

In the absorption of low-frequency radiation by matter, the principal process by which radiation is taken out of the beam is through the scattering of the radiation by the electrons in the material. Thomson calculated the cross section using classical electrodynamics. The result that he obtained,

$$\sigma = \frac{8\pi}{3}\left(\frac{e^2}{m_e c^2}\right)^2 \text{cm}^2/\text{electron}, \tag{14.13}$$

will be derived later in this chapter (Eq. 14.45). Except for the numerical factor in front, the result is really determined by dimensional analysis. The cross section, representing an area, must be of the form of (length)^2. The only lengths in the problem are the wavelength of the radiation, which approaches infinity in the low frequency limit, and the quantity that can be constructed out of the electron charge e, its mass m_e and c, the velocity of light, which enters naturally into radiation problems. Since e^2/R is an energy (cf. the Coulomb potential) and thus of the same structure as $m_e c^2$, we can define a length, called the *classical electron radius*

$$r_e = e^2/m_e c^2 \tag{14.14}$$

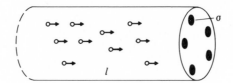

Figure 14.3. Radiation flux in a region with absorbing targets of area σ.

whose square must enter into the cross section. We are not saying that this must be the right answer: after all, the electron could be a little sphere of radius a unrelated to e and m_e, and it could be a that determines the absorption. We are saying, however, that a classical electrodynamic calculation in which the charge is treated as a point particle must have the above form. Note that in quantum theory we have another length, the Compton wavelength $h/m_e c$ that could be relevant. In this problem, where the frequency is low, we must be in the classical domain, as in the case of blackbody radiation, where the Rayleigh formula applies for low frequencies.

The magnitude of these quantities is

$$r_e \cong \frac{(4.8 \times 10^{-10})^2}{0.9 \times 10^{-27} \times 9 \times 10^{20}} \cong 2.8 \times 10^{-13} \text{ cm} \qquad (14.15)$$

and

$$\frac{h}{m_e c} \cong 2.4 \times 10^{-10} \text{ cm.} \qquad (14.16)$$

The Thomson cross section thus is

$$\sigma \cong 0.68 \times 10^{-24} \text{ cm}^2. \qquad (14.17)$$

A useful unit is the *barn*, defined by

$$1 \text{ barn } = 10^{-24} \text{ cm}^2 \qquad (14.18)$$

so that $\sigma_{\text{Thom}} \cong 2/3$ barns.

Thomson atomic model Measurements of the absorption of radiation showed that the number of electrons per atom was approximately $A/2$ where A is the atomic weight. We now know that the number of electrons Z is equal to the number of protons in the nucleus of the atom. In any case, it became clear that the mass of the atom resided in the positively charged component. Thomson devised a model of the atom which consisted of electrons embedded in a positively charged cloud and distributed in a way such that the structure was stable. The electrons had to be at rest, since otherwise they would radiate. Classical electrodynamics shows that an electrically charged particle, with charge q, moving with an instantaneous acceleration a, radiates instantaneous power P given by

$$P = \frac{2}{3} \frac{q^2}{c^3} a^2. \qquad (14.19)$$

The Thomson model had some difficulties which we will soon come to. It did, however, provide a coherent picture of electrons bound in atoms, and vibrating with simple harmonic motion if disturbed. Support for it came from the somewhat earlier discovery by Zeeman that a magnetic field affected the frequency of radiation emitted by an atom. Lorentz described the undisturbed atom as oscillating with some circular frequency ω_0. The amplitude of the oscillator (the displacement of the electron from its equilibrium position) thus satisfies the equation

$$\frac{d^2 x}{dt^2} = -\omega_0^2 x \qquad (14.20)$$

and the electromagnetic radiation is emitted with the same frequency ω_0 $(= 2\pi \nu_0)$.

In a magnetic field there is an additional force, the Lorentz force, so that

$$\frac{d^2\mathbf{x}}{dt^2} = -\omega_0^2\mathbf{x} - \frac{e}{mc}\frac{d\mathbf{x}}{dt} \times \mathbf{B}. \tag{14.21}$$

If we choose $\mathbf{B}$ as pointing in the z direction, then we have, with the notation

$$\omega_L = -\frac{e}{2mc}B \tag{14.22}$$

(ω_L is known as the Larmor frequency), the equations

$$\frac{d^2x}{dt^2} = -\omega_0^2 x + 2\omega_L\frac{dy}{dt},$$

$$\frac{d^2y}{dt^2} = -\omega_0^2 y - 2\omega_L\frac{dx}{dt}. \tag{14.23}$$

It is easy to check that the solutions

$$x = A\cos\Omega t,$$
$$y = \pm A\sin\Omega t \tag{14.24}$$

satisfy the equations, provided that

$$\Omega^2 \pm 2\omega_L\Omega - \omega_0^2 = 0, \tag{14.25}$$

that is,

$$\Omega = \mp\,\omega_L + (\omega_0^2 + \omega_L^2)^{1/2}. \tag{14.26}$$

Now typically, the characteristic frequencies for the vibration of electrons in atoms is of the order of 10^{15} Hz, whereas for a magnetic field of 10^4 gauss (1 tesla)

$$\omega_L = \frac{eB}{2m_ec} = \frac{4.8 \times 10^{-10} \times 10^4}{2 \times 0.9 \times 10^{-27} \times 3 \times 10^{10}} \approx 10^{11}.$$

Thus $\omega_L \ll \omega_0$ and therefore

$$\Omega = \omega_0 \mp \omega_L. \tag{14.27}$$

The above prediction of a splitting of a spectral line into two is in perfect agreement with the observations of Zeeman for some substances. In others Zeeman found what in terms of the above theory must be called an anomalous effect. The anomalous Zeeman effect was not explained until quantum mechanics was developed.

Lorentz electron theory The model of the oscillating charge also gave a good description of the interaction of light with matter, and even though quantum theory has superseded the *Lorentz Electron Theory*, the latter is still a good guide to the understanding of the dispersion of light, that is, the variation of the index of refraction with frequency.

In the presence of an external electric field oscillating with frequency ω, the equation describing the motion of the oscillating electron is

$$\frac{d^2\mathbf{r}}{dt^2} = -\omega_0^2\mathbf{r} + \frac{e\mathbf{E}}{m}\cos\omega t. \tag{14.28}$$

Actually, because the electron radiates in the course of acceleration, there is a damping built into the problem. Classical electromagnetic theory dealt in a rather complicated way with this problem, and that aspect of the solution has been completely superseded by quantum theory, so that we shall represent the damping in the simplest way possible, as a force proportional to the velocity. Thus Eq. (14.28) should be modified by the extra term in

$$\frac{d^2\mathbf{r}}{dt^2} = -\omega_0^2\mathbf{r} - \Gamma\frac{d\mathbf{r}}{dt} + \frac{e\mathbf{E}}{m}\cos\omega t. \tag{14.29}$$

The constant Γ has the dimensions of $\sec^{-1}$, and we shall soon be able to give it a physical interpretation. To solve the equation, let us try

$$\mathbf{r} = \mathbf{A}\cos\omega t + \mathbf{B}\sin\omega t. \tag{14.30}$$

When this is inserted into Eq. (14.29) we find that

$$\begin{aligned}
-\omega^2\mathbf{A}\cos\omega t - \omega^2\mathbf{B}\sin\omega t = {} & -\omega_0^2\mathbf{A}\cos\omega t - \omega_0^2\mathbf{B}\sin\omega t \\
& + \omega\Gamma\mathbf{A}\sin\omega t - \omega\Gamma\mathbf{B}\cos\omega t \\
& + (e\mathbf{E}/m)\cos\omega t.
\end{aligned}$$

This implies that

$$\begin{aligned}
(\omega_0^2 - \omega^2)\mathbf{A} + \omega\Gamma\mathbf{B} &= (e\mathbf{E}/m) \\
(\omega_0^2 - \omega^2)\mathbf{B} - \omega\Gamma\mathbf{A} &= 0
\end{aligned} \tag{14.31}$$

since the coefficients of the sine and cosine terms must vanish independently (for example, take $t = 0$, and $t = \pi/2\omega$). Thus we get the solution

$$\mathbf{B} = \mathbf{A}\frac{\omega\Gamma}{\omega_0^2 - \omega^2}$$

for the second equation. If this is inserted in the first one, we get

$$\left(\omega_0^2 - \omega^2 + \frac{\omega^2\Gamma^2}{\omega_0^2 - \omega^2}\right)\mathbf{A} = e\mathbf{E}/m.$$

Thus we finally obtain

$$\mathbf{r} = \frac{e\mathbf{E}}{m}\left[\frac{\omega_0^2 - \omega^2}{(\omega_0^2 - \omega^2)^2 + \omega^2\Gamma^2}\cos\omega t + \frac{\omega\Gamma}{(\omega_0^2 - \omega^2)^2 + \omega^2\Gamma^2}\sin\omega t\right]. \tag{14.32}$$

The electron therefore oscillates with the impressed frequency, and the amplitude is proportional to that of the electric field. The damping parameter has a magnitude that is always much smaller than ω_0, so that unless we are inducing oscillations with a frequency very near the natural frequency of the atom, that is, when $\omega \neq \omega_0$, we can neglect the second term. The dipole moment of the atom is given by $e\mathbf{r}(t)$, and in the case presently being discussed, it is proportional to the impressed electric field. If there are N atoms per unit volume, we may speak of a polarization per unit volume of magnitude $Ne\mathbf{r}(t)$. It follows from electromagnetic theory that the dielectric constant is given by

$$\epsilon(\omega)\mathbf{E}(t) = \mathbf{E}(t) + 4\pi\mathbf{P}(t) = \mathbf{E}(t) + 4\pi Ne\mathbf{r}(t)$$

so that

$$\epsilon(\omega) = 1 + 4\pi N\frac{e^2}{m}\frac{\omega_0^2 - \omega^2}{(\omega_0^2 - \omega^2)^2 + \omega^2\Gamma^2} \tag{14.33}$$

It follows from Maxwell's equations that the speed of propagation of radiation is $(\mu\epsilon(\omega))^{-1/2}$ where μ is the magnetic susceptibility generally close to 1, and $\epsilon(\omega)$ is the

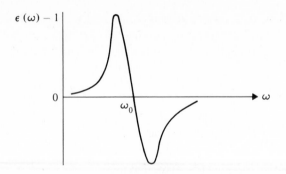

Figure 14.4. The dependence of the dielectric constant on frequency. The region where $\epsilon - 1 < 0$ is known as the anomalous dispersion region.

dielectric constant. Since the speed can also be written as $c/n(\omega)$, where $n(\omega)$ is the index of refraction, we have

$$n(\omega) = \sqrt{\epsilon(\omega)}. \tag{14.34}$$

The figure (Fig. 14.4) shows the shape of the curve, and shows that the dispersion, that is, the dependence of the dielectric constant on frequency ω increases as we approach a characteristic *resonance* frequency, and then changes very precipitously if Γ is small. If the atomic system is quite complicated, there may be many such normal modes in the system, and one might expect that there will be many characteristic frequencies ω_k for a system. In that case we would expect

$$\epsilon(\omega) = 1 + \frac{4\pi N e^2}{m} \sum_k f_k \frac{\omega_k^2 - \omega^2}{(\omega_k^2 - \omega^2)^2 + \omega^2 \Gamma_k^2} \tag{14.35}$$

where f_k represents the effective number of electrons that should be counted as oscillating with frequency ω_k. In the form taken so far, the quantities ω_k, Γ_k, and f_k are parameters to be determined from experiment, although in a complete theory, they would be predicted.

The term that we have neglected is called the *absorptive* contribution to the dielectric constant because it determines the rate of work done by the external electric field. The rate of work is given by

$$\frac{dW}{dt} = e\mathbf{E}(t) \cdot \frac{d\mathbf{r}}{dt}, \tag{14.36}$$

that is,

$$\frac{e^2 \mathbf{E}^2}{m} \cos \omega t \left[-\frac{\omega(\omega_0^2 - \omega^2)}{(\omega_0^2 - \omega^2)^2 + \omega^2 \Gamma^2} \sin \omega t + \frac{\omega^2 \Gamma}{(\omega_0^2 - \omega^2)^2 + \omega^2 \Gamma^2} \cos \omega t \right].$$

If we want to calculate a time average of the work done, we see that the average of

$$\langle \cos \omega t \sin \omega t \rangle = \tfrac{1}{2} \langle \sin 2\omega t \rangle = 0$$

vanishes, since the average of any oscillating quantity vanishes. On the other hand, since

$$\langle \cos^2 \omega t \rangle = \left\langle \frac{1 + \cos 2\omega t}{2} \right\rangle = \frac{1}{2},$$

only the first term averages to zero, so that

$$\left\langle \frac{dW}{dt} \right\rangle = \frac{e^2 E^2}{2m} \frac{\omega^2 \Gamma}{(\omega_0^2 - \omega^2)^2 + \omega^2 \Gamma^2}. \tag{14.37}$$

Thus it is the term that involves $\sin \omega t$, that is, $90°$ out of phase with the incident, external field involving $\cos \omega t$, that is responsible for the absorption of energy from the electric field. The term vanishes in the limit that $\Gamma \to 0$.

The power radiated is given by Eq. (14.19) which involves

$$\frac{d^2 \mathbf{r}}{dt^2} = -\omega^2 \mathbf{r} \tag{14.38}$$

with $\mathbf{r}$ given by Eq. (14.32). Again, averaging something of the form taken by $\mathbf{r}^2$ involves

$$\begin{aligned}
\langle (a \cos \omega t + b \sin \omega t)^2 \rangle &= a^2 \langle \cos^2 \omega t \rangle \\
&\quad + 2ab \langle \sin \omega t \cos \omega t \rangle \\
&\quad + b^2 \langle \sin^2 \omega t \rangle \\
&= \tfrac{1}{2}(a^2 + b^2)
\end{aligned} \tag{14.39}$$

so that

$$P = \frac{2}{3} \frac{e^2}{c^3} \omega^4 \left(\frac{e E}{m} \right)^2 \frac{1}{(\omega_0^2 - \omega^2)^2 + \omega^2 \Gamma^2}. \tag{14.40}$$

The cross section is defined as the ratio

$$\sigma = \frac{\text{Power radiated}}{\text{Incident energy flux}} = \frac{P}{F}. \tag{14.41}$$

This has dimensions of

$$[\sigma] = \frac{\text{Erg/sec}}{\text{Erg/cm}^2 \, \text{sec}} = [\text{cm}^2]. \tag{14.42}$$

In our calculation we must take the expression for the energy flux from electromagnetic theory without explanation. It is

$$F = \frac{c}{4\pi} E^2. \tag{14.43}$$

This leads to

$$\sigma = \frac{8\pi}{3} \left(\frac{e^2}{mc^2} \right)^2 \frac{\omega^4}{(\omega_0^2 - \omega^2)^2 + \omega^2 \Gamma^2}. \tag{14.44}$$

a) The shape of the cross section (Fig. 14.5) shows a peak at $\omega = \omega_0$. The width of the peak is approximately Γ in that the cross section falls to one half its peak value at $\omega = \omega_0 \pm \Gamma/2$, in the case of a narrow "resonance," that is, when $\Gamma \ll \omega_0$.

b) For free electrons $\omega_0 = 0$ and $\Gamma = 0$, so that

$$\sigma = \frac{8\pi}{3} \left(\frac{e^2}{mc^2} \right)^2, \tag{14.45}$$

the Thomson result.

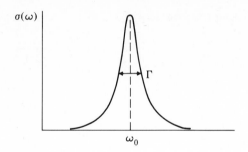

Figure 14.5. Cross section for the scattering of light as a function of frequency in the vicinity of a resonance at ω_0.

c) For scattering by an electron bound by an atom, for frequencies below any resonance absorption region, $\omega \ll \omega_0$, we get

$$\sigma = \frac{8\pi}{3} \left(\frac{e^2}{mc^2}\right)^2 \left(\frac{\omega}{\omega_0}\right)^4. \tag{14.46}$$

The formula was first derived by Rayleigh, and one speaks of scattering in this region as *Rayleigh Scattering*. It is characterized by the wavelength dependence $1/\lambda^4$, and it explains why blue light scatters better than red light, say.

The cross section can be used to calculate the mean free path of light in a medium. If there are N atoms/cm^3, then l, the mean free path is given by

$$l = \frac{1}{N\sigma} = \frac{3}{8\pi N (e^2/m\omega_0^2)^2} \left(\frac{c}{\omega}\right)^4$$

$$= \frac{3}{8\pi N (e^2/m\omega_0^2)^2} \left(\frac{\lambda}{2\pi}\right)^4.$$

The unknown parameter $(e^2/m\omega_0^2)$ can be determined from the index of refraction (14.33) in the case that its deviation from unity is small. In that case

$$n^2 - 1 = 4\pi N (e^2/m\omega_0^2)$$

$$n - 1 \simeq 2\pi N (e^2/m\omega_0^2)$$

$$l \simeq \frac{3\pi N}{(n-1)^2} \left(\frac{\lambda}{2\pi}\right)^4.$$

For air $n - 1 = 2.93 \times 10^{-4}$. Lord Rayleigh knew this number, and the story goes that on a visit to Darjeeling he observed Mount Everest from a terrace in his hotel. From the dimness of the outline of the mountain, and from the known distance of about 100 miles he determined the number of air molecules per cm^3 to be 3×10^{19}, which is quite close to the correct value.

The λ^{-4} dependence explains why the sky is blue, and why the sunset is red. In the latter case, the blue and violet components of the white light are taken out of the incident beam more efficiently than the reds and yellows, which come through more strongly. Why are clouds white? The reason is that in clouds, as in liquids and solids, the molecules take on a more orderly arrangement than in air. As soon as the molecules arrange themselves with a well-defined spacing, the pattern of scattering ceases to be random. There is constructive scattering for radiation going in the forward direction, and there is destructive interference in other directions. It is this effect that explains why a beam of light entering a liquid is not totally scattered, even though the density N is thousands of times larger than in a gas. When light is reflected from a surface, that light is not blue, even though the scattering cross section has the λ^{-4} dependence. The reason is that the reflected wave is a coherent effect. All the atoms in a volume that is approximately half a wave-length deep and covering an area of roughly (const) $\times \lambda$ (the first Fresnel zone, in optics) do the scattering *coherently*, so that the intensity is *proportional to the square of the number of scatterers*, that is, proportional to λ^4. Thus the wavelength dependence cancels, and reflected light has an unchanged spectral composition. Thus water appears colorless, though not quite. Because of its molecular structure, it happens to absorb red light weakly, and thus water looks greenish. At a depth of 30 meters or so, all of the red component of white light is absorbed, and everything looks green.

Difficulties of Thomson model The Lorentz electron theory was also successful in explaining in this qualitative way many properties of solids. Nevertheless, the Thomson model had flaws that made it ultimately unacceptable. The first had to do with the pattern of "resonant" frequencies ω_k. Experimental evidence accumulated through the last two decades of the nineteenth century that the frequencies did not behave as harmonics of a fundamental frequency. Rather, they appeared as differences of well-defined "terms," and for hydrogen, the formula empirically obtained by Rydberg was

$$\frac{1}{\lambda} = R_H \left(\frac{1}{n_1^2} - \frac{1}{n_2^2} \right) \qquad \begin{array}{l} n_1 = 1, 2, 3, \ldots \\ n_2 = 2, 3, 4, \ldots \\ n_1 < n_2 \end{array} \qquad (14.47)$$

with the Rydberg constant given as

$$R_H = 109677.575 \pm 0.012 \text{ cm}^{-1}. \qquad (14.48)$$

The Thomson model has a well-defined prediction for the frequency. For hydrogen, the single electron is at the center of a sphere of positive charge, when in its equilibrium position. If the electron is displaced a distance r through the excitation of the atom, there is a restoring force due to the attraction by the positive charge inside the sphere of radius r. The amount of that charge is e times the fraction of the volume of the atom that the sphere of radius r occupies, when the charge density is uniform (Fig. 14.6). Thus

$$q(r) = \frac{e}{4\pi R^3/3} \cdot 4\pi r^3/3 = e \left(\frac{r}{R} \right)^3 \qquad (14.49)$$

and therefore the restoring force has magnitude

$$F = \frac{eq(r)}{r^2} = \frac{e^2}{R^3} r. \qquad (14.50)$$

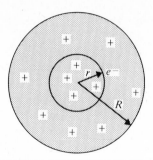

Figure 14.6. Electron at distance r from center of spherical uniform positive charge distribution of radius R.

This is linear in r, so that the motion is harmonic, with the frequency

$$\nu_0 = \frac{1}{2\pi}\omega_0 = \frac{1}{2\pi}\left(\frac{e^2}{mR^3}\right)^{1/2}.$$
(14.51)

This corresponds to a wavelength of ($R \simeq 10^{-8}$ cm)

$$\lambda = \frac{c}{\nu_0} = 2\pi c\left(\frac{mR^3}{e^2}\right)^{1/2} \cong 1180\,\text{Å}.$$
(14.52)

This number is of the correct order of magnitude (the α-Lyman line in hydrogen is 1215 Å), but there is no way of understanding the formula (14.47). Nonuniformities in the charge distribution will change Eq. (14.52) somewhat, but Eq. (14.47) eludes us completely.

Rutherford alpha scattering The death blow to the Thomson model came from the experiments of Rutherford in 1911. These experiments were made possible by the discovery of radioactivity by Becquerel and Maria and Pierre Curie, who found that heavy elements spontaneously emit radiation. The radiation consisted of three components, named α-, β- and γ-radiation. The γ-rays turned out to be high-frequency photons. The β-rays turned out to be electrons, and the α-rays, which were generally emitted with a well-defined energy, that varied from element to element, turned out to be particles of positive charge $2e$ and mass equal to four times that of the hydrogen atom. The α-particles were thus doubly ionized Helium. Rutherford scattered α-particles off a target consisting of thin gold foil, and he observed a certain amount of large angle scattering. This was totally incompatible with the expectations from the Thomson atomic model (Fig. 14.7). For that model, the maximum force on a positive charge occurs when it grazes the atom. Then the force is

$$F = \frac{(2e)(Ze)}{R^2}.$$
(14.53)

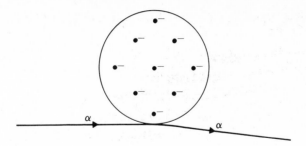

Figure 14.7. Deflection of alpha particle by Thomson atom.

To estimate the change in momentum of the α-particle (Fig. 14.8), we multiply the force by the time that the particle spends near the atom, that is, approximately $2R/v$ where v is the velocity of the α-particle. Hence

$$\frac{\Delta p}{p} = \frac{2FR/v}{M_\alpha v} = \frac{2Ze^2/R}{\frac{1}{2}M_\alpha v^2}. \tag{14.54}$$

Thus the deflection due to the positive charge distribution is at most

$$\frac{2Z(4.8 \times 10^{-10})^2}{1.6 \times 10^{-6}K(\text{Mev})} = 0.3 \times 10^{-4} \frac{Z}{K_{\text{Mev}}} \text{ radians} \tag{14.55}$$

where K is the kinetic energy of the incident α-particle. The scattering due to the electrons embedded in the positive charge is negligible, because the electrons are approximately 8000 times lighter than the α-particles. Even in a head-on collision, the maximum momentum transfer is of the order of $m_e v$ so that

$$\frac{\Delta p}{p} \simeq \frac{m_e}{M_\alpha} \sim \frac{1}{8000} \sim 10^{-4}. \tag{14.56}$$

Thus

$$\theta < 10^{-4} \frac{Z}{K_{\text{Mev}}} \tag{14.57}$$

is a safe estimate. For gold, with $Z = 79$ and with $K = 5$ MeV, the maximum deflection per collision is of the order of 10^{-3} radians. A deflection of as much as $1°$ can occur only as a result of many collisions; since these are random in direction the deviation from the root mean square becomes very improbable. One can estimate the probability of getting a $90°$ deflection, and for the Rutherford experiment that turned out to be 10^{-3500}! Experimentally it was found that one α-particle in 8,000 scattered through $90°$.

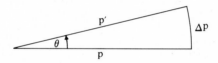

Figure 14.8. Momentum transfer in collision and associated angle of deflection.

The only way to increase the angle in the above estimate is to decrease R significantly. This, however, means that the positive charge occupies a volume that is much smaller than the volume of the atom. This led Rutherford to propose an alternative to the Thomson model. *Rutherford's planetary model* had the following features.

Rutherford's nuclear atom The positive charge occupies a very tiny region in space; it forms the nucleus of the atom, in which all of the mass (except for a small electronic contribution) resides. The electrons attracted to the nucleus with a Ze^2/r potential, circle the nucleus in Keplerian orbits.

For this model, the α-particles are again not scattered by the electrons. The deflection due to the point charge Ze is calculable by using methods of classical mechanics, and one finds that the cross section for scattering into a solid angle of magnitude $d\Omega$, making an angle θ with the initial direction is

$$d\sigma = \left(\frac{Ze^2}{\frac{1}{2}M_\alpha v^2}\right)^2 \frac{1}{\sin^4 \theta/2}\, d\Omega \qquad (14.58)$$

and this turns out to be in excellent agreement with experiment (Fig. 14.9). More crudely we can use our estimate for $\Delta p/p$ with a new value of R, which we now determine as the distance of closest approach. The total energy is given by

$$E = \frac{1}{2}M_\alpha v^2 + \frac{2Ze^2}{r}\, . \qquad (14.59)$$

Far away E is the kinetic energy of the α-particle, K. The point of closest approach is where the particle turns around, that is where $v = 0$. Thus at the point of closest approach

$$R = \frac{2Ze^2}{K}\, . \qquad (14.60)$$

This implies that $\Delta p/p \sim 1$, so that large angle scattering can occur.

The Rutherford atom had difficulties of its own. It also could not explain the Rydberg formula for the spectrum of radiation from the hydrogen atom. It had two other major flaws:

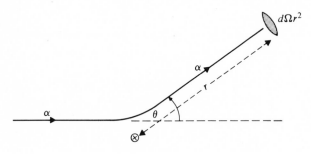

Figure 14.9. Alpha scattering into a solid angle $d\Omega$ with a deflection of θ.

a) The Rutherford atom could not explain why all atoms of a given element are alike. The planetary orbits depend on the initial conditions (position and velocity) and there is nothing in classical mechanics that can explain why all hydrogen atoms, for example, should have started out exactly the same way, and that even after excitation, and subsequent de-excitation, the orbits should be the same.

b) The fatal flaw of the Rutherford atom was that electrons in orbits are constantly accelerating, and they should therefore radiate. With this loss of energy going on, the electrons should finally spiral into the nucleus. How long would this process take? Consider circular orbits for simplicity. Balancing the centripetal and Coulomb forces gives

$$\frac{m_e v^2}{R} = \frac{Ze^2}{R^2}. \tag{14.61}$$

If we introduce the angular momentum

$$L = m_e v R, \tag{14.62}$$

we find that

$$v = \frac{Ze^2}{L}. \tag{14.63}$$

The radius is given by

$$R = \frac{L}{m_e v} = \frac{L^2}{m_e Ze^2}. \tag{14.64}$$

The time to go around a circular orbit once is

$$T = \frac{2\pi R}{v} = \frac{2\pi L^2}{m_e Ze^2} \frac{L}{Ze^2} = \frac{2\pi L^3}{m_e(Ze^2)^2}. \tag{14.65}$$

The acceleration is

$$\frac{v^2}{R} = \left(\frac{Ze^2}{L}\right)^2 \frac{m_e Ze^2}{L^2} = m_e \frac{(Ze^2)^3}{L^4}. \tag{14.66}$$

From classical radiation theory, the energy radiated per unit time is

$$P = \frac{2}{3}\frac{e^2}{c^3} a^2 = \frac{2}{3}\frac{e^2}{c^3} m_e^2 \frac{(Ze^2)^6}{L^8}. \tag{14.67}$$

Thus the fractional energy loss per traversal time is

$$\frac{\dfrac{2}{3}\dfrac{m_e^2 e^2}{c^3}\dfrac{(Ze^2)^6}{L^8}}{\dfrac{1}{2} m_e \left(\dfrac{Ze^2}{L}\right)^2} \frac{2\pi L^3}{m_e(Ze^2)^2} = \frac{8\pi}{3} Z^2 \left(\frac{e^2}{Lc}\right)^3. \tag{14.68}$$

This becomes unity in a time = (traversal time) × (number of traversals), that is

$$\tau = \frac{2\pi L^3}{m_e(Ze^2)^2} \bigg/ \frac{8\pi}{3} Z^2 \left(\frac{e^2}{Lc}\right)^3$$

$$= \frac{3}{4}\frac{1}{Z^4}\frac{L}{m_e c^2}\left(\frac{Lc}{e^2}\right)^5. \tag{14.69}$$

To go further, we need an estimate of L. If we take the radius of the orbit to be of the order of 10^{-8} cm, then

$$L \simeq (Ze^2 Rm_e)^{1/2} \simeq Z^{1/2} 4.8 \times 10^{-10} (10^{-8} \times 0.9 \times 10^{-27})^{1/2}$$
$$\simeq Z^{1/2} 4.8 \times 10^{-10} \times 3 \times 10^{-18}$$
$$\simeq Z^{1/2} 1.4 \times 10^{-27}. \qquad (14.70)$$

This leads to a spiraling-in time that for $Z = 1$ is of the order of

$$\tau = \frac{1.4 \times 10^{-27}}{0.9 \times 10^{-27} \times 9 \times 10^{20}} \left(\frac{1.4 \times 10^{-27} \times 3 \times 10^{10}}{(4.8 \times 10^{-10})^2} \right)^5$$

$$\simeq 1.2 \times 10^{-10} \sec. \qquad (14.71)$$

Thus the Rutherford model had a fatal flaw. All of the difficulties were swept aside by the bold proposal of Niels Bohr, whose atomic model led directly to the discovery of quantum mechanics and a complete understanding of the structure of atoms.

NOTES AND COMMENTS

1. A very detailed discussion of the work of Thomson may be found in F.L. Friedman and L. Sartori, *The Classical Atom*, Addison-Wesley, Reading, Mass., 1965.
2. A good discussion of the damped oscillator and the application of the mathematical formalism to electron theory may be found in F.S. Crawford, Jr., *Waves* (Berkeley Physics Course, vol. 3), McGraw-Hill, New York, 1968.
3. A very stimulating discussion of the interaction of light with matter, without any technical details, may be found in the essay "How light interacts with matter," in V.F. Weisskopf, *Physics in the Twentieth Century*, The MIT Press, Cambridge, Mass., 1972.
4. Dielectric constants are discussed in R.P. Feynman, R.B. Leighton, and M. Sands, *The Feynman Lectures on Physics*, vol. II, Addison-Wesley, Reading, Mass., 1964.

PROBLEMS

14.1 Consider scattering of particles in iron. Given that the atomic weight of iron is 57 and that the density is 7.9 gm/cm^3, calculate the number of target nuclei per cm^3. Calculate the mean free path of a projectile if the cross section is given by

i) 10^{-16} cm^2,

ii) 10^{-24} cm^2,

iii) 10^{-27} cm^2,

iv) 10^{-44} cm^2

per unit atomic weight (that is, $\sigma = A\sigma_0$, where σ_0 is what is given above).

14.2 Estimate the mean free path of a student crossing a field $50\,m \times 100\,m$ if there are $50, 100, 1000$ people on it, distributed at random.

14.3 Using the power absorbed, as given in Eq. (14.37), and the incident energy flux, as given in Eq. (14.43), we may define an absorption cross section, defined as in Eq. (14.41). Write down this cross section. Show that it has the correct dimensions. Write down the ratio of the absorptive to the total cross sections, where the latter is given by Eq. (14.44).

14.4 The index of refraction of air is 1.0002926 at 1 atmosphere pressure and $0°C$, at low frequencies. Use formula (14.33) to estimate the characteristic wavelength λ_0 $(= 2\pi c/\omega_0)$. You need to calculate N.

14.5 The index of refraction of water is 1.33 at low frequencies. What does this say about ω_0 (and therefore λ_0)?

14.6 The expression for the dielectric constant

$$\epsilon - 1 \cong 4\pi N \frac{e^2}{m\omega_0^2} \equiv N\alpha$$

relevant for gases, should be modified to

$$\epsilon - 1 = N\alpha/(1 - N\alpha/3)$$

for liquids (Clausius–Mossotti equation). The experimental data for several substances are

CS_2	$N\alpha = 1.11$	$\epsilon = 2.64$
O_2	$N\alpha = 0.435$	$\epsilon = 1.507$
A	$N\alpha = 0.441$	$\epsilon = 1.54$
CCl_4	$N\alpha = 0.977$	$\epsilon = 2.24$

a) Check the correctness of the Clausius–Mossotti equation.

b) If the densities of these substances are given as $1.293, 1.19, 1.44,$ and 1.59, respectively, in gm/cm^3, calculate the values of ω_0 for these substances. (The values of A for the various elements are given in Appendix C).

15
The Bohr Atom

Bohr Postulates Two years after Rutherford proposed his planetary model, Niels Bohr, in 1913, proposed a model of an atom which radically broke with classical physics, bypassed the flaws in the Rutherford model, and gave an enormous impetus to the revolution that was started by Planck and Einstein. Bohr postulated:

I. That an atomic system can only exist in certain stationary states with definite discrete energies, and that consequently any change of the energy of the system, including emission and absorption of radiation, must take place by a complete transition between such states.

II. That the radiation absorbed or emitted during such a transition between two stationary states has a frequency ν given by the relation

$$h\nu = E' - E'' \tag{15.1}$$

where h is Planck's constant and where E' and E'' are the values of the energy in the two states under consideration.

For circular orbits, the postulate determining the stationary states took a simple form:

III. The stationary states for circular orbits are determined by the condition that the angular momentum be an integral multiple of $\hbar = h/2\pi$,

that is,

$$L = n\hbar \quad n = 1, 2, 3, \ldots . \tag{15.2}$$

Since the various quantities associated with circular orbits of the Rutherford atom have been worked out in terms of L in Eq. (14.63) on, the calculations of the properties of the stationary states are straightforward.

The velocity of an electron in an orbit characterized by the quantum number n is, following Eq. (14.63),

$$v = \frac{Ze^2}{n\hbar} = \frac{1}{n} c \frac{Ze^2}{\hbar c}. \tag{15.3}$$

It is very convenient to introduce the dimensionless number

$$\alpha \equiv \frac{e^2}{\hbar c} \approx \frac{1}{137} \tag{15.4}$$

called the *fine structure constant*. The precise value of $1/\alpha$ is 137.035982 ± 0.000030. In terms of α we have

$$v = \frac{Z\alpha c}{n}. \tag{15.5}$$

For Z not too large, v/c is always much smaller than unity, so that the atomic motion is nonrelativistic. There are small $(v/c)^2$ corrections that we will run into later, but for the time being we will not worry about them. In fact, c does not really appear in any of the formulas like Eq. (14.63). It only enters through the use of α, that is, for the sake of convenience. The radius of the nth orbit of the Bohr atom is given by

$$R_n = \frac{n^2 \hbar^2}{mZe^2} = \frac{n^2}{Z\alpha}\frac{\hbar}{mc}. \tag{15.6}$$

The expression of the radius involves the reduced Compton wavelength of the electron, that is, h/mc divided by 2π. The radius of the lowest orbit is

$$R_1 = \frac{1}{Z\alpha}\frac{\hbar}{mc} \equiv \frac{a_0}{Z}. \tag{15.7}$$

Here

$$a_0 = \frac{\hbar}{mc\alpha} \simeq \frac{137 \times 1.05 \times 10^{-27}}{0.9 \times 10^{-27} \times 3 \times 10^{10}} \simeq 0.5 \times 10^{-8}\,\text{cm}$$
$$\simeq 0.5\,\text{Å}, \tag{15.8}$$

a number that should be remembered.

Let us now calculate the energy of the electron in the nth orbit. We expect that number to be negative, since the electron is bound. A positive energy electron could escape to infinity, where the potential energy vanishes, and still have some kinetic energy. We have

$$E_n = \frac{1}{2}mv_n^2 - \frac{Ze^2}{R_n} = \frac{1}{2}m\left(\frac{Z\alpha c}{n}\right)^2 - \frac{Ze^2(Z\alpha)mc}{\hbar n^2}$$

$$= -\frac{1}{2}m(Z\alpha c)^2/n^2. \tag{15.9}$$

To obtain the magnitude, we note that

$$mc^2 = 0.9 \times 10^{-27} \times 9 \times 10^{20}/1.6 \times 10^{-6}\ \text{MeV}$$
$$= 0.51\ \text{MeV}, \tag{15.10}$$

and therefore

$$\frac{m(\alpha c)^2}{2} = \frac{0.51 \times 10^6}{2(137)^2} = 13.6\ \text{eV}. \tag{15.11}$$

The frequency of radiation emitted in the transition from an orbit of quantum number n_2 to one of quantum number n_1 is

$$v(n_2 \to n_1) = \frac{m(\alpha cZ)^2}{2h}\left(\frac{1}{n_1^2} - \frac{1}{n_2^2}\right). \tag{15.12}$$

Now

$$\frac{mc^2\alpha^2}{2h} \approx \frac{13.6 \times 1.6 \times 10^{-12}}{6.53 \times 10^{-27}} \cong 3.3 \times 10^{15}\,\text{Hz}, \tag{15.13}$$

and therefore

$$\frac{1}{\lambda} = \frac{v}{c} \approx \frac{3.3 \times 10^{15}Z^2}{3 \times 10^{10}}\left(\frac{1}{n_1^2} - \frac{1}{n_2^2}\right)$$

$$\approx 1.1 \times 10^5 Z^2 \left(\frac{1}{n_1^2} - \frac{1}{n_2^2}\right)\text{cm}^{-1}. \tag{15.14}$$

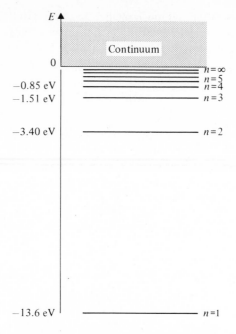

Figure 15.1. Energy spectrum of stationary states of Bohr atomic model with circular orbits.

A more precise evaluation of the constants shows complete agreement with the Rydberg formulas (14.47) and (14.48).

The energy spectrum that follows from these results is shown in Fig. 15.1. Note that the energy levels get closer and closer together as n increases. The zero level is called the *ionization* level. It takes $13.6Z^2$ eV energy to ionize the atom, and it takes $10.2Z^2$ eV to excite the electron from the ground state to the first excited state.

Two comments are in order at this point:

a) If the mass of the nucleus is M, rather than infinite, the nucleus can also move. As in the mechanics of two particles interacting with some potential that depends only on their separation, the two-body problem can be reduced to a one-body problem where the one body has the reduced mass

$$\mu = \frac{m_e M}{m_e + M} = \frac{m_e}{1 + m_e/M} .$$
(15.15)

(We do not derive this familiar result from classical mechanics, since it will come up again in quantum mechanics in Chapter 25.) In the hydrogen atom $m_e/M \approx 1/2000$ so that to a good approximation we may take $\mu \cong m_e$. Nevertheless, the correct formula for the energy is

$$E_n = -\tfrac{1}{2}\mu(Z\alpha c)^2/n^2 .$$
(15.16)

In 1932 Urey and collaborators noticed in the spectrum of hydrogen the presence of lines that were slightly displaced from hydrogen lines. It turned out that these were due to atoms of an isotope of hydrogen, *deuterium* $(Z = 1, A = 2)$, whose nuclear mass is approximately twice that of the hydrogen nucleus, so that the reduced mass for these atoms is

$$\mu' \cong \frac{m_e}{1 + m_e/2M}. \tag{15.17}$$

It was in this way that deuterium was discovered.

b) In our discussion we have considered only circular Keplerian orbits. The planetary model allows for elliptical orbits, and the Bohr quantization condition (15.2) can be generalized to cover the situation. In this way a second quantum number is introduced. The orbits that have maximum angular momentum for a given energy are the circular ones. We do not go into detail here, since the problem of the hydrogen atom will be solved quantum mechanically in Chapter 29, at which point we will study the full richness of the hydrogen atom. At this point we merely note that the Bohr rules were generalized by Sommerfeld and Wilson, to the condition that

$$\oint p_\theta \, rd\theta \ = \ 2\pi n_\theta \, \hbar$$

$$\oint p_r dr \ = \ 2\pi n_r \, \hbar \tag{15.18}$$

where p_θ and p_r are the momenta in the tangential and radial directions respectively, and the integrals are taken around closed orbits. The generalization allowed Sommerfeld to study finer details of the hydrogen spectrum such as arise from the relativistic change of the electron mass, but it was of no use for nonperiodic systems. There was general recognition, constantly stressed by Bohr, that the old quantum theory was a provisional one, consisting, as it did, of quantum rules grafted onto classical mechanical ideas. There was no understanding of when an electron decides to "jump" from one orbit to another, nor what the electron did between the orbits. The theory could not cope with atoms that were not hydrogenlike, although Bohr, with his extraordinary insight, managed to construct a very complete explanation of the Mendeleev periodic table. In doing this he anticipated many purely quantum-mechanical effects, and he provided crucial leadership in the search for the correct quantum theory.

The Correspondence Principle In the development of the old quantum theory, the Correspondence Principle formulated by Bohr was of paramount importance. The principle was based on the idea that the theory, whatever its final form, had to go over, in the limit in which quantum effects decrease in importance, into classical radiation theory. At first just an observation, the principle became an increasingly sharp tool in the early 1920s in guessing quantum formulae from the classical dispersion formulae and in studying absorption of light. In the hands of Bohr, Kramers, Heisenberg, and Van Vleck, the old quantum theory was formulated in a way that made the transition to quantum mechanics possible. We shall illustrate the use of the Correspondence Principle to (i) derive a *selection rule* for atomic transitions between stationary states with circular orbits, and (ii) to derive a value for the lifetime of the electron in the $n = 2$ state, that is, the length of time an electron remains in that state before making the transition to the ground state.

a) Consider the radiation emitted by a Bohr atom in a circular orbit, when the transition is made from the $(n + k)$ to the n orbit, with n very large, and k small, for example, $k = 1, 2, 3$, or so. The frequency of the radiation emitted is given by

$$\nu = \frac{m(Z\alpha c)^2}{Zh} \left(\frac{1}{n^2} - \frac{1}{(n+k)^2} \right)$$

$$\simeq \frac{1}{2} \frac{mZ^2 c^2}{2\pi\hbar} \left(\frac{e^2}{\hbar c} \right)^2 \frac{2nk}{n^2(n+k)^2}$$

$$\simeq \frac{e^4 Z^2 m}{2\pi L^3} k \tag{15.19}$$

where $L = n\hbar$ is the angular momentum. Let us compare this with the reciprocal of the time that an electron takes to go around the nucleus, that is, with the classical frequency of rotation. The frequency of the radiation in the classical limit is thus

$$\nu_{cl} = \frac{1}{T} = \frac{mZ^2 e^4}{2\pi L^3} \tag{15.20}$$

where we took T from eq. (14.65). These are to be equal for large L, since classical theory requires the frequency of radiation to be that of the radiator. This, therefore, implies that

$$k = 1. \tag{15.21}$$

Thus for large quantum numbers only transitions satisfying

$$\Delta n = 1 \tag{15.22}$$

take place. The Correspondence Principle is then used as a guide to what happens for small L, and the conjecture is that Eq. (15.22) is generally true. We shall later derive this *selection rule* from more general principles of quantum mechanics.

b) To calculate the lifetime of an electron in an orbit characterized by large quantum numbers, we take the formula (14.67) for the power radiated,

$$P = \frac{2}{3} \frac{e^2}{c^3} m^2 \frac{(Ze^2)^6}{L^8} . \tag{15.23}$$

This is an expression for the emission of radiation energy per unit time. Since the energy emitted consists of a single photon, with energy $h\nu$, where ν is calculated in Eq. (15.19) with $k = 1$, we get, from

$$\frac{\Delta E}{\Delta t} = P \tag{15.24}$$

the expression

$$\Delta t = \frac{\Delta E}{P} = \frac{h \dfrac{e^4 Z^2 m}{2\pi L^3}}{\dfrac{2}{3} \dfrac{e^2 m^2}{c^3} \dfrac{(Ze^2)^6}{L^8}}$$

$$= \frac{3}{4\pi} h \frac{e^2 Z^2 c^3 L^5}{m(Ze^2)^6} = \frac{3}{2} \left(\frac{\hbar}{mc^2} \right) \frac{1}{\alpha} \left(\frac{1}{Z\alpha} \right)^4 n^5 \text{ sec.} \tag{15.25}$$

In this expression we have written $L = n\hbar$. The formula must be correct in the limit of large n. If we take this formula as a guide for the situation for small n, we obtain, for $n = 2$ that is the $(2 \to 1)$ transition, the lifetime

$$\tau = \frac{3}{2}\left(\frac{1.05 \times 10^{-27}}{0.9 \times 10^{-27} \times 9 \times 10^{20}}\right) (137)^5 \, (2^5)$$

$$\cong 2.8 \times 10^{-9} \text{ sec.} \tag{15.26}$$

for hydrogen. The correct value, calculated using quantum mechanics, and measured in the laboratory, is

$$\tau(2 \to 1) = 1.6 \times 10^{-9} \text{ sec} \tag{15.27}$$

so that the estimate is not far off. This time appears to be very short. However, in terms of the time that it takes an electron to go round the nucleus once (the atomic "year"), which is

$$T = \frac{2\pi R}{v} \simeq \frac{2\pi \times 0.5 \times 10^{-8}}{3 \times 10^{10}/137} \simeq 4.3 \times 10^{-16} \text{ sec,}$$

this is really very long.

The way in which this lifetime manifests itself is the following: one knows from wave theory that a spread in the time-width of a pulse of radiation, Δt, manifests itself in the width of the frequency band, whose magnitude is

$$\Delta \nu = \frac{1}{2\pi} \Delta t \tag{15.28}$$

so that the energy emitted in the form of radiation is uncertain by an amount

$$\Delta E = h\Delta \nu = \frac{h}{2\pi\tau} = \frac{\hbar}{\tau} \cong \frac{2}{3}mc^2\alpha(Z\alpha)^4. \tag{15.29}$$

The spectral line thus has a "width" of the order of magnitude

$$\Delta \nu = \frac{1}{3\pi}\frac{mc^2\alpha}{\hbar}(Z\alpha)^4 \tag{15.30}$$

and this is just the quantity Γ_0 that appeared in Chapter 14 in the Lorentz theory of the (damped) oscillating electron.

We conclude by noting that the Bohr conditions can be applied to other systems that will be of interest to us.

The plane rotator The expression for the energy of a plane rotator of moment of inertia I about the axis of rotation, with angular momentum L is given by

$$E = \frac{L^2}{2I}. \tag{15.31}$$

If we now insert $L = m\hbar$ into the above, with $m = 0, \pm 1, \pm 2, \pm 3, \ldots$, we obtain

$$E = \frac{\hbar^2}{2I}m^2. \tag{15.32}$$

The two signs of the integers correspond to the two different senses of rotation about the axis. The energy levels depend only on the magnitude of m. The frequencies of radiation emitted in a transition are given by

$$\nu(m_1 \to m_2) = \frac{\hbar^2}{2Ih}(m_1^2 - m_2^2) = \frac{\hbar}{4\pi I}(m_1^2 - m_2^2). \tag{15.33}$$

If we take over the selection rule

$$\Delta m = 1 \tag{15.34}$$

derived for atomic transitions, then the dominant frequencies are those corresponding to

$$\nu(m + 1 \to m) = \frac{\hbar}{4\pi I}(2m + 1). \tag{15.35}$$

The above pattern of lines is characteristic of rotational spectra. They are clearly observed in the spectra of diatomic molecules such as CO, OH, H_2, and so on. For such molecules the moments of inertia are given by

$$I = \mu R^2 \tag{15.36}$$

where μ is the reduced mass of the diatomic molecule and R is the interatomic separation. Thus for an H_2 molecule, $\mu = M/2 = 0.8 \times 10^{-24}$ gm and with a separation of $1 \text{ Å} = 10^{-8}$ cm we get

$$I = 0.8 \times 10^{-4} \text{ cm}^2 \text{ gm}, \tag{15.37}$$

so that

$$\lambda = \frac{c}{\nu} = \frac{4\pi Ic}{\hbar}\frac{1}{2m + 1} = \frac{2.9 \times 10^{-2}}{2m + 1} \text{ cm}. \tag{15.38}$$

These wavelengths lie in the microwave region. The energy of the first excited state is given by

$$E_1 = \frac{\hbar^2}{2I} = \frac{(1.05 \times 10^{-27})^2}{1.6 \times 10^{-40}} = 6.9 \times 10^{-15} \text{ ergs}. \tag{15.39}$$

If we set this equal to kT, where k is Boltzmann's constant and T is the temperature, we find that

$$T = \frac{6.9 \times 10^{-15}}{1.38 \times 10^{-16}} \simeq 40°\text{K}. \tag{15.40}$$

Thus, using the Boltzmann factor $e^{-E_1/kT}$, we see that at temperatures as low as $40°$K molecular rotational states can be excited thermally.

The harmonic oscillator Consider a two-dimensional oscillator, with the spring constants in the x and y directions equal. In this case the potential energy can be written as

$$V(r) = \tfrac{1}{2}m\omega^2 r^2. \tag{15.41}$$

When the force, $-m\omega^2 r$ is equated to the centripetal force, we get

$$\frac{mv^2}{r} = m\omega^2 r \tag{15.42}$$

so that

$$v = \omega r. \tag{15.43}$$

The angular momentum quantization condition reads

$$L = mvr = m\omega r^2 = n\hbar \tag{15.44}$$

from which we find

$$r = (n\hbar/m\omega)^{1/2}$$
$$v = (n\hbar\omega/m)^{1/2} \tag{15.45}$$

and

$$E = \tfrac{1}{2}mv^2 + \tfrac{1}{2}m\omega^2 r^2$$
$$= n\hbar\omega \quad (n = 0, 1, 2, \ldots). \tag{15.46}$$

This answer agrees with what is derived from quantum mechanics except for the fact that in the correct theory the lowest state for a two-dimensional oscillator is not the zero energy state, but the state with $n = 1$.

We see that in both cases, as with atoms, there are discrete energy levels. The mere existence of these levels allows us to understand, at least qualitatively, the curious behavior of the specific heat curve pointed out in Fig. (12.9). The classical prediction for the specific heat, it will be recalled, is $7R/2$, because there are three translational degrees of freedom and two each, rotational and vibrational. The behavior of the curve can be understood as follows: at very low temperatures (according to our estimate of Eq. (15.40)), the only way in which a molecule can store energy is through the motion of its center of mass, that is, through its translational kinetic energy. Thus only the translational degrees of freedom count, and the specific heat is $3R/2$. At around $50°K$ it begins to be possible to excite rotational degrees of freedom, and energy can be stored in this way. Thus the specific heat rises to $5R/2$. At higher temperatures, the vibrational degrees of freedom get excited, and the specific heat rises to $7R/2$. At higher temperatures still, the electronic excitations become possible. This, however, occurs at temperatures corresponding to several electron volts. Since

$$1 \text{ eV} = 1200°K, \tag{15.47}$$

this does not occur until the temperature gets so high that the molecule has a good chance of dissociating.

The Bohr atomic model, the notion of stationary states, and the Correspondence Principle guided research in the period 1913–1925, when Heisenberg discovered the true laws of quantum mechanics. The Heisenberg path to quantum mechanics turned out to be a difficult one, and an alternative, but totally equivalent formulation, was developed a year later by Erwin Schrödinger. It is the latter approach, built upon the notion of matter waves due to DeBroglie, that we will follow in the next part of the book.

NOTES AND COMMENTS

1. Some historical background to the Bohr theory may be found in M. Jammer, *The Conceptual Development of Quantum Mechanics*, McGraw-Hill, New York, 1966, and also in the historical introduction to B.L. van der Waerden (Ed.), *Sources of Quantum Mechanics*, Dover Publications, New York, 1967. A number of chapters in E.T. Whittaker, *A History of the Theories of Aether and Electricity* (vol. 2), Philosophical Library, New York, 1954, deal with the matters discussed in this chapter, and in earlier chapters on the old quantum theory.

PROBLEMS

15.1 The law of attraction between two massive bodies is
$$V(r) = -Gm_1m_2/r$$
where $G = 6.67 \times 10^{-8}$ in c.g.s. units. Suppose the planetary orbits are quantized using the Bohr rules. What is the general expression for the magnitude of the Bohr radii for this problem?

15.2 Recently two new particles of mass $mc^2 = 3098\,\text{MeV}$ and $3684\,\text{MeV}$, respectively, were discovered. It is believed that they are bound states of two "fundamental" particles of equal mass, attracting each other with a potential $-q^2/r$. If the bound states are identified with the $n = 1$ and $n = 2$ states of the Bohr atom, what is the value of $q^2/\hbar c$ (the analog of α) and the mass of the constituent particle? (Do not forget the rest masses.)

15.3 A proton and an antiproton having equal and opposite charges e and with the proton mass $Mc^2 = 940\,\text{MeV}$ form an atom called protonium. What is the binding energy of protonium and what is the energy of the first excited state? How large is protonium?

15.4 Use the Sommerfeld–Wilson rule $\oint p\,dx = 2\pi n\hbar$ to calculate the energy levels of a particle confined to a box of width a (the box is one-dimensional). Suppose the box is $1\,\text{Å}$ wide, what is the energy difference between the ground state and the first excited state, if the particle is an electron?

15.5 Use the Bohr quantization rule to find the energy levels of a system with the potential energy
$$V(r) = V_0(r/a)^k$$
and circular motion in the orbit. Sketch the potential as a function of r for large k, and show that the energy levels agree with those obtained in Problem 15.4 when $k \to \infty$, as they should.

15.6 Consider a Bohr model of singly ionized Helium that is a hydrogen-like atom with $Z = 2$ and $A = 4$. Discuss the spectrum of this atom, and point out what difference the $A = 4$ makes.

15.7 Consider a single electron around a charge Z nucleus. For what value of Z will the binding energy exceed the rest mass of the electron?

15.8 One can show that for closed orbits (or, more generally, for motion confined to a finite region) that the average values of the kinetic energy is related to the average value of a function of the potential energy, in particular
$$\left\langle \frac{\mathbf{P}^2}{2m} \right\rangle = \frac{1}{2} \left\langle r\frac{dV}{dr} \right\rangle .$$
Use this for circular motion for the hydrogen atom, where the energy is given by $E = -mc^2\alpha^2/2n^2$ to calculate the magnitude of the momentum. Do the same for the harmonic oscillator, where the potential energy is $m\omega^2r^2/2$ and the energy is $n\hbar\omega$.

PART IV
Elementary
Quantum Mechanics

Quantum mechanics was discovered in two ways. In 1925 Heisenberg culminated his studies of optical dispersion, carried out within the Bohr Correspondence Principle program in collaboration with H. Kramers, in a remarkable paper that set forth the rules of treating a system "quantum mechanically." As was soon pointed out by Born, Heisenberg, and Jordan, and independently by Dirac, the crux of the new procedure was to treat dynamical variables such as the position x and the momentum p as *matrices* that are constrained by the rule that the two matrix products xp and px do not commute, but differ by a multiple of the unit matrix

$$xp - px = i\hbar I.$$

These matrices must be infinite-dimensional, and thus working with them is complicated. Although W. Pauli succeeded in solving the hydrogen atom spectrum using matrix mechanics in a very ingenious way, that approach never became particularly accessible to the average physicist. Fortunately, the second discovery of quantum mechanics, due to E. Schrödinger, presented quantum mechanics in the form of a wave equation. Since differential equations are more easily understood than matrix equations (in spite of their equivalence), much of the progress in the application of quantum mechanics to atomic systems came with the use of wave mechanics. In our introductory treatment of quantum mechanics we shall focus on wave mechanics, except when we discuss intrinsic *spin* in Chapter 30.

In this, the largest part of the book, we undertake a thorough introduction to quantum mechanics. We begin with a discussion of wave packets, to show how suitable superpositions of waves of different frequencies can give localized "disturbances" that could be interpreted as particles. This makes palatable the notion that a wave equation can describe atomic phenomena even when particles such as the electron are involved. The solution of the wave equation does not directly describe particle properties: rather, it is the absolute value squared of the complex wave function that has a physical interpretation as a probability density. This matter is discussed in Chapter 17, which is followed by a discussion of the uncertainty relations that provide the classical leeway to encompass both the particle and wave descriptions of the same quantal phenomenon. We follow with a general discussion of the Schrödinger equation, and the simplest physical system, a particle confined to a box. The nonclassical phenomenon of barrier penetration is treated rigorously for a simple one-dimensional system, and the manifestations of that phenomenon in the real world are discussed in some detail in Chapter 22. A discussion of bound states and the simple one-dimensional harmonic oscillator bring us to the end of the introductory aspects of quantum mechanics.

16
The Road to Wave Mechanics

De Broglie hypothesis The simplest approach to quantum mechanics makes use of the wave equation discovered by E. Schrödinger in 1926, which grew out of earlier work carried out in 1923 by Louis de Broglie. De Broglie was guided by certain analogies between geometrical optics (known to be a limiting case of "wave" optics) and dynamics to propose that if light could have both wave and particle properties, then matter should also share in this duality. He proposed that particles such as electrons should have wave properties associated with them, and through a study of wave packets, which will be discussed later in this chapter, he was led to propose the relationship between momentum and the associated wavelength that bears his name

$$\lambda = \frac{h}{p}. \tag{16.1}$$

Note that this relation also holds for radiation

$$\lambda = \frac{c}{\nu} = \frac{h}{(h\nu/c)} = \frac{h}{(E/c)} = \frac{h}{p}.$$

De Broglie's hypothesis provides a simple interpretation of the Bohr quantization condition

$$pr = n\hbar \tag{16.2}$$

by requiring that a stationary state be one in which an integral number of waves fit into an orbit. Such a pattern of standing waves might be expected to be nonradiating. The circumference of a circular orbit is $2\pi r$, and into it should fit n waves. This leads to the condition

$$2\pi r = n\lambda \tag{16.3}$$

which, with $\lambda = h/p$, allows us to write

$$pr = \frac{h}{\lambda}r = \frac{hn}{2\pi} = \hbar n. \tag{16.4}$$

Electron diffraction The thesis of de Broglie attracted much attention, and suggestions were made for its verification through the observation of electron diffraction. The experiments were carried out by C.J. Davisson and L.H. Germer in the United States and by G.P. Thomson in Great Britain in 1927. They found that in the scattering of electrons by a crystal surface (Fig. 16.1) there was preferential scattering in certain directions. The derivation of the interference conditions for waves, called the *Bragg*

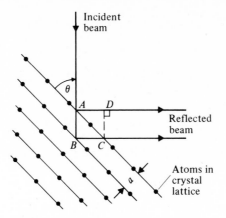

Figure 16.1. Scattering of ray at "scattering planes" in crystal lattice.

conditions, involves the phase difference between waves reflected from adjacent scattering "planes." That phase difference is given by

$$(2\pi/\lambda) \times (\text{difference in path length})$$

$$= (2\pi/\lambda) \times (ABC - AD). \tag{16.5}$$

The difference in paths, $ABC - AD$ can be worked out using the details of Fig. 16.2. We have

$$ABC - AD = AB + BC - (AX + XD)$$

and since

$$AX = BC,$$

the path difference is

$$AB - XD = \frac{a}{\sin \theta} - \frac{a}{\sin \theta} \cos 2\theta$$

$$= \frac{a}{\sin \theta} (1 - \cos 2\theta)$$

$$= 2a \sin \theta. \tag{16.6}$$

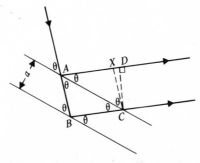

Figure 16.2. Details of optical path difference between rays reflected at first and at second scattering plane in crystal lattice.

Thus the phase difference is, using Eq. (16.5),

$$\frac{2\pi}{\lambda} 2a \sin \theta \; = \; \frac{4\pi a}{\lambda} \sin \theta. \tag{16.7}$$

There will be constructive interference whenever this phase difference is equal to $2\pi n$ where n is an integer, because two waves of the form

$$\cos\left(\frac{2\pi}{\lambda} x - 2\pi v t\right), \; \cos\left(\frac{2\pi}{\lambda}(x + d) - 2\pi v t\right)$$

are identical when $2\pi d/\lambda = 2\pi n$. Thus the *Bragg condition* is

$$\frac{4\pi a}{\lambda} \sin \theta \; = \; 2\pi n$$

that is,

$$\lambda \; = \; \frac{2a}{n} \sin \theta. \tag{16.8}$$

In the Davisson–Germer experiment, the spacing between the Bragg planes was measured to be 0.91 Å, using X-ray diffraction techniques. The diffraction maximum was observed at $65°$, and the incident electrons had kinetic energy of 54 eV. The application of Eq. (16.8) yields

$$\lambda \; = \; 2 \times 0.91 \times 10^{-8} \times 0.91 \; = \; 1.65 \times 10^{-8} \, \text{cm}$$

and the de Broglie relation implies that

$$p \; = \; \frac{h}{\lambda} \; = \; \frac{6.63 \times 10^{-27} \, \text{erg}}{1.65 \times 10^{-8} \, \text{cm}} \; = \; 4.02 \times 10^{-19} \, \text{gm cm/sec.}$$

This corresponds to an electron kinetic energy of

$$\frac{p^2}{2m_e} \; = \; \frac{(4.02 \times 10^{-19})^2}{2 \times 0.9 \times 10^{-27}} \; = \; 8.97 \times 10^{-11} \, \text{erg}$$

$$= \; 56.1 \, \text{eV}$$

in good agreement with what was measured.

The smallness of h guarantees that these wave aspects of particles do not manifest themselves for macroscopic objects. For a dust particle of mass 10^{-4} gm moving with a velocity of 10 cm/sec

$$\lambda \; = \; \frac{6.63 \times 10^{-27}}{10^{-3}} \; = \; 6.63 \times 10^{-24} \, \text{cm}$$

and this is so small that no wave properties become apparent. Nevertheless, the de Broglie discovery sharpened the need for a new theory that had to deal with the following difficulty.

When Thomson discovered the electron, he established the fact that it was a particle, that is, an object whose path could be described by a coordinate $\mathbf{r}(t)$. The fact that the equation of motion happens to take the simple form

$$m \frac{d^2 \mathbf{r}(t)}{dt^2} \; = \; -e \left(\mathbf{E} + \frac{1}{c} \frac{d\mathbf{r}(t)}{dt} \times \mathbf{B}\right)$$

is not of immediate significance. What is important is that an object that follows a well-defined path $\mathbf{r}(t)$ cannot give rise to interference phenomena. Instead of looking

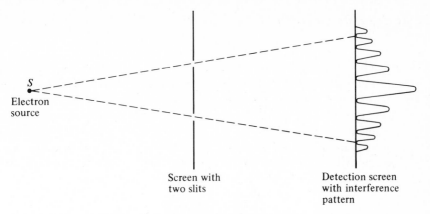

Figure 16.3. Thought experiment on electron diffraction by double slit.

at electron scattering, consider the transmission of electrons through a screen with a double slit. The same properties that give rise to diffraction will give rise to an interference pattern on the second screen in the "experimental" set up shown in Fig. 16.3. If the electrons come from the source at widely separated time intervals, they arrive at the slitted screen one by one. The path of a particular electron $\mathbf{r}_i(t)$ will go either through slit 1 or through slit 2. If that is so, then the unaffected slit could have been closed. With a sufficiently symmetric set up, the two-slit experiment could be divided into two one-slit experiments: slit 1 open half the time, slit 2 open the other half of the time. Under these circumstances, however, there is no way in which interference can take place, even if the electrons have wave-like attributes. Thus the description of an electron by a position coordinate $\mathbf{r}(t)$ *must be abandoned.* This must be reconciled with the evidence found by Thomson that electrons follow classical paths described by Newton's laws, and that they obey relations such as

$$E = \mathbf{p}^2/2m \tag{16.9}$$

and

$$\mathbf{p} = m\frac{d\mathbf{r}}{dt}. \tag{16.10}$$

Quantum mechanics explains this. The simplest description is through the fact that the properties of an electron are contained in a *wave function* $\psi(\mathbf{r}, t)$ that obeys the *Schrödinger equation*, of the form

$$i\hbar\frac{\partial \psi(\mathbf{r}, t)}{\partial t} = -\frac{\hbar^2}{2m}\left(\frac{\partial^2 \psi}{\partial x^2} + \frac{\partial^2 \psi}{\partial y^2} + \frac{\partial^2 \psi}{\partial Z^2}\right) + V(\mathbf{r})\psi(\mathbf{r}, t), \tag{16.11}$$

where m is the electron mass, $\hbar$ is Planck's constant divided by 2π, and $i = \sqrt{-1}$. This equation cannot be derived from classical physics. It was, in effect, guessed by Schrödinger who used brilliant intuition based on the work of de Broglie. In what follows we shall go through some arguments which might just make it plausible that the above equation somehow "follows" from what we have already learned. If the reader is not happy with the argument, he or she need not worry. What matters is that the equation makes predictions that can be compared with experiment, and it works, as we shall learn later.

Wave packets It is possible to add waves together in a way that instead of an ever-repeating pattern of crests and troughs they form a localized peak, and very little else. Consider, for example, the vibration of the ear drum in response to a clap of thunder. There is one large peak in the amplitude, and very little else. Nevertheless, one can analyze the sound pattern to show that it consists of a superposition of many waves, with a range of wavelengths and amplitudes that depend on the wavelength in such a way that outside of the time interval of the large noise, there is destructive interference.

To show this explicitly, consider a one-dimensional standing wave

$$A \cos \frac{2\pi}{\lambda} x = A \cos kx. \tag{16.12}$$

If we add together waves with a range of values of the wave number $k = 2\pi/\lambda$, with different amplitudes $A(k)$ that depend on the wave number, we obtain a wave packet

$$\psi(x) = \int_{-\infty}^{\infty} dk A(k) \cos kx. \tag{16.13}$$

For purposes of illustration we choose $A(k) = e^{-\alpha k^2}$. This choice is made because the integral (16.13) can be found in a table of definite integrals:

$$\psi(x) = \int_{-\infty}^{\infty} dk e^{-\alpha k^2} \cos kx = \left(\frac{\pi}{\alpha}\right)^{1/2} e^{-x^2/4\alpha}. \tag{16.14}$$

Clearly the wave packet is localized around $x = 0$. Figure 16.4 shows a plot of $A(k)$ and $\psi(x)$. It is easily seen that if the amplitude $A(k)$ is large for a wide range of k values, that is, when α is small, then the wave packet is sharply localized; conversely, if the amplitude $A(k)$ is sharply localized, then the wave packet is broad. The width of the amplitude $A(k)$, defined as the range in k within which the function drops to half its value is of the order of magnitude $(2 \times 0.69/\alpha)^{1/2}$, and the width of $\psi(x)$,

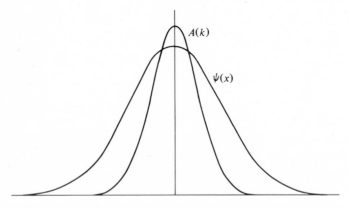

Figure 16.4. Plot of $A(k)$ and $\psi(x)$ for a Gaussian shape as in formula (16.14) evaluated with the parameter $\alpha = 1$.

similarly defined, is $(2 \times 0.69 \times 4\alpha)^{1/2}$. Thus the product of the widths, written in the form

$$\Delta k \, \Delta x \simeq 2.8 \qquad (16.15)$$

is independent of α. We write this in the form

$$\Delta k \, \Delta x \gtrsim 1. \qquad (16.16)$$

This reciprocal relation will play an important role later, when we discuss the uncertainty relations. It is not restricted to the gaussian shape that we have used above. To illustrate this, consider the amplitude to be of the form

$$A(k) = A_0 \frac{\sin ka}{k}. \qquad (16.17)$$

Now

$$\psi(x) = \int_{-\infty}^{\infty} \frac{dk}{k} \sin ka \cos kx$$

$$= \frac{1}{2} \int_{-\infty}^{\infty} \frac{dk}{k} \left[\sin k(x+a) - \sin k(x-a)\right].$$

Using the fact that

$$\frac{1}{2} \int_{-\infty}^{\infty} \frac{dk}{k} \sin k(x+a) = \quad \frac{\pi}{2} \quad x+a > 0$$

$$-\frac{\pi}{2} \quad x+a < 0$$

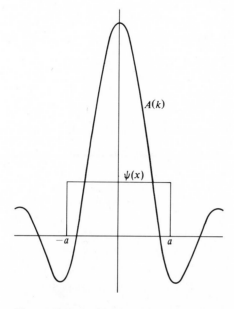

Figure 16.5. $A(k)$ and $\psi(x)$ for the example of Eq. (16.17) and Eq. (16.18).

and similarly for the second term, we see that

$$\psi(x) = \begin{cases} 0 & x > a \\ \pi A_0/2 & a > x > -a \\ 0 & -a > x. \end{cases}$$ (16.18)

The shapes of the amplitude $A(k)$ and of the resulting wave packet are shown in Fig. 16.5. The wave packet is sharply localized in space. It is an idealization of a wave train emitted by a source through a shutter that is instantaneously opened, and just as sharply closed again. The width of the wave train is $\Delta x = 2a$. The spread of the amplitude function (16.17) is roughly measured by the width of the central peak. This vanishes when $ka = \pm \pi$, and a rough measure of the width of the central peak is $\Delta k \cong \pi/a$. Thus

$$\Delta x \Delta k \simeq 2\pi \gtrsim 1$$ (16.19)

and we again observe the reciprocal relation, independent of a.

The motion of wave packets If such a wave packet were to describe a freely moving electron, how would we modify $\psi(x)$ to yield a time-dependent wave packet $\psi(x, t)$? Let us first consider this problem in the case of light. A single wave is described by

$$A(k) \cos \left(\frac{2\pi}{\lambda} x - 2\pi \nu t \right) = A(k) \cos (kx - \omega t)$$ (16.20)

and in the wave packet each of the waves that are superposed moves independently. Thus

$$\psi(x, t) = \int_{-\infty}^{\infty} dk A(k) \cos (kx - \omega t).$$ (16.21)

For light we have

$$\omega = 2\pi \nu = 2\pi \frac{c}{\lambda} = kc$$ (16.22)

so that

$$\psi(x, t) = \int_{-\infty}^{\infty} dk A(k) \cos k(x - ct) = \psi(x - ct).$$ (16.23)

Thus the superposition of light waves propagates with the velocity c, since $\psi(x - ct)$ is just the function we had before, now centered at $x - ct = 0$. If the light moves through a medium, then

$$\omega = \frac{kc}{n}$$

where n is the refractive index. If $n = n(k)$, we have a dispersive medium, and the propagation is much more complicated, in that the signal ψ does not keep its shape. For the propagation of electrons we have a situation analogous to dispersion. We proceed as follows:

1. We use the de Broglie relation between the wavelength and the momentum to write

$$k = \frac{2\pi}{\lambda} = 2\pi \frac{p}{h} = \frac{p}{\hbar}.$$ (16.24)

2. We use the relation between frequency and energy suggested by Planck:

$$\omega = \frac{E}{\hbar} = \frac{p^2}{2m\hbar} \tag{16.25}$$

and use these relations to provide the connection between ω and k. This gives for the propagating wave packet

$$\psi(x,t) = \int_{-\infty}^{\infty} dp\,\phi(p) \cos\frac{px - Et}{\hbar}. \tag{16.26}$$

We use the variable p instead of k inside the integral. This does not make any difference as far as the integration is concerned, since p is a dummy variable, but the new notation does suggest that there is a momentum associated with each component of the wave packet. The function $\phi(p)$ is a modulating function, just as $A(k)$ was.

3. Our next task is to prove that this wave packet really does propagate like a free particle in some approximation. To do so, let us consider $\phi(p)$ to be sharply peaked about some $p = p_0$. Then, to a good approximation we may write the integral for $\psi(x,t)$ in the form

$$\psi(x,t) \approx \phi(p_0) \int_{p_0-\Delta}^{p_0+\Delta} dp \cos\frac{px - Et}{\hbar} \tag{16.27}$$

(Fig. 16.6 shows the meaning of Δ). We now change the variable of integration to $q = p - p_0$ and write $p_0^2/2m = E_0$. Now if $\Delta \ll p_0$, then the largest value of q is always much less than p_0 and we may therefore write

$$E = \frac{(p_0 + q)^2}{2m} \cong E_0 + \frac{p_0 q}{m}. \tag{16.28}$$

Thus the wave packet takes the form

$$\psi(x,t) \simeq \phi(p_0) \int_{-\Delta}^{\Delta} dq \cos\left(\frac{p_0 x - E_0 t}{\hbar} + q\,\frac{x - p_0 t/m}{\hbar}\right). \tag{16.29}$$

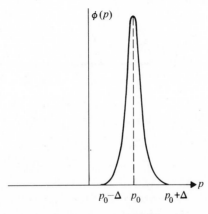

Figure 16.6. $\phi(p)$ sharply peaked about $p = p_0$, with width 2Δ.

With the notation $p_0/m = v_0$, we have

$$\cos\left(\frac{p_0 x - E_0 t}{\hbar} + q\frac{x - v_0 t}{\hbar}\right)$$

$$= \cos\frac{p_0 x - E_0 t}{\hbar}\cos q\frac{x - v_0 t}{\hbar} - \sin\frac{p_0 x - E_0 t}{\hbar}\sin q\frac{x - v_0 t}{\hbar}$$

$$= \cos\frac{p_0}{\hbar}\left(x - \frac{p_0 t}{2m}\right)\cos\frac{q}{\hbar}(x - v_0 t) - \sin\frac{p_0}{\hbar}\left(x - \frac{p_0 t}{2m}\right)\sin\frac{q}{\hbar}(x - v_0 t).$$

When this is inserted into the integral, the second product vanishes, since it is an odd function of q and the integral is over an interval that is symmetric about $q = 0$. Thus we are left with

$$\psi(x, t) \simeq \phi(p_0)\cos\frac{p_0}{\hbar}(x - p_0 t/2m)$$

$$\times \int_{-\Delta}^{\Delta} dq\cos\frac{q}{\hbar}(x - v_0 t)$$

$$= \phi(p_0)\cos\frac{p_0 x - E_0 t}{\hbar}\left(\frac{2\hbar}{x - v_0 t}\sin\frac{\Delta}{\hbar}(x - v_0 t)\right). \qquad (16.30)$$

Thus the wave packet has the form of a function evaluated at the sharp peak value of the momentum p_0, multiplied by an oscillating function with its peak value at

$$x = v_0 t. \qquad (16.31)$$

This modulating function forces the packet to move with velocity v_0. In general one can show that the velocity of propagation for a general $E(p)$ is given by

$$v_g = \frac{\partial E(p)}{\partial p}\bigg|_{p=p_0}. \qquad (16.32)$$

This is the *group velocity*. For example, with a relativistic connection between momentum and energy,

$$v_g = \frac{\partial}{\partial p}[p^2 c^2 + m^2 c^4]^{1/2}$$

$$= \frac{pc^2}{E} = \frac{\gamma m v c^2}{\gamma m c^2} = v. \qquad (16.33)$$

In summary, the wave packet of the form (16.26) does describe something that is localized and that moves with the group velocity $v_g = v_0$.

What we will do next is to study this wave packet and see what kind of equation could have led us to its form. Before doing that, however, two difficulties must be disposed of. One is the question of localization of the wave packet that we have been dealing with. We stated that the width of $\phi(p)$ is 2Δ, which is supposed to be very small. Does this mean that the spatial size of the wave packet is unreasonably large? The relation $\Delta k\Delta x \sim 1$, which with the help of the de Broglie relations takes the form

$$\Delta p\Delta x = \hbar\Delta k\Delta x \simeq \hbar, \qquad (16.34)$$

shows us that

$$\Delta x \sim \frac{\hbar}{2\Delta} \tag{16.35}$$

since $\Delta p \approx 2\Delta$. A typical value of p_0 that appears in atomic physics is

$$p_0 \simeq 0.01 \, m_e c.$$

If we take Δ to be some small fraction of that, say

$$\Delta \sim 10^{-2} p_0,$$

then

$$\Delta x \sim \frac{\hbar}{2 \times 10^{-4} m_e c} \sim \frac{1.05 \times 10^{-27}}{2 \times 10^{-4} \times 0.9 \times 10^{-27}} \frac{1}{3 \times 10^{10}}$$

$$\sim 19 \times 10^{-8} \, cm = 19 \, \text{Å}$$

which is still tiny on any macroscopic scale. Thus there is no problem of a contradiction with the evidence of the senses.

The second difficulty deals with the fact that in our expression for

$$E = p^2/2m = (p_0^2 + 2p_0 q + q^2)/2m$$

we neglected the last term. Since E is multiplied by t, for long times $tq^2/2m\hbar$ is not negligible. The effect of this is that the wave packet, whose initial width was of the order of $\hbar/2\Delta$, spreads to a larger width with time. Numerically this spread is not very rapid for short times, but the presence of this spread indicates that $\psi(x, t)$ cannot represent a distribution of matter in the most naive sense: we know that electrons have been around in metals since the early periods of the history of our universe, and these electrons are no different from the ones that are created in high-energy collisions or beta decays right now. We shall see that $\psi(x, t)$ will be interpreted in a way that removes this difficulty.

Wave equation for packet We are now ready to look for an equation that Eq. (16.26) is a solution of. To do so, we shall work out a number of derivatives of $\psi(x, t)$ with respect to x and t, and do so by interchanging differentiation and integration, which is permissible as long as $\phi(p)$ goes to zero sufficiently rapidly for large p. Thus

$$\hbar \frac{\partial \psi}{\partial x} = -\int dp\, p\, \phi(p) \sin(px - Et)/\hbar$$

$$\hbar^2 \frac{\partial^2 \psi}{\partial x^2} = -\int dp\, p^2 \phi(p) \cos(px - Et)/\hbar$$

$$\hbar^3 \frac{\partial^3 \psi}{\partial x^3} = \int dp\, p^3 \phi(p) \sin(px - Et)/\hbar$$

$$\hbar^4 \frac{\partial^4 \psi}{\partial x^4} = \int dp\, p^4 \phi(p) \cos(px - Et)/\hbar$$

$$\vdots$$

$$-\hbar \frac{\partial \psi}{\partial t} = -\int dp\, E\, \phi(p) \sin(px - Et)/\hbar$$

$$\hbar^2 \frac{\partial^2 \psi}{\partial t^2} = -\int dp\, E^2 \phi(p) \cos(px - Et)/\hbar \tag{16.36}$$

$$\vdots$$

If we use $E = p^2/2m$ and try to match up differentials with respect to x and differentials with respect to t, we see that the simplest equation that we can write down is

$$\frac{\hbar^4}{4m^2} \frac{\partial^4 \psi}{\partial x^4} = -\hbar^2 \frac{\partial^2 \psi}{\partial t^2} \tag{16.37}$$

which is a pretty complicated equation, and corresponds to

$$E^2 = \left(\frac{p^2}{2m}\right)^2. \tag{16.38}$$

This is not really what we want, since the equation describes not only the solution to the equation $E = p^2/2m$ but also that corresponding to $E = -p^2/2m$. Thus the equation contains redundant, nonphysical solutions, and that accounts for its complexity. Equation (16.37) is, in fact, not the correct equation. If we do want to deal with $E = p^2/2m$ alone, we want only a first-time derivative in the differential equation, and we want a second-time derivative with respect to x. This will not work with a cosine under the integral as in Eq. (16.26), nor will it work with a sine under the integral. If, however, we take a special combination of the two, that is, if we consider

$$\psi(x, t) = \int_{-\infty}^{\infty} dp\, \phi(p) e^{i(px - Et)/\hbar}, \tag{16.39}$$

then we can write a simple partial differential equation that (16.39) is a solution of, namely,

$$i\hbar \frac{\partial \psi(x, t)}{\partial t} = -\frac{\hbar^2}{2m} \frac{\partial^2 \psi(x, t)}{\partial x^2}. \tag{16.40}$$

This equation is the *Schrödinger Equation* for a free particle. Its solution, Eq. (16.39), has a real and imaginary part, and each part separately propagates with the group velocity, that is, like a free particle. This is satisfactory, but a new complication has appeared: the wave function $\psi(x, t)$ is necessarily complex and we must learn how to deal with that. The significance of a complex wave function, and its physical interpretation are the subject of the next chapter.

NOTES AND COMMENTS

1. The background to de Broglie's discovery is described in the interesting book: M. Jammer, *The Conceptual Development of Quantum Mechanics*, McGraw-Hill, New York, 1966.

2. Wave packets and associated concepts, such as the group velocity, are nicely discussed in much more detail in F.S. Crawford, Jr., *Waves* (Berkeley Physics course, vol. 3), McGraw-Hill, New York, 1966. The definition of the group velocity is obtained from noting that in a wave packet of the form

$$\int_{-\infty}^{\infty} dk\, A(k) \cos(kx - \omega(k)t)$$

the peak will be located at the point where the phase has an extremum (since there the waves will add coherently, and not tend to interfere destructively), and that occurs when

$$\frac{d}{dk}(kx - \omega(k)t) = x - \left(\frac{d\omega}{dk}\right)t$$

$$= 0.$$

3. The spread of the wave packet is qualitatively due to the fact that when $\omega(k)$ is not linear in k, the *phase velocity* of each wave, $v_p = \omega(k)/k$ is different for each wave, and the component waves will get out of phase with each other. This can be explicitly seen for a gaussian packet. See, for example, S. Gasiorowicz, *Quantum Physics*, John Wiley, New York, 1974.

PROBLEMS

16.1 What is the de Broglie wavelength of

a) an electron with kinetic energy of 1 eV?

b) an electron with kinetic energy of 1 keV?

c) an electron with kinetic energy of 10 MeV?

d) a neutron with energy kT, where $T = 300°K$?

e) a neutron with kinetic energy 10 MeV?

f) a molecule with molecular weight 1000 moving with kinetic energy kT with $T = 3000°K$?

16.2 For what energy will a particle's de Broglie wavelength equal its Compton wavelength?

16.3 Consider a crystal with planar spacing 2.5 Å. What energies would one need for (a) electrons, (b) neutrons, to observe up to three interference maxima?

16.4 The spacing between scattering planes in nickel is 2.15 Å. What is the maximum scattering angle of 80 eV electrons from a nickel crystal?

16.5 Consider a wavepacket of the form

$$\psi(x) = \int_{-\infty}^{\infty} dk A(k) \cos kx$$

with

$$A(k) = \frac{1}{k^2 + \alpha^2}.$$

Look up the integral and show that $\Delta k \Delta x$ is independent of α.

16.6 Consider the wavepacket with

$$A(k) = \begin{cases} \cos k\pi/2K & -K \leqslant k \leqslant K \\ 0 & \text{elsewhere} \end{cases}$$

and again show that $\Delta k \Delta x \sim 0(1)$.

16.7 The relation between the wavelength and the frequency of radiation propagating in a waveguide is given by

$$\lambda = \frac{c}{(\nu^2 - \nu_0^2)^{1/2}}$$

where ν_0 is a constant. What is the group velocity of such a wave? Sketch it as a function of λ, and as a function of ν.

16.8 The surface tension waves in shallow water obey the relation between frequency and wavelength

$$\nu = (2\pi T/\rho\lambda^3)^{1/2}$$

where T is the surface tension and ρ the density. What is the group velocity? What is the relation of the group velocity to the *phase velocity* defined by

$$\nu_p = \nu\lambda.$$

16.9 For gravity waves, that is, waves in deep water, the relation between frequency and wavelength is given by

$$\nu = (g/2\pi\lambda)^{1/2}.$$

What are the group and phase velocities?

16.10 The surface tension in water is given by $T = 72$ dynes/cm. Given that $g = 980$ cm/sec^2 and that

$$\nu = \left[\frac{2\pi T}{\rho\lambda^3} + \frac{g}{2\pi\lambda}\right]^{1/2},$$

calculate the group velocity numerically for a variety of wavelengths, and plot a curve. (The above formula is correct only when the depth of the water d is much larger than the wavelength under consideration. In the gravity wave limit, what is the group velocity for $\lambda = 20$ m?)

17
The Probability Interpretation

Given that the wave function $\psi(x, t)$ is a *complex* function of x and t, we must look more closely at what we mean when we say that the function is large in some region and small in another. A complex function can, in general, be written in the form

$$\psi(x, t) = R(x, t)e^{iS(x, t)} \tag{17.1}$$

where $R(x, t)$ is called the *modulus* and $S(x, t)$ is called the *phase* of the function. This is equivalent to

$$\text{Re } \psi = R \cos S,$$
$$\text{Im } \psi = R \sin S. \tag{17.2}$$

The magnitude of the vector represented by ψ (see Fig. 17.1) is determined by the size of the modulus, which is directly related to the absolute square of the wave function

$$|\psi(x, t)|^2 = \psi^*\psi = (Re^{-iS})(Re^{iS}) = R^2(x, t). \tag{17.3}$$

In analogy with the formula for the intensity of the electromagnetic radiation described by the field $\mathbf{E}(x, t)$,

$$I(x, t) \propto (\mathbf{E}(x, t))^2, \tag{17.4}$$

we would like to associate a large $|\psi(x, t)|^2$ with the presence of an electron in the region where that quantity is large. We shall see that $|\psi(x, t)|^2$ plays a central role in quantum mechanics.

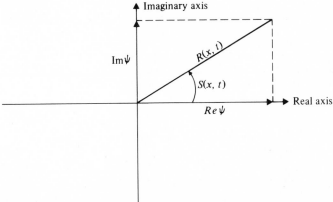

Figure 17.1. Geometrical representation of complex function as a vector in the two-dimensional space of the real and imaginary parts of functions.

Superposition Principle To see how interference phenomena can arise, we observe that the Schrödinger equation is *linear*. This implies that

a) if $\psi(x, t)$ is a solution, so is

$$\psi'(x, t) = a\psi(x, t)$$

where a is an arbitrary, possibly complex, constant.

b) If $\psi_1(x, t)$ and $\psi_2(x, t)$ are solutions, so is

$$\psi(x, t) = a_1\psi_1(x, t) + a_2\psi_2(x, t)$$

for arbitrary complex constants a_1 and a_2. Because of the linearity we thus get the *principle of superposition*, and we can understand interference phenomena. Let us discuss the two-slit experiment in a very schematic way. Suppose the solution of the Schrödinger equation with slit 1 open is $\psi_1(x, t)$; suppose the solution with slit 2 open is $\psi_2(x, t)$. The intensity pattern of electrons at the screen with only slit 1 open will be related to $|\psi_1(x, t)|^2_{\text{screen}}$; the intensity with only slit 2 open will similarly be related to $|\psi_2(x, t)|^2_{\text{screen}}$. If both slits are open, the solution is given by the superposition

$$\psi(x, t) = \psi_1(x, t) + \psi_2(x, t) \tag{17.5}$$

and the intensity will be related to $|\psi_1(x, t) + \psi_2(x, t)|^2_{\text{screen}}$. Let

$$\psi_1 = R_1 e^{iS_1},$$
$$\psi_2 = R_2 e^{iS_2}. \tag{17.6}$$

Then

$$|\psi_1(x, t)|^2 = R_1(x, t)^2 \tag{17.7}$$

and

$$|\psi_2(x, t)|^2 = R_2(x, t)^2. \tag{17.8}$$

When both slits are open, we have

$$|\psi_1(x, t) + \psi_2(x, t)|^2 = |R_1 e^{iS_1} + R_2 e^{iS_2}|^2.$$

This may be evaluated more explicitly:

$$
\begin{aligned}
(R_1 e^{-iS_1} &+ R_2 e^{-iS_2})(R_1 e^{iS_1} + R_2 e^{iS_2}) \\
&= R_1^2 + R_2^2 + R_1 R_2(e^{i(S_1 - S_2)} + e^{-i(S_1 - S_2)}) \\
&= R_1^2 + R_2^2 + 2R_1 R_2 \cos(S_1 - S_2).
\end{aligned} \tag{17.9}
$$

The first two terms are just the contributions from the individual slits, while the last term is the interference term. This term gives rise to variations in intensity due to the changing difference in the phases as we move along the screen, even when R_1 and R_2 remain fairly constant. The linearity of the equation is thus responsible for what we called the "wave properties" of the particles.

We still have the difficulty discussed in the last chapter. The interference pattern with two slits open can be understood, but if the electrons enter the system at a very slow rate, the pattern presumably builds up at a slow rate. In some way, each electron carries the interference pattern with it, that is, each individual electron, rather than a whole collection of electrons, satisfies the Schrödinger equation. If any one electron is described by the wave function $\psi_1(x, t) + \psi_2(x, t)$, what is the meaning of that wave function?

Probability density The answer, proposed by Max Born, and now accepted as the correct one, is that if a system is described by the wave function $\psi(x, t)$, then $|\psi(x, t)|^2$ represents the *probability density* for finding the system at the position x at the time t. More precisely, the probability of finding the system in the interval $(x, x + dx)$ at the time t is given by

$$P(x, t)dx = |\psi(x, t)|^2 dx. \tag{17.10}$$

The requirement that the system, the electron, say, be located somewhere, implies that

$$\int_{-\infty}^{\infty} dx P(x, t) = 1, \tag{17.11}$$

that is,

$$\int_{-\infty}^{\infty} dx |\psi(x, t)|^2 = 1. \tag{17.12}$$

Since the Schrödinger equation is linear, it is always possible to normalize the wave function to satisfy Eq. (17.12). Suppose it turns out that

$$\int_{-\infty}^{\infty} dx |\psi(x, t)|^2 = N.$$

Then the new wave function

$$\psi'(x, t) = \frac{1}{\sqrt{N}} \psi(x, t)$$

will be properly normalized. The condition (17.12) does imply that

$$\int_{-\infty}^{\infty} dx |\psi(x, t)|^2 < \infty. \tag{17.13}$$

If this were not so, if the integral were divergent, then we could not normalize the wave function. The restriction (17.13), called the *square integrability* condition, will turn out to be very important.

The probabilistic interpretation of the wave function removes the concern with the spreading of the wave packet with large times. It is not the electrons that spread, but rather, if an electron is initially localized in some region, then after a long time the probability of finding it outside that region increases with time.

The normalization of the wave function is possible only if the number N defined above is truly a constant. If it varied with t, then $\frac{1}{\sqrt{N}} \psi(x, t)$ would no longer be a solution of the Schrödinger equation. To show that N does not depend on time, let us write the Schrödinger equation in the form

$$\frac{\partial \psi}{\partial t} = -\frac{\hbar}{2im} \frac{\partial^2 \psi}{\partial x^2}. \tag{17.14}$$

The complex conjugate of the above equation reads

$$\frac{\partial \psi^*}{\partial t} = \frac{\hbar}{2im} \frac{\partial^2 \psi^*}{\partial x^2}. \tag{17.15}$$

Now

$$\frac{\partial}{\partial t} P(x, t) = \frac{\partial}{\partial t}(\psi^* \psi)$$

$$= \frac{\partial \psi^*}{\partial t} \psi + \psi^* \frac{\partial \psi}{\partial t}$$

$$= \frac{\hbar}{2im} \frac{\partial^2 \psi^*}{\partial x^2} \psi - \frac{\hbar}{2im} \psi^* \frac{\partial^2 \psi}{\partial x^2}$$

$$= -\frac{\partial}{\partial x} \left[\frac{\hbar}{2im} \left(\psi^* \frac{\partial \psi}{\partial x} - \psi \frac{\partial \psi^*}{\partial x} \right) \right]. \tag{17.16}$$

We may rewrite (17.16) in the form of a conservation law[†]

$$\frac{\partial}{\partial t} P(x,t) + \frac{\partial}{\partial x} j(x,t) = 0 \tag{17.17}$$

with the *probability current density*, or *probability flux*, given by the real quantity

$$j(x,t) = \frac{\hbar}{2im} \left(\psi^* \frac{\partial \psi}{\partial x} - \frac{\partial \psi^*}{\partial x} \psi \right). \tag{17.18}$$

The relation is called a conservation law, since the change in probability of finding an electron in the interval (a,b) is

$$\frac{d}{dt} \int_a^b P(x,t)dx = - \int_a^b dx \frac{\partial}{\partial x} j(x,t)$$
$$= j(a,t) - j(b,t), \tag{17.19}$$

that is, it is due to an in-flow at a and an out-flow at b. If we consider

$$\frac{d}{dt} N = \frac{d}{dt} \int_{-\infty}^{\infty} dx \, |\psi(x,t)|^2 = j(-\infty,t) - j(\infty,t), \tag{17.20}$$

we see that the right-hand side vanishes, because of the square-integrability condition. For the integral

$$\int_{-\infty}^{\infty} dx \, |\psi(x,t)|^2$$

to converge, $|\psi(x,t)|$ must go to zero faster than $|1/x^{1/2}|$ as $|x| \to \infty$. This means, however, that

$$j(x,t) \sim x^{-1/2} \frac{d}{dx} x^{-1/2} \sim x^{-2} \tag{17.21}$$

so that the flux vanishes at infinity.

As a final comment, we state without proof, that if the wave function $\psi(x,t)$ is properly normalized, and

$$\psi(x,t) = \frac{1}{\sqrt{2\pi\hbar}} \int_{-\infty}^{\infty} dp\phi(p,t)e^{ipx/\hbar}, \tag{17.22}$$

then $|\phi(p,t)|^2$ is the probability of finding an electron with momentum p at the time t,[‡] and, consistent with this,

$$\int_{-\infty}^{\infty} dp \, |\phi(p,t)|^2 = 1. \tag{17.23}$$

We will make use of the fact, established in the last chapter, that the width of the x-distribution and the width of the p-distribution are related by

$$\Delta x \, \Delta p \sim \hbar \tag{17.24}$$

in the next chapter.

[†] Eq. (17.17) is identical to the conservation law of electric charge where $P(x,t)$ is replaced by the charge density and $j(x,t)$ by the current density.

[‡] For a free particle, which is what we have been considering so far, $\phi(p,t) = \phi(p)e^{-iEt/\hbar}$. Both the interpretation and Eq. (17.23) are more general.

NOTES AND COMMENTS

Although the mathematical content of this chapter is rather straightforward, the philosophical implications are staggering. Until the advent of quantum mechanics, there was general agreement of what a theory was: certain initial conditions would be expected to have well-defined consequences, and this notion accorded well with how both mechanics and electrodynamics were formulated, namely, in terms of differential equations. In quantum mechanics, the differential equation, the Schrödinger equation, is not so different from the equations of classical physics, but the interpretation is fundamentally different; here the consequence of certain initial conditions is a probability statement. Both Einstein and Schrödinger were very unhappy with this aspect of quantum mechanics, and viewed the theory as an interim theory. Nowadays the philosophical discomfort with quantum mechanics is not expressed by physicists very much. There are still debates about how one should deal with the problem of an interpretation that is not contained in, but is superimposed on the mathematical apparatus of the theory. This is not the place to deal with these problems, but the reader should at least be aware that they are still being debated.

PROBLEMS

17.1 Insert the form $\psi(x, t) = R(x, t)e^{iS(x, t)}$ into the free particle Schrödinger equation and obtain the equations that describe the time development of $R(x, t)$ and $S(x, t)$. Show that one of the equations (there are two, obtained by equating the real and imaginary parts of the equations to zero) is just the flux conservation equation (17.17).

17.2 Consider the Schrödinger equation for a particle in the presence of a potential,

$$i\hbar \frac{\partial \psi(x, t)}{\partial t} = -\frac{\hbar^2}{2m} \frac{\partial^2 \psi(x, t)}{\partial x^2} + V(x)\psi(x, t).$$

Show that the conservation law (17.17) still holds, with the same definitions of $P(x, t)$ and $j(x, t)$ as were used for the free particle equations, provided $V(x)$ is real.

17.3 Suppose the potential in Problem 17.2 had the form

$$V(x) = U(x) + iW(x)$$

where U and W are real functions of x. What form does the conservation equation take? What is the sign of $W(x)$ if the potential is to act as a sink of probability flux, so that

$$\frac{d}{dt} \int_{-\infty}^{\infty} P(x, t)dx < 0.$$

17.4 It is sometimes useful to work with wave functions that do not go to zero at infinity, but that instead have the property that

$$\psi(x + L, t) = \psi(x, t).$$

Show that if one restricts oneself to a single "cell," that is, an interval (O, L), then

$$\frac{d}{dt} \int_0^L P(x, t) dx = 0$$

still holds.

17.5 At one time the so-called Klein–Gordon equation was studied as an alternative to the Schrödinger equation. This equation has the form

$$\frac{1}{c^2} \frac{\partial^2 \psi(x, t)}{\partial t^2} - \frac{\partial^2 \psi(x, t)}{\partial x^2} + \left(\frac{mc}{\hbar}\right)^2 \psi(x, t) = 0$$

and it, too, has a conservation law. Given that

$$j(x, t) = i \left(\psi^* \frac{\partial \psi}{\partial x} - \frac{\partial \psi^*}{\partial x} \psi\right),$$

find $P(x, t)$. Note that in contrast to $|\psi(x, t)|^2$ it is not necessarily a positive function, so that it may be difficult to give $P(x, t)$ a probability interpretation.

17.6 Consider a "momentum space" wave function

$$\phi(p) = N e^{-\alpha^2 p^2}.$$

Use the fact that

$$\int_{-\infty}^{\infty} dp e^{-\alpha^2 p^2} = \frac{\sqrt{\pi}}{\alpha}$$

to calculate N so that the normalization condition (17.23) is satisfied. Next, use Eq. (17.22) to calculate $\psi(x)$. [*Hint*: Complete squares in the exponent, use the above integral, and blithely go ahead!] . Show that $\psi(x)$ is properly normalized.

17.7 Use the results of Problem 17.6 to show that Eq. (17.24) is satisfied. Take the "widths" in p and in x from the shapes of $|\phi(p)|^2$ and $|\psi(x)|^2$.

18
The Uncertainty Relations

Uncertainty Relation and Classical Description The relation between the width of the probability distribution in x and the width of the probability distribution in momentum p

$$\Delta x \Delta p \gtrsim \hbar \tag{18.1}$$

plays a very important role in the analysis of the relation between classical mechanics and quantum mechanics. What the above *uncertainty relation* states is that in quantum mechanics it is not possible to define the position and the momentum of the particle with arbitrary precision. If the location of a path of a particle is determined with a certain degree of precision, then the momentum cannot be determined with a precision higher than that described by (18.1). This indeterminacy is not related to any indeterminacy in the Schrödinger equation. The equation is well defined, and given the initial condition $\psi(x, 0)$, the solution $\psi(x, t)$ can, in principle, be determined with arbitrary precision. Where the indeterminacy comes in is in the attempt to use classical concepts, such as the specification of a path by $\mathbf{r}(t)$ in describing phenomena that are really described by quantum mechanics. The smallness of $\hbar$ guarantees that no difficulty arises in the use of these concepts in the macroscopic domain. We can get an idea of the order of magnitude of the uncertainty by an example: consider a grain of dust, of mass 10^{-4} gm moving with velocity of 10^3 cm/sec. If the uncertainty in the velocity is 10^{-4} cm/sec, then $\Delta p = 10^{-8}$ gm cm/sec. This means

$$\Delta x \gtrsim \hbar/\Delta p \simeq 10^{-27}/10^{-8} \simeq 10^{-19} \text{cm};$$

that is, the classical specification of the position of the dust particle cannot be made to better than 10^{-19} cm. This, however, is 10^{11} times smaller than the size of one of the approximately 10^{19} atoms that make up the dust particle, so that the uncertainty cannot play any role in the physics of the dust particle. For atoms, on the other hand, the relation does have implications. Consider an electron in a Bohr atom. Suppose we consider circular orbits: since the velocity in an orbit is $\alpha c/n$, a lack of knowledge in which orbit an electron is to be found means that the momentum can lie anywhere between $mc\alpha$ and 0. This means that $\Delta p \sim mc\alpha$, and implies that

$$\Delta x \gtrsim \hbar/mc\alpha.$$

Thus the uncertainty in location is comparable to the radius of the Bohr atom, and is therefore in the range of relevant physical magnitudes. The uncertainty relation (18.1) can be derived using the general machinery of quantum mechanics. We shall

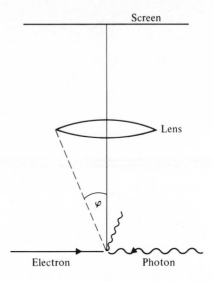

Figure 18.1. "Heisenberg microscope" used in the analysis of electron position measurement, and the establishment of the indeterminacy relation.

argue that it is forced upon us, independent of its origin from the wave-packet construction of wave functions, by the need to reconcile the complementary particle and wave properties of radiation and matter. The following example illustrates this.

Heisenberg microscope Consider the experimental setup shown in Fig. 18.1, designed to measure the position of an electron. The electrons move from left to right with momentum p_x. The microscope, consisting of a lens and a detecting screen (or photographic plate), is to be used to measure the position of an electron by looking at light scattered by the electron. We shine the light, of wavelength λ along the negative x axis. A particular electron will scatter a particular photon into the aperture of the lens. The resolution of the microscope is, according to classical optics,

$$\lambda/\sin \varphi \tag{18.2}$$

where φ is the angle subtended by the lens at the point of observation. That, therefore, is the precision with which the electron can be localized,

$$\Delta x \sim \lambda/\sin \varphi, \tag{18.3}$$

and it would appear that one could reduce Δx by making λ small enough. This, we will now show, can be done only at the expense of losing information about the x-component of the momentum of the electron after the collision. The scattered photon has a final momentum that is undetermined by

$$(\Delta p_\gamma)_x \simeq 2\frac{h\nu}{c}\sin \varphi \tag{18.4}$$

since it can reach the screen by entering the lens anywhere within the angle of the aperture. This, by momentum conservation, implies that there is a similar uncertainty in the electron momentum, so that

$$\Delta x \Delta p_x \simeq \frac{\lambda}{\sin \varphi} \, \frac{2h\nu}{c} \sin \varphi \simeq h. \tag{18.5}$$

The two-slit experiment The uncertainty relation permits us to remove the paradox associated with the two-slit experiment that we discussed before. If it were possible to tell which slit the electron came through, it would be possible to treat each electron transmittal as an event in which only one slit is open, and this would lead to the accumulation of electron arrivals on the screen as the "sum" of two experiments, in each of which one slit was open and the other closed. The two experiments do not, however, give a diffraction pattern. It seems that just knowing which slit the electron went through destroys the interference pattern! We shall use the uncertainty relation to show that the presence of a monitor that is good enough to identify the slit of passage disturbs the experiment sufficiently to destroy the interference pattern, and thus remove the logical inconsistency.

Let the slits be separated by a distance a and let the distance from the slits to the detection screen be d (Fig. 18.2). The condition for constructive interference is that

$$a \sin \theta_n = n\lambda \tag{18.6}$$

where θ_n is the angle that the beam leading to the maximum makes with the symmetry axis. The distance between adjacent maxima is

$$d \sin \theta_{n+1} - d \sin \theta_n = \lambda d/a. \tag{18.7}$$

Consider now a monitor that determines the position of the electron behind the slit screen to an accuracy sufficient to determine which slit the electron came through. This means that for the electron

$$\Delta y < \frac{a}{2}. \tag{18.8}$$

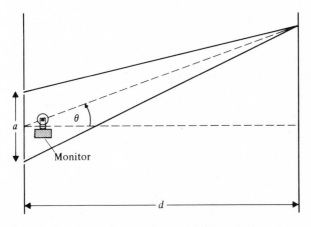

Figure 18.2. Two-slit experiment with monitor to determine which slit the electron goes through.

Such a monitor might be a source of photons, as in the last example. Such a position measurement will impart to the electron a momentum in the y direction, with an uncertainty given by

$$\Delta p_y \gtrsim \frac{h}{a/2}. \tag{18.9}$$

Hence

$$\frac{\Delta p_y}{p} \gtrsim \frac{2h}{ap} \simeq \frac{2\lambda}{a}. \tag{18.10}$$

This, however, introduces an indeterminacy in the position of arrival at the screen of magnitude $2\lambda d/a$, which is larger than the separation between the adjacent maxima. Thus the monitor wipes out the interference pattern and there is no logical inconsistency. We could, of course, have argued that logical consistency *demands* that $\Delta y \Delta p_y \gtrsim \hbar$.

Bohr atomic orbits We may use the uncertainty relation to show that the orbits of the Bohr atom are not really observable, in the sense that it is not possible to take "snapshots" of the electron as it goes around the orbit. The radius of the nth orbit is given by

$$R_n = \frac{\hbar n^2}{mc\alpha}. \tag{18.11}$$

An experiment designed to differentiate between this orbit and neighboring ones must yield position measurements with a precision

$$\Delta x \ll \frac{\hbar}{mc\alpha} (n^2 - (n-1)^2) \cong \frac{2\hbar n}{mc\alpha}. \tag{18.12}$$

This implies an uncontrollable momentum transfer to the electron whose magnitude is

$$\Delta p \gtrsim \frac{\hbar}{\Delta x} \gg \frac{mc\alpha}{2n}. \tag{18.13}$$

The change of momentum from p to $p + \Delta p$ implies a change in the energy of the order of magnitude $\Delta E \simeq \Delta p^2/2m \simeq p\Delta p/m$. Thus

$$\Delta E \gg \frac{1}{m} \left(\frac{mc\alpha}{n} \right) \frac{mc\alpha}{2n} = \frac{1}{2} mc^2 \alpha^2/n^2 . \tag{18.14}$$

This is much larger than the energy of the electron in the orbit, and therefore such a measurement will not, in general, leave the electron in the orbit. Thus taking a "film" of the motion is impossible.

Zero-point energy The uncertainty principle also explains a peculiarly quantum mechanical phenomenon, the *zero-point energy*. Consider, for example, a harmonic oscillator, for which the energy is given by

$$E = \frac{p^2}{2m} + \frac{1}{2} m\omega^2 x^2 . \tag{18.15}$$

Classically, the minimum energy is zero when the particle is at rest at the bottom of the potential. The uncertainty relation forbids this. If the position is uncertain by $\Delta x = a$, then the momentum will be given by

$$p = \Delta p \simeq \frac{\hbar}{a} \tag{18.16}$$

(the total momentum is just the amount by which it deviates from zero). The energy is then given by

$$E \cong \frac{\hbar^2}{2ma^2} + \frac{1}{2}m\omega^2 a^2.$$ (18.17)

This is a function of a, and it is minimum when

$$\frac{dE}{da} = -\frac{\hbar^2}{ma^3} + m\omega^2 a = 0,$$

that is, when $a^2 = \hbar/m\omega$. Substituting this into Eq. (18.17) gives

$$E \cong \tfrac{1}{2}\hbar\omega + \tfrac{1}{2}\hbar\omega \simeq \hbar\omega.$$ (18.18)

The zero-point energy is real, and it is in fact the effect that prevents helium from solidifying at low temperatures.

Estimates of ground state energies We may use a similar procedure to estimate the ground state energy for a hydrogenlike atom, for which

$$E = \frac{p^2}{2m} - \frac{Ze^2}{r}.$$ (18.19)

If the electron is localized within a distance R from the center, then the momentum is uncertain by $\hbar/R$ and the smallest value of p is $\hbar/R$. Thus

$$E(R) = \frac{\hbar^2}{2mR^2} - \frac{Ze^2}{R},$$ (18.20)

and this is minimum when

$$\frac{dE}{dR} = -\frac{\hbar^2}{mR^3} + \frac{Ze^2}{R^2} = 0,$$

that is, when

$$R = \frac{\hbar^2}{mZe^2}.$$ (18.21)

When this is substituted into Eq. (18.20),

$$E = -\frac{m(Ze^2)^2}{2\hbar^2}$$ (18.22)

results. This is in agreement with what is obtained from the Bohr model. The estimates that were made were not precise, and therefore the agreement with the exact value should not be taken too seriously. However, what is important is that the uncertainty relation provides us with a minimum energy. Classically, a particle at rest could have an arbitrarily large binding energy, depending on how small r was.

If we know to what extent a system is localized, we can estimate its kinetic energy. For example, it is known that the range of nuclear forces is of the order of 1 fermi $= 10^{-13}$ cm. The momentum of a particle localized in a region of that size must be of the order of magnitude

$$p \sim \frac{\hbar}{R} \sim \frac{10^{-27}}{10^{-13}} = 10^{-14}\,\text{gm cm/sec}.$$ (18.23)

Thus the kinetic energy of a nucleon of mass $M = 1.6 \times 10^{-24}$ gm is of the order of

$$K = \frac{p^2}{2M} = \frac{10^{-28}}{3.2 \times 10^{-24}} = 0.32 \times 10^{-4}\,\text{erg}$$

$$\simeq 20\,\text{MeV}.$$ (18.24)

From this we can deduce that the potential well which binds a nucleon to a nucleus must be at least of the order of magnitude 30 MeV, and attractive.

In addition to the uncertainty relation (18.1) there is a relation that reads

$$\Delta E \Delta t \gtrsim h. \tag{18.25}$$

It can be obtained by a plausibility argument, according to which

$$\Delta E = \frac{p \Delta p}{m} = v \Delta p \gtrsim v \frac{h}{\Delta x} = \frac{h}{(\Delta x/v)} \simeq \frac{h}{\Delta t},$$

where v is the velocity, and the uncertainty in time is related to the uncertainty in position. It is more general than the argument implies, and it asserts that a state of finite duration Δt cannot have a precisely defined energy: (18.25) specifies the uncertainty. Thus if an atomic state decays to the ground state, and the mean life of the excited state is τ, then the excited state does not have a precise energy, but rather one that is uncertain by $\Delta E \sim \hbar/\tau$. This implies that the frequency of the radiation emitted will not be precisely what the Bohr condition tells us, $\nu = (E_1 - E_0)/h$, but it will be fuzzy by an amount $\Delta \nu \cong 1/2\pi\tau$. The spectral line will thus be *broadened*. The broadening is an intrinsic quantum mechanical phenomenon, and the width of the spectral line is called the *natural line width*.

The uncertainty relations are an integral part of quantum mechanics, and they are extremely useful in estimating magnitudes of quantum mechanical effects. We shall have many occasions to use them in what follows.

NOTES AND COMMENTS

1.   The Uncertainty Principle is discussed in every textbook on quantum mechanics. A particularly detailed discussion may be found in W. Heisenberg, *The Physical Principles of the Quantum Theory*, Dover Publications, New York, 1947. This is a reprint of some lectures delivered in 1930, and is very readable. Another excellent reference is the essay by Niels Bohr, "Discussion with Einstein on Epistemological Problems in Atomic Physics," in *Albert Einstein: Philosopher-Scientist*, P. Schlipp (Ed.), Library of Living Philosophers, Evanston, Ill., 1949.

2.   There is an apparent loophole in the discussion of the Heisenberg microscope. One could argue that since the direction of the photon is correlated with the magnitude of its momentum, one could make the photon momentum (and therefore the electron momentum) measurement more precise by measuring the recoil momentum of the photographic plate. The difficulty with this procedure is that once the plate becomes part of the "observed" system, one must worry about specifying its location, since its momentum is now to be specified to an accuracy better than the Δp_x given in the text. The plate, too, obeys the uncertainty relation; and if its momentum needs to be specified with great accuracy, then its location and, therefore, that of the image on it will be "fuzzy." Thus there is no way of getting around the uncertainty relation. This argument also shows why quantum mechanics is different from other physical theories: the final measure-

ments are always specified in classical terms, but there is no clear rule of what the dividing line between the observed system and the measurement apparatus is.

PROBLEMS

18.1 Consider a particle of mass m in a potential of the form
$$V(x) = \infty \qquad x < 0$$
$$= V_0 x/a \quad x > 0$$
with V_0 and a constants of dimensions $[E]$ and $[L]$, respectively. Use the uncertainty relations to estimate the ground state energy of the particle.

18.2 The dimensions of a nucleus are given by $R = 1.3\,A^{1/3}\,10^{-13}$ cm. Use the uncertainty relation to estimate the energy of an electron that can be emitted from the nucleus, assuming that the electron was localized inside the nucleus before its emergence. Express your answer in MeV.

18.3 A beam of electrons is incident on a slit in a screen. The electrons have momentum p and impinge on a plate a distance L from the screen. What is the size of the slit that minimizes the width of the electron beam when it reaches the plate?

18.4 A particle is to be dropped from a height h on a line drawn below. Use the uncertainty relation to estimate the minimum spread forced by quantum mechanics, about the line. [*Hint*: Think of dropping the particle through a slit of width w, and calculate the spread, and its minimum value.]

18.5 In relativistic quantum mechanics one deals with the concept of a "virtual particle," a particle that is created on borrowed energy for a time of duration $\hbar/E$, consistent with the uncertainty relation. Write down an expression for the maximum length of time that a virtual particle of rest mass m can exist. If a particle of energy E is created, and it travels with momentum p, what is the largest distance that it can travel? If the distance of travel is 10^{-13} cm, what is the rest energy of the particle in MeV?

18.6 Use the uncertainty relation to estimate the ground state energy of a particle in a potential given by $V(x) = \lambda x^4$. Check the dimensions of your answer.

18.7 What is the natural line width of a state with a mean life of (a) 2.6×10^{-10} sec, (b) 2.2×10^{-6} sec, (c) 12 minutes. Express your answer in eV.

18.8 A resonant state in pion-nucleon scattering has a natural line width of 115 MeV. What is the lifetime of such a state?

18.9 Monochromatic light of wavelength $\lambda = 6000$ Å passes through a fast shutter that opens for 10^{-9} sec. The light emerging will no longer be monochromatic. What will be its spread in wavelengths?

19
Properties of
the Schrödinger Equation

The Schrödinger equation describing a free particle is given by

$$i\hbar \frac{\partial \psi(x,t)}{\partial t} = -\frac{\hbar^2}{2m} \frac{\partial^2 \psi(x,t)}{\partial x^2} \tag{19.1}$$

and the interpretation of the solution $\psi(x,t)$ is that

$$P(x,t)dx = \psi^*(x,t)\psi(x,t)dx \equiv |\psi(x,t)|^2 dx \tag{19.2}$$

is the probability of finding the particle between x and $x + dx$ at time t. This equation has the property of *linearity*: if $\psi_1(x,t)$ and $\psi_2(x,t)$ are solutions, so is

$$\psi(x,t) = a_1\psi_1(x,t) + a_2\psi_2(x,t) \tag{19.3}$$

with a_1 and a_2 arbitrary complex numbers. The wave packet, representing a sum (integral) of plane wave solutions of Eq. (19.1), is constructed by making use of this property.

Probability interpretation The probability interpretation requires that we normalize the solution so that it satisfies

$$\int_{-\infty}^{\infty} dx \, \psi^*(x,t)\psi(x,t) = 1. \tag{19.4}$$

This can be carried out provided that the solution is square integrable. If the integral over the absolute square of a solution exists, then it is a constant, as a consequence of

$$\frac{\partial}{\partial t} P(x,t) + \frac{\partial}{\partial x} j(x,t) = 0 \tag{19.5}$$

where

$$j(x,t) = \frac{\hbar}{2im} \left(\psi^* \frac{\partial \psi}{\partial x} - \frac{\partial \psi^*}{\partial x} \psi \right) \tag{19.6}$$

is called the probability current density or the probability flux. Note that this quantity is real for complex ψ, and that it vanishes when ψ is real.

A special property of equations that are first order in the time derivative is that the solution at a time t is completely specified if $\psi(x,t)$ is given at some initial time, say $t = 0$. Thus *if we know $\psi(x,0)$, then $\psi(x,t)$ is completely specified by the equation*. This can be made apparent by rewriting the time derivative as a difference. Writing

$$\frac{\partial \psi(x,t)}{\partial t} \cong \frac{1}{\Delta t}(\psi(x,t+\Delta t) - \psi(x,t)), \tag{19.7}$$

we get the Schrödinger equation in the form that one would use for integrating it on a digital computer

$$\psi(x, t + \Delta t) \cong \psi(x, t) + \Delta t \left(\frac{\hbar}{2im} \frac{\partial^2 \psi(x, t)}{\partial x^2} \right). \tag{19.8}$$

This shows that $\psi(x, t + \Delta t)$ can be obtained from ψ at the earlier time t. This can then be used to determine ψ at the time $t + 2\Delta t$ and so on.

Expectation values Given a probability density $P(x, t)$ one may calculate averages of functions of x. For example,

$$\langle x \rangle_t = \int_{-\infty}^{\infty} dx\, x P(x, t) = \int_{-\infty}^{\infty} dx\, \psi^*(x, t) x \psi(x, t)$$

$$\langle x^2 \rangle_t = \int_{-\infty}^{\infty} dx\, x^2 P(x, t) = \int_{-\infty}^{\infty} dx\, \psi^*(x, t) x^2 \psi(x, t)$$

and in general, for any function of x,[†]

$$\langle f(x) \rangle_t = \int_{-\infty}^{\infty} dx\, \psi^*(x, t) f(x) \psi(x, t). \tag{19.9}$$

The meaning of the average is the following: consider a large collection of identical systems, each described by the wave function $\psi(x, t)$. For each, measure x, or x^2, or $f(x)$. Each measurement will yield some value of $f(x)$, and the average of all the measurements is $\langle f(x) \rangle$. This average can change with time, because $\psi(x, t)$ changes with time. Note that the average of a function of x is not the same as a function of the average. For example,

$$\langle x^2 \rangle = \int_{-\infty}^{\infty} dx\, \psi^* x^2 \psi \neq \langle x \rangle^2 = \left(\int_{-\infty}^{\infty} dx\, \psi^* x \psi \right)^2$$

except under very special circumstances.

The expressions for the average or, more accurately stated, for the *expectation value* given above are generalizations of the discrete averages or expectation values that we learned to calculate in Chapter 9, such as

$$\langle n \rangle = \sum n P(n)$$

$$\langle n^2 \rangle = \sum n^2 P(n),$$

and so on.

The rule given in Eq. (19.9) does not tell us how to calculate the average value of the momentum. To get a hint of how to proceed, we calculate

$$\frac{d}{dt} \langle x \rangle = \frac{d}{dt} \int dx\, \psi^*(x, t) x \psi(x, t)$$

$$= \int dx \left(\frac{\partial \psi^*}{\partial t} x \psi + \psi^* x \frac{\partial \psi}{\partial t} \right). \tag{19.10}$$

[†] The reason for inserting the function f between ψ^* and ψ will become apparent in Eq. (19.13); sometimes derivatives appear instead of functions of x, and there the instruction is clear that these act on ψ.

We can manipulate this into a more suggestive form by using the Schrödinger equation. We write

$$\frac{\partial \psi^*}{\partial t} x \psi + \psi^* x \frac{\partial \psi}{\partial t} = \frac{\hbar}{2im} \left(\frac{\partial^2 \psi^*}{\partial x^2} x \psi - \psi^* x \frac{\partial^2 \psi}{\partial x^2} \right)$$

$$= \frac{\hbar}{2im} \frac{\partial}{\partial x} \left(\frac{\partial \psi^*}{\partial x} x \psi - \psi^* x \frac{\partial \psi}{\partial x} - \psi^* \psi \right)$$

$$+ \frac{\hbar}{im} \psi^* \frac{\partial \psi}{\partial x}. \tag{19.11}$$

Now note that because of square integrability

$$\int_{-\infty}^{\infty} dx \frac{\partial}{\partial x} \left(\frac{\partial \psi^*}{\partial x} x \psi - \psi^* x \frac{\partial \psi}{\partial x} - \psi^* \psi \right) = 0 \tag{19.12}$$

(remember that ψ has to go to zero at infinity faster than $1/\sqrt{x}$), so that

$$m \frac{d}{dt} \langle x \rangle = \int_{-\infty}^{\infty} dx \, \psi^*(x, t) \frac{\hbar}{i} \frac{\partial}{\partial x} \psi(x, t). \tag{19.13}$$

This result suggests the following procedure: if the expectation value of the momentum is required, just calculate the expectation value of the *differential operator* $\frac{\hbar}{i} \frac{\partial}{\partial x}$, inserted between ψ^* and ψ, as in

$$\langle p \rangle = \int_{-\infty}^{\infty} dx \, \psi^* p_{op} \psi = \int_{-\infty}^{\infty} dx \, \psi^* \left(\frac{\hbar}{i} \frac{\partial}{\partial x} \right) \psi. \tag{19.14}$$

This generalizes to higher powers of p. Thus

$$\langle p^2 \rangle = \int_{-\infty}^{\infty} dx \, \psi^* p_{op}^2 \psi$$

$$= \int_{-\infty}^{\infty} dx \, \psi^* \left(\frac{\hbar}{i} \frac{\partial}{\partial x} \right)^2 \psi = -\hbar^2 \int \psi^* \frac{\partial^2 \psi}{\partial x^2} dx, \tag{19.15}$$

and so on. The fact that physical quantities appear as *operators* in quantum mechanics differentiates it from classical mechanics. In classical mechanics the position and the momentum are independent dynamical variables that are determined by solving the equations of motion. They are constrained by the law of conservation of energy

$$E = \frac{p^2}{2m} + V(x) \tag{19.16}$$

whenever the potential does not have an *explicit* time dependence. In quantum mechanics the equations of motion are replaced by the Schrödinger equation, and the path of the particle is specified in terms of the solution by

$$\langle x \rangle_t = \int_{-\infty}^{\infty} dx \, \psi^* x \psi$$

$$\langle p \rangle_t = \int_{-\infty}^{\infty} dx \, \psi^* \frac{\hbar}{i} \frac{\partial \psi}{\partial x}. \tag{19.17}$$

The fact that both of these numbers can be calculated does not mean that it makes sense to speak of a simultaneous determination of position and momentum. The average values are just the central values of a spread of possible position and momentum determinations, and the spreads are such that the uncertainty relation

$$\Delta p \Delta x \gtrsim \hbar$$

is not violated.

Schrödinger equation for particle in potential The free particle Schrödinger equation may be rewritten as

$$i\hbar \frac{\partial \psi}{\partial t} = \frac{1}{2m} \left(\frac{\hbar}{i} \frac{\partial}{\partial x} \right)^2 \psi \qquad (19.18)$$

and this takes the form

$$i\hbar \frac{\partial \psi}{\partial t} = \frac{1}{2m} p_{op}^2 \psi. \qquad (19.19)$$

Remember that this is the Schrödinger equation for a free particle, for which the energy is

$$E = p^2/2m. \qquad (19.20)$$

It is conventional to denote the energy operator by the letter H and to call it the Hamiltonian. Thus the Hamiltonian for a free particle is

$$H = p_{op}^2/2m. \qquad (19.21)$$

The Schrödinger equation for a free particle has the form

$$i\hbar \frac{\partial \psi}{\partial t} = H\psi. \qquad (19.22)$$

This suggests the generalization of the Schrödinger equation to the case of a particle in a potential $V(x)$. In that case the energy is given by Eq. (19.16) and thus

$$H = \frac{p_{op}^2}{2m} + V(x). \qquad (19.23)$$

The equation (19.19) then generalizes to

$$i\hbar \frac{\partial \psi(x, t)}{\partial t} = H\psi = -\frac{\hbar^2}{2m} \frac{\partial^2 \psi(x, t)}{\partial x^2} + V(x)\psi(x, t), \qquad (19.24)$$

which is the correct equation. We shall leave it to the reader to show that if $V(x)$ is real, then the conservation law (19.5) is unaltered, as is the form for $P(x, t)$ and $j(x, t)$. Thus it is possible to maintain the interpretation of $\psi(x, t)$.

Hermitian operators The expectation value of the momentum operator should certainly be real, since momentum is a physical quantity. The appearance of i in its definition is disquieting. Let us check that the expectation value is indeed real. We have

$$\langle p \rangle = \frac{\hbar}{i} \int dx \, \psi^* \frac{\partial \psi}{\partial x}. \qquad (19.25)$$

Hence

$$\langle p \rangle^* = -\frac{\hbar}{i} \int dx \, \psi \frac{\partial \psi^*}{\partial x}. \qquad (19.26)$$

Therefore

$$\langle p\rangle - \langle p\rangle^* = \frac{\hbar}{i} \int dx \left(\psi^* \frac{\partial \psi}{\partial x} + \psi \frac{\partial \psi^*}{\partial x}\right)$$

$$= \frac{\hbar}{i} \int_{-\infty}^{\infty} dx \frac{\partial}{\partial x} (\psi^* \psi) = \frac{\hbar}{i} [|\psi(x,t)|^2]_{-\infty}^{\infty}$$

$$= 0 \qquad (19.27)$$

for square integrable functions.

An operator whose expectation value is real for all acceptable (that is, square integrable) wave function $\psi(x, t)$ is called a *hermitian operator*. Clearly all operators representing physical quantities are hermitian.

Linear operators An operator is defined by what it does to a function that it acts on. Let us list a number of freely invented operators:

i) $Of(x) = k$,

ii) $Of(x) = 2f(x)$,

iii) $Of(x) = 3xf(x)$,

iv) $Of(x) = f(x)/\sin x$,

v) $Of(x) = df(x)/dx$,   (19.28)

vi) $Of(x) = (f(x))^2$,

vii) $Of(x) = \lambda f(x)$,

viii) $Of(x) = 3f(x) - 2x^2$,

ix) $Of(x) = \int_0^x dy f(y)$,

x) $Of(x) = f(x + a)$.

An examination of these operators show that some of them have the property that

$$O(\alpha f_1(x) + \beta f_2(x)) = \alpha Of_1(x) + \beta Of_2(x). \qquad (19.29)$$

These are the *linear operators*. Thus (i) is a linear operator when $k = 0$, but not generally so; (ii) is linear, as are (iii) and (iv); (v) is linear; but (vi) is not, and so on. *Only linear operators play a role in quantum mechanics*. Operators acting on possible wave functions are supposed to yield other wave functions – they transform the original wave functions into new ones. Since both the old and the new wave functions are required to obey the superposition rules, we require that if

$$\psi(x, t) = \sum_n C_n u_n(x, t), \qquad (19.30)$$

then

$$O\psi(x, t) = \sum_n C_n (Ou_n(x, t)). \qquad (19.31)$$

Not all linear operators are admissible: only those operators that, when acting on square integrable functions, give square integrable functions are allowed. We thus require

$$\int_{-\infty}^{\infty} dx (Of(x))^* (Of(x)) < \infty. \qquad (19.32)$$

Thus, although (iv) is a linear operator, it is inadmissible since $1/\sin x$ blows up at the origin. We may make it admissible if we restrict ourselves to wave functions $f(x)$ that vanish at the origin (and at $n\pi$ generally). Similarly (ix) will generally not be admissible since with

$$\hat{\psi}(x) \;=\; \int_0^x dy\, \psi(y),$$

$$\mathrm{Lim}_{x \to \infty}\, \hat{\psi}(x) \;=\; \int_0^\infty dy\, \psi(y)$$

will not generally vanish.

If the potential $V(x)$ does not have an explicit time dependence, we can start to solve the Schrödinger equation by the technique of separation of variables. Into Eq. (19.24) insert the following form of solution:

$$\psi(x, t) \;=\; T(t)u(x). \tag{19.33}$$

This leads to

$$i\hbar \frac{dT(t)}{dt}\, u(x) \;=\; T(t)\left(-\frac{\hbar^2}{2m}\frac{d^2 u(x)}{dx^2} + V(x)u(x)\right)$$

which may be rewritten in the form

$$i\hbar \frac{dT(t)/dt}{T(t)} \;=\; \frac{-\dfrac{\hbar^2}{2m}\dfrac{d^2 u(x)}{dx^2} + V(x)u(x)}{u(x)}. \tag{19.34}$$

Here the left side depends on t alone, and the right side depends on x alone. This can hold only if each side is equal to a constant which we denote by E, a quantity with the dimensions of energy. Thus

$$i\hbar \frac{dT(t)}{dt} \;=\; ET(t) \tag{19.35}$$

and

$$-\frac{\hbar^2}{2m}\frac{d^2 u(x)}{dx^2} + V(x)u(x) \;=\; Eu(x). \tag{19.36}$$

Time dependence The first of these is easily solved. The solution is

$$T(t) \;=\; e^{-iEt/\hbar}. \tag{19.37}$$

The second equation, which reads

$$\left(\frac{p_{\mathrm{op}}^2}{2m} + V(x)\right) u(x) \;=\; Eu(x), \tag{19.38}$$

is not so easily solved, unless $V(x)$ is particularly simple. The equation, which we write in the form

$$Hu(x) \;=\; Eu(x), \tag{19.39}$$

where H as usual represents the energy operator, has the form of an *eigenvalue equation*.

When a linear operator acts on a function $f(x)$, it changes it into another function $Of(x) = g(x)$. Sometimes, for a special set of $f(x)$ the operator acting on $f(x)$ changes it into $f(x)$ multiplied by a numerical coefficient. Those special functions $f(x)$ are called *eigenfunctions* and the coefficients in front are called *eigenvalues*. In Eq. (19.39) the possible values of E for which Eq. (19.39) holds are the eigenvalues. We call them energy eigenvalues, because the operator is the energy operator. The functions $u(x)$ are called energy eigenfunctions. In the next chapter we shall solve a particularly simple eigenvalue problem and discuss properties of eigenvalues and eigenfunctions.

NOTES AND COMMENTS

1. An important property of operators is that when two operators are applied to a wave function, the order of application matters. For example,

$$p_{op}x\psi(x) - xp_{op}\psi(x) = \frac{\hbar}{i}\frac{d}{dx}(x\psi(x)) - \frac{\hbar}{i}x\frac{d}{dx}\psi(x) = \frac{\hbar}{i}\psi(x) \neq 0.$$

It is the fact that operators describing physical quantities *do not commute* that leads to the uncertainty relations. In more advanced treatments one shows that if the uncertainties are defined by

$$(\Delta A)^2 = \langle A^2 \rangle - \langle A \rangle^2$$
$$(\Delta B)^2 = \langle B^2 \rangle - \langle B \rangle^2,$$

then

$$(\Delta A)^2(\Delta B)^2 \geqslant \tfrac{1}{4}\langle i(AB - BA)\rangle^2.$$

See, for example, S. Gasiorowicz, *Quantum Physics*, John Wiley, New York, 1974, Appendix B.

2. In solving the time-dependent Schrödinger equation by a separation-of-variables method, it was essential that the potential did not depend on the time t. There is a good reason for that: only when the potential is independent of time is energy conserved in classical mechanics, and the argument carries over into quantum mechanics. This is really related to a deep principle, that energy conservation is a consequence of the invariance of the theory under a displacement in time: a solution depending on t still remains a solution when t is replaced by $t + T$, even if it is a different solution (for example, an orbit defined by the initial condition that it is at aphelion at time $t = 0$ will, when every t in the solution is replaced by $t + T$, still be an orbit, but not the one with the above initial condition). It is still possible (even though mathematically more complicated) to discuss the Schrödinger equation with a time-dependent potential, but the examples where this is of physical interest are beyond the scope of this book.

PROBLEMS

19.1 Consider a system described by the wave functions

a) $\psi(x) = Ae^{ikx} + Ae^{i\delta}e^{-ikx}$ A real

b) $\psi(x) = A\cos kx + B\sin kx$ A, B complex

c) $\psi(x) = R(x)e^{iS(x)/\hbar}$ $\qquad$ R, S real

d) $\psi(x) = Ae^{-x^2/2a^2}e^{ipx/\hbar}$ $\qquad$ A real.

Calculate the flux and the probability density in each case.

19.2 Consider the wave function

$$\psi(x) = \frac{\sqrt{2}}{\sqrt{b}}\sin\frac{n\pi x}{b} \qquad 0 \leqslant x \leqslant b$$

$$= 0 \quad \text{elsewhere.}$$

a) Calculate $\langle x \rangle$, $\langle x^2 \rangle$ and $(\Delta x)^2 \equiv \langle x^2 \rangle - \langle x \rangle^2$.

b) Calculate $\langle p \rangle$, $\langle p^2 \rangle$ and $(\Delta p)^2 \equiv \langle p^2 \rangle - \langle p \rangle^2$.

c) Use your results to calculate $\Delta p \Delta x$.

19.3 Consider the peculiar situation of wave functions that are only defined in a half space $x \geqslant 0$. Show that the momentum operator is hermitian only if all $\psi(x)$ satisfy the condition that $\psi(0) = 0$.

19.4 Consider the operator O defined by

$$O\psi(x) \equiv \int_{-\infty}^{x} dy\, y\, \psi(y).$$

What are the eigenfunctions, that is, the solutions of

$$O\psi(x) = \lambda\psi(x),$$

and what are the constraints on λ?

[*Hint*: (a) Differentiate; (b) the integral involved in the operator must exist.]

19.5 Consider the operator defined by

$$O\psi(x) = \psi(x + a).$$

a) Consider a to be infinitesimal, so that the right side in the defining equation may be expanded to first order in a. Show that O may be expressed in terms of the momentum operator.

b) Expand $\psi(x + a)$ in a formal Taylor series in a and thus show that the operator O may be written in the form

$$O = e^{iap_{op}/\hbar}.$$

19.6 Which of the following are linear operators?

a) $Of(x) = x + 3f(x)$

b) $Of(x) = f(x)/(x + 2)$

c) $Of(x) = kf(x)$

d) $Of(x) = d^2f(x)/dx^2$

e) $Of(x) = (f(x))^{1/2}$

f) $Of(x) = x^2 f(x) - 2df(x)/dx$

19.7 Take the linear operators and write down the expectation values of O. If the $f(x)$ for which the operators are defined as in Problem 19.6 are such that they go to zero at infinity, which of the expectation values of O are real for all (complex) $f(x)$?

19.8 Calculate the expectation values of p and p^2 for the wave functions

a) $\psi(x) = Ne^{-ax^2}$,

b) $\psi(x) = Ne^{-ax^2+ibx}$,

c) $\psi(x) = R(x)e^{iS(x)}$, with R and S real functions of x.

20
Simple Systems

Particle in a box The simplest eigenvalue problem is that for a particle in an infinite box, that is, for the potential shown in Fig. 20.1. The potential energy is

$$V(x) = 0 \quad -a \leqslant x \leqslant a$$
$$= \infty \quad \text{elsewhere.} \tag{20.1}$$

The particle is free, but confined to the region $-a \leqslant x \leqslant a$, and this confinement is represented by the potential energy given above. The classical motion is simple: the particle bounces back and forth between the two walls, with constant, arbitrary energy. There is no conserved momentum, since the walls are rigid, and upon each collision, the direction of the momentum changes. The form of the potential cannot be achieved in reality: it is impossible to construct a perfectly sharp discontinuity in the potential energy, and one should think of the edge as the limiting case of a smooth, rapidly varying function. The potential

$$V(x) = V_0(x/a)^{2N},$$

with N very large, approaches the infinite box. The reason for considering the idealized version (20.1) is that the equation for the motion becomes manageable.

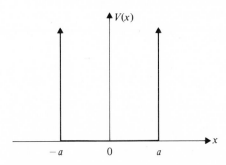

Figure 20.1. Infinite potential box of width $2a$.

193

Solution of the Schrödinger Equation The equation (19.36) may be written in the form

$$\frac{d^2 u(x)}{dx^2} + \left(\frac{2mE}{\hbar^2}\right) u(x) = \frac{2m}{\hbar^2} V(x) u(x) \tag{20.2}$$

and when $V(x)$ as given by Eq. (20.1) is inserted into this, one finds that in the region $-a \leqslant x \leqslant a$

$$\frac{d^2 u(x)}{dx^2} + k^2 u(x) = 0, \tag{20.3}$$

where

$$k^2 = \frac{2mE}{\hbar^2}. \tag{20.4}$$

Outside of that region $V(x) = \infty$, and this implies that for $x < -a$ and for $x > a$,

$$u(x) = 0, \tag{20.5}$$

since otherwise the energy would be infinite.

We shall require that $u(x)$ be continuous. This must be imposed so that a physical quantity such as the probability density $|u(x)|^2$ has this property. Equation (20.2) shows, when it is integrated from $x - \epsilon$ to $x + \epsilon$, with ϵ infinitesimal, that $du(x)/dx$ *must also be continuous*, unless $V(x)$ has an infinite jump. The latter is the case here, so that we impose the continuity requirement only on $u(x)$. When this is done, we get the boundary conditions

$$u(a) = u(-a) = 0. \tag{20.6}$$

The most general solution of Eq. (20.3) is

$$u(x) = A \cos kx + B \sin kx. \tag{20.7}$$

The condition (20.6) can be satisfied only if either

$$B = 0 \quad \cos ka = 0 \tag{20.8}$$

or

$$A = 0 \quad \sin ka = 0. \tag{20.9}$$

In the first case, the condition is satisfied provided that $ka = \pi/2$ (mod π), that is,

$$k = \frac{n\pi}{2a} \quad n \text{ an odd integer.} \tag{20.10}$$
$$(1, 3, 5, \ldots)$$

The corresponding eigenvalues for the energy are

$$E_n = \frac{\hbar^2 \pi^2 n^2}{8ma^2} \quad n \text{ odd} \tag{20.11}$$

and the eigenfunctions are

$$u_n(x) = A \cos \frac{n\pi x}{2a} \quad n \text{ odd.} \tag{20.12}$$

It is simple to work out that the requirement that

$$\int_{-a}^{a} dx \, |u_n(x)|^2 = 1 \tag{20.13}$$

leads to the eigenfunctions

$$u_n(x) = \frac{1}{\sqrt{a}} \cos \frac{n\pi x}{2a} \quad n \text{ odd.} \tag{20.14}$$

Note that these eigenfunctions do not change sign under the interchange $x \rightarrow -x$: they are *even* about the $x = 0$ axis.

In the second case, the condition $\sin ka = 0$ implies that $ka = 0 \pmod{\pi}$, or equivalently that

$$k = \frac{n\pi}{2a} \quad n \text{ an even integer.} \qquad (20.15)$$
$$(2, 4, 6, \ldots)$$

The energy eigenvalues are

$$E_n = \frac{\hbar^2 \pi^2 n^2}{8ma^2} \quad n \text{ even} \qquad (20.16)$$

and the normalized eigenfunctions are

$$u_n(x) = \frac{1}{\sqrt{a}} \sin \frac{n\pi x}{2a} \quad n \text{ even.} \qquad (20.17)$$

This set of eigenfunctions changes sign when $x \rightarrow -x$; that is, they are odd on reflection through the $x = 0$ axis.

We see that the energy eigenfunctions have the property that as the number of nodes (zeros, not counting $\pm a$) increases, so does the energy. This is a general feature of wave functions, since the kinetic energy is large if the wave function has a lot of curvature in it. This can be seen from

$$\langle p^2/2m \rangle = \frac{1}{2m} \int dx \, \psi^*(x)(-\hbar^2 d^2 \psi(x)/dx^2)$$

$$= -\hbar^2/2m \int dx \left[\frac{d}{dx} (\psi^* d\psi/dx) - (d\psi^*/dx)(d\psi/dx) \right]$$

$$= \frac{\hbar^2}{2m} \int dx \left| \frac{d\psi}{dx} \right|^2 \qquad (20.18)$$

where the first term in the second line vanishes because of boundary conditions, either at the wall, or at infinity. Parenthetically, the form of the energy can be determined from dimensional arguments. The energy can only depend on $\hbar$, m, and a. If we note that the dimensions of $\hbar$ are ergs/sec, that is, ML^2/T, those of m are M and those of a

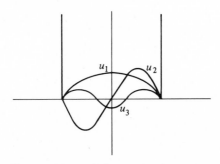

Figure 20.2. Sketch of the lowest three eigenfunctions $n = 1, 2, 3$ for the infinite box potential.

are L, then

$$\left(M\,\frac{L^2}{T^2}\right) = \left(\frac{ML^2}{T}\right)^\alpha M^\beta L^\gamma.$$

This requires that

$$\alpha + \beta = 1$$
$$2\alpha + \gamma = 2$$
$$\alpha = 2$$

which leads to

$$E \propto \frac{\hbar^2}{2ma^2}\,.$$

We could have obtained this more quickly by recalling that by the uncertainty principle $p \sim \hbar/a$, so that an energy of the form $p^2/2m$ involves the combination $\hbar^2/2ma^2$.

The time dependence of a given eigenfunction is $e^{-iEt/\hbar}$ so that the solutions even about the x axis are

$$u_n(x)e^{-iE_nt/\hbar} = \frac{1}{\sqrt{a}}\cos\frac{n\pi x}{2a}\,e^{-iE_nt/\hbar}\quad n\text{ odd}$$

and those odd about the x axis are

$$u_n(x)e^{-iE_nt/\hbar} = \frac{1}{\sqrt{a}}\sin\frac{n\pi x}{2a}\,e^{-iE_nt/\hbar}\quad n\text{ even.}$$

The most general solution of the Schrödinger equation that vanishes at the boundaries is of the form

$$\psi(x,t) = \sum_{n=1}^{\infty} A_n u_n(x)e^{-iE_nt/\hbar} \tag{20.19}$$

with the sum going over even as well as odd integers n. Note the absence of the $n = 0$ term. This has an eigenfunction that vanishes everywhere. The lowest state has $n = 1$, and thus the lowest energy is $\hbar^2\pi^2/8ma^2$. This is to be contrasted with the classical solution, for which the lowest energy is zero, corresponding to a particle at rest. In quantum mechanics there is always a finite energy, the zero point energy, when a system is localized.

Orthogonality of Eigensolutions We observe that since

$$\int_{-a}^{a} dx \cos\frac{n\pi x}{2a}\cos\frac{m\pi x}{2a} = \int_{-a}^{a} dx \sin\frac{n\pi x}{2a}\sin\frac{m\pi x}{2a} = 0, m \neq n$$

$$\int_{-a}^{a} dx \cos\frac{n\pi x}{2a}\sin\frac{m\pi x}{2a} = 0$$

we have

$$\int_{-a}^{a} dx u_n(x)u_m(x) = 0\quad m \neq n. \tag{20.20}$$

Whenever

$$\int dx f^*(x)g(x) = 0 \tag{20.21}$$

we say that $f(x)$ and $g(x)$ are *orthogonal*. This is because an integral of the type appearing in Eq. (20.21) is a continuous version of the discrete relation

$$\sum_{i=1}^{n} f_i^* g_i = 0, \qquad (20.22)$$

and here the left side is recognized as the scalar (dot) product of two n-dimensional vectors in a complex vector space, which vanishes when the vectors are orthogonal. Thus the result of Eq. (20.20) can be stated in the form

Eigenfunctions corresponding to different eigenvalues are orthogonal.

This result, too, is quite general, and not restricted to the potential under consideration.

Parity The general form of the solution (20.19) leads us to the interesting notion of *inversion symmetry* or *parity*. The most general form of an initial wave function leading to Eq. (20.19) at an arbitrary time is

$$\psi(x,0) = \sum_{n \text{ odd}} A_n u_n(x) + \sum_{n \text{ even}} A_n u_n(x) \qquad (20.23)$$

where the first sum involves only functions that are even under the reflection in the x axis, and the second sum involves only odd functions. If the initial state is an even function, then

$$A_n = 0 \quad n \text{ even} \qquad (20.24)$$

and $\psi(x, t)$ is, according to Eq. (20.19), even under reflection at all times. Similarly, if the initial state is odd under reflection in the x axis, then the wave function $\psi(x, t)$ is odd at all times. Thus evenness and oddness are "conserved." Since conservation laws are always interesting, it is important to find out under what circumstances we may expect this to be true more generally.

Let us introduce an operator that reflects any function in the $x = 0$ axis. Such a *parity operator* is defined by the rule that

$$Pf(x) = f(-x) \qquad (20.25)$$

for all functions $f(x)$. What are the eigenvalues and eigenfunctions of this new operator? The equation

$$Pu(x) = \lambda u(x) \qquad (20.26)$$

implies that

$$u(-x) = \lambda u(x). \qquad (20.27)$$

Changing x into $-x$ or, equivalently, applying P again yields

$$u(x) = \lambda u(-x) \qquad (20.28)$$

so that

$$u(x) = \lambda^2 u(x). \qquad (20.29)$$

Thus the eigenvalues must have the property that

$$\lambda^2 = 1. \qquad (20.30)$$

For $\lambda = +1$ we have

$$u(x) = u(-x); \qquad (20.31)$$

that is, $u(x)$ is any even function. For $\lambda = -1$ we have

$$u(x) = -u(-x); \qquad (20.32)$$

that is, $u(x)$ is any odd function. Thus *even and odd functions of x are eigenfunctions of the parity operator with eigenvalues ± 1, respectively*.

Consider now the Schrödinger equation written most generally in the form

$$i\hbar \frac{\partial}{\partial t} \psi(x, t) = H\psi(x, t). \tag{20.33}$$

Let us apply P to both sides. On the left side we get $i\hbar \dfrac{\partial}{\partial t} P\psi$ since P does not affect the time derivative. On the right side we get $PH\psi(x, t)$. Thus P acts on H as well as on the wave function. The kinetic energy term in H is proportional to d^2/dx^2, which is even under the interchange $x \to -x$. In the special case that the potential energy $V(x)$ has the property that

$$V(-x) = V(x), \tag{20.34}$$

that is, when H is an even function of x, then we get the equation

$$i\hbar \frac{\partial}{\partial t} P\psi(x, t) = HP\psi(x, t). \tag{20.35}$$

Thus both the wave function and its space-inverted (reflected in the origin) form obey the same equation. The sum of the two,

$$\tfrac{1}{2}(\psi(x, t) + P\psi(x, t)) = \psi_e(x, t), \tag{20.36}$$

which is the even part of $\psi(x, t)$, and the difference,

$$\tfrac{1}{2}(\psi(x, t) - P\psi(x, t)) = \psi_o(x, t), \tag{20.37}$$

which is the odd part of $\psi(x, t)$, separately obey the Schrödinger equation. Thus, if, say, $\psi(x, 0)$ is even, then this implies

$$\psi_o(x, 0) = 0. \tag{20.38}$$

Using Eq. (19.8) we can then show that

$$\psi_o(x, t) = 0. \tag{20.39}$$

Lest the reader think that we have spent too much space on a minor point, it should be pointed out that the notion of invariance under reflections, which in three dimensions involves

$$P\psi(x, y, z) = \psi(-x, -y, -z), \tag{20.40}$$

is a very important one in atomic, nuclear, and elementary particle physics, and we shall have many occasions to return to it later.

The free particle Another simple eigenvalue problem is that of a free particle for which $V(x) = 0$ everywhere. One way of treating it is to let the box $(-a, a)$ become very large, as large as we wish. The solutions

$$\frac{1}{\sqrt{a}} \cos \frac{n\pi x}{2a} \; ; \quad \frac{1}{\sqrt{a}} \sin \frac{n\pi x}{2a} \tag{20.41}$$

for n of moderate size become trivial as a becomes very large. We must therefore let $n \to \infty$ as $a \to \infty$. If we define

$$k = \frac{n\pi}{a} \tag{20.42}$$

and keep k finite as a becomes very large, the solutions (20.41) become

$$\frac{1}{\sqrt{a}} \cos kx; \quad \frac{1}{\sqrt{a}} \sin kx. \tag{20.43}$$

Note that the energy is no longer quantized. The difference between adjacent energy eigenvalues

$$E_n - E_{n-1} = \frac{\hbar^2 \pi^2}{8ma^2} (n^2 - (n-1)^2) = \frac{\hbar^2 \pi^2}{8ma^2} (2n-1) \approx \frac{\hbar^2 k^2}{4mn} \tag{20.44}$$

approaches zero as $n \to \infty$ and k is fixed.

The solutions (20.43) can be independently seen to be solutions of the free particle Schrödinger equation

$$\frac{d^2 u(x)}{dx^2} + k^2 u(x) = 0. \tag{20.45}$$

There is the problem of the normalizing factor $1/\sqrt{a}$ in front, since that factor tends to zero as $a \to \infty$. This is just a reflection of the fact that we have normalized our solutions to describe one particle, and if the box becomes infinitely large, and the particle is not forced to be localized in a particular region (say in a hydrogen atom), then it is not surprising that the probability of finding it in any finite region should go to zero. If we ask only sensible questions, such as involve relative probabilities (will the particle be more likely to be here than there .... ?) we will never get into trouble. In the answer to any sensible question, the size of the box will drop out.

Degeneracy The energy is now

$$E = \frac{\hbar^2 k^2}{2m} \tag{20.46}$$

and we notice something new: there are *two* solutions, $\cos kx$ and $\sin kx$ corresponding to the same energy. We say that there is a *two-fold degeneracy*. In contrast to the finite box, knowledge of the energy is not enough to specify the state of the system. To distinguish between the two solutions we must use other information: in this case we can use *parity*. The solution $\cos kx$ is an energy eigenfunction that is simultaneously a parity eigenfunction with eigenvalue $+1$; the solution $\sin kx$ is an energy eigenfunction that is simultaneously a parity eigenfunction with eigenvalue -1.

Another solution to Eq. (20.45) is a linear combination of the above, namely

$$\frac{1}{\sqrt{a}} e^{ikx}; \quad \frac{1}{\sqrt{a}} e^{-ikx}. \tag{20.47}$$

These solutions were not admissible for a box, but here there is no boundary, and therefore no boundary condition $u(\pm a) = 0$. The above solutions also have the same energy, but they differ in that they are simultaneously eigenfunctions of the momentum operator, with eigenvalues $\pm \hbar k$:

$$p_{op} e^{\pm ikx} = \frac{\hbar}{i} \frac{d}{dx} e^{\pm ikx} = \pm \hbar k e^{\pm ikx}. \tag{20.48}$$

Again, with more information, the solutions can be uniquely specified.

For a solution of the form $u(x) = Ne^{ikx}$, the probability flux, defined in Eq. (19.6) can be calculated. One finds

$$j = \frac{\hbar}{2im} |N|^2 (e^{-ikx}(ik)e^{ikx} - (-ik)e^{-ikx}e^{ikx})$$

$$= \frac{\hbar k}{m} |N|^2 = \frac{p}{m} |N|^2 = v|N|^2. \tag{20.49}$$

Thus there are $|N|^2 v$ (v is the velocity) particles moving to the right and passing a given point per unit time. A beam of such particles contains $|N|^2$ particles per unit length in it; when these move with velocity v, then in one unit of time a piece of the beam v units long crosses a given point and thus the flux is $|N|^2 v$. For the solution Ne^{-ikx} the only change is $k \to -k$ and this corresponds to the same flux moving to the left, in the negative x direction. The choice of N is at our disposal, in that it is determined by the experimental conditions. We are free to change N as long as we remember to calculate only relative probabilities. We shall see how this freedom is used in the problems to be discussed in the next chapter.

NOTES AND COMMENTS

1. One aspect of the infinite box problem that we did not discuss, but that is quite important in the interpretation of quantum mechanics, is the physical interpretation of the coefficients A_n in the general expansion of a wave function in terms of the eigenstates $u_n(x)$ of the energy operator. One can show (and this is the substance of Problem 20.3) that $|A_n|^2$ may be interpretated as the probability that a measurement of the energy for the state

$$\psi(x) = \sum_n A_n u_n(x)$$

yields the eigenvalue E_n. It is only for states where $\psi(x)$ is an eigenstate of the energy operator that the energy measurement will always give the same number. In general, the average energy need not be one of the eigenvalues, but the result of a particular measurement must yield an eigenvalue.

This is one of the peculiar aspects of quantum mechanics: once a measurement is made, the system is in an eigenstate, since otherwise a repetition of the measurement could give a totally different answer, and we could never check an experiment. This, however, means that in making a measurement, we "collapse" the general wave function to a particular eigenstate. The peculiar aspect is that this notion does not reside in the Schrödinger equation, but is part of the interpretative machinery. This has led to a great deal of unhappiness with quantum mechanics on the part of some philosophers of science, an unhappiness compounded by the fact that quantum mechanics is such an extraordinarily successful theory that it must be right, or that any other theory must look a great deal like it, if it is to be a successful candidate for replacing it.

PROBLEMS

20.1 Consider a particle in the potential well

$$V(x) = 0 \quad 0 \leqslant x \leqslant b$$
$$= \infty \quad \text{elsewhere.}$$

Show that the eigenfunctions are given by

$$u_n(x) = \frac{\sqrt{2}}{\sqrt{b}} \sin \frac{n \pi x}{b}.$$

What are the energy eigenvalues?

20.2 Show that the eigenfunctions satisfy the orthonormality condition

$$\int_0^b dx u_m^*(x) u_n(x) = \delta_{mn} = \begin{cases} 0 & m \neq n \\ 1 & m = n. \end{cases}$$

20.3 Use the above relation to calculate the expectation value of the energy, $\langle H \rangle$, for the state described by

$$\psi(x) = \sum_{n=1}^{\infty} a_n u_n(x)$$

where the a_n are arbitrary numbers. Show that if $\psi(x)$ is normalized to one,

$$\sum_{n=1}^{\infty} |a_n|^2 = 1.$$

20.4 Use the orthonormality condition to obtain an expression for a_n in terms of $\psi(x)$ and $u_n(x)$:

$$a_n = \int_0^b dx u_n^*(x) \psi(x).$$

20.5 Show that the parity operator is hermitian.

20.6 Consider a particle in a one-dimensional box, with mass $m = 0.9 \times 10^{-27}$ gm and box size $b = 1$ Å. If the transitions between levels are made to satisfy the $\Delta n = 1$ selection rule, and the energy change in a transition appears in the form of radiation according to the Bohr rules, list the wavelengths of the three lowest transitions.

20.7 Consider the potential of Problem 20.1. Calculate the motion of a classical particle in such a well; that is, obtain an expression for the position $x(t)$ as a function of time, given that the particle starts at $x = 0$ when $t = 0$ and it has energy E.

20.8 Calculate the classical expectation value of a function $f(x)$, defined by

$$\langle f(x) \rangle_{cl} = \frac{1}{T} \int_0^T dt f(x(t))$$

where T is the period of a single to-and-fro passage of the particle. Show that it may be written in the form

$$\langle f(x) \rangle_{cl} = \frac{1}{b} \int_0^b dy f(y)$$

independent of the energy of the particle. (b is the size of the box.)

20.9 Calculate the quantum mechanical expectation value of a function $f(x)$ in a state that is an eigenstate of H with quantum number n. Show that in the limit of large n, the result of this problem approaches that of Problem 20.8. This is an illustration of the *Correspondence Principle*.

[*Hint*: Use the fact that $\sin^2\theta = (1 - \cos 2\theta)/2$ and in the next step, integrate by parts to show that the remainder goes to zero as $1/n^2$.]

20.10 Calculate the expectation value of p for the $\psi(x)$ given in Problem 20.3.

21
Barrier Penetration

The Square Barrier The examples discussed in the last chapter show that the Schrödinger equation is simple to solve when the potential is constant. Many physical problems can be simulated by discontinuous flat potentials. Let us consider a potential barrier of the form (Fig. 21.1)

$$
\begin{aligned}
V(x) &= 0 & -\infty < x < -a & \quad \text{(region I)} \\
&= V_0 & -a < x < a & \quad \text{(region II)} \\
&= 0 & a < x < \infty & \quad \text{(region III)}.
\end{aligned}
\tag{21.1}
$$

As in the case of the infinite box considered in the last chapter, the sharp edges represent an idealization that is never seen in the real world, but that does approximate sudden changes in potential energy.

The classical motion is simple. A particle approaching the barrier from the left will clear it if the energy is large enough for the particle to have positive kinetic energy in the potential region, and it will be reflected if the energy lies below the critical value. We shall see that in quantum mechanics a particle can penetrate a barrier, provided the barrier is not too high or too wide. We want to solve the Schrödinger equation in the three regions. The equation may be written in the form

$$
\frac{d^2 u(x)}{dx^2} + k^2 u(x) = \frac{2mV(x)}{\hbar^2} u(x)
\tag{21.2}
$$

with $k^2 = 2mE/\hbar^2$.

In *region I* there is no potential, so that the most general solution is of the form

$$
u(x) = Me^{ikx} + Ne^{-ikx}.
\tag{21.3}
$$

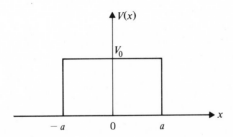

Figure 21.1. Shape of potential barrier in Eq. (21.1).

203

This solution represents a superposition of particles going to the right with flux $v|M|^2$ and particles going to the left with flux $v|N|^2$.

In *region III* there is again no potential, and the general solution is of the form

$$u(x) = Fe^{ikx} + Ge^{-ikx}. \tag{21.4}$$

This solution again represents a flux $v|F|^2$ traveling to the right and a flux $v|G|^2$ traveling to the left. If there were no potential *anywhere*, we would have the same solution everywhere, that is, $F = M$ and $G = N$. Thus the flux to the right as $x \to +\infty$ would be the same as the flux that started out toward the right at $x = -\infty$, and the flux going to the left as $x \to -\infty$ would be the same as the flux sent in toward the left starting at $x = +\infty$. The numbers M and G thus represent the flux being sent in, and they are determined by the experimental setup. If we decide that only particles from the left are to be sent in, and this with flux v, then the solutions in regions I and III are, with a change in notation

$$\begin{aligned} u(x) &= e^{ikx} + Re^{-ikx} \quad \text{(region I)} \\ u(x) &= Te^{ikx}. \quad\quad\quad\ \text{(region III)} \end{aligned} \tag{21.5}$$

For zero potential, $R = 0$ and $T = 1$. With the potential present this will no longer be the case. When $R \neq 0$ we have some flux going to the left. Since no flux was sent in from the right, this is due to reflection from the barrier. Flux conservation suggests that $|T|^2 < 1$, since only part of the incident flux is transmitted. We expect to find

that is
$$j_{\text{incident}} = j_{\text{reflected}} + j_{\text{transmitted}};$$
$$v = v|R|^2 + v|T|^2, \tag{21.6}$$

since the total flux is conserved.

To solve the Schrödinger equation in *region II*, we note that the equation takes the form

$$\frac{d^2u(x)}{dx^2} + \frac{2m}{\hbar^2}(E - V_0)u(x) = 0; \tag{21.7}$$

that is,

$$\begin{aligned} \frac{d^2u}{dx^2} + q^2u &= 0 \quad \text{if } E > V_0 \\ \frac{d^2u}{dx^2} - q^2u &= 0 \quad \text{if } E < V_0 \end{aligned} \tag{21.8}$$

with

$$q^2 = \frac{2m}{\hbar^2}|E - V_0|. \tag{21.9}$$

In the first case, we have a situation in which classically only transmission occurs. The particle slows down when crossing the barrier (since there is less kinetic energy, K.E. $= E - V_0$) but continues on. In the second case, the kinetic energy in region II is negative, so that classically there is only reflection. Let us consider this case in detail. The solution to

$$\frac{d^2u}{dx^2} - q^2u = 0$$

has the general form

$$u(x) = A\cosh qx + B\sinh qx. \tag{21.10}$$

Let us match wave functions and derivatives at $x = -a$ and at $x = a$, as required by the continuity of $u(x)$ and $du(x)/dx$. We obtain four equations:

$$e^{-ika} + Re^{ika} = A \cosh qa - B \sinh qa, \qquad (21.11a)$$

$$ike^{-ika} - ikRe^{ika} = -qA \sinh qa + qB \cosh qa, \qquad (21.11b)$$

$$Te^{ika} = A \cosh qa + B \sinh qa, \qquad (21.11c)$$

$$ikTe^{ika} = qA \sinh qa + qB \cosh qa. \qquad (21.11d)$$

Dividing the fourth by the third equation yields

$$ik = q \, \frac{A \sinh qa + B \cosh qa}{A \cosh qa + B \sinh qa} \, .$$

We may solve this for B and find

$$B = A \, \frac{ik \cosh qa - q \sinh qa}{q \cosh qa - ik \sinh qa} \, . \qquad (21.12)$$

Taking the ratio of the second to the first equation, we get

$$ik \, \frac{e^{-ika} - Re^{ika}}{e^{-ika} + Re^{ika}} = q \, \frac{\sinh qa + (B/A) \cosh qa}{\cosh qa - (B/A) \sinh qa} \, .$$

This may be solved for R

$$R = e^{-2ika} \, \frac{ik (\cosh qa - (B/A) \sinh qa) - q (\sinh qa + (B/A) \cosh qa)}{q (\sinh qa + (B/A) \cosh qa) + ik (\cosh qa - (B/A) \sinh qa)} \, . \qquad (21.13)$$

We can simplify this by working out

$$\sinh qa + \frac{B}{A} \cosh qa$$

$$= \frac{1}{q \cosh qa - ik \sinh qa} \, [q \sinh qa \cosh qa - ik \sinh^2 qa$$
$$+ ik \cosh^2 qa - q \sinh qa \cosh qa]$$

$$= \frac{ik \cosh^2 qa - ik \sinh^2 qa}{\text{denominator}} = \frac{ik}{\text{denominator}}$$

and

$$\cosh qa - \frac{B}{A} \sinh qa$$

$$= \frac{1}{q \cosh qa - ik \sinh qa} \, [q \cosh^2 qa - ik \sinh qa \cosh qa$$
$$- ik \cosh qa \sinh qa + q \sinh^2 qa]$$

$$= \frac{q \cosh 2qa - ik \sinh 2qa}{\text{denominator}} \, .$$

When this is inserted in Eq. (21.13) we get

$$R = e^{-2ika} \, \frac{(q^2 + k^2) \sinh 2qa}{(k^2 - q^2) \sinh 2qa + 2ikq \cosh 2qa} \, . \qquad (21.14)$$

This may now be used to solve for A, and after a little algebra we get

$$T = e^{-2ika} \, \frac{2ikq}{(k^2 - q^2) \sinh 2qa + 2ikq \cosh 2qa} \, . \qquad (21.15)$$

Let us check that

$$|R|^2 + |T|^2 = 1 \qquad (21.16)$$

as anticipated from flux conservation. We have, using

$$\left|\frac{1}{x+iy}\right|^2 = \frac{1}{x+iy}\frac{1}{x-iy} = \frac{1}{x^2+y^2}$$

$$|e^{-2ika}|^2 = e^{-2ika} \times e^{2ika} = 1,$$

$$|T|^2 + |R|^2 = \frac{(q^2+k^2)^2 \sinh^2 2qa + 4k^2q^2}{(k^2-q^2)^2 \sinh^2 2qa + 4k^2q^2 \cosh^2 2qa}$$

$$= \frac{(q^2+k^2)^2 \sinh^2 2qa + 4k^2q^2(\cosh^2 2qa - \sinh^2 2qa)}{[(k^2+q^2)^2 - 4k^2q^2]\sinh^2 2qa + 4k^2q^2 \cosh^2 2qa}$$

$$= 1. \tag{21.17}$$

If we define the *reflection coefficient* by

$$r = \frac{\text{reflected flux}}{\text{incident flux}} = \frac{v|R|^2}{v} = |R|^2, \tag{21.18}$$

and the *transmission coefficient* as

$$t = \frac{\text{transmitted flux}}{\text{incident flux}} = \frac{v|T|^2}{v} = |T|^2, \tag{21.19}$$

then for $E < V_0$,

$$r = \frac{(q^2+k^2)^2 \sinh^2 2qa}{(q^2+k^2)^2 \sinh^2 2qa + 4k^2q^2} \tag{21.20}$$

and

$$t = \frac{4k^2q^2}{(q^2+k^2)^2 \sinh^2 2qa + 4k^2q^2}. \tag{21.21}$$

Before analyzing these expressions in detail, we must deal with an apparent difficulty. The wave function does not vanish in region II and therefore there is some probability of actually finding a particle with negative kinetic energy. How would such a particle look? How can this make sense? The uncertainty relation again rescues from a conflict with classical physics. If we want to observe the particle inside the barrier, we must measure its position with an accuracy

$$\Delta x \ll 2a. \tag{21.22}$$

Such a measurement transfers to the particle an uncontrollable amount of momentum given by

$$\Delta p \gg \frac{\hbar}{a}. \tag{21.23}$$

This corresponds to a transfer of energy

$$\Delta E \gg \frac{\hbar^2}{2ma^2}. \tag{21.24}$$

There will be a chance of observing a negative kinetic energy particle only if this energy transfer is much less than $|E - V_0|$; that is, if

$$\frac{\hbar^2 q^2}{2m} \gg \Delta E \gg \frac{\hbar^2}{2ma^2}. \tag{21.25}$$

This implies that

$$qa \gg 1. \tag{21.26}$$

Under these circumstances

$$t \simeq \frac{16k^2 q^2}{(q^2 + k^2)^2} e^{-4qa} ; \qquad (21.27)$$

that is, the probability of there being penetration into the barrier becomes vanishingly small.[†]

Let us look at the transmission coefficient, using the fact that

$$q^2 + k^2 = \frac{2mV_0}{\hbar^2} . \qquad (21.28)$$

It has the form

$$t = \frac{1}{1 + \left(\dfrac{mV_0}{kq\,\hbar^2}\right)^2 \sinh^2 qL} \qquad (21.29)$$

where $L = 2a$ is the width of the barrier. In the only case that we want to consider in detail, namely, when $qL \gg 1$, the case of a wide and/or high barrier, we have

$$t \cong \left(\frac{2kq\,\hbar^2}{mV_0}\right)^2 e^{-2qL} = 16 \frac{E}{V_0} \left(1 - \frac{E}{V_0}\right) e^{-2qL} \qquad (21.30)$$

since

$$\sinh qL = \tfrac{1}{2}(e^{qL} - e^{-qL}) \approx \tfrac{1}{2} e^{qL}.$$

The transmission coefficient is the fraction of the flux that comes through the barrier, and we may therefore interpret it as the *probability* for a single particle to get through the barrier. If we double the thickness of the barrier we do not quite get the square of the probability of getting through a single barrier. The reason is that there is usually some reflection at the $x = a$ end of the barrier, followed by reflection at the $x = -a$ end which adds to the transmitted flux, and so on, so that the simple law of tunneling through a very thick barrier as a sequence of rare independent events of getting through parts of the thick barrier is not quite correct. Nevertheless, to the extent that the exponential dominates, it is sensible to write the transmission coefficient for a sum of thick barriers in the form

$$t \cong A e^{-2\sum_i q_i L_i} \qquad (21.31)$$

We shall return to this subject in the next chapter.

To illustrate the impossibility of classical tunneling, consider the value of qL for a 10 gm marble confronted with a step 0.1 cm high and 1 cm wide. Let us take the kinetic energy to be negligible. With $g = 980 \text{ gm/sec}^2$,

$$qL = \frac{1}{\hbar}(2mV_0)^{1/2}L = \frac{1}{\hbar}(2mgH)^{1/2}L$$

$$\cong 10^{27}(2 \times 10 \times 980 \times 0.1)^{1/2} \times 1 \simeq 10^{28}.$$

The exponential of this is absolutely enormous, and thus the transition probability is effectively zero. On the other hand, consider an electron with energy $-E_n = -\tfrac{1}{2}mc^2\alpha^2/n^2$ in a Bohr atom. What is the probability that it will tunnel to a neighboring atom (in a crystal, say), a distance of 1 Å away? If we treat the potential barrier as a

---

† For $qa \simeq 10$, $e^{-4qa} \simeq 10^{-18}$.

square barrier of height E_n and width $1\,\text{Å}$, then, again ignoring kinetic energy, and using $1\,\text{Å} \simeq 2a_0 = 2\hbar/mc\alpha$,

$$qL \cong \left(\frac{2m}{\hbar^2} \frac{mc^2\alpha^2}{2n^2} \right)^{1/2} \left(2\frac{\hbar}{mc\alpha} \right) \cong \frac{2}{n}$$

so that for electrons bound only a little below the continuum threshold, the transmission probability $e^{-2/n}$ is not small at all. In fact, in metals, the valence electrons find it easy to tunnel, so that one may describe them as effectively free.

Transmission resonances To work out the transmission problem when $E > V_0$ in Eq. (21.8) it is not necessary to repeat the whole calculation. It is enough to replace q^2 by $-q^2$ in our expressions or, to be more specific, to write

$$q = i\kappa$$

$$\kappa^2 = \frac{2m}{\hbar^2} (E - V_0). \tag{21.32}$$

Using the fact that $\sinh i\kappa L = i \sin \kappa L$ we obtain

$$t = \frac{1}{1 + \dfrac{m^2 V_0^2}{\hbar^2 k^2 \kappa^2} \sin^2 \kappa L}. \tag{21.33}$$

If we fix V_0 and plot t as a function of energy, then if $\sin^2 \kappa L$ were replaced by unity we would get a smooth approach to the asymptotic value of 1. The $\sin^2 \kappa L$ term modulates this approach (Fig. 21.2). Note that whenever

$$\kappa L = n\pi \quad n = 1, 2, \ldots \tag{21.34}$$

we obtain total transmission. This is a pure wave phenomenon, and arises because the waves that are reflected at $x = a$ and $x = -a$ once, twice, ..., and so on, interfere destructively, giving no reflected flux. This phenomenon occurs in optics. There a wave in a vacuum, when entering a medium of refractive index n, has its wavelength reduced by a factor n. This is just what happens in wave mechanics (see Fig. 21.3). The correspondence between n and the potential energy is seen from

$$n = \frac{\lambda_{\text{free}}}{\lambda_{\text{medium}}} = \frac{\lambda_{\text{free}}}{\lambda_{\text{barrier}}} = \frac{p_{\text{barrier}}}{p_{\text{free}}} = \frac{(2m(E - V_0))^{1/2}}{(2mE)^{1/2}} = \left(1 - \frac{V_0}{E} \right)^{1/2}. \tag{21.35}$$

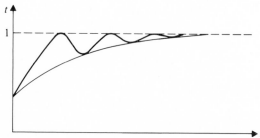

Figure 21.2. Transmission coefficient as a function of energy.

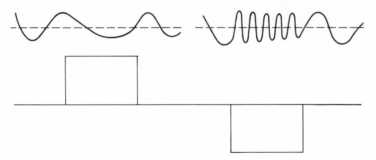

Figure 21.3. Variation of wavelength of particle waves in repulsive and attractive potentials.

The effect that we observed in quantum mechanics has as its counterpart the existence of optical systems which are totally transmitting in the neighborhood of certain frequencies. This is sometimes used in the construction of nonreflecting shop windows.

The so-called *transmission resonance*, as the above effect has been named, has been observed in the scattering of low energy electrons (~ 0.1 eV) by noble gases such as Neon and Argon. The effect was first observed by Ramsauer and Townsend and has been explained in this way.

Finally, if the potential barrier is replaced by a potential hole, then we just change $V_0 \rightarrow - V_0$. The only change is that

$$\kappa^2 = \frac{2m}{\hbar^2}(E - V_0) = \frac{2m}{\hbar^2}(E + |V_0|) \tag{21.36}$$

is now larger than k^2. Figure 21.3 gives a rough picture of the shape of the wave function above a barrier or potential hole. Since the wavelength is inversely proportional to the wave number, the wavelength decreases for an attractive potential. It should be stressed that such drawings are only suggestive, since the solution is really complex. They are useful, however, in the case of bound states.

PROBLEMS

21.1 Consider the potential given by

$$V(x) = 0 \qquad x < 0$$
$$= V_0 \qquad x > 0.$$

a) Solve the Schrödinger equation with the condition that $E > V_0$ and that the flux incident from the left is $\hbar k/m$. Calculate the reflection coefficient and the transmission coefficient. Consider the behavior of these as E increases.

b) Repeat the calculation with $E < V_0$. Show that the transmission coefficient vanishes.

c) Repeat the calculation with flux incident from the right side instead of the left. Show that there will be reflection as well as transmission.

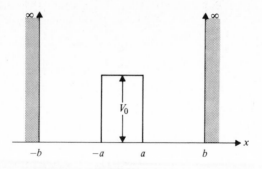

Figure 21.4.

21.2 Consider a particle in the potential shown in Fig. 21.4. Set up the eigenvalue problem in the three regions for the even and the odd solutions separately, and write down the matching conditions at the various points of discontinuity in most convenient form, but do not solve them.

Examine your matching conditions in the limiting case that V_0 is much larger than the energy, and show that the solution approaches that of two independent potential holes.

Examine the matching conditions in the limiting case that V_0 is much smaller than the energy.

21.3 Consider the potential shown in Fig. 21.5. Write down the solutions in the various regions for the case that $E > V_0$. Assume that a flux of $\hbar k/m$ enters from the left.

Write down the matching conditions in most convenient form but do not solve them. If the solution for large negative x is of the form

$$e^{ikx} + Re^{-ikx}$$

what is a special characteristic of R?

21.4 Consider the potential shown in Fig. 21.6. Write down the solutions of the Schrödinger equation in the various regions and the matching conditions corresponding to a flux of 1 unit coming in from the far right.

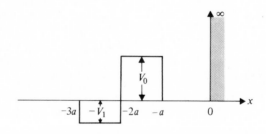

Figure 21.5.

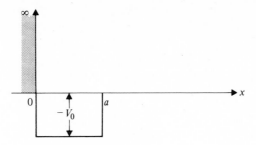

Figure 21.6.

21.5 Consider the potential given by

$$V(x) = 0 \qquad x < -a$$
$$= U(x) \qquad -a < x < a$$
$$= 0 \qquad a < x.$$

If the solution for $x < -a$ is written in the form $Ae^{ikx} + Be^{-ikx}$ and if the solution for $x > a$ is written in the form $Ce^{ikx} + De^{-ikx}$, show that if we relate the constants B and C to the *incoming* flux constants A and D by

$$C = S_{11}A + S_{12}D$$
$$B = S_{21}A + S_{22}D,$$

then flux conservation implies that

$$|S_{11}|^2 + |S_{12}|^2 = 1$$
$$|S_{21}|^2 + |S_{22}|^2 = 1$$
$$S_{11}S_{12}^* + S_{21}S_{22}^* = 0.$$

21.6 Calculate the coefficients $S_{11}, \ldots$ for the potential given by

$$V(x) = -V_0 \qquad -a < x < a$$
$$= 0 \qquad \text{elsewhere}$$

and verify the conditions established in Problem 21.5.

21.7 Write a relation of the form

$$C = T_{11}A + T_{12}B$$
$$D = T_{21}A + T_{22}B$$

with the $T_{11}, \ldots$ expressed in terms of the $S_{11}, \ldots$.

21.8 Assume that the one-dimensional attractive well provides a good model for the Ramsauer–Townsend effect. Given that the energy for the first resonance in the transmission is at 0.1 eV, and that the size of the atom is an Angstrom, estimate the magnitude of the depth of the potential well. How does this compare with the kind of energies familiar from atomic physics?

22
Tunneling

The expression, derived in the last chapter, for the probability of tunneling through a series of barriers,

$$t \approx Ae^{-2\Sigma q_i L_i} \tag{22.1}$$

may be used to calculate the tunneling probability for an arbitrary potential barrier $V(x)$. Figure 22.1 shows such a barrier, which we divide up into individual square barriers. If each barrier has width ΔL, then the probability of getting through all of them is

$$t \cong Ae^{-2\Sigma q_i \Delta L}$$
$$\cong Ae^{-2\int dx q(x)}$$
$$\cong Ae^{-2\int_{x_1}^{x_2} dx[2m(V(x)-E)/\hbar^2]^{1/2}} \tag{22.2}$$

The constant in front of the exponent is not determined by our approximation, so that the expression is most useful for determining the dependence of the penetration probability on the energy and other parameters that appear in $V(x)$. The range of integration is from the point where the kinetic energy first becomes zero, that is, $V(x_1) = E$, to the same point on the other side of the barrier, where $V(x_2) = E$. These are just the points where the integrand vanishes, and they are called *turning points* since that is where the classical particles would have to stop and turn around. Near the turning point $q = 0$, and hence the basic approximation ($q\Delta L$ is large) is not valid. A proper mathematical treatment of this, which is obtained by solving the Schrödinger equation in the limit that the wavelength changes slowly, must take the end points

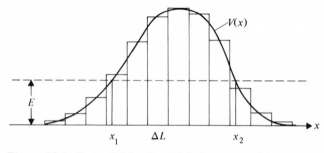

Figure 22.1. Potential barrier $V(x)$ decomposed into square barriers. The turning points x_1 and x_2 corresponding to the energy E are sketched in.

into account. This is done in the so-called WKB (Wentzel–Kramers–Brillouin) approximation. For our purposes Eq. (22.2) is good enough. We apply the formula to several examples from the physics of metals, nuclei, and molecules.

Tunneling in Metals Metals are characterized by the property that they contain free electrons. As we learnt in the discussion of the photoelectric effect, these electrons may be viewed as being held in a metal by a potential, which to first approximation may be described by a potential well of finite depth. A box of finite depth and macroscopic size has energy levels that are very closely spaced (to first approximation this looks like an infinite box with the size $2a$ of the order of centimeters instead of Angstroms). It is a deep law of nature, which we will discuss later, that *no more than two electrons can occupy a single energy level*. A consequence of this law, the Pauli Exclusion Principle, is that the free electrons, which number one or two per atom, fill the lowest levels up to some highest energy, known as the *Fermi energy*. The difference between the top of the potential and the Fermi energy is the *work function* that appeared in our discussion of the photoelectric effect. Electrons can get out of the well by acquiring an energy that is at least as large as the work function W from a photon, or when the metal is heated up, by thermionic emission. When an external electric field is applied, *field emission* occurs, because the potential seen by the electron is changed from W to $W - e\mathscr{E}x$, where $\mathscr{E}$ is the applied electric field, for electrons near the top of the Fermi "sea" of the energy levels. As Fig. 22.2 shows, these electrons can now tunnel through the triangular barrier. The electrons near the top of the "sea" are the most likely to tunnel, since the barrier is lowest and narrowest there. The transmission probability is

$$|T|^2 = Ae^{-\frac{2}{\hbar}\int_0^{W/e\mathscr{E}} dx[2m(W - e\mathscr{E}x)]^{1/2}} \tag{22.3}$$

which, after a little algebra, leads to

$$|T|^2 = Ae^{-\frac{4}{3}\left(\frac{2mW}{\hbar^2}\right)^{1/2}\frac{W}{e\mathscr{E}}}. \tag{22.4}$$

This formula, called the *Fowler–Nordheim formula*, is in qualitative agreement with experiment. To improve it, one could include the additional attraction back to the

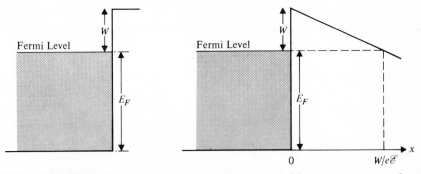

Figure 22.2. Potential seen by electrons in a metal (a) in the absence of an external potential, (b) in the presence of an external potential. The Fermi energy and the work function W are indicated in the figures.

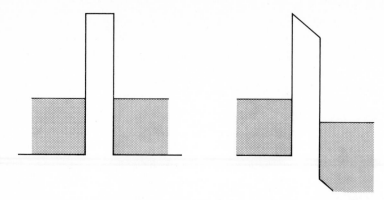

Figure 22.3. The energy levels for two metals separated by a thin barrier (a) in the absence of an electric field, and (b) in the presence of an electric field. The lowering of the Fermi level by the potential has been enormously exaggerated in the second figure.

metal plate (which has now acquired a charge $+e$), but a much more important effect, one much harder to treat, is the fact that there are surface imperfections in the metal that change the electric field locally, and since $\mathscr{E}$ appears in the exponent, these local fluctuations can change the cold emission rate substantially, and somewhat unpredictably. Note that if we write $|T|^2$ in the form $e^{-2\langle q \rangle L}$, where L is the thickness of the barrier at the top of the Fermi sea, then $\langle q \rangle$ is just 2/3 of the q we would have if the barrier were rectangular instead of triangular.

   A different effect occurs when two metal plates are placed in contact, with a thin oxide film between them. Tunneling could occur, except for the effect of the Pauli Exclusion Principle. The levels on both sides of the barrier are filled, and there is no place for the electrons to go. The effect of even a weak electric field is to depress the top of the Fermi sea on one side of the barrier relative to that on the other side, and thus empty levels will be brought in correspondence with filled ones (Fig. 22.3). The transition probability is of the rough form

$$|T|^2 \propto e^{-2a(2mW/\hbar^2)^{1/2}}, \tag{22.5}$$

where a is the gap thickness. This number is very small, and no flow is possible unless a is of the magnitude of 5–20 Å. (The value of $|T|^2$ is of the order of 10^{-5} when the work function is 2 eV and $a = 7$ Å.) As the electric field is increased, the number of empty states made available for electrons to flow into increases, and thus one gets a current that is proportional to the voltage over a certain range of voltage values. Measurements confirm the above manifestation of tunneling and the rough dependence on the gap width, though the latter is difficult to check, since it is very difficult to make such extremely thin films.

Metal-superconductor junction An interesting effect occurs when the metal on one side of the barrier is in a *superconducting state*. A characteristic of a superconductor is that just above the Fermi energy, and just below it, there is a gap in the level density; that is, roughly speaking, the energy levels above E_F are squeezed upward to a level

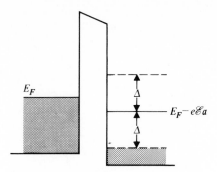

Figure 22.4. The energy level structure of a contact between a metal on the left and a superconductor on the right, when the electric field is too small to lower the level beyond the width of the gap Δ.

$E_F + \Delta$, and the levels below E_F are squeezed downwards. If the electric field is too small, that is, if $e\,\mathscr{E}a$ is less than Δ, then there will still be no tunneling, because no empty levels will be placed in correspondence with the filled levels in the metal (Fig. 22.4). This leads one to expect the relation between current and voltage applied to be as shown in Fig. 22.5. Figure 22.6 shows the actual experimental data for tunneling in an aluminum – aluminum oxide – lead sandwich. At room temperature there is no evidence for a gap. Below $7.2°$K, at which temperature lead becomes superconducting,

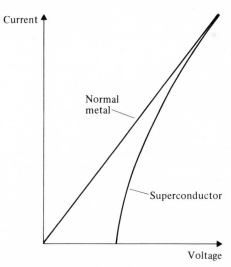

Figure 22.5. Current-voltage characteristic for a metal-oxide-superconductor junction compared with the metal-oxide-metal junction. The slope of the curves depends on $|T|^2$.

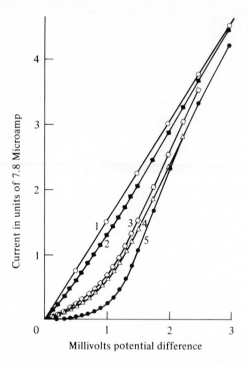

Figure 22.6. Current-voltage curve for an Al-Al$_2$O$_3$-Pb junction for various temperatures. Curves 3, 4, and 5 refer to the lead in the superconducting state. (Source: I. Giaever, *Phys. Rev. Letters*, **5**: 147 (1960). Reprinted by permission.)

one sees the onset of the effect outlined above. The experiments were first carried out by I. Giaever in 1960. Tunneling experiments are very useful in the study of superconductivity. The "superconducting gap" in the energy level density near the top of the Fermi sea is believed due to an attraction between electron pairs, which form something like bound states, with the binding energy of the order of magnitude of the energy Δ. If this is the case, one might expect that irradiating a "sandwich" made of two superconductors separated by an insulating oxide layer could break up some of the "bound states." The electrons that are no longer bound can acquire enough energy from the photon to move into the region of unoccupied levels above the gap and then tunnel across the barrier. This increases the current. Such "photon-assisted" tunneling has been observed by placing the sandwich inside a cavity in which microwaves of frequency 4×10^{10} Hz existed.

Tunneling in Nuclei One of the first applications of the notion of barrier penetration was to the phenomenon of alpha particle emission by nuclei. The study of scattering of alpha particles by nuclei showed that to fairly close distances the potential energy was just the repulsive Coulomb potential, and the height of the potential was some-

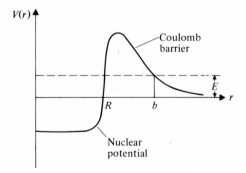

Figure 22.7. Potential energy of alpha particle in nucleus.

times several times larger than the kinetic energy of an alpha particle emitted by the target nuclei. This was mysterious, since if the alpha particle came from the inside of the nucleus, and it had enough energy to cross the barrier (see Fig. 22.7), its kinetic energy would have had to be much larger than observed. Tunneling, allowed by quantum mechanics, can explain the effect. To obtain a simple quantitative treatment of the process, one must assume that somehow individual nucleons inside the nucleus cluster together to form alpha particles, which emerge, leaving behind the residual nucleus, with atomic number $Z - 2$. The total potential felt by a particle of charge $Z_1 e$ is the sum of the nuclear potential due to the presence of the other nuclei, and the Coulomb repulsion due to the remaining charge $Z_2 e$, which is

$$V(r) = \frac{Z_1 Z_2 e^2}{r}. \tag{22.6}$$

The transmission probability for the particle is of the form

$$|T|^2 \simeq e^{-G} \tag{22.7}$$

where

$$G = 2 \left(\frac{2m}{\hbar^2}\right)^{1/2} \int_R^b dr \, [Z_1 Z_2 e^2/r - E]^{1/2}. \tag{22.8}$$

Here R is the nuclear radius, and b is the turning point determined by the vanishing of the integrand. Thus

$$b = Z_1 Z_2 e^2/E. \tag{22.9}$$

The integral can be done exactly. We shall leave this as an exercise for the reader and merely write down the answer

$$\int_R^b dr \, [Z_1 Z_2 e^2/r - E]^{1/2} = (Z_1 Z_2 e^2)^{1/2} \int_R^b dr (1/r - 1/b)^{1/2}$$

$$= b(Z_1 Z_2 e^2/b)^{1/2} (\cos^{-1} \sqrt{R/b} - \sqrt{R/b(1 - R/b)}). \tag{22.10}$$

At energies that are low compared to the height of the Coulomb barrier at $r = R$, we have $b/R \gg 1$ and then $\cos^{-1} \sqrt{R/b} \simeq \pi/2$ so that approximately,

$$G \simeq 2 \left(\frac{2m}{\hbar^2}\right)^{1/2} \cdot \frac{Z_1 Z_2 e^2}{\sqrt{E}} \cdot \frac{\pi}{2} \simeq \pi\sqrt{2} \left(\frac{mc^2}{E}\right)^{1/2} Z_1 Z_2 \alpha, \tag{22.11}$$

where m is the mass of the particle of charge Z_1, that is, $mc^2 = 3750\,\text{MeV}$ for the alpha particle. Also, as usual $\alpha = e^2/\hbar c \simeq 1/137$. Thus with $Z_1 = 2$ we have

$$G \simeq \frac{2Z_1 Z_2}{\sqrt{E(\text{MeV})}} = \frac{4(Z-2)}{\sqrt{E(\text{MeV})}}, \tag{22.12}$$

where Z is the charge of the decaying nucleus and E is the kinetic energy of the alpha particle in MeV.

The time that it takes the alpha particle to get out of the nucleus may be estimated as follows: consider the α-particle as bouncing back and forth inside the nuclear well. With each encounter with the barrier, the particle has a probability of e^{-G} of getting out. Thus the number of encounters needed is

$$n = e^G. \tag{22.13}$$

Each encounter takes $2R/v$ seconds, where v is the velocity of the alpha particle inside the well. Hence the lifetime is of the order of

$$\tau \simeq \frac{2R}{v} e^G. \tag{22.14}$$

The factor in front of the exponential cannot be taken as more than a very rough approximation, because it is based on a very classical picture of what goes on. It is still worth estimating what its value is for a 1 MeV alpha particle. The velocity is given by

$$v = (2E/m)^{1/2} = c(2E/mc^2)^{1/2} \simeq 6.9 \times 10^8 \text{ cm/sec}. \tag{22.15}$$

The radius of the nucleus is approximately

$$R = 1.3\, A^{1/3} \times 10^{-13} \text{ cm}. \tag{22.16}$$

This number was, in fact, first obtained from consideration of alpha decay for energies in which the whole expression (22.10) had to be used. It has as a consequence the fact that the nuclear density

$$\rho \cong \frac{A M_n c^2}{4\pi R^3/3} \cong \frac{3 M_n c^2}{4\pi (1.3 \times 10^{-13})^3} \frac{\text{gm}}{\text{cm}^3}, \tag{22.17}$$

where M_n is the nuclear mass, is independent of A, a crucial factor in understanding nuclear forces. With that value of R we get

$$\tau \cong 3.5 \times 10^{-22} A^{1/3} \frac{1}{\sqrt{E(\text{MeV})}} e^{4(Z-2)/\sqrt{E(\text{MeV})}}. \tag{22.18}$$

If we plot

$$\log_{10} \frac{1}{\tau} \simeq 21.5 + \log_{10} \frac{\sqrt{E}}{A^{1/3}} - 1.74 \frac{(Z-2)}{\sqrt{E(\text{MeV})}} \tag{22.19}$$

against $(Z-2)/\sqrt{E(\text{MeV})}$, we expect to obtain a straight line. A plot of the data for a number of alpha emitters shows that nuclei, with lifetimes ranging from less than one microsecond to more than 10^{10} years, lie on the same curve, given by approximately

$$\log_{10} \frac{1}{\tau} = 28.9 + 1.6\, (Z-2)^{2/3} - 1.61 \frac{Z-2}{\sqrt{E(\text{MeV})}} \tag{22.20}$$

according to one analysis (Fig. 22.8). Given the uncertainties of the model, the agreement between Eq. (22.19) and Eq. (22.20) is impressive, and shows that the description of alpha decay in terms of tunneling is basically correct. This description

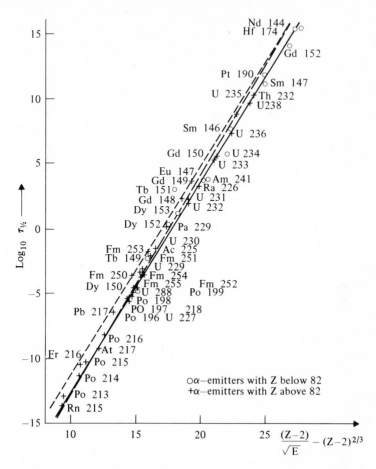

Figure 22.8. Experimental data for alpha decays. The logarithm of the half-life is plotted against $(Z-2)/\sqrt{E} - (Z-2)^{2/3}$. (Source: E. K. Hyde, I. Perlman, and G. T. Seaborg, *The Nuclear Properties of the Heavy Elements*, vol. 1, Fig. 4.12. Reprinted by permission.)

was one of the first applications of quantum mechanics to nuclear physics, carried out in 1928 by G. Gamow and by E. Condon and R. Gurney, independently.

Nuclear reactions between nuclei of charge Z_1 and Z_2, respectively, are attenuated by the barrier factor

$$B = e^{-2Z_1 Z_2 /\sqrt{E(\text{MeV})}} \tag{22.21}$$

and this has important implications in astrophysics and elsewhere. Thus in stellar nucleosynthesis, the burning of hydrogen can occur at a temperature much lower than that required for the burning of helium. The connection with the temperature occurs because the nuclei have approximately a Maxwellian distribution, that is, the number of nuclei with a given energy is roughly proportional to $e^{-E/kT}$. It is because of the presence of the barrier factor that fission reactors use neutrons for the penetration of heavy nuclei. Neutrons are electrically neutral and do not encounter the barrier.

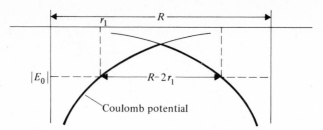

Figure 22.9. Barrier seen by electron in hydrogen atom when a proton is a distance R away.

Attempts at constructing thermonuclear reactors also focus on the burning of isotopes of hydrogen, as in the reactions

$$^2H + {}^2H \rightarrow {}^3H\dot{e} + n + 3.27\,\text{MeV}$$

$$^2H + {}^2H \rightarrow {}^3H + p + 4.03\,\text{MeV}$$

$$^2H + {}^3H \rightarrow {}^4He + n + 17.6\,\text{MeV}$$

since the burning of higher Z elements requires much higher temperatures and correspondingly more difficulties with confinement.

Tunneling in Molecules Tunneling plays a role in the structure of molecules. To illustrate this, consider the simplest of all molecules, the H_2^+ molecule, consisting of two protons and an electron. We take a hydrogen atom and a proton widely separated. The electron is originally localized about one of the protons, and cannot travel to the other proton, because it sees a barrier of height of the order of magnitude $mc^2\alpha^2/2$, since the binding energy is $E_0 = -mc^2\alpha^2/2$, and of width

$$R - 2r_1 = R - 2e^2/|E| = R - 2e^2/\tfrac{1}{2}mc^2\alpha^2$$

$$= R - \frac{4e^2}{\hbar c}\frac{\hbar c}{mc^2\alpha^2} = R - \frac{4\hbar}{mc\alpha}, \tag{22.22}$$

where r_1 is the classical turning point inside the potential. As soon as the protons are within a couple of Ångstroms of each other, tunneling becomes possible (Fig. 22.9), and the electron can move back and forth between the protons, effectively shared by both of them. We shall see in the next chapter, when we discuss bound states, that a particle moving in a double potential has a lower energy than if it were in a single potential. To the extent that this gain in binding overcomes the proton-proton Coulomb repulsion, one gets a stable molecule.

The notion of tunneling is very important in quantum mechanics, and we shall run into it in a variety of contexts.

NOTES AND COMMENTS

1. A very interesting account of the discovery of electron tunneling into super-conductors may be found in *Adventures in Experimental Physics*, Bogdan Maglich, Ed., Vol. 4, World Science Education, Princeton, N.J.

2. A more accurate discussion of alpha decay takes into account the wave function of the particle inside the potential. A more advanced text-book that deals with this subject is M.A. Preston, *Physics of the Nucleus*, Addison-Wesley, Reading, Mass., 1962.

3. The formula (22.2) can be justified more rigorously, and an expression for the constant in front can be obtained using the so-called W.K.B. (Wentzel-Kramers-Brillouin) approximation to the solution of the Schrödinger equation. That solution is obtained by writing

$$\psi(x) = R(x)e^{iS(x)/\hbar},$$

inserting this into the equation, and solving the equations for R and S with the assumption that these vary slowly. The treatment around the endpoints, where the variation is not slow, is somewhat complicated, and that is why we have chosen to derive the formula in the nonrigorous way that we did.

PROBLEMS

22.1 Consider the potential shown in Fig. 22.10.

a) Write down an approximate expression for the transmission probability for a low energy electron $(E \ll V)$.

b) Calculate the transmission probability for $E = V/2$ using the approximate expression.

c) Estimate this numerically, if $V = 10\,\text{eV}$ and $a = 0.5\,\text{Å}$ for an electron of mass $m = 0.9 \times 10^{-27}\,\text{gm}$.

22.2 What is the probability that a 1000 kg automobile of zero energy will tunnel through a bump in the road that is 1 m wide and 10 cm high? To be able to use the approximate formula (22.2), represent the potential by an appropriately chosen integrable form such as

$$V(x) = A \sin^2 wx$$

in the region where the bump is located. You will surely expect a small number, but to appreciate just how small it is, make a guess before starting the calculation and then compare.

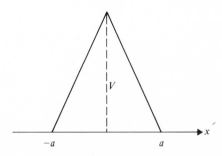

Figure 22.10

22.3 The nucleus $_{88}$Ra^{226} undergoes alpha decay, with the energy of the alpha particle equal to $E = 4.78$ MeV. Should one use Eq. (22.10) or Eq. (22.11) for the computation of the mean life? Calculate the half-life using the appropriate formula, and compare it with the experimental value of 5×10^{10} sec.

22.4 When a particle has angular momentum that is different from zero, the decay is inhibited by what is called a centrifugal barrier, which is represented by the potential
$$V(r) = l(l+1)\hbar^2/2mr^2.$$
Calculate the transmission probability through such a centrifugal barrier (Eq. (22.7)), assuming that the barrier starts at $r = R$ and extends to infinity. Show that for low energies the decay rate is suppressed by approximately $(kR)^{2l}$.

22.5 The following alpha emitters, all with $Z = 84$, may be used to study the energy dependence of the rate of decay, and the nuclear radius.

A	E (MeV)	$T_{1/2}$
210	5.4	138 days
212	8.95	3×10^{-7} sec
214	7.83	1.5×10^{-4} sec
215	7.5	1.8×10^{-3} sec
216	6.89	0.16 sec
218	6.12	3.05 min

Use Eq. (22.8) to check the tunneling relation between half-life and energy.

22.6 Compute the reflection and transmission coefficients for the square barrier and square potential hole in the limit that the width goes to zero and the magnitude of the potential goes to infinity so that
$$\int_{-\infty}^{\infty} dx \, V(x) = \lambda.$$

23
Bound States

Bound states in square well In Chapter 22 we discussed solutions of the Schrödinger equation with positive energy. Such solutions were seen to describe states which were not localized: the wave functions did not vanish at infinity. For attractive potentials there are localized solutions. Such solutions have $E < 0$, and they describe bound states. In classical mechanics, too, states with $E < 0$ are localized, since in the distant regions, where the potential energy vanishes, and only the kinetic energy remains, solutions with $E < 0$ are inadmissible.

Let us consider the potential (Fig. 23.1)

$$V(x) = -V_0 \qquad -a \leqslant x \leqslant a$$
$$= 0 \qquad \text{elsewhere.} \qquad (23.1)$$

Such a potential represents a sharp-edge idealization of finite range potentials that occur in nuclear physics and in solid-state physics. The classical motion of a particle in such a potential is just the to-and-fro motion of a particle bouncing off the rigid walls. We might expect the quantum mechanical states to be quantized, as for the infinite box, at least for a deep enough potential well. The situation is different from the case of the infinite box, in that we can no longer expect the wave function to vanish at the

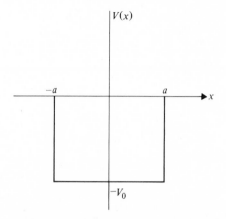

Figure 23.1. Square well potential described by Eq. (23.1).

223

boundaries; we have learned from our study of tunneling that a penetration of particles into forbidden regions is possible.

The sharp edges of the potential allow us to divide space into regions. Outside of the potential region, the wave equation reads

$$\frac{d^2 u(x)}{dx^2} + \frac{2mE}{\hbar^2} u(x) = 0 \tag{23.2}$$

and if we want to consider $E < 0$, we write

$$\frac{2mE}{\hbar^2} = -\alpha^2 \tag{23.3}$$

and look at solutions of

$$\frac{d^2 u}{dx^2} - \alpha^2 u = 0. \tag{23.4}$$

These must be of the form $e^{\pm \alpha x}$ and they must vanish at infinity so as to be square integrable, or at least finite. Thus

$$\begin{aligned} u(x) &= A e^{\alpha x} & x < -a \\ &= B e^{-\alpha x} & x > a. \end{aligned} \tag{23.5}$$

In the potential region we have

$$\frac{d^2 u}{dx^2} + \frac{2m}{\hbar^2} (E - (-V_0)) u(x) = 0. \tag{23.6}$$

We write

$$\frac{2m}{\hbar^2} (E + V_0) = \frac{2mV_0}{\hbar^2} - \alpha^2 \equiv q^2 \tag{23.7}$$

so that the equation becomes

$$\frac{d^2 u}{dx^2} + q^2 u = 0 \tag{23.8}$$

and the most general solution is

$$u(x) = C \cos qx + D \sin qx. \tag{23.9}$$

There is a relation between the coefficients A, B, C, and D that comes from requiring continuity of $u(x)$ and $du(x)/dx$ at $x = \pm a$. We do this by a shortcut, in which we match $(du/dx)(1/u)$. This eliminates the quantities A and B, but we shall see that they can later be determined by the normalization condition, that is, by the requirement that

$$\int_{-\infty}^{\infty} dx \, |u(x)|^2 = 1. \tag{23.10}$$

Matching the ratio at $x = -a$ gives us

$$\alpha = \frac{qC \sin qa + qD \cos qa}{C \cos qa - D \sin qa}. \tag{23.11}$$

Matching the ratio at $x = a$ gives

$$-\alpha = \frac{-qC \sin qa + qD \cos qa}{C \cos qa + D \sin qa}. \tag{23.12}$$

Comparing these relations gives us a relation between C and D. A little algebra gives the remarkable condition

$$CD = 0. \tag{23.13}$$

If $D = 0$, the solution in the potential region is a cosine, that is, it is even under exchange $x \to -x$. In that case the inside solution is the same at $x = a$ and at $x = -a$, so that $A = B$ there. If $C = 0$, the inside solution is odd under reflection $x \to -x$. This corresponds to a solution in which $B = -A$. It is generally true for bound state problems, in contrast to the kind of "scattering" problems discussed in Chapter 21, that eigenfunctions are characterized by a parity quantum number (evenness or oddness) if the potential is symmetric under the interchange $x \to -x$. We shall make use of this in discussing more complicated potentials. Consider the two cases in detail:

1. Let $D = 0$. In that case the two conditions (23.11) and (23.12) are the same, and they read

$$\alpha = q \tan qa. \qquad (23.14)$$

This is an *eigenvalue condition*, and is satisfied only for certain discrete values of α and hence $E = -\hbar^2 \alpha^2 / 2m$. To analyze the condition, we introduce the notation

$$qa = y; \quad 2mV_0 a^2 / \hbar^2 = \lambda \qquad (23.15)$$

in terms of which the condition becomes

$$\tan y = \frac{(\lambda - y^2)^{1/2}}{y}. \qquad (23.16)$$

Figure (23.2) shows a plot of the two functions in the equation. The points of intersection yield the *eigenvalues*. The graph yields some interesting information:

a) No matter how small λ is, there will always be an intersection point. This means that no matter how weak the attractive potential, there will always be one bound state. This is characteristic of one-dimensional problems. In three dimensions there is a minimum depth for the potential of a given range for a bound state to develop.

b) As λ increases, a second bound state develops as soon as λ becomes as large as π. With increasing λ, that is, with increasing depth of the potential, more and more bound states become possible.

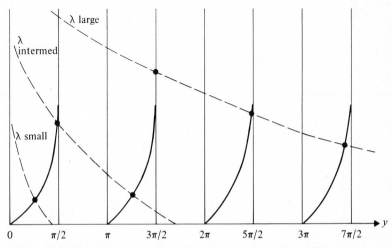

Figure 23.2. Eigenvalue condition for even solutions. Heavy lines represent $\tan y$ and dotted lines represent $(\lambda - y^2)^{1/2}/y$.

c) For a very deep potential, that is, for λ very large, the points of intersection of the two curves become very close to $y = (n + \frac{1}{2})\pi$, since the tangent curve becomes steep rather quickly. The condition

$$q^2 a^2 = (n + \tfrac{1}{2})^2 \pi^2 \qquad n = 0, 1, 2, \ldots \tag{23.17}$$

shows that the spacing between levels follows the same rule as was found for the infinite box in Chapter 20. This should come as no surprise: conditions at the bottom of a deep well should not depend very much on whether the potential is infinitely deep or not.

2. When $C = 0$, the condition reads

$$\alpha = -q \cot q\alpha = q \tan\left(qa + \frac{\pi}{2}\right). \tag{23.18}$$

The same figure can be used for studying the eigenvalue conditions, since the only difference is that the tangent curve is displaced to the right by $\pi/2$ (Fig. 23.3). Here there is a condition for the appearance of the first bound state: it is $\lambda \geqslant \pi/2$, that is,

$$\frac{2mV_0 a^2}{\hbar^2} \geqslant \frac{\pi^2}{4}. \tag{23.19}$$

Other than that, the discussion is essentially the same as for the even solution.

The odd solutions all vanish for $x = 0$, and therefore the solution is the same as for a potential of the form

$$\begin{aligned} V(x) &= \infty & x &< 0 \\ &= -V_0 & 0 &\leqslant x \leqslant a \\ &= 0 & a &< x \end{aligned} \tag{23.20}$$

where we would impose the condition $u(0) = 0$ as for the infinite box. We shall see that this is the condition that appears for a square well in three dimensions, so that the odd solutions solve that problem as well.

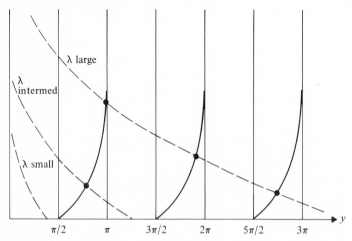

Figure 23.3. Eigenvalue condition for odd solutions. Heavy lines represent $\tan y$ and dotted lines represent $(\lambda - y^2)^{1/2}/y$.

The square-well problem is sufficiently informative to allow for some generalizations: The constraint that gives rise to discrete eigenvalues is the need to have a wave function that vanishes at infinity, and this is directly related to the probability interpretation of the wave function. Had we allowed a wave function of the form

$$u(x) = A_1 e^{-\alpha x} + A_2 e^{\alpha x} \tag{23.21}$$

for $x > a$, the continuity conditions would have given us only relations between the coefficients of the various solutions, but no eigenvalue condition. Why an eigenvalue condition arises can be seen graphically (Fig. 23.4). For the even-parity ground state, the wave function inside is of the form $\cos qx$ and it is to be tied continuously to a falling exponential $e^{-\alpha x}$ with $\alpha^2 = 2mE_B/\hbar^2$. Large binding means a rapidly falling exponential. Since $q^2 = 2m(V_0 - E_B)/\hbar^2$, a large binding energy means that q is small, that is, the wave inside the potential has a long wavelength. This means that matching is impossible. As we reduce the trial value of E_B the exponential falls off less steeply, and the inside wave function curves more, so that a match is possible. For the first excited bound state, with odd parity, the wave function vanishes at the origin, and it can only tie onto a falling exponential if it has a chance to turn over (at least once) inside the potential. The condition that it turns over just at the edge of the well, and ties on to a straight line ($\alpha = 0$) is that $\sin qa = 1$, so that

$$qa = \pi/2. \tag{23.22}$$

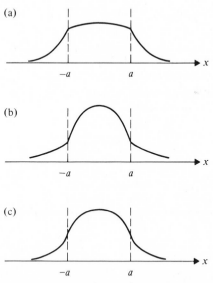

Figure 23.4. Matching of wave function inside potential to exponentially falling outside wave function. (a) Mismatch due to too large a choice of (negative) energy parameter; (b) mismatch due to too small a choice of energy parameter; (c) proper match at the correct value of energy parameter.

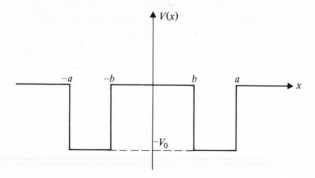

Figure 23.5. Double potential well.

Double potential well We may use these considerations to discuss an electron bound in a double potential well (Fig. 23.5). The ground state is again of even parity. The reason for that is the fact that the kinetic energy contribution to the total energy is smallest for wave functions with the smallest number of nodes (zeros) as we saw in Eq. (20.18). Next we notice that for the double potential the binding can be stronger than for a single potential. For a single potential the wave function inside the well can match exponential fall-offs on the two sides only if it is appropriately curved. For a double well, the wave function must match a falling exponential on the outside of the system, but on the interior region it can tie onto a much flatter function cosh αx. This can be seen in Fig. 23.6. Note also that the wave function is large in the region between the two potentials. The electron in a single well must spend a good part of the time outside and this extended probability distribution means that the wave function outside falls off less sharply. For the double well, the electron can spend its time between the wells, instead of on the outside. Very crudely, the stronger binding is connected with the extended structure of the double well system. By the uncertainty

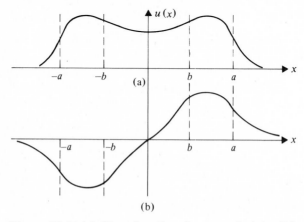

Figure 23.6. (a) Wave function for ground state. (b) Wave function for next highest bound state of double potential well.

relation, a more extended structure implies less kinetic energy, and therefore the negative potential energy is more effective in binding the particle. The existence of a second bound state of odd parity is also enhanced with a double well. Figure 23.6 shows the wave functions for the ground state and the first excited state of a double well. Note that if the two wells were far apart, then the even wave function and the odd wave function would tend to become degenerate, that is, correspond to the same energy, because the even function would have the form

$$u_E(x) \cong u_A(x) + u_B(x) \tag{23.23}$$

where $u_A(x)$ is the ground-state wave function for a well centered at A and the function $u_B(x)$ is the same wave function centered at B, and the odd function has the form

$$u_O(x) \cong u_A(x) - u_B(x). \tag{23.24}$$

The degeneracy is not quite perfect, since the total wave function really does "know" about the other potential, even in the region around one of them.

It is amusing that one can actually estimate the energy difference in the limiting case by invoking the notion of tunneling. Consider an electron localized around potential A. The wave function is, according to Eq. (23.23) and Eq. (23.24), of the form

$$u(x) \cong \frac{1}{\sqrt{2}}(u_E(x) + u_O(x)). \tag{23.25}$$

This is not an energy eigenstate, since u_E and u_O have different time dependences. At a later time the form of the wave function is

$$u(x, t) \cong \frac{1}{\sqrt{2}}(u_E(x)e^{-iE_0 t/\hbar} + u_O(x)e^{-iE_1 t/\hbar})$$

$$\cong \frac{1}{\sqrt{2}}(u_E(x) + u_O(x)e^{-i(E_1 - E_0)t/\hbar})e^{-iE_0 t/\hbar} \tag{23.26}$$

and this shows that after a time t such that

$$(E_1 - E_0)t/\hbar = \pi,$$

that is, after a time

$$t = \frac{\hbar \pi}{E_1 - E_0} \tag{23.27}$$

the electron will actually be localized around potential B. It got there by tunneling through a barrier whose width is $2b$ and whose height is approximately V_0. We may use the arguments of the preceding chapter to estimate the time that it takes to get from A to B, and equate it to t in Eq. (23.27).

Ammonia clock Although our considerations are only approximate, because we have considered only the two lowest eigenfunctions of the system, they are applicable to a variety of physical systems. A good example is the ammonia molecule: NH_3 has a tetrahedral shape (Fig. 23.7) with the three H nuclei forming a triangular base, and with the nitrogen at one of two symmetrically located apexes of the pyramid. The location of the N nucleus is determined by the condition that there the energy is a minimum. This implies that in the region between the two minima there is a potential

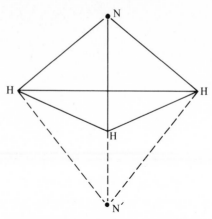

Figure 23.7. Schematic representation
of localized sites of nuclei in ammonia
molecule. The nitrogen nucleus oscillates between N and N'.

barrier. The two-well problem models this, and we might expect two nearly degenerate
states, with the electron, if slightly disturbed, oscillating between the two minima with
a well-defined frequency

$$\nu = \frac{E_1 - E_0}{h} = 23.87 \times 10^9 \text{ Hz.} \tag{23.28}$$

This frequency happens to lie in the microwave region of the spectrum, and it can be
studied by putting ammonia samples in resonance cavities; these can be tuned to high
precision to match the characteristic frequency of the ammonia molecule, and thus
provide us with an "atomic clock."

NOTES AND COMMENTS

1. The method of making an accurate "ammonia clock" using the energy splitting
    between the ground state and the lowest vibrational excitation level is the follow-
    ing: a beam of ammonia molecules, heated in an oven to 300°K contains a large
    number of molecules in the upper state. The beam is passed through an electro-
    static separator. Because the molecules in the two states have different electric
    dipole moments, such a separator can get rid of the molecules in the lower state,
    and send a beam of pure "upper state" molecules into a microwave cavity reso-
    nator. When the electric field in the cavity oscillates with a frequency that just
    matches the transition frequency of 23.87×10^9 Hz, it induces transitions to the
    ground state. The energy goes in part to make up for cavity losses, and the rest
    appears as electrical energy. The ammonia maser (*m*icrowave *a*mplification by
    *s*timulated *e*mission of *r*adiation) is not a good power source, since there are
    limitations on the flux of upper-state molecules. It does, however, make a good
    clock in that one can achieve high precision in cavity tuning to the transition
    frequency. The natural linewidth for the transition is 4×10^3 Hz, so that it
    appears that the limit of accuracy is $\Delta\nu/\nu \cong 1.6 \times 10^{-7}$. Actually it turns out that

because of the interaction with the cavity, the effective width of the spectrum of radiation is reduced to $\Delta \nu \sim 10^{-2}$ Hz, so that the "clock" is accurate to one part in 10^{12}, that is, of the order of seconds per 10^5 years. The measurements were made by taking two oscillators with a frequency difference of the order of 20– 30 Hz and looking at the beat frequency; that beat frequency was found to be stable, with fluctuations of less than 0.1 Hz.

PROBLEMS

23.1 Consider the potential shown in Fig. (23.8). Write down the solutions to the Schrödinger equation in the various regions and the matching conditions. Do not solve.

23.2 Consider the limiting case of a very deep short-range potential, with $V_0 \to \infty$ and $a \to 0$, so that
$$2aV_0 \;=\; \lambda,\, \text{fixed}.$$
Work out the limiting cases of the eigenvalue conditions for the even and odd solutions. Show that there is always just one even eigenvalue, and there is no odd eigenvalue. Sketch the wave function for the even solution in the limiting case.

23.3 Consider the positive energy solution involving an attractive potential. The solution is of the form(21.5). If we let the energy E become negative, this corresponds to letting $k \to i\alpha$. The solution (21.5) does not make sense, unless R and T become infinite, so that the bad part of the solution can be neglected. Show that the condition that R and T become infinite for $k = i\alpha$ is identical to the eigenvalue condition for the attractive potential.

23.4 Consider the following argument: If we have an electron in a potential of width a, then the kinetic energy is, by the uncertainty principle, of order of magnitude $\hbar^2/2ma^2$, and therefore the potential energy must be sufficiently negative to overcome this, in order to give a bound state. Thus one expects a condition of the type $V_0 \gtrsim \hbar^2/ 2ma^2$ to hold. Contrast this with the result that for a one-dimensional system, there is always a bound state, no matter how small V_0 is, for a given a. What is wrong with the above argument?

23.5 Consider a one-dimensional square-well potential of width 2 Å, and depth 10 eV. Given that the electron mass is $m = 0.9 \times 10^{-27}$ gm, calculate the binding energy for

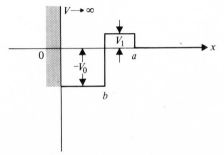

Figure 23.8. Potential for Problem 23.1.

an electron in the ground state of the potential. Is the potential deep enough to give rise to a second bound state?

23.6 Consider a square well that is deep enough to give rise to a weakly bound second eigenstate. The potential has depth $V = V_0 + W$, where

$$2mV_0a^2/\hbar^2 = \pi^2/4.$$

Express the binding energy in terms of W, neglecting terms of order W^2.
(One possibility is to use the graphical method and approximate the tangent curve by a straight line near the axis.)

23.7 Consider a particle in the double well shown in Fig. 23.5. Write down the solutions in the various regions and match them at the boundaries. Prove that the eigenvalue condition may be written in the form

$$\tan q(a - b) = \frac{q\alpha(1 + \tanh \alpha b)}{q^2 - \alpha^2 \tanh \alpha b}$$

for the even solution, and

$$\tan q(a - b) = \frac{q\alpha(1 + \coth \alpha b)}{q^2 - \alpha^2 \coth \alpha b}$$

for the odd solution.

23.8 Consider the expressions derived in Problem 23.7.

a) Show that the eigenvalue conditions approach the single-well conditions when $b \to 0$.

b) Consider the case where $a - b = c$ is fixed, and b becomes very large. Show that the even and odd eigenvalues approach each other in value. Can you show how the difference in energy is related to the tunneling time? [*Hint*: Recall that for large z, $\tanh z = 1 - 2e^{-2z}$ and $\coth z = 1 + 2e^{-2z}$, and work to lowest order in e^{-2z}.]

c) Consider the limiting case $c \to 0$, $V_0 \to \infty$, so that

$$V_0 c = \lambda \qquad \text{fixed}$$

and study the eigenvalue conditions. Show that there is always a single bound state for the even solution, and that there may be (no more than) one odd eigenfunction.

23.9 Consider a potential given by

$$V(x) = \infty \text{ for } x < 0$$
$$= 0 \text{ for } x > a$$
$$= \text{unknown in between.}$$

What is known is that the interior wave function is such that

$$\frac{1}{u}\frac{du}{dx} = f(k)$$

at $x = a$. Assuming that $f(k)$ varies slowly, so that it may be taken to be a constant, f_0, calculate the binding energy in terms of f_0. If a plane wave e^{-ikx} is sent in from $+\infty$, it will be reflected by the potential. Calculate the reflection coefficient in terms of f_0, and show that $|R|^2 = 1$. Relate R to the binding energy.

24
The Harmonic Oscillator

The harmonic potential The harmonic oscillator is described by the energy operator
$$H = p^2/2m + kx^2/2 \tag{24.1}$$
where k is the spring constant for the oscillator. Classically the motion in such a potential is sinusoidal, $x(t) = x_0 \sin(\omega t + \delta)$ with the angular frequency given by
$$\omega = (k/m)^{1/2} \tag{24.2}$$
The classical energy is given by
$$E = \tfrac{1}{2}kx_0^2 \tag{24.3}$$
where x_0 is the location at which the particle has zero kinetic energy (the classical "turning point"). This can take on any value. We shall find, as we did for the infinite box, that in the quantum treatment, the energy is quantized.

There are two reasons for discussing this problem: one is that we are for the first time dealing with a potential that is not (sectionally) constant. This will teach us something about how to deal with differential equations. The other is that the harmonic oscillator is very important. For any potential, the behavior in the vicinity of the classical equilibrium x_c point defined by
$$\left(\frac{dV(x)}{dx}\right)_{x=x_c} = 0 \tag{24.4}$$
is given by
$$V(x) \cong V(x_c) + \tfrac{1}{2}k(x-x_c)^2 \tag{24.5}$$
where
$$k = \left(\frac{d^2V(x)}{dx^2}\right)_{x=x_c}. \tag{24.6}$$

Thus for many systems the low-lying quantum mechanical states are just simple harmonic oscillator states. For example, the potential energy of two atoms interacting has the shape shown in Fig. 24.1. The long range attraction arises from electrostatic polarization forces. As the atoms approach each other, the attraction is increased because the electrons can redistribute themselves in a way to maximize the attractive Coulomb force between electrons and nuclei. At short distances there is a repulsion caused by a purely quantum mechanical effect forbidding large overlaps of electron wave functions. Thus the interatomic distance is the equilibrium radius R_0. The potential in that vicinity is parabolic and one expects that the molecule should have

233

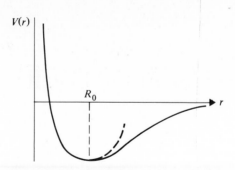

Figure 24.1. Potential energy of diatomic molecule as a function of the separation between the nuclei. R_0 represents the equilibrium separation. The potential energy is harmonic in the immediate vicinity of the classical equilibrium separation.

vibrational degrees of freedom that are properly described by the harmonic oscillator equation. This is fully borne out by experiment.

Let us replace k by $m\omega^2$ in the Hamiltonian, and write the Schrödinger equation as

$$-\frac{\hbar^2}{2m}\frac{d^2u(x)}{dx^2} + \frac{1}{2}m\omega^2 x^2 u(x) = Eu(x). \tag{24.7}$$

We write it in dimensionless form for convenience. Let $y = \alpha x$: then

$$\frac{d^2u}{dy^2} - \frac{m^2\omega^2}{\alpha^4\hbar^2}y^2u + \frac{2mE}{\hbar^2\alpha^2}u = 0.$$

The coefficient of the y^2 term becomes unity if we choose

$$\alpha^2 = \frac{m\omega}{\hbar} \tag{24.8}$$

and the energy term becomes

$$\frac{2mE}{\hbar^2\alpha^2} = \frac{2E}{\hbar\omega} \equiv \epsilon. \tag{24.9}$$

The equation thus takes the form of the eigenvalue equation

$$\frac{d^2u}{dy^2} + (\epsilon - y^2)u = 0 \tag{24.10}$$

with

$$y = \alpha x. \tag{24.11}$$

We leave it to the reader to check that both y and ϵ are now dimensionless quantities.

Behavior at infinity A standard procedure in solving equations of this type is to deal separately with the behavior at infinity and at the origin. To begin with, for large y, we can neglect ϵ and we get the approximate equation

$$\frac{d^2u}{dy^2} = y^2u. \tag{24.12}$$

We want a solution that goes to zero as $y \to \infty$. As a trial solution, anticipating some sort of exponential decay, let us try

$$u(y) = e^{-f(y)}.$$ (24.13)

Then

$$du/dy = -f'e^{-f}; \quad d^2u/dy^2 = f'^2 e^{-f} - f''e^{-f}$$

where the prime denotes differentiation with respect to y. Thus

$$f'^2 - f'' = y^2$$ (24.14)

and for large y it is easy to see that $f' = y$ is a good approximation. This implies that

$$f = y^2/2.$$ (24.15)

Let us now write

$$u(y) = F(y)e^{-y^2/2}.$$ (24.16)

Then

$$\frac{du}{dy} = F'e^{-y^2/2} - yFe^{-y^2/2}$$

and

$$\frac{d^2u}{dy^2} = F''e^{-y^2/2} - 2yF'e^{-y^2/2} + y^2 Fe^{-y^2/2} - Fe^{-y^2/2}$$

so that Eq. (24.10) takes the form

$$F''e^{-y^2/2} - 2yF'e^{-y^2/2} + y^2 Fe^{-y^2/2} + (\epsilon - 1 - y^2)Fe^{-y^2/2} = 0,$$

that is,

$$\frac{d^2F}{dy^2} - 2y\frac{dF}{dy} + (\epsilon - 1)F = 0.$$ (24.17)

The Eigenvalue Condition To solve this, let us try the series solution

$$F(y) = \sum_{n=0}^{\infty} a_n y^n.$$ (24.18)

This will be a solution provided that

$$\sum_{n=0}^{\infty} n(n-1)a_n y^{n-2} - 2\sum_{n=0}^{\infty} na_n y^n + (\epsilon - 1)\sum_{n=0}^{\infty} a_n y^n = 0.$$

The first sum actually starts with $n = 2$, since the $n = 0$ and $n = 1$ terms vanish. If we write $n - 2 = m$ in that sum, we obtain the expression

$$\sum_{m=0}^{\infty} ((m+2)(m+1)a_{m+2} + (\epsilon - 1 - 2m)a_m)y^m = 0.$$ (24.19)

The vanishing of such a series can be accomplished only if the coefficient of every power of y vanishes. This implies that

$$a_{m+2} = \frac{2m + 1 - \epsilon}{(m+1)(m+2)} a_m.$$ (24.20)

Such an equation is known as a *recursion relation*.

In this particular case, a_0 may be used to determine the coefficients of all the even powers of y, and a_1 yields all the odd powers of y. The terms proportional to a_0 and a_1, respectively, may be treated separately. This should come as no surprise: We saw in our discussion of the infinite box that for a potential invariant under reflections

$(x \rightarrow -x)$, the eigensolutions may be characterized by the parity quantum number. Let us now consider the even powers. For large m, we have

$$a_{m+2} \simeq \frac{2}{m} a_m \qquad (24.21)$$

and this is just the recursion relation that the successive coefficients in the expansion of

$$e^{y^2} = \sum_{k=0}^{\infty} \frac{1}{k!} y^{2k} \qquad (24.22)$$

satisfy:

$$\frac{a_{2k+2}}{a_{2k}} = \frac{1}{k+1} \sim \frac{2}{2k} .$$

One can show that the series generated by the a_m satisfying the recursion relation (24.21) consists of two terms: one proportional to e^{y^2} and the other smaller than that for large y. This, however, is a disastrous situation, since such an $F(y)$, even when multiplied by $e^{-y^2/2}$ grows unacceptably for large y. The only solution is if the series terminates. This happens if some a_M does not vanish, but a_{M+2} does. This can happen only if

$$\epsilon = 2M + 1. \qquad (24.23)$$

Thus we have found the eigenvalues: the energy must be of the form

$$E = \hbar\omega(M + \tfrac{1}{2}) \quad M = 0, 1, 2, \ldots . \qquad (24.24)$$

When $M = 0, 2, 4, 6, \ldots$ the series is even under the interchange $y \rightarrow -y$; when $M = 1, 3, 5, 7, \ldots$ the series is odd under the reflection. In either case, the function $F(y)$ is a polynomial.

Let us calculate a few of the polynomials, choosing a_0 and a_1 equal to unity. For $M = 0$, we have

$$F(y) = 1 \qquad (24.25)$$

so that the eigenfunction for $\epsilon = 1$ is just $e^{-y^2/2}$. This must still be normalized. For $M = 1$, we have

$$F(y) = y \qquad (24.26)$$

so that the eigensolution for $\epsilon = 3$ is of the form

$$u(y) = y e^{-y^2/2}. \qquad (24.27)$$

When $M = 2$, we have $a_2 = -2a_0$, so that the eigensolution for $\epsilon = 5$ is of the form

$$u(y) = (1 - 2y^2) e^{-y^2/2}. \qquad (24.28)$$

Hermite Polynomials There is no need to proceed in this tedious way, because the solutions of the equation (24.17) with $\epsilon = 2M + 1$, that is of

$$\frac{d^2 F}{dy^2} - 2y \frac{dF}{dy} + 2MF = 0 \qquad (24.29)$$

with M integral, are known in the mathematical literature. They are the so-called Hermite polynomials, $H_M(y)$. Many of their properties can be worked out directly from Eq. (24.29). We do not do this here, and limit ourselves to a list of useful properties:

1. $$H_{M+1} - 2yH_M + 2MH_{M-1} = 0 \qquad (24.30)$$

2.
$$H_{M+1} = 2yH_M - \frac{d}{dy}H_M \qquad (24.31)$$

3.
$$\sum_{M=0}^{\infty} H_M(y)\frac{z^M}{M!} = e^{2zy-z^2} \qquad (24.32)$$

4.
$$H_M(y) = (-1)^M e^{y^2}\frac{d^M}{dy^M}(e^{-y^2}) \qquad (24.33)$$

The normalization of the Hermite polynomials is such that

5.
$$\int_{-\infty}^{\infty} e^{-y^2} H_M^2(y)dy = 2^M M! \sqrt{\pi} . \qquad (24.34)$$

The reason for this complicated result is that the Hermite polynomials are defined such that the coefficient of the highest power of y in $H_M(y)$ is 2^M.

We list a few of the Hermite polynomials:

$$H_0(y) = 1$$
$$H_1(y) = 2y$$
$$H_2(y) = 4y^2 - 2$$
$$H_3(y) = 8y^3 - 12y$$
$$H_4(y) = 16y^4 - 48y^2 + 12$$
$$H_5(y) = 32y^5 - 160y^3 + 120y \qquad (24.35)$$

Finally, we may write the properly normalized eigenfunctions corresponding to the energy

$$E_M = \hbar\omega(M + 1/2) \qquad (24.36)$$

as

$$u_M(x) = \left(\frac{1}{2^M M! \sqrt{\pi}}\right)^{1/2} e^{-m\omega x^2/2\hbar} H_M\left(\sqrt{\frac{m\omega}{\hbar}}x\right). \qquad (24.37)$$

Figure 24.2 shows the shape of the first few eigenfunctions. We may use these eigenfunctions to calculate expectation values of powers of x and p.

First of all, we note that odd powers of x or p have vanishing expectation values. This happens because the eigenfunctions have well-defined transformation properties under reflections $x \rightarrow -x$. Using

$$u_M(-x) = (-1)^M u_M(x), \qquad (24.38)$$

we find that

$$\langle x^{2p+1}\rangle = \int_{-\infty}^{\infty} dx (u_M(x))^2 x^{2p+1}$$

$$= \int_{-\infty}^{\infty} dx (u_M(-x))^2 (-x)^{2p+1}$$

$$= (-1)^{2M+2p+1}\int_{-\infty}^{\infty} dx (u_M(x))^2 x^{2p+1}$$

$$= (-1)^{2M+2p+1}\langle x^{2p+1}\rangle$$

$$= 0 \qquad (24.39)$$

since a quantity that is equal to its negative must vanish.

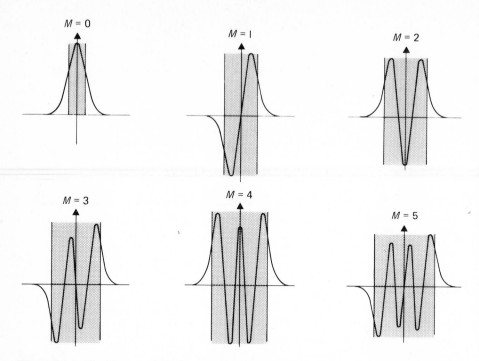

Figure 24.2. The shapes of the first six eigenfunctions for the harmonic oscillator. The shaded areas correspond to the classically accessible regions for the corresponding energy.

Virial theorem The expectation value of x^2 is easily calculated using the *Virial Theorem*, which states that

$$\langle T \rangle = \left\langle \frac{p^2}{2m} \right\rangle = \frac{1}{2} \left\langle x \frac{dV(x)}{dx} \right\rangle. \tag{24.40}$$

The proof is sufficiently long that we forgo presenting it here. What matters is that when the potential is of the form $V(x) = \alpha x^2$, then

$$\frac{1}{2} \left\langle x \frac{dV}{dx} \right\rangle = \langle V \rangle$$

so that

$$\langle T \rangle = \langle V \rangle = \tfrac{1}{2}\langle T + V \rangle = \tfrac{1}{2}\langle H \rangle. \tag{24.41}$$

This is what was obtained using equipartition, and here it implies that

$$\tfrac{1}{2}m\omega^2 \langle x^2 \rangle_M = \tfrac{1}{2}\hbar\omega(M + \tfrac{1}{2})$$

so that

$$\langle x^2 \rangle_M = \frac{\hbar}{m\omega}(M + \tfrac{1}{2}). \tag{24.42}$$

From this we can calculate

$$\langle p^2 \rangle_M = 2m \cdot \tfrac{1}{2}\hbar\omega(M + \tfrac{1}{2})$$

$$= \hbar\omega m(M + \tfrac{1}{2}). \tag{24.43}$$

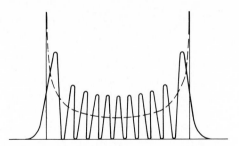

Figure 24.3. Plot of $|u_{10}|^2$ for the harmonic oscillator. The dashed line represents the classical probability distribution.

Using the definitions of the uncertainties

$$\Delta p = \sqrt{\langle p^2 \rangle - \langle p \rangle^2} = \sqrt{\langle p^2 \rangle}$$
$$\Delta x = \sqrt{\langle x^2 \rangle - \langle x \rangle^2} = \sqrt{\langle x^2 \rangle}, \tag{24.44}$$

we find

$$\Delta p \Delta x = \hbar(M + \tfrac{1}{2}). \tag{24.45}$$

Thus the uncertainty relation is satisfied, and the ground state has the smallest uncertainty. The ground state has a nonvanishing energy $\tfrac{1}{2}\hbar\omega$, which is also a quantum mechanical effect. A state of zero energy would have to have both

$$\langle p^2 \rangle = 0$$
$$\langle x^2 \rangle = 0 \tag{24.46}$$

which would violate the uncertainty relation. The crude estimate made in Eq. (18.18) is not precise enough to yield the correct result, but it does point out the origin of the zero point energy.

One may use the results given in Eq. (24.37) to calculate the probability distributions corresponding to different values of the quantum number M. In particular the properties of the Hermite polynomials for large M show that the square of Eq. (24.37) approaches the classical distribution, which is just proportional to the time that a particle spends in a given region, that is, proportional to

$$\frac{dt}{T} \propto \frac{dx}{(x_0^2 - x^2)^{1/2}} \tag{24.47}$$

where x_0 is the value of x at the classical turning point. Figure 24.3 shows how the classical probability distribution averages the quantum mechanical one, for $M = 10$.

NOTES AND COMMENTS

1. The harmonic oscillator is the simplest nontrivial system that can be treated just by using the abstract properties of x and p_{op}, constrained by the commutation relations

$$p_{op}x - xp_{op} = -i\hbar.$$

This is treated in many quantum mechanics text-books, and used as an illustration of the power of operator methods. See, for example, S. Gasiorowicz, *Quantum Physics*, John Wiley, New York, 1974.

2.   The zero point energy manifests itself clearly in the fact that at low temperatures helium remains a liquid at normal pressure. Normally at low temperatures a crystal is formed because the potential energy (attractive) together with the kinetic energy $\hbar^2/2mR^2$ (repulsive) add up to an energy dependence on the separation that has a rather deep minimum for a separation $R = R_0$. For liquid helium the mass m of the atom is very small, so that the kinetic energy is relatively large, and the energy dependence has a very shallow minimum. In that net potential a lot of motion is possible, and helium remains liquid. One might expect the same effect to hold for hydrogen, since hydrogen atoms are even lighter. In the case of hydrodgen, however, it turns out that the potential energy is much stronger than for helium. This has to do with the fact that neighboring hydrogen atoms easily polarize each other, so that there is a strong (induced) dipole-dipole interaction. The interaction, known as the van der Waals force, is an order of magnitude stronger than the corresponding helium-helium force, and thus there is a sharp minimum, leading to little room for motion of the ions, that is, a rigid lattice structure.

PROBLEMS

24.1 A charged particle in a harmonic oscillator is placed in an external electric field, so that the energy operator is

$$H = p^2/2m + kx^2/2 - eEx.$$

What is the energy spectrum for this problem, and what is the wave function for the ground state of the system?

24.2 Consider the potential

$$V(x) = \infty \quad 0 > x$$
$$= V_0(e^{-2(x-a)/b} - 2e^{-(x-a)/b}) \qquad x > 0.$$

Sketch the potential. For what values of the parameter b will the potential be wide? What is your estimate of the energy of the first excited state, given that the potential can be approximated by a harmonic oscillator about the equilibrium position?

24.3 Work out the first four Hermite polynomials.

24.4 Consider the problem described by the energy operator

$$H = p^2/2m + gx^4.$$

a) Make the transformation needed to write the Schrödinger equation in dimensionless form.

b) Find the large x behavior of the solutions of the equation. Do not go any further, since the rest of the problem is not as easily solved as for the x^2 potential!

24.5 Consider the potential
$$V(x) = m\omega^2 x^2/2 + \lambda x^3.$$

a) Sketch the potential.

b) Estimate the energy of the ground state, and the first excited state in the region around the minimum.

c) Note that the ground state cannot be stable, because the particle can tunnel into the region where $V(x) \to -\infty$.

24.6 Calculate $\langle x^4 \rangle - \langle x^2 \rangle^2$ for the first three states ($M = 0, 1, 2$) of the harmonic oscillator. Use this to calculate $\langle p^4 \rangle - \langle p^2 \rangle^2$.

24.7 Calculate the "dipole transition" amplitude defined by

$$\langle M + 1 | y | M \rangle \equiv \int_{-\infty}^{\infty} dy\, u_{M+1}(y) y u_M(y)$$

for $M = 0, 1$, and 2. A more challenging problem, left for more advanced students, is to calculate the above for arbitrary M. [*Hint*: Use Eq. (24.31) and Eq. (24.32) to show that

$$2M \int_{-\infty}^{\infty} dy\, e^{-y^2} H_M(y) y H_{M-1}(y) = -\int_{-\infty}^{\infty} dy\, e^{-y^2} H_M(y) y \frac{d}{dy} H_M(y).$$

Next show that

$$\int_{-\infty}^{\infty} dy\, H_M(y) y \frac{dH_M}{dy} e^{-y^2} = \frac{1}{2} \int_{-\infty}^{\infty} dy\, e^{-y^2} \frac{d}{dy} (y H_M^2(y)) - \frac{1}{2} \int_{-\infty}^{\infty} dy\, e^{-y^2} H_M^2(y)$$

and proceed from there.]

24.8 Show that
$$\langle M + K | y | M \rangle = 0$$

unless $K = \pm 1$. This is known as a *selection rule*. Use Eq. (24.30), and so on, to establish this result.

24.9 Check the Virial Theorem for the ground state of the harmonic oscillator.

24.10 This problem is for the mathematically more advanced student: Prove the Virial Theorem for energy states of an arbitrary potential. Proceed as follows:

a) Show that

$$\int_{-\infty}^{\infty} dx\, u(x) x \frac{dV}{dx} u(x) = -\langle V \rangle + 2 \int_{-\infty}^{\infty} dx \frac{du}{dx} x V u(x).$$

b) Use the time independent Schrödinger equation to show that

$$2 \int_{-\infty}^{\infty} dx \frac{du}{dx} x V u(x) = E + \frac{\hbar^2}{2m} \int_{-\infty}^{\infty} dx \left(\frac{du}{dx}\right)^2$$

$$= E + \langle T \rangle.$$

24.11 Show that the classical probability distribution function for a particle in a harmonic oscillator potential is given by

$$P(x)dx = \frac{1}{\pi} \frac{dx}{(x_0^2 - x^2)^{1/2}} .$$

[*Hint*: The probability distribution is given by dt/T where dt is the time spent by the particle in the interval dx and T is the total period. Use this, and the solution in the form $x = x_0 \cos \omega t.$]

PART V
Advanced Topics in Quantum Mechanics

The following seven chapters take us deeper into quantum mechanics. Although the extension of the Schrödinger equation to a description of more than one particle is straightforward, a very deep principle, totally quantum mechanical, dictates new constraints when the particles are identical. The constraint, discovered by W. Pauli, is that the wave function must be either totally symmetric or totally antisymmetric under the interchange of any two of the identical particles, with the nature of the symmetry depending only on the nature of the particles. The implications of this principle, discussed in Chapter 25, are vast. The first application, discussed in Chapter 26, is the existence of the so-called Fermi sea of energy levels when many identical particles are present.

Angular momentum is also more complicated than a simple correspondence argument would indicate, and Chapters 27 and 28 show how the fact that the position and momentum operators do not commute, determines the structure of the angular momentum eigenstates. These play an important role in the treatment of the hydrogen atom in Chapter 29. The results obtained by a proper quantum mechanical treatment are in agreement with those obtained by Bohr, but the real richness of the level structure that manifests itself in the complexity of atomic spectra can be understood only with quantum mechanics. The fine details of the hydrogen spectrum can be understood only with the help of the notion that the electron itself carries an intrinsic angular momentum, a spin. This is another example of a purely quantum mechanical phenomenon, and it is discussed in Chapter 30. The section ends with a discussion of the role of spin in atoms, molecules, and nuclei in Chapter 31.

This is probably the most difficult part of the book, and it may be desirable to omit certain topics, taking care to note only the results. Chapter 31 could be left for "cultural" reading. There is, however, no way of avoiding the difficult concepts associated with the exclusion principle, spin, and the addition of angular momenta: they all play an important role in atomic molecular, solid state, nuclear, and particle physics, as will be seen in the last part of the book.

25
The Quantum Mechanics of Multiparticle Systems and the Pauli Exclusion Principle

N-Particle Schrödinger Equation The extension of quantum mechanics to many particle systems appears to be fairly straightforward, if we now generalize the wave function to a function of N coordinates and the time t, for a one-dimensional N-particle system. The wave function $\psi(x_1, x_2, \ldots, x_N; t)$ is again expected to develop in time according to the equation

$$i\hbar \frac{\partial}{\partial t} \psi(x_1, x_2, \ldots, x_N; t) = H\psi(x_1, x_2, \ldots, x_N; t) \tag{25.1}$$

with the energy operator taking the form

$$H = \frac{p_1^2}{2m_1} + \frac{p_2^2}{2m_2} + \cdots + \frac{p_N^2}{2m_N} + V(x_1, x_2, \ldots, x_N) \tag{25.2}$$

and with the replacement

$$p_i = \frac{\hbar}{i} \frac{\partial}{\partial x_i} \quad i = 1, 2, \ldots, N. \tag{25.3}$$

When the potential energy $V(x_1, x_2, \ldots, x_N)$ does not depend on the time, one may again write

$$\psi(x_1, x_2, \ldots, x_N; t) = e^{-iEt/\hbar} u(x_1, x_2, \ldots, x_N) \tag{25.4}$$

to obtain the energy eigenvalue equation

$$\left(-\frac{\hbar^2}{2m_1} \frac{\partial^2}{\partial x_1^2} + \cdots - \frac{\hbar^2}{2m_N} \frac{\partial^2}{\partial x_N^2} + V(x_1, \ldots, x_N) \right) u(x_1, \ldots, x_N)$$

$$= Eu(x_1, \ldots, x_N). \tag{25.5}$$

For a two-particle system this reads

$$\left(-\frac{\hbar^2}{2m_1} \frac{\partial^2}{\partial x_1^2} - \frac{\hbar^2}{2m_2} \frac{\partial^2}{\partial x_2^2} + V(x_1, x_2) \right) u(x_1, x_2) = Eu(x_1, x_2). \tag{25.6}$$

In general such an equation, being a partial differential equation, is a great deal more complicated than the one-particle ordinary differential equation we studied before. There are a few circumstances when the equation simplifies. One is when the potential energy has the form

$$V(x_1, x_2) = V_1(x_1) + V_2(x_2). \tag{25.7}$$

In this case, the energy operator separates into two independent terms, one for each particle. The two particles do not interact, and the method of separation of variables may be used to solve the problem. It is easy to check that

$$u(x_1, x_2) = u_1(x_1)u_2(x_2) \tag{25.8}$$

where

$$\left(-\frac{\hbar^2}{2m_1}\frac{d^2}{dx_1^2} + V_1(x_1)\right)u_1(x_1) = E_1 u_1(x_1)$$

$$\left(-\frac{\hbar^2}{2m_2}\frac{d^2}{dx_2^2} + V_2(x_2)\right)u_2(x_2) = E_2 u_2(x_2) \tag{25.9}$$

and

$$E_1 + E_2 = E \tag{25.10}$$

is a solution.

Another, more interesting simplification arises when the potential energy depends only on the separation between the two particles, that is,

$$V(x_1, x_2) = V(x_1 - x_2). \tag{25.11}$$

All real potentials in nature have this form, because there is no "fixed origin" in the universe, or at least in our part of the universe. Thus forces can depend only on the relative separation of particles, and not on where the particles are relative to some absolutely determined coordinate frame. A succinct way of describing this is by the statement that *the laws of nature are invariant under spatial translation*, that is, under a displacement of every position coordinate by the same, fixed amount. When such a potential enters into the problem, one may separate out center of mass motion, just as in classical mechanics.

Center of mass motion The procedure is to introduce the center of mass coordinate

$$X = \frac{m_1 x_1 + m_2 x_2}{m_1 + m_2} \tag{25.12}$$

and the relative coordinate

$$x = x_1 - x_2. \tag{25.13}$$

In terms of these

$$\frac{\partial}{\partial x_1} = \frac{\partial X}{\partial x_1}\frac{\partial}{\partial X} + \frac{\partial x}{\partial x_1}\frac{\partial}{\partial x} = \frac{m_1}{M}\frac{\partial}{\partial X} + \frac{\partial}{\partial x}$$

$$\frac{\partial}{\partial x_2} = \frac{\partial X}{\partial x_2}\frac{\partial}{\partial X} + \frac{\partial x}{\partial x_2}\frac{\partial}{\partial x} = \frac{m_2}{M}\frac{\partial}{\partial X} - \frac{\partial}{\partial x}$$

where we have introduced the total mass

$$M = m_1 + m_2.$$

A straightforward calculation leads to

$$
-\frac{\hbar^2}{2m_1}\frac{\partial^2}{\partial x_1^2} - \frac{\hbar^2}{2m_2}\frac{\partial^2}{\partial x_2^2} = -\frac{\hbar^2}{2m_1}\left(\frac{m_1^2}{M^2}\frac{\partial^2}{\partial X^2} + \frac{2m_1}{M}\frac{\partial^2}{\partial X\partial x} + \frac{\partial^2}{\partial x^2}\right)
$$

$$
-\frac{\hbar^2}{2m_2}\left(\frac{m_2^2}{M^2}\frac{\partial^2}{\partial X^2} - \frac{2m_2}{M}\frac{\partial^2}{\partial X\partial x} + \frac{\partial^2}{\partial x^2}\right)
$$

$$
= -\frac{\hbar^2}{2M}\frac{\partial^2}{\partial X^2} - \frac{\hbar^2}{2}\left(\frac{1}{m_1} + \frac{1}{m_2}\right)\frac{\partial^2}{\partial x^2}
$$

$$
= -\frac{\hbar^2}{2M}\frac{\partial^2}{\partial X^2} - \frac{\hbar^2}{2\mu}\frac{\partial^2}{\partial x^2}. \tag{25.14}
$$

In the last step we have introduced the *reduced mass* defined by

$$
\frac{1}{\mu} = \frac{1}{m_1} + \frac{1}{m_2},
$$

that is,

$$
\mu = \frac{m_1 m_2}{m_1 + m_2}. \tag{25.15}
$$

The kinetic energy separates into the sum of a kinetic energy of the center of mass, with total mass M, and a kinetic energy of relative motion, with mass μ. The Schrödinger equation (25.6) now takes the simple form

$$
\left(-\frac{\hbar^2}{2M}\frac{\partial^2}{\partial X^2} - \frac{\hbar^2}{2\mu}\frac{\partial^2}{\partial x^2} + V(x)\right) u(x, X) = Eu(x, X) \tag{25.16}
$$

This equation may be solved by the method of separation of variables. Since the only place that X appears is in the kinetic energy term, a trial solution

$$
u(x, X) = e^{iPX/\hbar}\,\varphi(x) \tag{25.17}
$$

gives

$$
\left(-\frac{\hbar^2}{2\mu}\frac{d^2}{dx^2} + V(x)\right)\varphi(x) = (E - P^2/2M)\varphi(x) \equiv \epsilon\varphi(x) \tag{25.18}
$$

The total energy is thus seen to consist of the kinetic energy of motion of the two-particle system with mass M, and the internal energy of relative motion ϵ. It thus appears that no new features appear when we deal with more than one particle in quantum mechanics.

Identical particles   This expectation is not correct, since in quantum mechanics we must deal with systems in which several *identical particles* appear, such as atoms, in which there are Z electrons, and nuclei, containing many protons and neutrons. The first question is: Why is quantum mechanics different from classical physics on this issue? Let us imagine that we have two identical billiard balls and scatter them off each other, as shown in Fig. 25.1. The "counter" A will receive billiard balls that have undergone small-angle scattering, as well as those that have undergone the complementary large-scale scattering. Thus the number of counts of billiard balls in A per second will be $N_1 + N_2$, where the first number refers to those that got there via process "1," and the second to those that got there via process "2." Had the billiard balls been different, and the selected ones started out on the left side, then only process "2" would give us counts. Nevertheless, calling the billiard balls identical is fake. What we are saying is that we are willing to count all the billiard balls that come into A, without asking whether they came from the right or from the left. We still *know* where they came from, since we can take a film of their paths, for example, and we could divide the billiard balls entering A into two classes. This is like counting bullets that pass through a wall with two slits, and not asking which slit they came through: not asking the question does not mean that the information could not be obtained without altering the character of the experiment. In quantum mechanics, however, we cannot trace the path of a particle (as already discussed for the two-slit experiment) and thus, when an electron arrives at counter A *there is no way of knowing whether it arrived there via the process "1" or the process "2."*

Could we put a little spot on one electron (in principle) and thus tag it? Are all electrons identical? The evidence from spectroscopy is that there is only one hydrogen, one helium, and so on, which strongly suggests that there is only one kind of

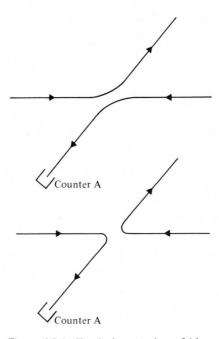

Figure 25.1. Classical scattering of identical billiard balls. In the first process, the billiard balls undergo small-angle scattering; in the second process they undergo large-angle scattering. Different, but indistinguishable billiard balls end up in the counter.

electron. Similar evidence from nuclear physics points to the indistinguishability of protons, and that of neutrons. Furthermore, the simplicity of the electron (and of the other fundamental particles) indicates that there is no way of tagging them. Thus the indistinguishability of these particles appears to be a fact of nature, and we must see how it is reflected in the structure of the wave function for two (or more) identical particles.

First of all, in the energy operator (note the equal masses)

$$H = \frac{p_1^2}{2m} + \frac{p_2^2}{2m} + V(x_1, x_2),$$ (25.19)

we must have

$$V(x_1, x_2) = V(x_2, x_1),$$ (25.20)

since H must not distinguish between the two particles. In addition, the eigensolutions may be chosen to be either symmetric

$$\psi(x_2, x_1) = \psi(x_1, x_2)$$ (25.21)

or antisymmetric

$$\psi(x_2, x_1) = -\psi(x_1, x_2)$$ (25.22)

under the interchange of the two particles. So far this is reminiscent of our discussion of parity in Chapter 20. There we argued that for a Hamiltonian invariant under reflection $x \to -x$, the wave function *could* be chosen even or odd under reflection. The choice of parity was determined by the way that the initial state was prepared, but once determined, it was conserved.

The Pauli Principle In the case of the exchange of two identical particles the situation is different. The choice of symmetry or antisymmetry is not up to the experimenter who prepares the initial state, but is decided by a very deep law of nature. According to that law, discovered by W. Pauli, and called the *Pauli Principle*, all particles fall into two classes. One class of particles contains the *fermions* (after Fermi, who with Dirac discovered the symmetry of these particles). The electron, the proton, the neutron, the muon, and the neutrino belong to this class. The other class contains the *bosons* (after S. Bose, who with Einstein discovered the symmetry of these particles). The photon, the alpha particle, the pion, the deuteron, and any bound state of an even number of fermions belong to this class. The Pauli Principle states that

> the wave function of a many particle system
> is antisymmetric under the interchange of
> two identical fermions, and is symmetric
> under the interchange of two identical
> bosons.

This seemingly innocuous restriction has very important consequences. As will become evident later, the entire richness in the structure of matter is a direct consequence of the operation of the Pauli Principle.

To illustrate this, consider two noninteracting electrons, such as are described by a potential energy of the form given in Eq. (25.7), with the requirement that $V_1 = V_2$, imposed by the indistinguishability of the two electrons. The unsymmetrized wave function for two electrons is given by

$$\psi(x_1, x_2) = u_{E_1}(x_1) u_{E_2}(x_2) \tag{25.23}$$

where

$$\left(-\frac{\hbar^2}{2m} \frac{d^2}{dx^2} + V(x) \right) u_{E_n}(x) = E_n u_{E_n}(x) \tag{25.24}$$

and

$$E = E_1 + E_2. \tag{25.25}$$

Thus, if the potential $V(x) = V_1(x) = V_2(x)$ is an infinite box that extends from $x = 0$ to $x = b$, then the unsymmetrized wave function is

$$\psi(x_1, x_2) = \frac{2}{b} \sin \frac{n_1 \pi x_1}{b} \sin \frac{n_2 \pi x_2}{b} \tag{25.26}$$

while the antisymmetrized wave function has the form

$$\psi_A(x_1, x_2) = \frac{1}{\sqrt{2}} \frac{2}{b} \left(\sin \frac{n_1 \pi x_1}{b} \sin \frac{n_2 \pi x_2}{b} - \sin \frac{n_1 \pi x_2}{b} \sin \frac{n_2 \pi x_1}{b} \right). \tag{25.27}$$

More generally, the antisymmetrized form of Eq. (25.23) is

$$\psi_A(x_1, x_2) = \frac{1}{\sqrt{2}} (u_{E_1}(x_1) u_{E_2}(x_2) - u_{E_1}(x_2) u_{E_2}(x_1)). \tag{25.28}$$

It is evident that if $E_1 = E_2$, this vanishes. This is a manifestation of a corollary of the Pauli Principle, known as the *Pauli Exclusion Principle*:

> *No more than one electron (fermion)*
> *can be in a given quantum mechanical*
> *state.*

One might be inclined to worry that the antisymmetrization requirement will lead us into technical problems due to the presence of other electrons in the system. If we are talking about an electron on earth, do we really have to antisymmetrize its wave function with that of an electron on the moon? To see just when antisymmetrization is necessary, let us consider the probability density for two electrons without antisymmetrization. We then have

$$P(x_1, x_2) = |\psi_a(x_1) \psi_b(x_2)|^2. \tag{25.29}$$

With antisymmetry we have

$$P(x_1, x_2) = \left| \frac{1}{\sqrt{N}} (\psi_a(x_1) \psi_b(x_2) - \psi_a(x_2) \psi_b(x_1)) \right|^2$$

$$= \frac{1}{N} |\psi_a(x_1) \psi_b(x_2)|^2 + \frac{1}{N} |\psi_a(x_2) \psi_b(x_1)|^2$$

$$- \frac{1}{N} (\psi_a^*(x_1) \psi_b^*(x_2) \psi_a(x_2) \psi_b(x_1)$$

$$+ \psi_a(x_1) \psi_b(x_2) \psi_a^*(x_2) \psi_b^*(x_1)). \tag{25.30}$$

The normalization constant N is determined by the requirement that

$$\int_{-\infty}^{\infty} dx_1 \int_{-\infty}^{\infty} dx_2 P(x_1, x_2) = 1$$

and since the $\psi_a(x)$ and $\psi_b(x)$ are normalized, this means that

$$1 = \frac{1}{N} + \frac{1}{N} + \frac{2}{N} \left| \int_{-\infty}^{\infty} dx \, \psi_a^*(x) \psi_b(x) \right|^2$$

that is,

$$N = 2 \left(1 + \left| \int_{-\infty}^{\infty} dx \, \psi_a^*(x) \psi_b(x) \right|^2 \right). \tag{25.31}$$

Suppose we now ask for the probability that the electron with label "a" ("a" may be the energy eigenvalue for the electron bound to a proton in a hydrogen atom) is in some spatial region R. For the unsymmetrized case we have

$$P_a(R) = \int_R dx_1 \, |\psi_a(x_1)|^2 \int_{-\infty}^{\infty} dx_2 \, |\psi_b(x_2)|^2. \tag{25.32}$$

We integrate over the whole range of the coordinates of the electron with the "b" label, since we do not care where it is. For the antisymmetrized wave function we have

$$P_a(R) = \frac{1}{N} \int_R dx_1 \, |\psi_a(x_1)|^2 \int_{-\infty}^{\infty} dx_2 \, |\psi_b(x_2)|^2$$

$$+ \frac{1}{N} \int_R dx_2 \, |\psi_a(x_2)|^2 \int_{-\infty}^{\infty} dx_1 \, |\psi_b(x_1)|^2$$

$$- \frac{1}{N} \int_R \int_R dx_1 \, dx_2 \, \psi_a^*(x_1) \psi_b^*(x_2) \psi_a(x_2) \psi_b(x_1)$$

$$- \frac{1}{N} \int_R \int_R dx_1 \, dx_2 \, \psi_a(x_1) \psi_b(x_2) \psi_a^*(x_2) \psi_b^*(x_1)$$

$$= \frac{2}{N} \int_R dx \, |\psi_a(x)|^2 \int_{-\infty}^{\infty} dy \, |\psi_b(y)|^2$$

$$- \frac{2}{N} \int_R \int_R dx \, dy \, \psi_a^*(x) \psi_a(y) \psi_b^*(y) \psi_b(x). \tag{25.33}$$

In the last step we made use of the fact that in the integrals the labels x_1 and x_2 are just dummy variables, and that by a change of labeling it is easy to see that the second term is equal to the first, and the fourth equal to the third. Note that in the second "interference" term, the integration for both variables is over the range R, since in the expression we must keep the variables associated with the quantum numbers "a" within that range. We see that except for the factor $2/N$, the first term is the same as if there were no antisymmetrization. The second term is only important if

$$\int_R dx \, \psi_a^*(x) \psi_b(x) \tag{25.34}$$

is not negligible. Now if the label "*a*" describes an electron in the ground state of hydrogen on earth, and "*b*" an electron in the ground state of hydrogen on the moon, for example, then

$$\psi_a(\mathbf{r}) = Ce^{-\alpha|\mathbf{r}-\mathbf{R}_e|}$$

$$\psi_b(\mathbf{r}) = Ce^{-\alpha|\mathbf{r}-\mathbf{R}_m|} \tag{25.35}$$

where $\mathbf{R}_e$ is the coordinate of the proton (hydrogen nucleus) on the earth and $\mathbf{R}_m$ that of the proton on the moon. Since these wave functions extend over the dimensions of a Bohr radius they do not overlap at all, and (25.34) is totally negligible. We can thus ignore antisymmetrization for all practical purposes. Even in crystal lattices, where the spacing between nuclei is several Ångstroms, the overlap is often very small, and antisymmetrization can be neglected. In the case of the two electrons in the Helium atom, on the other hand, the overlap is important and antisymmetrization is crucial.

In our discussion we illustrated the effect of antisymmetrization with noninteracting electrons in given quantum states, labeled with "*a*" and "*b*" or E_1 and E_2. In three-dimensional problems there will in general be more labels: an electron in a Coulomb potential has its quantum state specified by the energy, the total angular momentum, and the *z*-component of the angular momentum. This does not change the need to antisymmetrize, nor the form of the Pauli Exclusion Principle. The latter, then, implies that there cannot be more than one electron with all three of the quantum numbers the same. When the electrons interact, as they do in the real world via the Coulomb interaction, the labeling may get modified. Nevertheless, the requirement that the total wave function be antisymmetric under the interchange of the coordinates of the two electrons still holds.

Spin quantum number for electron There is one modification that must be brought in at this point: As we shall learn later, the electron has an additional quantum number that has no classical counterpart, which we shall call its *spin state*. The electron, like the proton and the neutron, is known to have an intrinsic angular momentum, called the spin, as a consequence of which, it may be in one of two states, which we call the spin-up and the spin-down states, respectively. All of our considerations hold if the electrons are in the same spin states. If they are in different spin states, then there is one quantum label that is different, and the two electrons can be in the same energy state. Formally, the existence of the additional state is denoted by writing the wave function of the electron in the form

$$u_{E,\sigma}(x) = u_E(x)\chi_\sigma \tag{25.36}$$

where σ may take on two values, ↑ (up) or ↓ (down). An unsymmetrized two-particle wave function is

$$u_{E_1}(x_1)u_{E_2}(x_2)\chi_{\sigma_1}^{(1)}\chi_{\sigma_2}^{(2)} \tag{25.37}$$

and an antisymmetrized one is

$$\frac{1}{\sqrt{2}}\left(u_{E_1}(x_1)u_{E_2}(x_2)\chi_{\sigma_1}^{(1)}\chi_{\sigma_2}^{(2)} - u_{E_1}(x_2)u_{E_2}(x_1)\chi_{\sigma_1}^{(2)}\chi_{\sigma_2}^{(1)}\right). \tag{25.38}$$

This vanishes only when both $E_1 = E_2$ *and* $\sigma_1 = \sigma_2$. The statement of the exclusion principle is unaltered, if it is understood that a "quantum mechanical state" includes the description of the spin state.

The connection between spin and the Pauli Principle is actually very deep. The classification of particles into fermion and boson classes is determined by the spins of the particles. Electrons, protons, neutrons, and so on, have intrinsic spin 1/2 (in units of $\hbar$); particles that have spin 3/2, 5/2, ... are also fermions. On the other hand, particles that have integral spin, such as the photon (spin 1), the pion (spin 0), the deuteron (spin 1), and so on are bosons. The connection between the spin of the particles, and the symmetry or antisymmetry of the wave function was first studied by Pauli, and is known as the *spin-statistics theorem*.

NOTES AND COMMENTS

1. Pauli proposed the exclusion principle in late 1924, before quantum mechanics had actually been discovered. The idea arose out of an attempt to understand the kind of effects that are now known to be associated with the electron spin (doublets, anomalous Zeeman effect; see Chapter 31). For a fascinating account see the essays by R. Kronig and B.L. van der Waerden in *Theoretical Physics in the Twentieth Century* (a Memorial Volume to Wolfgang Pauli), edited by M. Fierz and V.F. Weisskopf, Interscience Publishers, New York, 1960.

    Pauli pursued the subject and finally was able to explain the spin-statistics connection as arising out of some very general properties of relativistic quantum theory, in which the absence of arbitrarily large negative energy states, and the limitation of signal velocities to those that are smaller than the velocity of light, play an important role, together with the requirement that all probabilities be nonnegative.

PROBLEMS

25.1 Show that the total momentum operator P given by

$$P = p_1 + p_2$$

may be written in the form

$$\frac{\hbar}{i}\frac{\partial}{\partial X}.$$

From this show that in the wave function (25.17) the first factor represents the center of mass motion of the two-particle system with total momentum P.

25.2 Show that the relative momentum

$$p = \frac{m_2 p_1 - m_1 p_2}{m_1 + m_2}$$

may be written in the form

$$\frac{\hbar}{i}\frac{\partial}{\partial x}.$$

25.3 Consider two electrons in the infinite box given by

$$V(x) = 0 \quad 0 \leqslant x \leqslant L$$
$$= \infty \quad \text{elsewhere.}$$

Suppose the two electrons do not interact, and they are in different spin states. What is the lowest energy eigenstate for the two electron system? What is the lowest energy eigenstate if the two electrons are in the same spin state?

25.4 Consider two bosons, and consider them to have spin 0, so that there is no spin wave function. What is the lowest energy state wave function? What is the wave function for the state in which one boson is in the lowest state and the other in the first excited state?

25.5 Suppose we put two electrons in the infinite box. Given the freedom to arrange the spins as we wish, what is the lowest energy for the two electron state? What is it for three electrons? What is it for seven electrons? What is it for N electrons?

25.6 Suppose we put N spinless bosons in the infinite box. What is the lowest energy for the system?

25.7 Consider N bosons in the infinite box, with $N-1$ in the lowest state, and one in the first excited state. What is the wave function of that state?

25.8 Consider an electron in interaction with a proton. What is the reduced mass? What is the reduced mass when an electron is in interaction with another electron?

25.9 Consider two identical bosons described by the energy operator

$$H = \frac{p_1^2}{2m} + \frac{p_2^2}{2m} + \tfrac{1}{2}m\omega^2 x_1^2 + \tfrac{1}{2}m\omega^2 x_2^2.$$

What is the energy spectrum of the system? Do this in two ways: (i) by separation of variables x_1 and x_2; and (ii) by separating out the center of mass motion and the internal motion, and getting the energy in that way. Show that the spectrum, including the degeneracy, is the same however we solve this problem.

25.10 Suppose the two particles in Problem 25.9 are electrons. What is the explicit form of the wave function when the two electrons are both in the ground state of their respective one-particle energy operators? What is the energy in the lowest state for which the two spin states are the same?

26

The Schrödinger Equation
in Three Dimensions.
I. Noninteracting Fermions

The Wave Equation in Three Dimensions In three dimensions the energy takes the form

$$E = \frac{\mathbf{p}^2}{2m} + V(\mathbf{r}) \tag{26.1}$$

with $\mathbf{r} = (x, y, z)$ and $\mathbf{p} = (p_x, p_y, p_z)$ now vectors. The momentum operator now becomes

$$\mathbf{p} = \frac{\hbar}{i}\nabla = \left(\frac{\hbar}{i}\frac{\partial}{\partial x}, \frac{\hbar}{i}\frac{\partial}{\partial y}, \frac{\hbar}{i}\frac{\partial}{\partial z}\right) \tag{26.2}$$

so that the Schrödinger equation takes the form

$$\frac{1}{2m}\left(\frac{\hbar}{i}\nabla\right) \cdot \left(\frac{\hbar}{i}\nabla\right)\psi(\mathbf{r}) + V(\mathbf{r})\psi(\mathbf{r}) = E\psi(\mathbf{r})$$

that is,

$$-\frac{\hbar^2}{2m}\nabla^2\psi(\mathbf{r}) + V(\mathbf{r})\psi(\mathbf{r}) = E\psi(\mathbf{r}) \tag{26.3}$$

where we denote the energy eigenvalue by E, the eigenfunction by $\psi(\mathbf{r})$, and where

$$\nabla^2 = \frac{\partial^2}{\partial x^2} + \frac{\partial^2}{\partial y^2} + \frac{\partial^2}{\partial z^2}$$

in Cartesian coordinates. The normalization condition reads

$$\iiint dx\,dy\,dz\,|\psi(\mathbf{r})|^2 \equiv \int d^3r\,|\psi(\mathbf{r})|^2 = 1. \tag{26.4}$$

In the next chapter we will consider potentials that depend only on the separation between two particles, that is,

$$V(\mathbf{r}) = V(|\mathbf{r}|) = V(r) \tag{26.5}$$

in which case we will also replace m by μ, the reduced mass.

At this stage we consider the potential of the form

$$V(\mathbf{r}) = U_1(x) + U_2(y) + U_3(z). \tag{26.6}$$

The form of the potential suggests that the motion in the x-, y-, and z-directions is independent, since the force in the three directions, $-\partial U_1/\partial x, -\partial U_2/\partial y, -\partial U_3/\partial z$ are quite unrelated. Indeed, if we write the eigenfunction $\psi(\mathbf{r})$ in the form

$$\psi(\mathbf{r}) = X(x)Y(y)Z(z) \tag{26.7}$$

it is a rather simple exercise to show that the equation decomposes into

$$-\frac{\hbar^2}{2m}\frac{d^2X}{dx^2} + U_1(x)X(x) = \epsilon_1 X(x)$$

$$-\frac{\hbar^2}{2m}\frac{d^2Y}{dy^2} + U_2(y)Y(y) = \epsilon_2 Y(y) \tag{26.8}$$

$$-\frac{\hbar^2}{2m}\frac{d^2Z}{dz^2} + U_3(z)Z(z) = \epsilon_3 Z(z)$$

with

$$\epsilon_1 + \epsilon_2 + \epsilon_3 = E. \tag{26.9}$$

A particularly interesting case is the motion of a particle in a cubical box of side L, so that $U_1, U_2,$ and U_3 all have the same form

$$U(x) = 0 \quad 0 \leqslant x \leqslant L$$
$$= \infty \quad \text{elsewhere} \tag{26.10}$$

and similarly for $U(y)$ and $U(z)$. Drawing on our knowledge of the solution of this problem in one dimension, we can immediately write down the solution

$$\psi_{n_1 n_2 n_3}(x,y,z) = \left(\frac{2}{L}\right)^{3/2} \sin\frac{n_1 \pi x}{L} \sin\frac{n_2 \pi y}{L} \sin\frac{n_3 \pi z}{L} \tag{26.11}$$

with

$$n_1 = 1, 2, 3, \ldots$$
$$E_{n_1 n_2 n_3} = \frac{\hbar^2 \pi^2}{2mL^2}(n_1^2 + n_2^2 + n_3^2). \quad n_2 = 1, 2, 3, \ldots \tag{26.12}$$
$$n_3 = 1, 2, 3, \ldots$$

In contrast to the one-dimensional case, there is a great deal of degeneracy. There is only one ground state, with $(n_1, n_2, n_3) = (1, 1, 1)$, but the first excited states $(2, 1, 1), (1, 2, 1),$ and $(1, 1, 2)$ all have the same energy. As the energy increases, so does the degeneracy: for example, $(1, 2, 6), (1, 6, 2), (2, 6, 1), (2, 1, 6), (6, 1, 2), (6, 2, 1), (3, 4, 4), (4, 3, 4),$ and $(4, 4, 3)$ all have the same energy. In general, the number of ways in which we can get three integers n_1, n_2, n_3 such that $n_1^2 + n_2^2 + n_3^2$ add up to some large integer grows very quickly.

The Fermi Energy The problem that we have just solved becomes particularly interesting when we consider electrons in a box, and take into account the *exclusion principle* discussed in the last chapter. In filling up the box, we make use of the fact that, at most, two electrons can go into each state labeled by the integers (n_1, n_2, n_3). Thus if we try to fill the box with electrons in a way to minimize the energy, we put two electrons into $(1, 1, 1)$. two into each of $(2, 1, 1), (1, 2, 1),$ and $(1, 1, 2)$. The next electron must go into one of the states $(2, 2, 1), (2, 1, 2),$ or $(1, 2, 2),$ and so on. We see the picture that was briefly described in Chapter 22 when we discussed tunneling in metals. Electrons fill up energy levels, two at a time, until the top level, the *Fermi energy* level, is reached. Note that bosons, which do not obey the exclusion principle, would all fill the $(1, 1, 1)$ level.

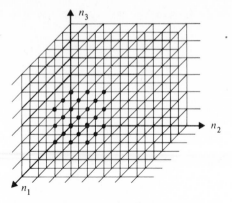

Figure 26.1. Counting of states for independent particle system.

To find out the magnitude of the Fermi energy E_F, we ask the following question: How many states are there with energy less than or equal to some given E? This is the same as asking: How many triplets of integers (n_1, n_2, n_3) are there such that

$$n_1^2 + n_2^2 + n_3^2 \leqslant \frac{2mEL^2}{\hbar^2 \pi^2} . \tag{26.13}$$

If the right-hand side is small, we just have to count. If the number is large, one can get a very accurate result by doing the following (Fig. 26.1):

Plot the integers on a cubic lattice. Each point on the lattice is denoted by some triplet of integers (n_1, n_2, n_3). The condition (26.13) may be written as

$$n_1^2 + n_2^2 + n_3^2 \leqslant R^2 \tag{26.14}$$

where

$$R^2 = \frac{2mEL^2}{\hbar^2 \pi^2} \tag{26.15}$$

Thus all the states of a given energy lie on a sphere of radius R, and this clarifies the statement that the degeneracy grows with energy. The degeneracy is the number of states lying on the surface of a sphere of radius R, and this is to a good approximation proportional to R^2, that is, the energy.

To answer our question, we need just calculate the number of lattice points inside the sphere of radius R. Since the integer-triplets form cubes of unit volume, we can do this by calculating the volume. Since the integers must all be positive, we actually need only to calculate the volume of an octant of the sphere. Thus the *number of states* is given by

$$\frac{1}{8} \frac{4\pi}{3} R^3 = \frac{\pi}{6} \left(\frac{2mEL^2}{\hbar^2 \pi^2} \right)^{3/2} , \tag{26.16}$$

and since the number of electrons that can be accommodated is twice that, (spin "up" and spin "down"), the number of electrons N_e is twice the above,

$$N_e = \frac{\pi}{3} \left(\frac{2mEL^2}{\hbar^2 \pi^2} \right)^{3/2} = \frac{\pi}{3} \frac{L^3}{\pi^3 \hbar^3} (2mE)^{3/2} . \tag{26.17}$$

Given a number of electrons N_e, we can find the Fermi energy from the above:

$$(2mE_F)^{3/2} = 3\pi^2\hbar^3 \frac{N_e}{L^3}$$

so that

$$E_F = \left(\frac{3\pi^2\hbar^3 N_e}{L^3}\right)^{2/3} \frac{1}{2m}. \tag{26.18}$$

In terms of the electron density

$$n_e = \frac{N_e}{L^3} \tag{26.19}$$

this reads

$$E_F = \frac{\hbar^2}{2m}(3\pi^2 n_e)^{2/3}. \tag{26.20}$$

To estimate the magnitude of this, consider the free electrons in a metal, copper, say. There are roughly $n_e = 8.8 \times 10^{22}$ electrons per cm^3, because there is about one free electron per atom, and with a density of $\rho = 8.95$ gm/cm^3 and atomic weight $A = 63.5$, the number of atoms is

$$\frac{\rho}{1.6 \times 10^{-24}A} \simeq 8.8 \times 10^{22} \frac{\text{atoms}}{\text{cm}^3}. \tag{26.21}$$

Hence

$$E_F = \frac{(1.05 \times 10^{-27})^2}{2 \times 0.9 \times 10^{-27}} (3 \times 9.87 \times 8.8 \times 10^{22})^{2/3}$$

$$\simeq 1.1 \times 10^{-11} \text{ ergs} \simeq 7 \text{ eV} \tag{26.22}$$

of the same order of magnitude as the work function.

We see from our formula that the wave number k, defined by $E = \hbar^2 k^2/2m$ is given by

$$k_F = (3\pi^2 n_e)^{1/3} \tag{26.23}$$

for an electron at the top of the Fermi sea. Since $k = 2\pi/\lambda$ we get, for the De Broglie wavelength,

$$\lambda_F \simeq 2.03 \, n_e^{-1/3}. \tag{26.24}$$

Since $n_e^{-1/3}$ is approximately the interparticle spacing, a, we can state our result in the easily remembered form that

$$a \approx \tfrac{1}{2}\lambda_F. \tag{26.25}$$

The exclusion principle, by forbidding the packing of electrons into the same state, effectively gives rise to a repulsion between them. It becomes very difficult to compress an electron (or generally a fermion) gas, because decreasing a means decreasing the De Broglie wavelength, that is, increasing the kinetic energy of the electrons.

Stellar evolution The resistance to compression has important applications in stellar evolution. It is now believed that stars "burn" by undergoing a succession of thermonuclear reactions. The sequence of reactions in which hydrogen converts to helium with energy generation

$$^1\text{H} + {}^1\text{H} \rightarrow {}^2\text{D} + e^+ + \nu + 1.44 \text{ MeV}$$

$$^2\text{D} + {}^1\text{H} \rightarrow {}^3\text{He} + \gamma + 5.49 \text{ MeV}$$

$$^3\text{He} + {}^3\text{He} \rightarrow {}^4\text{He} + {}^1\text{H} + {}^1\text{H} + 12.85 \text{ MeV}$$

ends when all of the hydrogen is gone. Gravitational contraction compresses the helium until it begins to burn in the reactions

$$^4\text{He} + {}^4\text{He} \rightarrow {}^8\text{Be} + \gamma - 95\,\text{keV}$$

$$^8\text{Be} + {}^4\text{He} \rightarrow {}^{12}\text{C} + \gamma + 7.4\,\text{MeV}.$$

The variety of possible processes gets larger and larger and in an extraordinary collaboration between nuclear physicists and astrophysicists, the details of nucleosynthesis have been worked out. At some point, when stars consist largely of iron, silicon, and neighboring elements, burning stops. The material then begins to contract under its own weight. Let us calculate the gravitational pressure, with the simplifying assumption that the density is independent of radius, and that the shape is spherical. Figure 26.2 shows the division of the material into shells. The potential energy of the shell of material lying between radius r and $r + dr$ is given by

$$-G\frac{(M(r))(4\pi r^2 dr\rho)}{r} = -G\frac{\left(\dfrac{4\pi}{3}r^3\rho\right)(4\pi r^2\rho dr)}{r} = -\frac{(4\pi)^2 G}{3}\rho^2 r^4 dr. \tag{26.26}$$

If the radius of the sphere is R, then the total potential energy is given by

$$V_G = -\frac{(4\pi)^2}{3}G\rho^2 \int_0^R r^4\,dr = -\frac{(4\pi)^2}{15}G\rho^2 R^5. \tag{26.27}$$

The relation between density, mass, and R is

$$M = \int_0^R 4\pi r^2\rho\,dr = \frac{4\pi}{3}\rho R^3. \tag{26.28}$$

Thus for N nucleons (initially in the form of iron nuclei), each of mass $m = 1.6 \times 10^{-24}$ gm, we have $M = Nm$ and thus

$$V_G = -\frac{3}{5}\left(\frac{4\pi}{3}\right)^2 G\rho^2\frac{R^6}{R} = -\frac{3}{5}G\frac{(Nm)^2}{R} = -\frac{3}{5}\left(\frac{4\pi}{3}\right)^{1/3}G(Nm)^2\,\Omega^{-1/3} \tag{26.29}$$

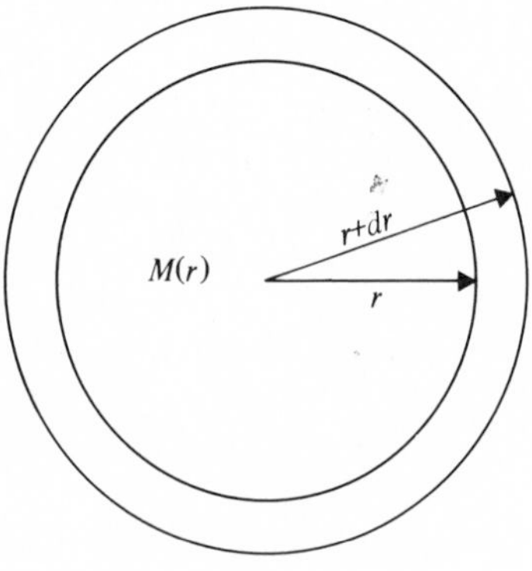

Figure 26.2. Shells of material for gravitational pressure calculation.

in terms of the volume of the star

$$\Omega = \frac{4\pi}{3} R^3. \tag{26.30}$$

The gravitational pressure is the force/unit area

$$= -\frac{\partial V_G/\partial R}{4\pi R^2} = -\frac{\partial V_G}{\partial \Omega} \tag{26.31}$$

so that

$$p_G = -\frac{1}{5} \left(\frac{4\pi}{3}\right)^{1/3} G(Nm)^2 \Omega^{-4/3}. \tag{26.32}$$

The minus sign indicates that the pressure acts to collapse matter.

Degeneracy Pressure As the matter is compressed, the electrons are pushed closer and closer together, and this compression is resisted by the exclusion principle. To work this out quantitatively, we need to know the total energy of N_e electrons in the lowest state in a box. The number of electrons that have energy less than or equal to E was calculated before

$$N = \frac{\pi}{3} (2m)^{3/2} \frac{L^3}{\pi^3 \hbar^3} E^{3/2}. \tag{26.33}$$

Thus the number of electrons that have energy between E and $E + dE$ is

$$dN = \frac{\pi}{2} (2m)^{3/2} \frac{L^3}{\pi^3 \hbar^3} E^{1/2} dE. \tag{26.34}$$

We need

$$\begin{aligned} E_{tot} &= \int_0^{E_F} E \, dN = \frac{\pi}{2} (2m)^{3/2} \frac{L^3}{\pi^3 \hbar^3} \int_0^{E_F} E^{3/2} \, dE \\ &= \frac{2}{5} \cdot \frac{\pi}{2} (2m)^{3/2} \frac{L^3}{\pi^3 \hbar^3} E_F^{5/2} \\ &= \frac{\hbar^2}{10m} 3^{5/3} \pi^{4/3} N_e^{5/3} L^{-2} \\ &= \frac{3^{5/3} \pi^{4/3}}{5} \frac{\hbar^2}{2m} N_e^{5/3} \Omega^{-2/3}. \end{aligned} \tag{26.35}$$

In the last step we changed from L^{-2} to $\Omega^{-2/3}$. This is permitted because the "counting of states," though most easily done for free particles in a rectangular box, gives a result that is independent of the shape of the container holding the free particles, if there are very many of them.

The pressure resisting compression, the *degeneracy pressure*, is

$$p_e = -\frac{\partial E_{tot}}{\partial \Omega} = \frac{6}{5} \left(\frac{\pi^4}{3}\right)^{1/3} \frac{\hbar^2}{2m} N_e^{5/3} \Omega^{-5/3}. \tag{26.36}$$

If the gravitational mass is not too large, equilibrium is obtained for some Ω. This occurs when the two pressures balance each other. If the mass is larger than some critical value, complications set in. The electrons get compressed into a volume small enough that their motion becomes relativistic. Then the electron energy, instead of being $p^2/2m$, approaches the relativistic pc. This, however, changes the dependence of E_{tot} on the volume from $\Omega^{-2/3}$ to $\Omega^{-1/3}$, and the electrons can no longer resist the gravitational pressure, since both p_G and p_e have the same volume dependence, and the first one wins out. Thus instead of the formation of a *white dwarf* (as an electron-degeneracy stabilized star is called), a collapse occurs. It becomes energetically favorable for the electrons to be squeezed into the protons

$$e^- + p \rightarrow n + \nu$$

to form neutrons. In the final stage we get a *neutron star*. Neutrons are also fermions, and they have spin $1/2$. The neutrons in the lowest state will exert a pressure

$$p_n = \frac{6}{5} \left(\frac{\pi^4}{3}\right)^{1/3} \frac{\hbar^2}{2m_n} N^{5/3} \Omega^{-5/3} \tag{26.37}$$

where now the number of neutrons is N, the number of nucleons originally present. Thus equilibrium is obtained when

$$\frac{6}{5} \left(\frac{\pi^4}{3}\right)^{1/3} \frac{\hbar^2}{2m_n} N^{5/3} \Omega^{-5/3} = \frac{G}{5} \left(\frac{4\pi}{3}\right)^{1/3} N^2 m_n^2 \Omega^{-4/3},$$

that is, when

$$\Omega^{1/3} = \frac{3\pi}{2^{2/3}} \frac{\hbar^2}{N^{1/3} m_n^3 G}. \tag{26.38}$$

This corresponds to a radius of

$$R = \frac{3^{4/3} \pi^{2/3}}{2^{4/3}} \frac{\hbar^2}{N^{1/3} m_n^3 G}. \tag{26.39}$$

For a star of two solar masses, $M = 4 \times 10^{33}$ gm, so that

$$N = \frac{4 \times 10^{33}}{1.6 \times 10^{-24}} = 2.5 \times 10^{57}.$$

With $m = 1.6 \times 10^{-24}$ and $G = 6.67 \times 10^{-8}$ in c.g.s. units we end up with

$$R \approx 10 \, \text{km}!$$

Our primitive calculation is actually in quite good agreement with more sophisticated ones.

If the mass of the star is much larger, then again the neutrons become relativistic and the neutron degeneracy pressure can no longer resist the gravitational pressure. Collapse again occurs, but there is no further mechanism to stop it. It is generally believed that under these circumstances a *black hole* is formed. A discussion of this topic involves general relativity. Whereas the existence of black holes is still somewhat controversial, there is unanimous agreement that the pulsars, first discovered by A. Hewish and collaborators in 1968, are indeed rotating neutron stars.

NOTES AND COMMENTS

1. The assumption that the electrons in a metal and the neutrons in a neutron star do not interact with each other is a gross simplification, in that one *knows* that in the first case there is the Coulomb repulsion between electrons, as well as attraction to the ions that form the crystal, and in the second case there are the strong nuclear forces whose range is of the same order of magnitude as the particle spacing. Nevertheless, the exclusion principle also accounts in part for the fact that the simplification does not lead to an invalid approximation. The reason is that when two particles interact, they scatter, that is, the momentum of each particle changes. Since all of the states in a given region in momentum space are occupied, such scattering is strongly inhibited. This argument will reappear when we discuss the independent particle model of the nucleus.

2. An interesting account of the discovery of the first optical pulsar (neutron star) may be found in *Adventures in Experimental Physics*, vol. α, World Science Communications, Princeton, N.J. (1972), B. Maglich (Ed.) A discussion of black holes may be found in *Cosmology + 1*, a collection of *Scientific American* reprints, W.H. Freeman, San Francisco, 1977.

PROBLEMS

26.1 Solve the Schrödinger equation for a particle in a box with sides L_1, L_2, and L_3. Calculate the energy eigenvalues. Compute the Fermi energy for a system of noninteracting fermions in such a box. Note that the volume of an ellipsoid whose surface is given by

$$\frac{x^2}{a^2} + \frac{y^2}{b^2} + \frac{z^2}{c^2} = 1$$

is $4\pi abc/3$.

26.2 Consider a particle in a three-dimensional harmonic oscillator potential, with

$$V(x, y, z) = \tfrac{1}{2} m\omega^2 (x^2 + y^2 + z^2).$$

Write down an expression for the energy eigenfunctions and for the energy eigenvalues. Consider noninteracting fermions in such a potential. Outline the procedure for calculating the Fermi energy.

26.3 The compressibility of material may be defined in terms of the pressure required to give a certain fractional change in volume. Conventionally one defines the *bulk modulus* by

$$\Delta P = -B \frac{\Delta V}{V}$$

or, equivalently,

$$B = -V \frac{\partial P}{\partial V}.$$

a) Show that $B = 5P/3$ where P is the degeneracy pressure.
b) Calculate the bulk modulus for the metals listed below: the table lists the metals, the electron density, and the experimental value of B.

Metal	n_e	B_{exp}
Na	2.65×10^{22} cm^{-3}	6.4×10^{10} dynes/cm^2
K	1.40×10^{22} cm^3	2.8×10^{10} dynes/cm^2
Cu	8.47×10^{22} cm^3	134.3×10^{10} dynes/cm^2
Al	$18.1 \ \times 10^{22}$ cm^3	76.0×10^{10} dynes/cm^2

Observe that the answers obtained from the noninteracting fermion model are of the right order of magnitude, which implies that the exclusion principle plays a significant part in the compressibility of matter.

26.4 Calculate the Fermi energy for noninteracting fermions that are massless, so that the energy is related to the momentum by $E = pc$. Explain the dependence on the fermion density n on purely dimensional grounds. (To warm up for this exercise, do the dimensional analysis for the nonrelativistic fermion gas.)

26.5 A nucleus consists of Z protons and $N = A - Z$ neutrons. The radius of the nucleus is given by

$$R = 1.2 \times 10^{-13} A^{1/3} \text{ cm}.$$

Assuming a box model for the nucleus, and assuming that the nucleons do not interact, obtain an expression for the Fermi energy of the proton gas, and for the neutron gas. Calculate the energies for lead, where $Z = 82$ and $N = 126$. Express your answer in MeV.

26.6 What is the velocity of an electron at the top of the Fermi sea in Na, K, Cu, Al? If the electron kinetic energy (Fermi energy) is equated to kT_F, what is the Fermi temperature for the four metals listed above?

26.7 Calculate the Fermi energy for particles that move relativistically, so that the energy – momentum relation is

$$E^2 = p^2 c^2 + m^2 c^4.$$

In a neutron star, for what densities would one begin to require the use of the relativistic relation?

26.8 Calculate the size of a white dwarf star making the same assumptions that were used in the neutron star calculation. Take the mass of the star to be one solar mass $(2 \times 10^{33} \text{ gm})$ and take the number of electrons to be approximately one half of the total number of nucleons.

27
The Schrödinger Equation in Three Dimensions. II. Central Potentials and Angular Momentum

The most interesting potentials in nature depend only on the separation between the interacting particles, so that

$$V(\mathbf{r}) = V(r). \qquad (27.1)$$

The force between the particles is then directed along the radius vector that connects them, and therefore it exerts no torque on the two-particle system. A consequence of this is the conservation of angular momentum. In quantum mechanics the argument proceeds somewhat differently (it is somewhat beyond the scope of this book), but the consequences are the same.

Angular momentum There exists an orbital angular momentum (*orbital*, because it has to do with the motion of the particle), whose operator form is

$$\mathbf{L}_{op} = \mathbf{r} \times \mathbf{p}_{op} \qquad (27.2)$$

and the angular momentum is a constant of the motion. One should be wary in the use of Eq. (27.2). We have already seen that the position operator $\mathbf{r}$ and the momentum operator $\mathbf{p}_{op}$ cannot have eigenfunctions such that $\mathbf{r}$ and $\mathbf{p}$ are simultaneously specified — that would violate the uncertainty relations. To see the difficulties, let us look at Eq. (27.2) using the representation $\mathbf{p} = (\hbar/i)\nabla$, to get

$$L_x = yp_z - zp_y = \frac{\hbar}{i}\left(y\frac{\partial}{\partial z} - z\frac{\partial}{\partial y}\right)$$

$$L_y = zp_x - xp_z = \frac{\hbar}{i}\left(z\frac{\partial}{\partial x} - x\frac{\partial}{\partial z}\right) \qquad (27.3)$$

$$L_z = xp_y - yp_x = \frac{\hbar}{i}\left(x\frac{\partial}{\partial y} - y\frac{\partial}{\partial x}\right).$$

With these forms we can show that different components of the angular momentum do not commute. Let us compute

$$[L_x, L_y] = L_xL_y - L_yL_x = -\hbar^2\left(y\frac{\partial}{\partial z} - z\frac{\partial}{\partial y}\right)\left(z\frac{\partial}{\partial x} - x\frac{\partial}{\partial z}\right)$$

$$+ \hbar^2\left(z\frac{\partial}{\partial x} - x\frac{\partial}{\partial z}\right)\left(y\frac{\partial}{\partial z} - z\frac{\partial}{\partial y}\right)$$

$$= -\hbar^2 \left(yz \frac{\partial^2}{\partial z \partial x} + y \frac{\partial}{\partial x} - xy \frac{\partial^2}{\partial z^2} - z^2 \frac{\partial^2}{\partial y \partial x} \right.$$

$$+ zx \frac{\partial^2}{\partial y \partial z} \right) + \hbar^2 \left(zy \frac{\partial^2}{\partial x \partial z} - z^2 \frac{\partial^2}{\partial y \partial x} \right.$$

$$\left. - xy \frac{\partial^2}{\partial z^2} + xz \frac{\partial^2}{\partial z \partial y} + x \frac{\partial}{\partial y} \right)$$

$$= -\hbar^2 \left(y \frac{\partial}{\partial x} - x \frac{\partial}{\partial y} \right) \ = \ i\hbar L_z . \tag{27.4}$$

Similarly,

$$[L_y, L_z] \ = \ i\hbar L_x \tag{27.5}$$

$$[L_z, L_x] \ = \ i\hbar L_y . \tag{27.6}$$

We shall see in Chapter 28 that as a consequence of these relations not all components of the angular momentum can simultaneously be specified. In quantum mechanics, it is the commutation relations (27.4)–(27.6) that *define* the angular momentum operators.

The representation in terms of differentials of the Cartesian coordinates (27.3) are not so useful when we consider a central potential, so that we need these operators in terms of spherical coordinates (r, θ, ϕ) (Fig. 27.1), defined by

$$x \ = \ r \sin \theta \cos \phi$$

$$y \ = \ r \sin \theta \sin \phi$$

$$z \ = \ r \cos \theta . \tag{27.7}$$

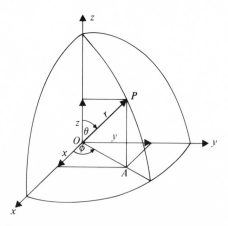

Figure 27.1.   Spherical coordinates: $z = r \cos \theta, x = (OA) \cos \phi = r \sin \theta \cos \phi, y = (OA) \sin \phi = r \sin \theta \sin \phi.$

We may write the differential form of the above as

$$dx = \sin\theta\cos\phi\, dr + r\cos\theta\cos\phi\, d\theta - r\sin\theta\sin\phi\, d\phi$$

$$dy = \sin\theta\sin\phi\, dr + r\cos\theta\sin\phi\, d\theta + r\sin\theta\cos\phi\, d\phi$$

$$dz = \cos\theta\, dr - r\sin\theta\, d\theta \qquad (27.8)$$

and solve for dr, $d\theta$, and $d\phi$. A little algebra that we leave to the reader gives

$$dr = \sin\theta\cos\phi\, dx + \sin\theta\sin\phi\, dy + \cos\theta\, dz$$

$$r\,d\theta = \cos\theta\cos\phi\, dx + \cos\theta\sin\phi\, dy - \sin\theta\, dz$$

$$r\sin\theta\, d\phi = -\sin\phi\, dx + \cos\phi\, dy. \qquad (27.9)$$

We can now calculate

$$\frac{\partial}{\partial x} = \frac{\partial r}{\partial x}\frac{\partial}{\partial r} + \frac{\partial \theta}{\partial x}\frac{\partial}{\partial \theta} + \frac{\partial \phi}{\partial x}\frac{\partial}{\partial \phi} \qquad (27.10)$$

$$= \sin\theta\cos\phi\frac{\partial}{\partial r} + \frac{1}{r}\cos\theta\cos\phi\frac{\partial}{\partial \theta} - \frac{1}{r\sin\theta}\sin\phi\frac{\partial}{\partial \phi},$$

$$\frac{\partial}{\partial y} = \frac{\partial r}{\partial y}\frac{\partial}{\partial r} + \frac{\partial \theta}{\partial y}\frac{\partial}{\partial \theta} + \frac{\partial \phi}{\partial y}\frac{\partial}{\partial \phi}$$

$$= \sin\theta\sin\phi\frac{\partial}{\partial r} + \frac{1}{r}\cos\theta\sin\phi\frac{\partial}{\partial \theta} + \frac{1}{r\sin\theta}\cos\phi\frac{\partial}{\partial \phi}, \qquad (27.11)$$

$$\frac{\partial}{\partial z} = \frac{\partial r}{\partial z}\frac{\partial}{\partial r} + \frac{\partial \theta}{\partial z}\frac{\partial}{\partial \theta} + \frac{\partial \phi}{\partial z}\frac{\partial}{\partial \phi} = \cos\theta\frac{\partial}{\partial r} - \frac{1}{r}\sin\theta\frac{\partial}{\partial \theta}. \qquad (27.12)$$

We may use these expressions to calculate further. Thus

$$L_z = \frac{\hbar}{i}\left(x\frac{\partial}{\partial y} - y\frac{\partial}{\partial x}\right)$$

$$= \frac{\hbar}{i}\left[r\sin\theta\cos\phi\left(\sin\theta\sin\phi\frac{\partial}{\partial r} + \frac{1}{r}\cos\theta\sin\phi\frac{\partial}{\partial \theta} + \frac{\cos\phi}{r\sin\theta}\frac{\partial}{\partial \phi}\right)\right. \qquad (27.13)$$

$$\left. - r\sin\theta\sin\phi\left(\sin\theta\cos\phi\frac{\partial}{\partial r} + \frac{1}{r}\cos\theta\cos\phi\frac{\partial}{\partial \theta} - \frac{\sin\phi}{r\sin\theta}\frac{\partial}{\partial \phi}\right)\right] = \frac{\hbar}{i}\frac{\partial}{\partial \phi}.$$

This is a particularly simple result. More complicated are

$$L_x = \frac{\hbar}{i}\left(y\frac{\partial}{\partial z} - z\frac{\partial}{\partial y}\right) = \frac{\hbar}{i}\left[r\sin\theta\sin\phi\left(\cos\theta\frac{\partial}{\partial r} - \frac{\sin\theta}{r}\frac{\partial}{\partial \theta}\right)\right.$$

$$\left. - r\cos\theta\left(\sin\theta\sin\phi\frac{\partial}{\partial r} + \frac{\cos\theta\sin\phi}{r}\frac{\partial}{\partial \theta} + \frac{\cos\phi}{r\sin\theta}\frac{\partial}{\partial \phi}\right)\right]$$

$$= \frac{\hbar}{i}\left(-\sin\phi\frac{\partial}{\partial \theta} - \frac{\cos\theta\cos\phi}{\sin\theta}\frac{\partial}{\partial \phi}\right) \qquad (27.14)$$

and

$$L_y = \frac{\hbar}{i}\left(z\frac{\partial}{\partial x} - x\frac{\partial}{\partial z}\right) = \frac{\hbar}{i}\left[r\cos\theta\left(\sin\theta\cos\phi\frac{\partial}{\partial r} + \frac{\cos\theta\cos\phi}{r}\frac{\partial}{\partial\theta} - \frac{\sin\phi}{r\sin\theta}\frac{\partial}{\partial\phi}\right)\right.$$

$$\left. - r\sin\theta\cos\phi\left(\cos\theta\frac{\partial}{\partial r} - \frac{\sin\theta}{r}\frac{\partial}{\partial\theta}\right)\right] = \frac{\hbar}{i}\left(\cos\phi\frac{\partial}{\partial\theta} - \frac{\cos\theta\sin\phi}{\sin\theta}\frac{\partial}{\partial\phi}\right).$$

$$(27.15)$$

Note that all three angular momentum operators depend only on angles and not on the radial coordinate r.

There is one more operator that plays a major role in the development of angular momentum theory, and that is the operator

$$\mathbf{L}^2 = L_x^2 + L_y^2 + L_z^2. \tag{27.16}$$

This too can be worked out, and a certain amount of tedious algebra leads to the expression

$$\mathbf{L}^2 = -\hbar^2\left(\frac{\partial^2}{\partial\theta^2} + \frac{\cos\theta}{\sin\theta}\frac{\partial}{\partial\theta} + \frac{1}{\sin^2\theta}\frac{\partial^2}{\partial\phi^2}\right). \tag{27.17}$$

The eigenvalue problem associated with angular momentum will occupy us later. Our next task is to calculate $\mathbf{p}^2$ using Eqs. (27.13)–(27.15). We might expect this operator to involve the operator $\mathbf{L}^2$, as can be seen from Fig. 27.2: the momentum vector $\mathbf{p}$ can be decomposed into a radial component p_r and a component $\mathbf{p}_\perp$ perpendicular to the radius vector. Since $\mathbf{L} = r\mathbf{p}_\perp$, we have the classical relation

$$\mathbf{p}^2 = \left(\hat{\mathbf{i}}_r p_r + \frac{\mathbf{L}}{r}\right)\cdot\left(\hat{\mathbf{i}}_r p_r + \frac{\mathbf{L}}{r}\right) = p_r^2 + \frac{\mathbf{L}^2}{r^2}. \tag{27.18}$$

This can also be seen from the vector identity

$$(\mathbf{r}\times\mathbf{p})\cdot(\mathbf{r}\times\mathbf{p}) = r^2\mathbf{p}^2 - (\mathbf{r}\cdot\mathbf{p})^2. \tag{27.19}$$

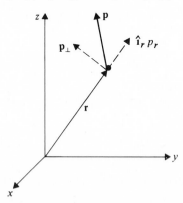

Figure 27.2. The decomposition of the momentum $\mathbf{p}$ into a radial and a transverse component. The transverse component is related to the angular momentum by $\mathbf{p}_\perp = \mathbf{L}/\mathbf{r}$.

In quantum mechanics the commutation relations involving $\mathbf{p}$ and $\mathbf{r}$ complicate this a little, so that the final result is

$$\mathbf{p}^2 = \frac{1}{r^2}\left(\frac{\hbar}{i}r\frac{\partial}{\partial r}\right)\left(\frac{\hbar}{i}r\frac{\partial}{\partial r}\right) + \frac{1}{r^2}\frac{\hbar}{i}\left(\frac{\hbar}{i}r\frac{\partial}{\partial r}\right) + \frac{\mathbf{L}^2}{r^2} = -\hbar^2\frac{\partial^2}{\partial r^2} - \frac{2\hbar^2}{r}\frac{\partial}{\partial r} + \frac{\mathbf{L}^2}{r^2}.$$

(27.20)

Thus we see that solving the angular momentum eigenvalue problem will also help with the separation of the Schrödinger equation into an angular part and a radial part. Since the potential does not depend on the variables θ and ϕ, but only on r, we expect that only the radial part will depend on the potential. If we know the eigenfunctions of $\mathbf{L}^2$,

$$\mathbf{L}^2 Y_\lambda(\theta, \phi) = \lambda\hbar^2 Y_\lambda(\theta, \phi),$$

(27.21)

then a trial form

$$\psi(\mathbf{r}) = R(r)Y_\lambda(\theta, \phi)$$

(27.22)

inserted into the equation

$$\frac{1}{2m}\left(-\hbar^2\frac{\partial^2}{\partial r^2} - \frac{2\hbar^2}{r}\frac{\partial}{\partial r} + \frac{\mathbf{L}^2}{r^2}\right)\psi(\mathbf{r}) + V(r)\psi(\mathbf{r}) = E\psi(\mathbf{r})$$

(27.23)

yields the *ordinary differential equation* for the *radial* wave function $R(r)$:

$$-\frac{\hbar^2}{2m}\left(\frac{d^2}{dr^2} + \frac{2}{r}\frac{d}{dr} - \frac{\lambda}{r^2}\right)R(r) + V(r)R(r) = ER(r).$$

(27.24)

The normalization condition in three dimensions is

$$\int d^3r\,|\psi(\mathbf{r})|^2 = 1$$

(27.25)

and since the volume element in spherical coordinates is

$$d^3r = r^2 \sin\theta\,dr\,d\theta\,d\phi, \quad \begin{array}{l} 0 \leqslant r \leqslant \infty \\ 0 \leqslant \theta \leqslant \pi \\ 0 \leqslant \phi \leqslant 2\pi \end{array}$$

(27.26)

we shall require separately that

$$\int_0^\infty r^2\,|R(r)|^2\,dr = 1$$

(27.27)

and

$$\int_0^\pi \sin\theta\,d\theta \int_0^{2\pi} d\phi\,|Y_\lambda(\theta, \phi)|^2 = 1.$$

(27.28)

Radial equation The radial equation can be written in a form that resembles the one-dimensional equation. Introduce

$$u(r) = rR(r).$$

(27.29)

Then

$$\frac{d^2}{dr^2}\left(\frac{u}{r}\right) + \frac{2}{r}\frac{d}{dr}\left(\frac{u}{r}\right) = \frac{1}{r}\frac{d^2u}{dr^2} - 2\frac{1}{r^2}\frac{du}{dr} + \frac{2u}{r^3} + \frac{2}{r}\frac{1}{r}\frac{du}{dr} - \frac{2}{r^3}u$$

$$= \frac{1}{r}\frac{d^2u}{dr^2}$$

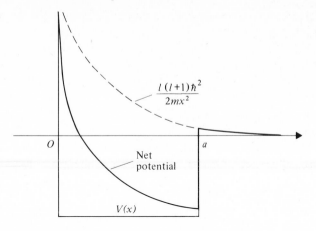

Figure 27.3. Equivalent one-dimensional well corresponding to attractive square well in three dimensions.

so that the equation takes the form

$$-\frac{\hbar^2}{2m}\frac{d^2u(r)}{dr^2} + \left(V(r) + \frac{\hbar^2\lambda}{2mr^2}\right)u(r) = Eu(r) \tag{27.30}$$

subject to the condition

$$u(0) = 0. \tag{27.31}$$

This is equivalent (Fig. 27.3) to a one-dimensional problem with potential

$$V(x) = \infty \qquad\qquad x < 0$$

$$V(x) = V(x) + \frac{\hbar^2\lambda}{2mx^2} \qquad 0 < x \tag{27.32}$$

since the infinite barrier at the origin imposes Eq. (27.31). We will see in the next chapter that the number λ is of the form $l(l+1)$ where $l = 0, 1, 2, \ldots$. Thus for $l = 0$, the three-dimensional radial equation looks exactly like the one-dimensional Schrödinger equation with solutions that vanish at the origin.

NOTES AND COMMENTS

1. The conservation of angular momentum for systems in which the potential energy depends on r only is a consequence of a deep connection between conservation laws and symmetry. Here the symmetry states that the energy operator does not depend on a *direction*. There is no preferred direction in the system. All the potentials that occur in nature have this property. The conservation law can be violated in a subsystem, if an external potential is set up: An atom may be placed between the pole pieces of a magnet or between the plates of a condenser,

in which case that atom is in an environment in which there is a preferred direction. (This will be discussed in Chapter 29 in connection with the Zeeman effect.) The existence of angular momentum operators as a consequence of the transformation laws under rotations is discussed in more advanced quantum mechanics textbooks: see, for example, S. Gasiorowicz, *Quantum Physics*, John Wiley, New York, 1974, Chapter 9.

2. The connection between a symmetry—here the invariance of the equation under rotations — and the separability of the equation into ordinary differential equations, is no accident. If the equation has an invariance, there is a quantity that is conserved, and one may then find eigenfunctions of the energy operator that are simultaneously eigenfunctions of the conserved operator. Thus, if the potential in the Schrödinger equation were cylindrically symmetric, that is, of the form $V(x^2 + y^2)$ alone, then (a) there is symmetry under displacement along the z axis, and therefore conservation of momentum along the z axis, p_z; and (b) there is symmetry under rotations about the z axis, and therefore conservation of angular momentum about the z axis, L_z. One may thus look for solutions of the Schrödinger equation of the form

$$e^{ipz/\hbar} e^{im\phi} f(x^2 + y^2)$$

where the first is an eigenfunction of the operator p_z, the second is an eigenfunction of the angular momentum in the z direction L_z, and the third depends on $x^2 + y^2$.

PROBLEMS

27.1 Use the expressions (27.3) to calculate a differential operator for $\mathbf{L}^2$, and use that

a) to show that
$$[\mathbf{L}^2, L_x] = [\mathbf{L}^2, L_y] = [\mathbf{L}^2, L_z] = 0,$$

b) to derive Eq. (27.20).

27.2 Solve the eigenvalue problem

$$L_z \Phi(\phi) = l_z \hbar \Phi(\phi)$$

subject to the condition that the wave function is unaltered when the system is rotated about the z axis by 2π, that is, when $\phi \to \phi + 2\pi$. Why is the eigenvalue a multiple of $\hbar$?

27.3 Use the expressions Eq. (27.3) to calculate the commutation relations

$$[L_x, p_x], [L_x, p_y], [L_x, p_z], \ldots, [L_z, p_z].$$

27.4 Show that the expectation values of L_x, L_y, and L_z vanish when the system is in a state described by a spherically symmetric wave function $u(r)$.

27.5 Consider a free particle propagating in the z direction. Its wave function is proportional to e^{ikz}. Show that this state is not an eigenstate of L_x and L_y, but is an eigenstate of L_z. What is the eigenvalue in the last case?

28
Angular Momentum

Angular momentum eigenfunctions In classical mechanics a system has a well-defined angular momentum, that is, all three components of $\mathbf{L}$, L_x, L_y, and L_z have definite values. In quantum mechanics, the corresponding situation would be described by the statement that a system is in a state, described by a wave function $Y(\theta, \phi)$, such that

$$L_x Y(\theta, \phi) = l_1 Y(\theta, \phi)$$

$$L_y Y(\theta, \phi) = l_2 Y(\theta, \phi)$$

$$L_z Y(\theta, \phi) = l_3 Y(\theta, \phi). \tag{28.1}$$

We will now show that the angular momentum commutation relations (27.4)–(27.6) make this impossible. Let us have Eq. (27.4) act on the wave function $Y(\theta, \phi)$:

$$
\begin{aligned}
i\hbar L_z Y(\theta, \phi) &= L_x L_y Y(\theta, \phi) - L_y L_x Y(\theta, \phi) \\
&= L_x l_2 Y(\theta, \phi) - L_y l_1 Y(\theta, \phi) \\
&= l_2 L_x Y(\theta, \phi) - l_1 L_y Y(\theta, \phi) \\
&= l_2 l_1 Y(\theta, \phi) - l_1 l_2 Y(\theta, \phi) = 0.
\end{aligned}
\tag{28.2}
$$

This implies that l_3 must vanish. If we repeat the calculation with Eq. (27.5), we find that l_1 must vanish, and Eq. (27.6) shows that l_2 must vanish! *In quantum mechanics it is impossible to prepare a state that is a simultaneous eigenstate of all three components of angular momentum*, except for the state of zero angular momentum, where all components vanish. This is not unprecedented: we found that it is impossible to find a state of simultaneously well-defined momentum and position. We can find an eigenstate of one component of angular momentum. The simplest one is L_z, and it is easy to solve for

$$\frac{\hbar}{i} \frac{\partial}{\partial \phi} Y(\theta, \phi) = l_3 Y(\theta, \phi). \tag{28.3}$$

The solution is of the form

$$Y(\theta, \phi) \propto e^{il_3 \phi/\hbar}$$

and single-valuedness for the wave function requires that

$$e^{il_3(\phi + 2\pi)/\hbar} = e^{il_3 \phi/\hbar}. \tag{28.4}$$

This means that $l_3/\hbar$ must be an integer, so that the eigenfunction, properly normalized, becomes

$$Y(\theta, \phi) = \frac{f(\theta)}{\sqrt{2\pi}} e^{im\phi} \qquad (m \text{ integer}) \qquad (28.5)$$

with

$$\int_0^\pi \sin \theta \, d\theta \, |f(\theta)|^2 = 1 \qquad (28.6)$$

and with the eigenvalues

$$L_z = m\hbar \quad m = 0, \pm 1, \pm 2, \pm 3, \ldots. \qquad (28.7)$$

To determine the function $f(\theta)$ we shall require that $Y(\theta, \phi)$ be an eigensolution of

$$\mathbf{L}^2 Y(\theta, \phi) = \lambda \hbar^2 Y(\theta, \phi). \qquad (28.8)$$

Using Eq. (27.17) we get

$$-\left(\frac{\partial^2}{\partial\theta^2} + \frac{\cos\theta}{\sin\theta}\frac{\partial}{\partial\theta} + \frac{1}{\sin^2\theta}\frac{\partial^2}{\partial\phi^2}\right) f(\theta)e^{im\phi} = \lambda f(\theta)e^{im\phi}$$

so that

$$\left(\frac{d^2}{d\theta^2} + \frac{\cos\theta}{\sin\theta}\frac{d}{d\theta} - \frac{m^2}{\sin^2\theta} + \lambda\right) f(\theta) = 0. \qquad (28.9)$$

This may be written in the form

$$\frac{1}{\sin\theta}\frac{d}{d\theta}\left(\sin\theta\frac{df}{d\theta}\right) + \left(\lambda - \frac{m^2}{\sin^2\theta}\right)f(\theta) = 0. \qquad (28.10)$$

A detailed study of this differential equation yields the following results:

a) The eigenvalues of the equation are of the form

$$\lambda = l(l+1) \qquad (28.11)$$

with $l = 0, 1, 2, 3, \ldots$.

b) For a given value of l, not all values of m are allowed. The values of m are restricted to be

$$m = l, l-1, l-2, \ldots, -l. \qquad (28.12)$$

c) The $f(\theta)$ are polynomials in $\sin\theta$ and $\cos\theta$ with the order depending on l and m. They are the *associated Legendre polynomials*, discussed in many textbooks on mathematical physics and special functions. When these are combined with the $e^{im\phi}$ factors, and properly normalized, we get the so-called *spherical harmonics* $Y_{lm}(\theta, \phi)$. The first few of the spherical harmonics are listed in Table 28.1. They satisfy

$$\int_0^\pi \sin\theta \, d\theta \int_0^{2\pi} d\phi (Y_{lm}(\theta, \phi))^* Y_{l'm'}(\theta, \phi) = \delta_{ll'}\delta_{mm'}, \qquad (28.13)$$

that is, the simultaneous eigenfunctions of $\mathbf{L}^2$ and L_z are orthogonal when either the eigenvalue of $\mathbf{L}^2$ or the eigenvalue of L_z is different.

Table 28.1

Spherical harmonics

$$Y_{00} = (1/4\pi)^{1/2}$$

$$Y_{11} = -(3/8\pi)^{1/2} e^{i\phi} \sin\theta$$

$$Y_{10} = (3/4\pi)^{1/2} \cos\theta$$

$$Y_{22} = (15/32\pi)^{1/2} e^{2i\phi} \sin^2\theta \qquad (28.14)$$

$$Y_{21} = -(15/8\pi)^{1/2} e^{i\phi} \sin\theta \cos\theta$$

$$Y_{20} = (5/16\pi)^{1/2} (3\cos^2\theta - 1)$$

$$Y_{l,-m} = (-1)^m (Y_{lm})^*$$

A general formula that may be used to compute higher-order spherical harmonics is

$$Y_{lm}(\theta,\phi) = (-1)^m \left[\frac{2l+1}{4\pi} \frac{(l-m)!}{(l+m)!} \right]^{1/2} e^{im\phi} P_l^m(\cos\theta)$$

$$P_l^m(u) = (-1)^m \frac{(l+m)!}{(l-m)!} \frac{(1-u^2)^{-m/2}}{2^l l!} \left(\frac{d}{du} \right)^{l-m} (u^2-1)^l. \qquad (28.15)$$

The functions $P_l^m(\cos\theta)$ are known as associated Legendre polynomials. For $m=0$ they reduce to the more familiar Legendre polynomials.

The fact that the values of m, for a given l should be restricted, is reasonable. After all, in an eigenstate of $\mathbf{L}^2$ and L_z,

$$\langle \mathbf{L}^2 \rangle = \langle L_x^2 + L_y^2 + L_z^2 \rangle = \hbar^2 l(l+1) \qquad (28.16)$$

and

$$\langle L_z^2 \rangle = \hbar^2 m^2, \qquad (28.17)$$

so that we expect

$$l(l+1) \geqslant m^2.$$

What is surprising is that, even when L_z takes on its maximum value l, then

$$\langle \mathbf{L}^2 \rangle - \langle L_z^2 \rangle = l\hbar^2 \neq 0. \qquad (28.18)$$

Classically it is certainly true that

$$\langle \mathbf{L}^2 \rangle = \langle L_z^2 \rangle_{max}$$

but quantum mechanically this would imply that

$$\langle L_x^2 + L_y^2 \rangle = 0,$$

and since both are positive,

$$\langle L_x^2 \rangle = \langle L_y^2 \rangle = 0, \qquad (28.19)$$

and this would imply that $Y_{lm}(\theta,\phi)$ are eigenfunctions of L_x and L_y with eigenvalues zero, which, from the commutation relations implies that L_z also has zero eigenvalue. The fact that Eq. (28.18) holds is taken into account in a semiclassical graphical representation of angular momentum (Fig. 28.1). On a circle of radius $\sqrt{l(l+1)}$ (angular momentum is represented in units of $\hbar$), the vertical axis represents the z direction,

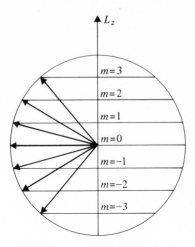

Figure 28.1. Semiclassical representation of an angular momentum $l = 3$ vector. The circle represents a sphere of radius $\sqrt{3 \times 4}$.

and the horizontal axis represents a projection of the $x - y$ plane. The angular momentum vector is constrained to lie on the sphere, with the values of L_z determining the latitude of the circles on which the vector lies. We may view the angular momentum vector as precessing about the z axis. A calculation of $\langle L_x \rangle$ and $\langle L_y \rangle$ in eigenstates of $\mathbf{L}^2$ and L_z yields zero, whereas

$$\langle L_x^2 \rangle = \langle L_y^2 \rangle = \tfrac{1}{2} \hbar^2 (l(l+1) - m^2). \tag{28.20}$$

For large angular momenta, the state with $m = l$ (or $m = -l$) represents an almost classical angular momentum vector pointing in the z direction. Classically such an angular momentum would arise from a distribution of matter rotating in the $x - y$ plane. The quantum mechanical wave function mimics this. One can show that

$$|Y_{ll}(\theta, \phi)|^2 \propto (\sin \theta)^{2l} \tag{28.21}$$

and this function is confined to a small angular region around the equatorial plane. The other distributions $|Y_{lm}(\theta, \phi)|^2$ also do not depend on the azimuthal angle ϕ. They are less transparent and more complicated; some of them are shown in Fig. 28.2.

We chose to find eigenfunctions and eigenvalues of L_z (and $\mathbf{L}^2$) because of the simplicity of L_z in spherical coordinates. There is nothing special about that choice. We could equally well have chosen to find mutual eigenfunctions of $\mathbf{L}^2$ and L_x, say. These eigenvalues would still be $l(l + 1)\hbar^2$ for $\mathbf{L}^2$ and the number $l\hbar$, $(l - 1)\hbar$, ..., $-l\hbar$ for L_x. The eigenfunctions would be some particular linear combinations of the

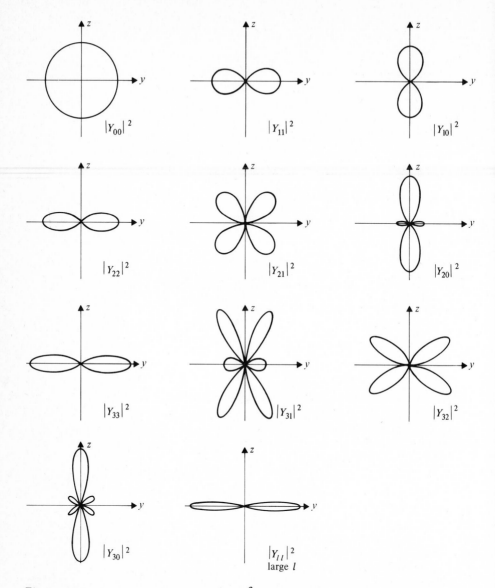

Figure 28.2. Distributions of $|Y_{lm}(\theta, \phi)|^2$. The sketches represent sections of the distributions made in the z-y plane. It should be understood that the three-dimensional distributions are obtained by rotating the figures about the z axis.

Y_{lm}: We would find, if we chose to do the algebra, that for a given l we can find a unique set of coefficients a_{mn} such that

$$L_x \left(\sum_{n=-l}^{l} a_{mn} Y_{ln} \right) = m\hbar \left(\sum_{n=-l}^{l} a_{mn} Y_{ln} \right). \qquad (28.22)$$

The choice of L_z as the sole angular momentum component is so convenient that we will stick with it.

Connection with magnetic moment The fact that angular momentum is quantized can be demonstrated directly because of the connection between angular momentum and the magnetic dipole moment of atoms. The connection can be obtained in the classical limit. There the magnetic moment of a current loop is

$$\mu = \frac{IA}{c} \tag{28.23}$$

where A is the area and I is the current. For an electron of charge e, moving with velocity v in a circle of radius r, this becomes

$$|\mu| = \frac{ev}{2\pi r}\frac{\pi r^2}{c} = \frac{1}{2}\frac{e}{c}\frac{mvr}{m} = \frac{e}{2mc}|\mathbf{L}|. \tag{28.24}$$

Since the electron has a negative charge, the magnetic moment points in a direction opposite to that of the angular momentum **L**. For more complicated systems the coefficient in front may change but the proportionality to the angular momentum remains, so that we have

$$\mu = -g\frac{e}{2mc}\mathbf{L}. \tag{28.25}$$

The factor g is called the *gyromagnetic ratio*.

Classical electrodynamics tells us that if a system with magnetic dipole moment μ is placed in a magnetic field **B**, then the energy is changed by a term

$$V = -\mu \cdot \mathbf{B}. \tag{28.26}$$

This is smallest when the magnetic dipole moment points in the direction of the external magnetic field **B**. There is also a torque $\mu \times \mathbf{B}$ that acts on the moment, and it changes the angular momentum **L**. Classically the moment precesses in a way such that the energy V in Eq. (28.26) remains constant.

In quantum mechanics Eq. (28.26) represents an addition to the total energy operator H of a term

$$V = \frac{eg}{2mc}\mathbf{B} \cdot \mathbf{L}_{\text{op}} \tag{28.27}$$

which has a particularly simple form if the direction of **B** is defined as the z axis. If the magnetic field is not uniform, but depends on **r**, then it exerts a force on the system, given by

$$\mathbf{F} = -\nabla V = +\nabla(\mu_x B_x + \mu_y B_y + \mu_z B_z). \tag{28.28}$$

Since L_x, L_y have an average value of zero, so do μ_x and μ_y, and hence the force reduces to the average value $\mu_z \, \partial B_z/\partial z$ in the z direction.

Stern and Gerlach in 1922 set up the apparatus shown in Fig. (28.3). According to classical theory, the beam of atoms should be spread out by the field, since μ can point in any direction. In fact, the experiment showed that the beam splits into a discrete number of beams, indicating that the magnetic dipole moment, and hence the

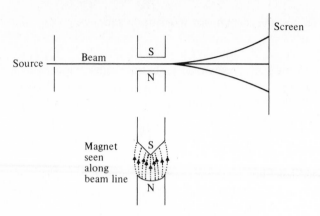

Figure 28.3. Apparatus for Stern–Gerlach experiment.

angular momentum is quantized. With knowledge of the field and the measured splitting, the gyromagnetic ratio can also be measured.

According to the theory of angular momentum discussed above, the number of possible values of $\mathbf{L} \cdot \mathbf{B}$ is $(2l + 1)$, one for each value of m $(-l \leqslant m \leqslant l)$; this is most easily seen by choosing $\mathbf{B}$ to define the z direction, $\mathbf{B} = (0, 0, B)$. Thus the number of beams should always be odd. It turns out that sometimes the number of beams is even, which suggests the possibility of angular momentum $1/2, 3/2, \ldots$ in units of $\hbar$. This important phenomenon is associated with the existence of *spin*, and will be discussed in Chapter 30.

As a final comment, we note that for historical reasons, orbital angular momentum states are frequently denoted by letters. Thus states with values of $l = 0, 1, 2, 3, 4, 5, \ldots$ are denoted with the letters $S, P, D, F, G, H, \ldots$. The nomenclature no longer makes sense, but it has struck, and is used in atomic, nuclear, and even in elementary particle physics.

NOTES AND COMMENTS

1. The Stern-Gerlach experiment is discussed in great detail in many modern physics textbooks, and I have therefore felt free to limit myself to a very brief discussion. For more details see, for example, R. Eisberg and R. Resnick, *Quantum Physics*, John Wiley, New York, (1974). The Stern-Gerlach apparatus is used by Feynman in the *Feynman Lectures on Physics* (vol. III) Addison-Wesley, Reading, Mass., 1963 to introduce the basic notions of quantum mechanical states to the reader.

PROBLEMS

28.1 The solution of Eq. (28.10) must be regular at $\theta = 0$ and π, when $\sin \theta = 0$. Show that the behavior of $f(\theta)$ near $\sin \theta = 0$ is given by

$$f(\theta) \sim (\sin \theta)^{|m|}.$$

28.2 Write

$$f(\theta) = (\sin\theta)^m P(\theta), \quad m > 0$$

and obtain the differential equation obeyed by $P(\theta)$. Rewrite the differential equation in terms of the variable $u = \cos\theta$, and examine the solution, assuming that it is a polynomial, so that a power series expansion terminates. Examine the recursion relation.

28.3 Consider a system described by the energy operator

$$H = \mathbf{L}^2/2I.$$

Sketch the spectrum, and label the degeneracy, that is, the number of states that have a given energy. Suppose the system is put in an external magnetic field, chosen to point in the z direction. Show that the energy operator becomes

$$H = \mathbf{L}^2/2I + \alpha L_z.$$

Work out the energy spectrum, and show that the degeneracy is now lifted, that is, each energy level has only one state associated with it.

28.4 A plane rotator is described by the energy operator

$$H = L_z^2/2I.$$

What are the eigenvalues and eigenfunctions of this system? There is a degeneracy, in that there are two states for (almost) each energy eigenvalue. What is the physical reason for this degeneracy?

28.5 Consider now a two-particle system described by the energy operator

$$H = L_z^2/2I.$$

The two particles lie at opposite ends of a diameter rotating in a plane about the midpoint. If the two particles are *identical*, the singlevaluedness of the wave function under interchange implies that

$$\psi(\phi + \pi) = \psi(\phi).$$

How does this affect the spectrum?

28.6 Repeat the above calculation with N identical, equally spaced particles on a ring. How does the spectrum look when the singlevaluedness condition

$$\psi(\phi + 2\pi/N) = \psi(\phi)$$

is imposed? How much energy does it take to excite the system from the ground state to the first excited state? What happens when one particle is removed, and a gap left?

29
The Hydrogen Atom

The radial equation The hydrogen-like atom, consisting of a nucleus of charge Ze and a single electron of charge $-e$ bound to it, is described by the eigenvalue equation

$$-\frac{\hbar^2}{2\mu} \nabla^2 \psi(\mathbf{r}) - \frac{Ze^2}{r} \psi(\mathbf{r}) = E\psi(\mathbf{r}) \tag{29.1}$$

with $E < 0$ for the bound state solutions. For hydrogen, $Z = 1$ and the parameter μ is the reduced mass, given by

$$\mu = \frac{m_e M}{m_e + M} \tag{29.2}$$

where M is the mass of the proton. When the wave function is written in the form

$$\psi(\mathbf{r}) = R(r)Y_{lm}(\theta, \phi), \tag{29.3}$$

we get the radial equation (27.24)

$$\frac{d^2R}{dr^2} + \frac{2}{r}\frac{dR}{dr} - \frac{l(l+1)}{r^2}R + \frac{2\mu}{\hbar^2}\left(\frac{Ze^2}{r} - |E|\right)R = 0 \tag{29.4}$$

with the normalization condition

$$\int_0^\infty dr\, r^2 (R(r))^2 = 1. \tag{29.5}$$

Given the probability interpretation of $\psi(\mathbf{r})$, the quantity $r^2(R(r))^2\, dr$ is the probability of finding the electron in a spherical shell lying between the radii r and $r + dr$. As in our discussion of the harmonic oscillator, it is convenient to use dimensionless variables. We introduce

$$\rho = (8\mu|E|/\hbar^2)^{1/2} r \tag{29.6}$$

and the dimensionless parameter

$$\eta = \left(\frac{Z^2 e^4 \mu}{2\hbar^2 |E|}\right)^{1/2} = Z\alpha \left(\frac{\mu c^2}{|E|}\right)^{1/2} ; \alpha = \frac{e^2}{\hbar c} \tag{29.7}$$

in terms of which the radial equation takes the form

$$\frac{d^2R}{d\rho^2} + \frac{2}{\rho}\frac{dR}{d\rho} - \frac{l(l+1)}{\rho^2}R + \frac{\eta}{\rho}R - \frac{1}{4}R = 0. \tag{29.8}$$

The parameter η is the eigenvalue that is to be determined.

To solve this equation we proceed as we did for the harmonic oscillator. First we look for the behavior for large ρ. In that limit the equation simplifies to

$$\frac{d^2R}{d\rho^2} - \frac{1}{4}R = 0. \tag{29.9}$$

This has solutions $R \sim e^{\pm \rho/2}$, but only the $e^{-\rho/2}$ can be bounded at infinity. For small ρ, the leading terms in the equation are

$$\frac{d^2R}{d\rho^2} + \frac{2}{\rho}\frac{dR}{d\rho} - \frac{l(l+1)}{\rho^2}R = 0. \tag{29.10}$$

If we try the solution of the form ρ^s, we find that

$$(s(s-1) + 2s - l(l+1))\rho^{s-2} = 0$$

so that $s = l$. The solution that behaves as ρ^{-l-1} is called the *irregular* solution and blows up at the origin. It plays no role here. We combine the information obtained so far and write

$$R(\rho) = e^{-\rho/2}\rho^l F(\rho). \tag{29.11}$$

When this is substituted into Eq. (29.8) a differential equation for $F(\rho)$ is obtained. A page or so of algebra, left to the reader, leaves us with the equation

$$\frac{d^2F}{d\rho^2} + \left(\frac{2l+2}{\rho} - 1\right)\frac{dF}{d\rho} + \frac{\eta - l - 1}{\rho}F = 0 \tag{29.12}$$

and it is this equation that we will attempt to solve by a series expansion, as was already done for the harmonic oscillator problem. Write

$$F(\rho) = \sum_{n=0}^{\infty} a_n \rho^n. \tag{29.13}$$

This gives

$$\sum_{n=2}^{\infty} n(n-1)a_n\rho^{n-2} + \sum_{n=1}^{\infty} n(2l+2)a_n\rho^{n-2}$$

$$- \sum_{n=0}^{\infty} na_n\rho^{n-1} + (\eta - l - 1)\sum_{n=0}^{\infty} a_n\rho^{n-1} = 0.$$

In the first two terms we take note of the fact that because of the differentiation, the series starts with $n = 2$ and $n = 1$, respectively. We can rearrange this to get

$$\sum_{m=0}^{\infty} m(m+1)a_{m+1}\rho^{m-1} + \sum_{m=0}^{\infty} 2(m+1)(l+1)a_{m+1}\rho^{m-1}$$

$$- \sum_{m=0}^{\infty} ma_m\rho^{m-1} + (\eta - l - 1)\sum_{m=0}^{\infty} a_m\rho^{m-1} = 0$$

and this recursion relation is satisfied provided that

$$a_{m+1} = \frac{m+l+1-\eta}{(m+1)(2l+m+2)}a_m. \tag{29.14}$$

Again, as for the harmonic oscillator, we find that the series must terminate, that is, $F(\rho)$ *must be a polynomial.* Otherwise, the recursion relation indicates that $a_m \sim 1/m!$ for large m and the series approaches e^ρ.

The spectrum   The simplest polynomial is a constant. In that case the equation is satisfied, provided that

$$\eta = l + 1, \quad l = 0, 1, 2, \ldots . \tag{29.15}$$

The next simple case is a first-order polynomial. If we set $F(\rho) = \rho + c$, then we get

$$\frac{2(l+1)}{\rho} - 1 + (\rho + c)\frac{\eta - l - 1}{\rho} = 0$$

so that

$$\eta = l + 2, \quad l = 0, 1, 2, \ldots \tag{29.16}$$

and the terms proportional to $1/\rho$ determine c. In general, *if the polynomial is of degree* n_r, so that $a_{n_r+1} = 0$, the recursion relation yields

$$\eta = n_r + l + 1. \tag{29.17}$$

Thus η is an integer, and hence

$$E = -\frac{1}{2}\mu c^2 (Z\alpha)^2 \frac{1}{n^2}, \quad n = 1, 2, 3, \ldots , \tag{29.18}$$

a result familiar from the Bohr model. The integer is of the form

$$n = n_r + l + 1 \tag{29.19}$$

so that $n \geqslant l + 1$. For $n = 1$ we must have $n_r = 0$ and $l = 0$. The wave function is spherically symmetric, and it has the form

$$R(\rho) \sim e^{-\rho/2}. \tag{29.20}$$

For $n = 2$ there are two possibilities:

a) $n_r = 1, l = 0$. This is again a spherically symmetric function, but the polynomial is of first order. It is easy to check that it has a zero, so that the wave function has a node.

b) We can have $n_r = 0$ and $l = 1$. In that case the wave function has no radial nodes, except for the origin, and it has a $Y_{lm}(\theta, \phi)$ angular dependence.

The same arguments, allowing for more possibilities, hold for $n = 3, 4, 5, \ldots$. In all cases we have the surprising result that the energy does not depend on the parameter l, even though the Schrödinger equation does contain that parameter. The fact that for a given l (for example, $l = 1$, above) all the states with $-l \leqslant m \leqslant l$ have the same energy is not surprising: the radial equation that determines the eigenvalue contains no reference to the L_z eigenvalue. The additional degeneracy in l is a very special feature of the $1/r$ potential. To see this in a specific example, suppose that we

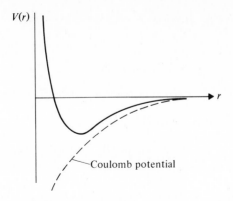

Figure 29.1. Potential considered in Eq. (29.21).

had a potential that had the Coulomb form far out, but a $1/r^2$ repulsive part close in (Fig. 29.1),

$$V(r) = -\frac{Ze^2}{r} + \frac{\hbar^2}{2\mu}\frac{f^2}{r^2}.$$ (29.21)

If this is inserted into the radial equation, nothing changes except that where we had

$$l(l+1)/r^2$$

we now have

$$l^*(l^*+1)/r^2,$$

where we have written

$$l(l+1)+f^2 = l^*(l^*+1).$$

Thus the energy is now of the form

$$E = -\frac{1}{2}\mu c^2 (Z\alpha)^2 \frac{1}{(n_r + l^* + 1)^2}$$ (29.22)

where l^* can be solved for:

$$l^* = -\tfrac{1}{2} + \sqrt{(l+1/2)^2 + f^2}.$$ (29.23)

If we now write the energy in terms of $n = n_r + l + 1$, we get

$$E = -\tfrac{1}{2}\mu c^2 (Z\alpha)^2 [n - l - 1/2 + \sqrt{(l+1/2)^2 + f^2}]^{-2}.$$ (29.24)

Thus for fixed n, the *principal quantum number*, there is an explicit l-dependence in the expression for the energy.

The orbits of maximum angular momentum in the classical Kepler problem are the circular orbits. Here they are the solutions with $n_r = 0$, that is,

$$l = n-1.$$ (29.25)

In that case, the square of the radial function is

$$|R(\rho)|^2 \sim \rho^{2(n-1)}e^{-\rho}$$ (29.26)

and the probability density in the ρ variable is

$$\rho^2 |R(\rho)|^2 \sim \rho^{2n} e^{-\rho}. \tag{29.27}$$

This peaks when

$$2n\rho^{2n-1} e^{-\rho} - \rho^{2n} e^{-\rho} = 0, \tag{29.28}$$

that is, for

$$\rho = 2n.$$

When this is expressed in terms of r, it is found that

$$r = a_0 n^2 \tag{29.29}$$

where $a_0 = \hbar/\mu c(Z\alpha)$. This again is a result familiar from the Bohr model.

Radial eigenfunctions The radial functions may be written quite generally in terms of known mathematical functions, the associated Laguerre polynomials, and for detailed calculations involving the hydrogen-like atoms, one must know their properties. Table 29.1 lists a few of the radial eigenfunctions.

Table 29.1

n	l	$R_{nl}(r)$
1	0	$2a_0^{-3/2} e^{-r/a_0}$
2	0	$(2a_0)^{-3/2}(2 - r/a_0)e^{-r/2a_0}$
2	1	$\dfrac{1}{\sqrt{3}}(2a_0)^{-3/2}\dfrac{r}{a_0}e^{-r/2a_0}$
3	0	$\dfrac{2}{81\sqrt{3}}a_0^{-3/2}\left(27 - 18\dfrac{r}{a_0} + 2\dfrac{r^2}{a_0^2}\right)e^{-r/3a_0}$
3	1	$\dfrac{4}{81\sqrt{6}}a_0^{-3/2}\left(6 - \dfrac{r}{a_0}\right)\dfrac{r}{a_0}e^{-r/3a_0}$
3	2	$\dfrac{4}{81\sqrt{30}}a_0^{-3/2}\dfrac{r^2}{a_0^2}e^{-r/3a_0}$

With the help of these, quantities of physical interest can be calculated. Figure 29.2 shows some radial eigenfunctions. There are also expectation values of powers of r that are sometimes useful. We list a few of them:

i)

$$\langle r \rangle = a_0 n^2 \left[1 + \frac{1}{2}\left(1 - \frac{l(l+1)}{n^2}\right)\right]$$

ii)

$$\left\langle \frac{1}{r^2} \right\rangle = \frac{1}{a_0^2}\frac{1}{n^3(l+1/2)}$$

iii)

$$\left\langle \frac{1}{r^3} \right\rangle = \frac{1}{a_0^3}\frac{1}{n^3 l(l+1/2)(l+1)} \tag{29.30}$$

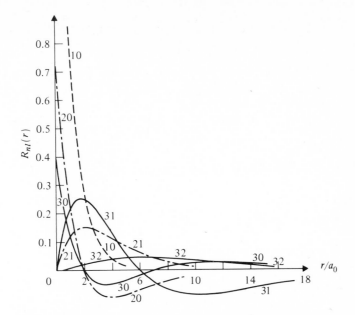

Figure 29.2. The radial wave functions $R_{nl}(r)$ for hydrogenic atoms for $n = 1, 2, 3$. Each curve is labeled with two integers, representing the corresponding n and l values. Note the effect of the centrifugal force in "pushing out" the wave function from the center of the atom. Note also that the functions have $n - 1 - l$ nodes. (Source: R. H. Dicke and J. P. Wittke, *Introduction to Quantum Mechanics*, Addison-Wesley, 1960, p. 163. Reprinted by permission.)

iv)
$$\left\langle \frac{1}{r} \right\rangle = \frac{1}{a_0} \frac{1}{n^2}$$

v)
$$\langle r^2 \rangle = \tfrac{1}{2} a_0^2 n^2 (5n^2 + 1 - 3l(l + 1))$$

These are calculated using

$$\langle r^k \rangle = \int_0^\infty dr\, r^2 (R(r))^2 r^k. \tag{29.31}$$

We observe that $\langle r \rangle$ is not quite at the same location as the peak in the probability distribution. The result (iv) can also be easily calculated from the *Virial Theorem* which states that for any stationary, confined state, the expectation values of the kinetic energy and a function related to the potential are equal

$$\langle \mathbf{p}^2/2\mu \rangle_{n,l} = \frac{1}{2} \left\langle r \frac{dV}{dr} \right\rangle_{n,l} \tag{29.32}$$

which here means that

$$\langle \mathbf{p}^2/2\mu \rangle_{n,l} = -\tfrac{1}{2} \langle V \rangle_{n,l}. \tag{29.33}$$

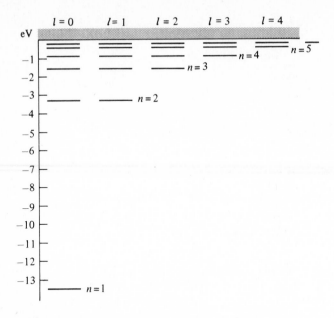

Figure 29.3. Spectrum of hydrogen atom. The energy levels for a given l have a degeneracy of $(2l + 1)$. The shaded region on top represents the continuum.

The spectrum of hydrogen, according to the idealized version discussed above, and shown in Fig. 29.3, shows a great deal of degeneracy. For a given n, the number of states that have the same energy are the states of angular momentum $0, 1, 2, \ldots$ $(n - 1)$. Since for a given l the degeneracy is $(2l + 1)$, we end up with

$$\sum_{l=0}^{n-1} (2l + 1) = n^2. \tag{29.34}$$

In reality, the degeneracy that sets all the angular momentum states from $l = 0$ to $l = n - 1$ on the same level is broken. This arises from relativistic effects, which destroy the pure $1/r$ nature of the potential. These will be discussed in Chapter 31.

The degeneracy of the various m − values for a given l, $-l \leqslant m \leqslant l$, can only be broken by an external mechanism, which picks out a preferred direction. Only in that case will different L_z states respond differently. The external mechanism is the presence of a magnetic field, and the splitting of spectral lines, which reflects the splitting of energy levels, was first discovered by Zeeman.

Zeeman effect The *Zeeman effect* can be understood by examining the effect of an external magnetic field on an atom. Equation (28.26) shows that the energy operator is changed from

$$\mathbf{p}^2/2\mu + V(r)$$

to

$$\mathbf{p}^2/2\mu + V(r) - \boldsymbol{\mu} \cdot \mathbf{B} \tag{29.35}$$

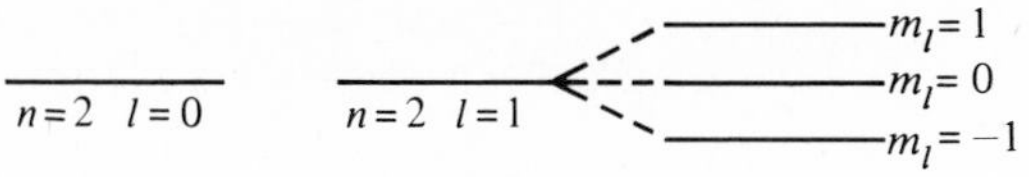

Figure 29.4. Effect of magnetic field on $n = 1$ and $n = 2$ levels in hydrogen spectrum.

which, for the *normal Zeeman effect*, just becomes

$$\mathbf{p}^2/2\mu + V(r) + \frac{eB}{2m_e c} L_z \qquad (29.36)$$

since $\boldsymbol{\mu} = -\dfrac{e}{2m_e c} \mathbf{L}$. Thus we are asked to solve the eigenvalue problem

$$\left(\frac{\mathbf{p}^2}{2\mu} + V(r) + \frac{eB}{2m_e c} L_z \right) \psi(\mathbf{r}) = E\psi(\mathbf{r}) \qquad (29.37)$$

obtained by defining the z-axis by B. The substitution

$$\psi(\mathbf{r}) = R(r) Y_{im}(\theta, \phi), \qquad (29.38)$$

where $R(r)$ is the radial eigenfunction for $B = 0$, immediately leads to

$$E = E_n + \mu_B B m \qquad (29.39)$$

where $\mu_B = e\hbar/2m_e c$ is called the *Bohr magneton*, and where

$$L_z Y_{lm} = \hbar m Y_{lm}. \qquad (29.40)$$

Thus each energy level is split into its $(2l + 1)$ components, and the size of the splitting is the same for each level. Figure 29.4 shows the split spectrum. This altered structure clearly has an effect on the spectrum. For example, in the transition from the $n = 2, l = 1$ levels to the ground state, instead of one spectral line, of frequency

$$\nu_{21} = \frac{E_2 - E_1}{h} \qquad (29.41)$$

there will be three lines, of frequency

$$\nu_{21} + \frac{eB}{4\pi m_e c}, \nu_{21}, \nu_{21} - \frac{eB}{4\pi m_e c}. \qquad (29.42)$$

One might expect that in the transition $n = 3, l = 2$ to $n = 2, l = 1$ there are more than three lines. It turns out, however, that as a consequence of a *selection rule*, which we discuss in more detail later on, transitions are restricted to those that satisfy

$$\Delta m = 1, 0, -1, \qquad (29.43)$$

and thus there are also only three spectral lines (See Fig. 29.5).

Relativistic fine structure The normal Zeeman effect is frequently observed, but there are also many instances of the presence of an *anomalous effect*, in which the number

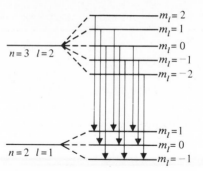

Figure 29.5. Zeeman effect in transition $(n = 3, l = 2) \rightarrow (n = 2, l = 1)$.

of lines differs from three. In particular, the presence of an even number of lines was very troublesome, since for any integer l, there is always an odd number of m_l-substates. Even in the absence of an external magnetic field, some *fine structure* in the spectra was observed. Some of that was easily understood as kinematic relativistic effects. The kinetic energy of an electron, taken to be $\mathbf{p}^2/2m$, is only an approximation to

$$(\mathbf{p}^2 c^2 + m^2 c^4)^{1/2} - mc^2 \cong \frac{\mathbf{p}^2}{2m} - \frac{1}{8} \frac{(\mathbf{p}^2)^2}{m^3 c^2}$$

so that the Schrödinger equation ought to read

$$\left(\frac{\mathbf{p}^2}{2\mu} + V(r) - \frac{1}{8} \frac{(\mathbf{p}^2)^2}{m^3 c^2} \right) \psi(\mathbf{r}) = E\psi(\mathbf{r}). \qquad (29.44)$$

The extra term is a small correction. It is of order of magnitude

$$\left(\frac{\mathbf{p}^2}{2m} \right)^2 \frac{1}{2mc^2} \sim \frac{\mathbf{p}^2}{2m} \left(\frac{p}{2mc} \right)^2 \sim \frac{\mathbf{p}^2}{2m} \left(\frac{mc(Z\alpha)}{2mc} \right) \sim \frac{\mathbf{p}^2}{2m} (Z\alpha)^2$$

that is a 10^{-4} factor effect, of the right order of magnitude. This contribution acts to split the l-degeneracy, because with the notation

$$\mathbf{p}^2/2\mu - \frac{Ze^2}{r} \equiv H_0, \qquad (29.45)$$

the extra term has the form

$$-\frac{1}{2} \frac{1}{mc^2} \left(\frac{\mathbf{p}^2}{2m} \right)^2 \simeq -\frac{1}{2mc^2} \left(H_0 + \frac{Ze^2}{r} \right) \left(H_0 + \frac{Ze^2}{r} \right) \qquad (29.46)$$

one effect of which is to introduce a term of the form A/r^2 which, as discussed earlier, definitely splits the lines of different l for a given n.

Electron spin This relativistic splitting, although definitely present, does not tell the whole story. Often a doublet structure appears — a pair of close-lying spectral lines —

and this does not fit in with any "classical" perturbations. The explanation came in two steps: First, Wolfgang Pauli postulated a new two-valued quantum number for electrons: the electrons were to be labeled by (n, l, m_l, σ) where σ was the new quantum mechanical variable. Then S. Goudsmit and G. Uhlenbeck suggested that the new degree of freedom be associated with an internal angular momentum of the electron, the *electron spin*. It was suggested that $s = 1/2$ so that the square of the angular momentum of the electron in the absence of orbital angular momentum (that is, in an $l = 0$ state) was $\frac{1}{2}(\frac{1}{2} + 1)\hbar^2$. Furthermore, the exact correspondence with orbital angular momentum tells us that the electron can be in one of two states,[†] of $m_s = \hbar/2$ and $-\hbar/2$. Such a spin had to be quantum mechanical. One could not associate it with an extended structure for the electron, and make an analogy with the planets spinning about their own axes in addition to orbiting about the sun. To see the difficulty, consider a sphere of radius R, spinning with angular velocity ω. If the sphere is uniform in mass, the angular momentum is

$$L = I\omega$$

with $I = 2MR^2/5$. Thus if the radius of the electron is taken to be the classical electron radius

$$R = \frac{e^2}{mc^2},$$

then the velocity at the equator is

$$v = R\omega = R\frac{L}{I} = \frac{5}{4}\frac{\hbar}{mR} = \frac{5}{4}\frac{\hbar c}{e^2}c = \frac{5}{4} \times 137c!$$

which violates the special theory of relativity. Even with the radius taken to be the Compton wavelength,

$$R = \frac{\hbar}{mc},$$

we get $v = 5/4c$. Thus the electron spin is not a quantity that has a classical limit or counterpart. It must be treated quantum-mechanically, and that will be done in the next chapter.

PROBLEMS

29.1 An electron is in the state

$$\psi(\mathbf{r}) = \frac{1}{\sqrt{30}}(2\psi_{100}(\mathbf{r}) - 3\psi_{211}(\mathbf{r}) + 4\psi_{200}(\mathbf{r}) - \psi_{320}(\mathbf{r}))$$

where the $\psi_{nlm}(\mathbf{r})$ are eigenstates of the hydrogen atom characterized by the quantum number n, l and m.

a) What is the expectation value of the energy?

---

† The Pauli label σ is thus identified with $2m_s/\hbar$.

   b) What is the expectation value of L^2?

   c) What is the expectation value of L_z?

[*Hint*: Use the orthogonality of the eigenfunctions corresponding to different quantum numbers.]

29.2 The uncertainty in position and momentum are defined by

$$(\Delta x)^2 = \langle x^2 \rangle - \langle x \rangle^2 ; \quad (\Delta p_x)^2 = \langle p_x^2 \rangle - \langle p_x \rangle^2.$$

Use (29.30) on one hand, and the Virial Theorem (29.33) on the other, to calculate

$$\Delta p_x \Delta x$$

for an arbitrary n and l. Note that $\langle p_x^2 \rangle = \langle \mathbf{p}^2 \rangle/3$ and $\langle x^2 \rangle = \langle r^2 \rangle/3$. (Why is this the case?) Show that the uncertainty increases for the excited states.

29.3 Calculate the eigensolutions of the radial equation for $n = 2$, $l = 1, 0$ and for $n = 3, l = 2, 1$ and 0.

29.4 Compare the frequency of the α-Lyman line ($n = 2$, $l = 1 \to n = 1$, $l = 0$ transition) for the hydrogen atom with that for the atom in which the nucleus is a deuteron, with $Z = 1$ and mass twice that of the hydrogen nucleus.

29.5 Apply the techniques used in this chapter to the three-dimensional harmonic oscillator

$$V(r) = \tfrac{1}{2} m\omega^2 r^2.$$

   a) Separate out the asymptotic behavior of the wave function.

   b) Separate out the behavior at the origin.

   c) Assume that the remaining function is a polynomial. What is the spectrum in terms of l and the number of radial nodes? What is the degeneracy?

Check the degeneracy in another way, namely by solving the problem in Cartesian coordinates. The solution is known from Chapter 24.

30
Spin and the Addition of Angular Momenta

Spin A major step in the understanding of the structure of matter came with the discovery of a new, purely quantum phenomenon, the *intrinsic spin* of a system. The spin of a system is the angular momentum that it has when it is at rest, but the new discovery was that even systems that are structureless, such as the elementary particles, the electron, the proton, the neutron, and so on, have an intrinsic angular momentum. Furthermore, the range of values of this angular momentum was enlarged: whereas angular momentum associated with the motion (orbital angular momentum) could only take on integer values l, intrinsic spin could be half-integral. Thus the electron, just like the proton and the neutron, has spin 1/2, by which we mean that the square of the spin $\mathbf{S}^2$ has the eigenvalue $\frac{1}{2}(\frac{1}{2} + 1)\hbar^2$, and the possible values of S_z start at $\frac{1}{2}\hbar$, and decrease in integer steps to $-\frac{1}{2}\hbar$. There are only two states that are independent for a spin $\frac{1}{2}$ object (Fig. 30.1), with

$$S_z \chi_{+1/2} = \tfrac{1}{2}\hbar\chi_{+1/2}$$

$$S_z \chi_{-1/2} = -\tfrac{1}{2}\hbar\chi_{-1/2}.$$

(30.1)

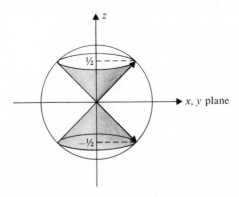

Figure 30.1. Graphical representation of spin 1/2 system. The $m_s = +\frac{1}{2}\hbar$ eigenstate lies on the upper cone, intersecting a sphere of radius $\sqrt{3/4}\,\hbar$; the $m_s = -\frac{1}{2}\hbar$ eigenstate lies on the lower cone. These diagrammatic representations must be viewed only as visual aids, since a spin $\frac{1}{2}$ system has no classical limiting case, as would be the case for a large value of the orbital angular momentum.

Furthermore,

$$S^2 \chi_{\pm 1/2} = s(s+1)\hbar^2 \chi_{\pm 1/2} = \tfrac{3}{4}\hbar^2 \chi_{\pm 1/2}. \tag{30.2}$$

For the orbital angular momentum, various components could be represented by differential operators, such as

$$L_z = \frac{\hbar}{i}\left(x\frac{\partial}{\partial y} - y\frac{\partial}{\partial x} \right) = \frac{\hbar}{i}\frac{\partial}{\partial \phi}, \tag{30.3}$$

but here we must represent the operators in some other way, still consistent with the basic rules that

$$S_x S_y - S_y S_x = i\hbar S_z$$
$$S_y S_z - S_z S_y = i\hbar S_x$$
$$S_z S_x - S_x S_z = i\hbar S_y. \tag{30.4}$$

It turns out to be possible to satisfy the above rules if the operators are represented by *matrices*. If we take, following Pauli, the representation

$$S = \tfrac{1}{2}\hbar\boldsymbol{\sigma} \tag{30.5}$$

where the *Pauli matrices* $\boldsymbol{\sigma}$ are given by

$$\sigma_x = \begin{pmatrix} 0 & 1 \\ 1 & 0 \end{pmatrix}; \quad \sigma_y = \begin{pmatrix} 0 & -i \\ i & 0 \end{pmatrix}; \quad \sigma_z = \begin{pmatrix} 1 & 0 \\ 0 & -1 \end{pmatrix}, \tag{30.6}$$

then it is easily checked that

$$\sigma_x \sigma_y - \sigma_y \sigma_x = \begin{pmatrix} 0 & 1 \\ 1 & 0 \end{pmatrix}\begin{pmatrix} 0 & -i \\ i & 0 \end{pmatrix} - \begin{pmatrix} 0 & -i \\ i & 0 \end{pmatrix}\begin{pmatrix} 0 & 1 \\ 1 & 0 \end{pmatrix}$$

$$= \begin{pmatrix} i & 0 \\ 0 & -i \end{pmatrix} - \begin{pmatrix} -i & 0 \\ 0 & i \end{pmatrix}$$

$$= 2i\begin{pmatrix} 1 & 0 \\ 0 & -1 \end{pmatrix} = 2i\sigma_z \tag{30.7}$$

and so on, which with Eq. (30.5) implies that (30.4) is satisfied. With this representation of the operators in terms of 2×2 matrices, the eigenstates are necessarily two-component column vectors. The first of the equations (30.1) now reads

$$\tfrac{1}{2}\hbar\begin{pmatrix} 1 & 0 \\ 0 & -1 \end{pmatrix}\chi_{+1/2} = \tfrac{1}{2}\hbar\chi_{+1/2}$$

which implies that

$$\chi_{+1/2} = \begin{pmatrix} 1 \\ 0 \end{pmatrix}. \tag{30.8}$$

The second one implies that

$$\chi_{-1/2} = \begin{pmatrix} 0 \\ 1 \end{pmatrix}. \tag{30.9}$$

Eq. (30.2) is automatically satisfied for any linear combination of the $\chi_{\pm 1/2}$, since

$$S_x^2 = (\tfrac{1}{2}\hbar)^2 \begin{pmatrix} 0 & 1 \\ 1 & 0 \end{pmatrix} \begin{pmatrix} 0 & 1 \\ 1 & 0 \end{pmatrix} = \frac{\hbar^2}{4} \begin{pmatrix} 1 & 0 \\ 0 & 1 \end{pmatrix}$$

$$S_y^2 = (\tfrac{1}{2}\hbar)^2 \begin{pmatrix} 0 & -i \\ i & 0 \end{pmatrix} \begin{pmatrix} 0 & -i \\ i & 0 \end{pmatrix} = \frac{\hbar^2}{4} \begin{pmatrix} 1 & 0 \\ 0 & 1 \end{pmatrix} \tag{30.10}$$

$$S_z^2 = (\tfrac{1}{2}\hbar)^2 \begin{pmatrix} 1 & 0 \\ 0 & -1 \end{pmatrix} \begin{pmatrix} 1 & 0 \\ 0 & -1 \end{pmatrix} = \frac{\hbar^2}{4} \begin{pmatrix} 1 & 0 \\ 0 & 1 \end{pmatrix}$$

so that

$$\mathbf{S}^2 = S_x^2 + S_y^2 + S_z^2 = \tfrac{3}{4}\hbar^2 \begin{pmatrix} 1 & 0 \\ 0 & 1 \end{pmatrix}. \tag{30.11}$$

A unit matrix acting on any two-component object

$$\begin{pmatrix} a \\ b \end{pmatrix}$$

reproduces it, so that all two-component column vectors (we call them *spinors*) are eigenstates of $\mathbf{S}^2$.

A general spin state is described by the spinor

$$\begin{pmatrix} a \\ b \end{pmatrix}.$$

Its complex conjugate is described by the row vector $(a^* \, b^*)$ and the scalar product is defined to be

$$\chi^\dagger \chi = (a^* \, b^*) \begin{pmatrix} a \\ b \end{pmatrix} = |a|^2 + |b|^2. \tag{30.12}$$

Spinors are usually normalized such that

$$\chi^\dagger \chi = |a|^2 + |b|^2 = 1. \tag{30.13}$$

As with orbital angular momentum, the choice of S_z as the component of spin for which the spinors are eigenstates is a matter of convenience, but not really necessary. If we want to find the eigenstates of $S_x \cos\alpha + S_y \sin\alpha$, for example, we ask to solve

$$(S_x \cos\alpha + S_y \sin\alpha)\chi = \lambda\chi. \tag{30.14}$$

The matrix on the left is

$$\tfrac{1}{2}\hbar \begin{pmatrix} 0 & \cos\alpha \\ \cos\alpha & 0 \end{pmatrix} + \tfrac{1}{2}\hbar \begin{pmatrix} 0 & -i\sin\alpha \\ i\sin\alpha & 0 \end{pmatrix} = \tfrac{1}{2}\hbar \begin{pmatrix} 0 & e^{-i\alpha} \\ e^{i\alpha} & 0 \end{pmatrix} \qquad (30.15)$$

and we shall write $\lambda = \tfrac{1}{2}\hbar m_s$. Thus

$$\begin{pmatrix} 0 & e^{-i\alpha} \\ e^{i\alpha} & 0 \end{pmatrix} \begin{pmatrix} a \\ b \end{pmatrix} = m_s \begin{pmatrix} a \\ b \end{pmatrix} \qquad (30.16)$$

therefore

$$e^{-i\alpha}b = m_s a$$

$$e^{i\alpha}a = m_s b.$$

From this it follows that

$$m_s^2 = 1,$$

that is,

$$m_s = \pm 1.$$

Thus again the eigenvalues are $\pm\,\hbar/2$. For the eigenvalue $+\,\hbar/2$, we have

$$b = e^{i\alpha}a$$

so that

$$\chi_{+1/2} = a \begin{pmatrix} 1 \\ e^{i\alpha} \end{pmatrix}.$$

The square of this spinor is given by

$$\chi_{+1/2}^{\dagger}\chi_{+1/2} = |a|^2 (1\ e^{-i\alpha}) \begin{pmatrix} 1 \\ e^{i\alpha} \end{pmatrix} = 2|a|^2$$

and if we set this equal to unity, we get

$$\chi_{+1/2} = \frac{1}{\sqrt{2}} \begin{pmatrix} 1 \\ e^{i\alpha} \end{pmatrix}. \qquad (30.17)$$

Similarly,

$$\chi_{-1/2} = \frac{1}{\sqrt{2}} \begin{pmatrix} 1 \\ -e^{i\alpha} \end{pmatrix}. \qquad (30.18)$$

Note that

$$\chi_{+1/2}^{\dagger}\chi_{-1/2} = \frac{1}{\sqrt{2}}(1\ e^{-i\alpha})\frac{1}{\sqrt{2}} \begin{pmatrix} 1 \\ -e^{i\alpha} \end{pmatrix} = \tfrac{1}{2}(1-1) = 0. \qquad (30.19)$$

The eigenstates corresponding to different eigenvalues are orthogonal — a familiar result.

Although there are circumstances in which one may restrict one's attention to the spin of an electron, to the exclusion of its motion, many applications of spin involve electrons that also have an orbital angular momentum, and the coupling of spin and orbital angular momentum. To be able to deal with such problems, we must learn how to add angular momenta.

Addition of angular momenta In classical mechanics, angular momenta are vectors, and the sum of two angular momenta, the total angular momentum

$$\mathbf{J} = \mathbf{L_1} + \mathbf{L_2} \qquad (30.20)$$

can be a vector, whose length ranges from $L_1 + L_2$ to $|L_1 - L_2|$. In quantum mechanics this is not quite true, since angular momentum is quantized. Thus the sum of two angular momenta, if it is to be an angular momentum, must also be quantized. To see the problems that arise, let us use our pictorial representation of angular momentum (cf. Fig. 28.1) to find out what happens when we add two angular momenta each with $l = 1$. Figure 30.2 indicates a possible strategy for the addition of angular momenta, using vector addition of two vectors that are constrained to have well-defined z-components. If we add one unit of angular momentum to the vector representing one unit of angular momentum with z-component 1, the possible values of the z component of the sum are 2, 1, and 0 (Fig. 30.2 c); if we add it to the unit angular momentum with z-component 0, the possible values of the total z-component are 1, 0, -1 (Fig. 30.2 d); if we add it to the unit angular momentum with z-component -1, the possible values of the total z-component are 0, -1, and -2 (Fig. 30.2 e). The lengths of the vectors are $\sqrt{1(1+1)} = \sqrt{2}$ in units of $\hbar$, and this points up the difficulty of using the vectorial addition. The vector with the maximum value of L_z (2 units of $\hbar$) appears to lie on a sphere with radius $2\sqrt{2}\hbar$, whereas a vector of total angular momentum 2 lies, in the sense of the above diagrams, on a sphere of radius $\sqrt{2(2+1)} = \sqrt{6}$ in units of $\hbar$, and there is no integer L such that $\sqrt{L(L+1)} = 2\sqrt{2}$.

The z-component addition indicates that a state with total $L_z = 2$ appears once, $L_z = 1$ appears twice, $L_z = 0$ appears thrice, $L_z = -1$ appears twice, and $L_z = -2$ appears once. This suggests that a set of L_z values $(2, 1, 0, -1, -2)$ form a total angular momentum state 2 (with $\sqrt{L(L+1)} = \sqrt{6}$); another set of L_z values $(1, 0, -1)$ form a state of total angular momentum 1, and the remaining $L_z = 0$ state belongs to a state with total angular momentum 0. This decomposition is at least consistent with the idea that the eigenvalues of L_z start with the value of the angular momentum and decrease in unit steps till they reach the minimum value of L_z which is minus the total angular momentum, and with the classical limit that the largest value of the total angular momentum should be $L_1 + L_2$, and the smallest one $|L_1 - L_2|$.

The suggested idea is the correct one. We state without proof that when two angular momenta are to be added, then:

i) The operator

$$\mathbf{J} = \mathbf{L_1} + \mathbf{L_2} \qquad (30.21)$$

is an angular momentum, in that it satisfies commutation relations (27.4)–(27.6).

ii) The solution of the eigenvalue problem

$$\mathbf{J}^2 \, \mathscr{Y}_{j,m_j} \equiv (\mathbf{L_1} + \mathbf{L_2})^2 \, \mathscr{Y}_{j,m_j} = j(j+1)\hbar^2 \, \mathscr{Y}_{j,m_j} \qquad (30.22)$$

yields

$$j = l_1 + l_2, l_1 + l_2 - 1, l_1 + l_2 - 2, \ldots, |l_1 - l_2| \qquad (30.23)$$

for the values of j, where the eigenvalues of the operators $\mathbf{L}_1^2$ and $\mathbf{L}_2^2$ are $l_1(l_1 + 1)$ and $l_2(l_2 + 1)$ in units of $\hbar^2$.

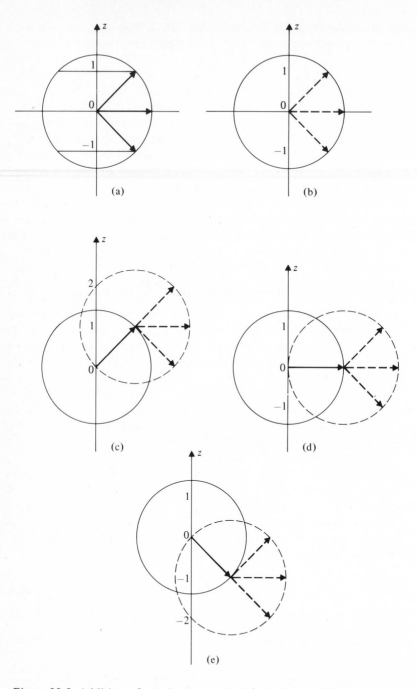

Figure 30.2. Addition of angular momenta $1 + 1$ using graphical representation of angular momentum quantization; (a), (b) represent the individual angular momenta; (c), (d), and (e) represent the nine different ways in which the addition can be carried out.

iii) The eigenvalues of J_z, defined by

$$J_z = L_{1z} + L_{2z} \tag{30.24}$$

are given by m_j, where for a given j, the values of m_j are

$$m_j = j, j-1, j-2, \ldots, -j. \tag{30.25}$$

iv) The eigenfunctions of Eq. (30.22) and

$$J_z \mathscr{Y}_{j,m_j} = m_j \hbar \mathscr{Y}_{j,m_j} \tag{30.26}$$

are linear combinations of products of the eigenfunctions $Y_{l_1 m_1}$ and $Y_{l_2 m_2}$ for angular momentum $\mathbf{L}_1$ and $\mathbf{L}_2$, respectively. It is easy to see that

$$J_z Y_{l_1 m_1} Y_{l_2 m_2} = (L_{1z} + L_{2z}) Y_{l_1 m_1} Y_{l_2 m_2}$$

$$= \hbar(m_1 + m_2) Y_{l_1 m_1} Y_{l_2 m_2} \tag{30.27}$$

but beyond the scope of our treatment to show that linear combinations of products of $Y_{l_1 m_1} Y_{l_2 m_2}$ with $m_1 + m_2 = m_j$ fixed can be constructed to satisfy Eq. (30.22) with the j values listed in Eq. (30.23). We can, at least, make the counting come out right.

Suppose we want to add angular momentum 2 and angular momentum 1. The two systems are represented by eigenfunctions $Y_{22}^{(a)}, Y_{21}^{(a)}, \ldots, Y_{2,-2}^{(a)}$ and $Y_{11}^{(b)}, Y_{10}^{(b)}, Y_{1,-1}^{(b)}$, respectively. There is a total of fifteen products of the two. The eigenvalues of J_z range from 3 to -3 (in units of $\hbar$), as given in Table 30.1.

Table 30.1

$m_j = m_a + m_b$	Products
3	$Y_{22}^{(a)} Y_{11}^{(b)}$
2	$Y_{22}^{(a)} Y_{10}^{(b)}$, $Y_{21}^{(a)} Y_{11}^{(b)}$
1	$Y_{22}^{(a)} Y_{1,-1}^{(b)}$, $Y_{21}^{(a)} Y_{1,0}^{(b)}$, $Y_{20}^{(a)} Y_{11}^{(b)}$
0	$Y_{21}^{(a)} Y_{1,-1}^{(b)}$, $Y_{20}^{(a)} Y_{10}^{(b)}$, $Y_{2,-1}^{(a)} Y_{11}^{(b)}$
-1	$Y_{20}^{(a)} Y_{1,-1}^{(b)}$, $Y_{2,-1}^{(a)} Y_{10}^{(b)}$, $Y_{2,-2}^{(a)} Y_{11}^{(b)}$
-2	$Y_{2,-1}^{(a)} Y_{1,-1}^{(b)}$, $Y_{2,-2}^{(a)} Y_{1,0}^{(b)}$
-3	$Y_{2,-2}^{(a)} Y_{1,-1}^{(b)}$

Of the fifteen products, seven, including the first, the last, and appropriate linear combinations from each line, form the eigenfunctions corresponding to $j = 3$. The remaining independent (orthogonal) linear combination from the $m_j = 2$ and $m_j = -2$ lines, together with appropriate linear combinations from the $1, 0, -1$ lines, yield the five eigenfunctions corresponding to $j = 2$; the remaining linear combinations, orthogonal to the other two from the lines corresponding to $m_j = 1, 0, -1$ form the $j = 1$ eigenfunctions. The total $7 + 5 + 3 = 15$ add up to the total number of products.

Addition of two spins The addition of two spin 1/2 angular momenta yields 1 and 0. This system is simple enough to exhibit the choosing of linear combinations explicitly. We will choose

$$\mathbf{S}_1^2 \xi_\pm = \tfrac{3}{4} \hbar^2 \xi_\pm$$

$$S_{1z} \xi_\pm = \pm \tfrac{1}{2} \hbar \xi_\pm \tag{30.28}$$

and

$$\mathbf{S}_2^2 \eta_\pm = \tfrac{3}{4} \hbar^2 \eta_\pm$$

$$S_{2z} \eta_\pm = \pm \tfrac{1}{2} \hbar \eta_\pm. \tag{30.29}$$

There are four products

$$\xi_+\eta_+, \xi_+\eta_-, \xi_-\eta_+, \xi_-\eta_- \tag{30.30}$$

which we are instructed to combine into linear combinations of total angular momentum 1 and 0. The total angular momentum will be denoted by $\mathbf{S}$:

$$\mathbf{S} = \mathbf{S}_1 + \mathbf{S}_z. \tag{30.31}$$

We note that

$$S_z (\xi_+\eta_+) = (S_{1z} + S_{2z})\xi_+\eta_+$$

$$= (S_{1z} \xi_+) \eta_+ + (S_{2z} \eta_+)\xi_+$$

$$= \tfrac{1}{2} \hbar \xi_+\eta_+ + \tfrac{1}{2} \hbar \eta_+\xi_+$$

$$= \hbar \xi_+\eta_+. \tag{30.32}$$

Similarly,

$$S_z \xi_+\eta_- = 0 \tag{30.33}$$

$$S_z \xi_-\eta_+ = 0 \tag{30.34}$$

$$S_z \xi_-\eta_- = -\hbar \xi_-\eta_-. \tag{30.35}$$

Thus the combinations $\xi_+\eta_+$ and $\xi_-\eta_-$ cannot have angular momentum 0, since S_z has eigenvalues different from 0. We surmise that they have angular momentum 1. With $\xi_+\eta_-$ and $\xi_-\eta_+$ we cannot tell which linear combination is the $S = 0$ and which is the $S = 1$, $S_z = 0$ state. To work this out, we must ask for which combination we have an eigenstate of $\mathbf{S}^2 = \mathbf{S}_1^2 + \mathbf{S}_2^2 + 2\mathbf{S}_1 \cdot \mathbf{S}_2$ with eigenvalue 0 and for which the eigenvalue will be $2\hbar^2$. If we wish to apply the operator $\mathbf{S}^2$ to the products, we must know what $\mathbf{S}_1 \cdot \mathbf{S}_2$ does, when applied to them, and so we need to know $S_{1x} \xi_+$, and so on. We now use the explicit representations in terms of the Pauli matrices. Thus

$$S_{1x} \xi_+ = \frac{\hbar}{2} \begin{pmatrix} 0 & 1 \\ 1 & 0 \end{pmatrix} \begin{pmatrix} 1 \\ 0 \end{pmatrix} = \frac{\hbar}{2} \begin{pmatrix} 0 \\ 1 \end{pmatrix} = \frac{\hbar}{2} \xi_-$$

$$S_{1x} \xi_- = \frac{\hbar}{2} \begin{pmatrix} 0 & 1 \\ 1 & 0 \end{pmatrix} \begin{pmatrix} 0 \\ 1 \end{pmatrix} = \frac{\hbar}{2} \begin{pmatrix} 1 \\ 0 \end{pmatrix} = \frac{\hbar}{2} \xi_+$$

$$S_{1y} \xi_+ = \frac{\hbar}{2} \begin{pmatrix} 0 & -i \\ i & 0 \end{pmatrix} \begin{pmatrix} 1 \\ 0 \end{pmatrix} = \frac{\hbar}{2} \begin{pmatrix} 0 \\ i \end{pmatrix} = i\frac{\hbar}{2} \xi_-$$

$$S_{1y} \xi_- = \frac{\hbar}{2} \begin{pmatrix} 0 & -i \\ i & 0 \end{pmatrix} \begin{pmatrix} 0 \\ 1 \end{pmatrix} = \frac{\hbar}{2} \begin{pmatrix} -i \\ 0 \end{pmatrix} = -i\frac{\hbar}{2} \xi_+ \tag{30.36}$$

with identical results when S_2 components act on the $\eta_\pm$. Thus

$$\mathbf{S}_1 \cdot \mathbf{S}_2 \, \xi_+ \eta_- = (S_{1x} S_{2x} + S_{1y} S_{2y} + S_{1z} S_{2z}) \xi_+ \eta_-$$

$$= (S_{1x} \xi_+)(S_{2x} \eta_-) + (S_{1y} \xi_+)(S_{2y} \eta_-)$$

$$+ (S_{1z} \xi_+)(S_{2z} \eta_-)$$

$$= \left(\frac{\hbar}{2}\right)^2 \xi_- \eta_+ + \left(\frac{\hbar}{2}\right)^2 \xi_- \eta_+ - \frac{\hbar^2}{4} \xi_+ \eta_-$$

$$= \frac{\hbar^2}{2} \xi_- \eta_+ - \frac{\hbar^2}{4} \xi_+ \eta_-. \tag{30.37}$$

Hence

$$\mathbf{S}^2 \xi_+ \eta_- = (\mathbf{S}_1^2 \xi_+) \eta_- + \xi_+ \mathbf{S}_2^2 \eta_- + 2 \mathbf{S}_1 \cdot \mathbf{S}_2 \, \xi_+ \eta_-$$

$$= \frac{3}{4} \hbar^2 \xi_+ \eta_- + \frac{3}{4} \hbar^2 \xi_+ \eta_- + \hbar^2 \xi_- \eta_+ - \frac{\hbar^2}{2} \xi_+ \eta_-$$

$$= \hbar^2 \xi_+ \eta_- + \hbar^2 \xi_- \eta_+. \tag{30.38}$$

In the same way we show that

$$\mathbf{S}^2 \xi_- \eta_+ = \hbar^2 \xi_- \eta_+ + \hbar^2 \xi_+ \eta_-. \tag{30.39}$$

Thus adding

$$\mathbf{S}^2 (\xi_+ \eta_- + \xi_- \eta_+) = 2\hbar^2 (\xi_+ \eta_- + \xi_- \eta_+), \tag{30.40}$$

and subtracting

$$\mathbf{S}^2 (\xi_+ \eta_- - \xi_- \eta_+) = 0. \tag{30.41}$$

We have thus obtained the result that (with proper normalization) the linear combinations of products for total angular momentum 1 are

$$\xi_+ \eta_+$$

$$\frac{1}{\sqrt{2}} (\xi_+ \eta_- + \xi_- \eta_+)$$

$$\xi_- \eta_- \tag{30.42}$$

while the linear combination with total angular momentum 0 is

$$\frac{1}{\sqrt{2}} (\xi_+ \eta_- - \xi_- \eta_+). \tag{30.43}$$

The procedure for adding more general angular momenta, while much more complicated in detail, is in principle the same. Thus if we add spin S (with $S = 1/2$) to orbital angular momentum L, we get a total of $2(2l + 1)$ products $Y_{lm}(\theta, \phi)\chi_+$, $Y_{lm}(\theta, \phi)\chi_-$ which decompose into $2(l + 1/2) + 1$ linear combinations of angular momentum $j = l + 1/2$ and $2(l - 1/2) + 1$ linear combinations of angular momentum $j = l - 1/2$. The particular linear combinations happen to be

$$\mathcal{Y}_{l+1/2, m_j} = \left(\frac{l + 1/2 + m_j}{2l + 1}\right)^{1/2} Y_{lm} \chi_+ + \left(\frac{l + 1/2 - m_j}{2l + 1}\right)^{1/2} Y_{lm} \chi_- \tag{30.44}$$

and

$$\mathcal{Y}_{l-1/2, m_j} = \left(\frac{l + 1/2 - m_j}{2l + 1}\right)^{1/2} Y_{lm} \chi_+ - \left(\frac{l + 1/2 + m_j}{2l + 1}\right)^{1/2} Y_{lm} \chi_-. \tag{30.45}$$

Note that neither is an eigenstate of S_z, for example, nor L_z, but the eigenstates are eigenstates of J_z. In the absence of external torques, that is, in the absence of a preferred direction, the total angular momentum is conserved, even though orbital and spin angular momenta are not. This "classical" statement of the conservation of angular momentum is equivalent to the statement that if a system is initially in an eigenstate of $\mathbf{J}^2$ and J_z, then it will always remain in such an eigenstate.

Spectroscopic notation We conclude this chapter with the introduction of a standard notation that appears in the discussion of energy levels, both atomic and nuclear. In both cases one frequently deals with situations in which, to first approximation, spin and orbital angular momentum are independently conserved. Although spin-orbit coupling† makes meaningless the assignment of any quantum number other than total angular momentum, the coupling does not, in many cases, completely break down the original characterization of the state. For example, if we have an electron with spin $1/2$ in an orbital state of $L = 1$, then we have a doublet P state, denoted by ^2P, and since the total angular momentum may, according to the addition rules be $1/2$ or $3/2$, we describe the possible states as $^2P_{1/2}$ and $^2P_{3/2}$. In general the characterization is

$$^{(2S+1)}L_J$$

with the orbital angular momentum labeled by the appropriate letter $S(L = 0)$, $P(L = 1)$, $D(L = 2)$, $F(L = 3)$, and so on. As another example, if we have two electrons in a total spin 0 state, there is only a singlet, and the superscript will be 1. Two electrons in a singlet S state can only be in the $^1S_0$ state, since by the addition rules, $J = 0$. When the electrons are in a spin 1 state, we have a $^3S_1$ state. When the orbital angular momentum of the two-electron state is $L = 1$, then the singlet state is $^1P_1$; when they are in a spin 1 state, we have $^3P_2$, $^3P_1$, and $^3P_0$ as possible states, since spin 1 and orbital angular momentum 1 can add up to $J = 2, 1$, or 0. It should be noted that the spin does not affect the *parity* of the state; that property is determined by the orbital wave function. Thus an S state has a spherically symmetric wave function of the form $R(r)$. Since under an inversion $x \to -x$, $y \to -y$, $z \to -z$ $r = \sqrt{x^2 + y^2 + z^2}$ is unchanged, the S state wave function is even. For the P state, the $m = 0$ wave function is proportional to

$$\cos\theta = z/r$$

and this is *odd*. The $m = 1$ component is proportional to

$$\sin\theta\, e^{i\phi} = \frac{x + iy}{r}$$

which is also odd. The D state, for which the $m = 0$ component is proportional to

$$3\cos^2\theta - 1 = \frac{3z^2 - r^2}{r^2},$$

the parity is even again. *The parity of the orbital angular wave function is in general* $(-1)^L$. Thus a $^3P_1$ state is odd, while a $^3D_1$ state is even, and so on.

---

† This interaction between the spin of the electron and its motion in a potential is discussed in Chapter 31.

NOTES AND COMMENTS

1. The idea that the double-valuedness associated with an electron state description be associated with an intrinsic spin was first proposed by R. Kronig, and rediscovered and published less than a year later by S. Goudsmit and G. Uhlenbeck. The history of this discovery is discussed in the articles by Kronig and van der Waerden mentioned in the notes and comments following Chapter 25.

2. The description of angular momentum in terms of matrices is not restricted to spin 1/2 (or perhaps other odd half-integer spins). Any spin can be described by a set of matrices, provided that the angular momentum commutation relations are satisfied. For example, the three components of angular momentum for spin 1 can be represented by the matrices

$$
L_x = \frac{\hbar}{2}\begin{pmatrix} 0 & \sqrt{2} & 0 \\ \sqrt{2} & 0 & \sqrt{2} \\ 0 & \sqrt{2} & 0 \end{pmatrix}, \quad L_y = \frac{i\hbar}{2}\begin{pmatrix} 0 & -\sqrt{2} & 0 \\ \sqrt{2} & 0 & -\sqrt{2} \\ 0 & \sqrt{2} & 0 \end{pmatrix};
$$

$$
L_z = \hbar\begin{pmatrix} 1 & 0 & 0 \\ 0 & 0 & 0 \\ 0 & 0 & -1 \end{pmatrix},
$$

which satisfy the commutation relations and the condition that $L^2 = 2\hbar^2 \mathbf{1}$. The eigenstates corresponding to $L_z = \hbar, 0, -\hbar$ are given by

$$
\begin{pmatrix} 1 \\ 0 \\ 0 \end{pmatrix}, \begin{pmatrix} 0 \\ 1 \\ 0 \end{pmatrix}, \begin{pmatrix} 0 \\ 0 \\ 1 \end{pmatrix}.
$$

3. When three angular momenta are to be added,

$$
\mathbf{J} = \mathbf{L}_1 + \mathbf{L}_2 + \mathbf{L}_3,
$$

the procedure is to add two of them first, and then add the third to the sum of the first two. The counting and the final number of states of total angular momentum is independent of the sequence in which the addition is done, but the eigenfunctions reflect the choice made. This is discussed in advanced texts on this subject.

PROBLEMS

30.1 Find the properly normalized eigenstates of S_x.

30.2 Find the properly normalized eigenstates of $3S_x/5 - 4S_z/5$. Show that the eigenvalues are $\pm \hbar/2$.

30.3 Prove that for the Pauli matrices

$$\sigma_x \sigma_y + \sigma_y \sigma_x = \sigma_x \sigma_z + \sigma_z \sigma_x = \sigma_y \sigma_z + \sigma_z \sigma_y = 0.$$

30.4 Prove that

$$(\mathbf{A} \cdot \boldsymbol{\sigma})(\mathbf{B} \cdot \boldsymbol{\sigma}) = \mathbf{A} \cdot \mathbf{B} + i\boldsymbol{\sigma} \cdot (\mathbf{A} \times \mathbf{B}).$$

30.5 The expectation value of an operator in spin space is defined by

$$\langle M \rangle = \chi^\dagger M \chi$$

where M is the matrix representing the operator. Calculate the expectation value of S_x for the eigenstates of $S_x \cos \alpha + S_y \sin \alpha$ calculated in the text.

30.6 An isolated "spin" interacts with an external magnetic field so that the energy operator is given by

$$H = \frac{eg}{2mc} \mathbf{S} \cdot \mathbf{B} = \frac{egB}{2mc} S_z$$

where $\mathbf{B}$ is chosen to point in the z direction $\mathbf{B} = (0, 0, B)$. Show that the solution of the Schrödinger equation

$$i\hbar \frac{\partial \psi}{\partial t} = H\psi$$

is of the form

$$\psi = \begin{pmatrix} ae^{-i\omega t} \\ be^{i\omega t} \end{pmatrix}$$

where

$$a^2 + b^2 = 1 \quad \text{and} \quad \omega = \frac{egB}{4mc}.$$

30.7 Suppose in Problem 30.6, at time $t = 0$, the spinor is in an eigenstate of S_x with eigenvalue $+ \hbar/2$. Show that at a later time

$$\langle S_x \rangle = \cos 2\omega t$$

$$\langle S_y \rangle = \sin 2\omega t,$$

which suggests that in the presence of an external magnetic field the spin precesses about the direction of the field with a frequency (known as the Larmor frequency)

$$\omega_L = 2\omega.$$

30.8 What are the total angular momentum states obtained by adding

   i) $L_1 = 2$ and $L_2 = 3$

   ii) $L = 2$ and $S = 1/2$

   iii) $S_1 = 1/2$ and $S_2 = 1/2$

   iv) $L = 1$ and $S = 3/2$.

In each case check that the total number of products adds up to the total number of angular momentum states.

30.9 Suppose we have a system describing two spinning systems, and the energy operator has the form

$$H = \mathbf{S}^{(1)} \cdot \mathbf{S}^{(2)} E_1/\hbar^2 + (S_z^{(1)} + S_z^{(2)})E_2/\hbar.$$

Calculate the energy spectrum (i) when both spins are 1/2; (ii) when particle 1 has spin 1/2 and particle 2 has spin 3/2.

[*Hint*: Use the fact that the total angular momentum satisfies

$$\mathbf{J}^2 = (\mathbf{S}^{(1)})^2 + (\mathbf{S}^{(2)})^2 + 2\mathbf{S}^{(1)} \cdot \mathbf{S}^{(2)}$$

and the fact that eigenstates of $\mathbf{J}^2$ and J_z are linear combinations of products of eigenstates of $(\mathbf{S}^{(1)})^2$ and $(\mathbf{S}^{(2)})^2$. Thus

$$\mathbf{S}^{(1)} \cdot \mathbf{S}^{(2)} = \frac{\hbar^2}{2}(J(J+1) - S_1(S_1 + 1) - S_2(S_2 + 1)).$$

What are the possible values of J in the two cases?]

30.10 Make a list of all the possible states that can be made out of two spin 1/2 particles, using $L = 0$ to $L = 3$, in spectroscopic notation, such as $^1S_0$ and so on. Suppose the particles are identical. Because of the exclusion principle, certain states cannot occur. Given that $S, D, G, \ldots$ (even L) states are symmetric under the interchange of the particles, and $P, F, H, \ldots$ (odd L) states are odd under the interchange of the particles, and recalling the symmetry of the singlet and triplet spin wave functions, indicate which states cannot exist because of the exclusion principle. Which states can have angular momentum 1?

30.11 It is believed that the "elementary particles" studied by high-energy physicists are made up of fundamental constituents called *quarks* and their antiparticles *antiquarks*. Given that antiquarks have opposite parity to that of quarks, list the total angular momentum and parity of quark-antiquark combinations in their possible spin 0 and 1 states, with orbital angular momentum $L = 0, 1, 2, 3$. The spin of a quark (and an antiquark) is 1/2. Are there any J^P states that are forbidden? ($P = \pm$ refers to the parity.)

30.12 What are the possible states that can be made of three quarks, given that the total orbital angular momentum of the three-quark state is 0, 1, 2, 3. What are the parities of the various states, given that the parity of the spatial wave function is given by $(-1)^L$ where L is the total orbital angular momentum.

30.13 Add angular momenta 3, 2, and 1, by first adding two of them and then adding the third. Show that the total number of states with angular momentum $6, 5, \ldots 0$ is the same, no matter which two are added first. Check that the total number of eigenstates adds up to $7 \times 5 \times 3 = 105$ products of the eigenstates of the individual angular momenta.

31
Spin in Atoms, Molecules, and Nuclei

In this chapter we discuss a few manifestations of the role that spin plays in various physical systems.

Spin-orbit coupling Spin was discovered through attempts to understand the existence of and the splitting of doublets of spectral lines in alkali metals such as sodium. In the alkali metals only the outer electron is easily excited; the inner electrons effectively form an inert core and thus the system behaves very much like a hydrogen-like atom. If the potential is not pure $1/r$, we expect the structure of the spectrum discussed in Chapter 29 to be modified by the splitting of the l-degeneracy. We shall use the conventional nomenclature calling the states by letters that have their roots in the history of spectroscopy,

$$l = 0 \quad S \text{ state}$$
$$l = 1 \quad P \text{ states}$$
$$l = 2 \quad D \text{ states}$$
$$l = 3 \quad F \text{ states}$$
$$l = 4 \quad G \text{ states}$$

and so on. Figure 31.1 is labeled accordingly.

Since the energy operator is of the form

$$H = \frac{p^2}{2m} - \frac{Ze^2}{r} - \frac{1}{2mc^2}\left(\frac{Ze^2}{r}\right)^2 + \cdots \tag{31.1}$$

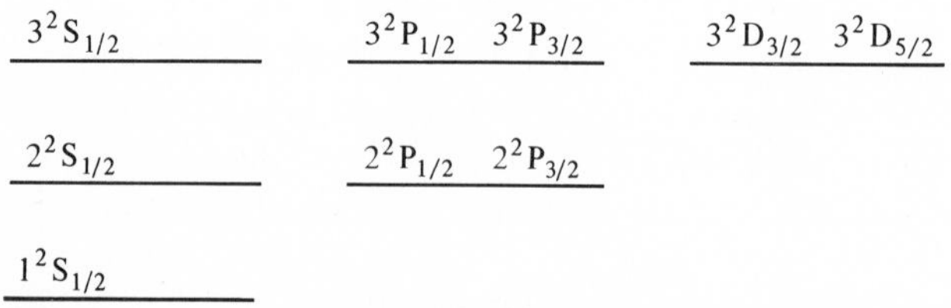

$$3^2 S_{1/2} \underline{\hspace{2cm}} \qquad 3^2 P_{1/2} \quad 3^2 P_{3/2} \underline{\hspace{2cm}} \qquad 3^2 D_{3/2} \quad 3^2 D_{5/2} \underline{\hspace{2cm}}$$

$$2^2 S_{1/2} \underline{\hspace{2cm}} \qquad 2^2 P_{1/2} \quad 2^2 P_{3/2} \underline{\hspace{2cm}}$$

$$1^2 S_{1/2} \underline{\hspace{2cm}}$$

Figure 31.1. Low-lying levels in hydrogen-like spectrum. The states are labeled using the spectroscopic notation, with the principal quantum number preceding the angular momentum labeling.

with the $1/r^2$ and other such terms coming from the relativistic corrections to the kinetic energy, it does not involve spin. Thus the fact that each electron can be in a spin "up" and a spin "down" state merely doubles the degeneracy of each state from $(2l + 1)$ to $2(2l + 1)$. For a given l state, we can either view the states as products of $Y_{lm}\chi_+$ and $Y_{lm}\chi_-$ with the radial wave functions, or, quite equivalently, rearrange them into linear combinations of the type (30.44), (30.45), that is, eigenstates of the total angular momentum $J = L + S$. With the Hamiltonian (31.1) the $j = l \pm \frac{1}{2}$ states are degenerate, since spin does not appear anywhere. In reality, because the electron has a magnetic dipole moment (by virtue of its spin), the degeneracy *is* lifted. This comes about as follows:

In first approximation, the electron only experiences an electric field that arises because of the charge of the nucleus. The field is

$$\mathbf{E} = -\boldsymbol{\nabla} V(r) = -\frac{\mathbf{r}}{r}\frac{dV}{dr} = \mathbf{r}\frac{Ze}{r^3}. \tag{31.2}$$

Because the electron is not at rest, it effectively "sees" a magnetic field too. If the electron were moving with uniform velocity $\mathbf{v}$, it would "see" a magnetic field of magnitude

$$\mathbf{B} = -\frac{1}{c}\mathbf{v} \times \mathbf{E} = -\frac{1}{mc}\mathbf{p} \times \mathbf{E}$$

$$= \frac{1}{mc}\mathbf{p} \times \mathbf{r}\frac{1}{r}\frac{dV}{dr}$$

$$= -\frac{1}{mc}\mathbf{L}\frac{1}{r}\frac{dV}{dr} \tag{31.3}$$

(see Eq. (7.24)). Because the motion is not rectilinear, this is not the correct expression. The proper result, first calculated by Thomas, is just half of the above, so that the magnetic field experienced by the electron is

$$\mathbf{B} = -\frac{1}{2mc}\mathbf{L}\frac{1}{r}\frac{dV}{dr}. \tag{31.4}$$

The electron has a magnetic dipole moment

$$\boldsymbol{\mu} = -\frac{e}{mc}\mathbf{S}. \tag{31.5}$$

(More precisely, it is $-\dfrac{eg}{2mc}\mathbf{S}$ with

$$\frac{g-2}{2} = 1159656.7 \pm 3.5 \times 10^{-9}.$$

The deviation from $g = 2$ is accounted for by quantum electrodynamics.) Thus the interaction of the electron dipole moment with the magnetic field changes the energy operator by the additional term

$$H' = -\boldsymbol{\mu} \cdot \mathbf{B} = \frac{1}{2m^2c^2} \mathbf{S} \cdot \mathbf{L} \frac{1}{r} \frac{d}{dr}(-eV(r))$$

$$= \frac{Ze^2}{2m^2c^2} \frac{1}{r^3} \mathbf{S} \cdot \mathbf{L}. \tag{31.6}$$

We may rewrite $\mathbf{S} \cdot \mathbf{L}$ in terms of $\mathbf{J}^2$ thus:

$$\mathbf{J}^2 = \mathbf{L}^2 + \mathbf{S}^2 + 2\mathbf{L} \cdot \mathbf{S}$$

so that

$$\mathbf{L} \cdot \mathbf{S} = \tfrac{1}{2}(\mathbf{J}^2 - \mathbf{L}^2 - \mathbf{S}^2). \tag{31.7}$$

When this operator acts on an eigenstate of $\mathbf{J}^2$ with eigenvalue $j(j+1)\hbar^2$, constructed out of eigenstates of $\mathbf{L}^2$ with eigenvalue $l(l+1)\hbar^2$ and spin eigenstates with eigenvalue of $\mathbf{S}^2$ equal to $\tfrac{1}{2}(\tfrac{1}{2}+1)\hbar^2$, we get a *number* instead of an operator:

$$\mathbf{L} \cdot \mathbf{S} \, \mathscr{Y}_{j,m_j} = \tfrac{1}{2}\hbar^2[j(j+1) - l(l+1) - \tfrac{3}{4}] \, \mathscr{Y}_{j,m_j}$$

$$= \tfrac{1}{2}\hbar^2 \left\{ \begin{array}{c} l \\ -l-1 \end{array} \right\} \mathscr{Y}_{j,m_j} \text{ for } j = l \pm \tfrac{1}{2}. \tag{31.8}$$

Thus the relativistic term arising from the electron magnetic dipole moment gives rise to a spin-orbit coupling (L-S coupling) that has different forms for $j = l + \tfrac{1}{2}$ and $j = l - \tfrac{1}{2}$ states, or equivalently, for fixed j, it has a different coefficient for $l = j - \tfrac{1}{2}$ and for $l = j + \tfrac{1}{2}$. The energy difference that gives rise to the splitting may be estimated by calculating the expectation value of the difference in the perturbing potentials

$$\Delta E = \frac{Ze^2}{2m^2c^2} \frac{\hbar^2}{2} l \left\langle \frac{1}{r^3} \right\rangle - \frac{Ze^2}{2m^2c^2} \frac{\hbar^2}{2} (-l-1) \left\langle \frac{1}{r^3} \right\rangle$$

$$= \frac{Ze^2\hbar^2}{4m^2c^2} (2l+1) \left\langle \frac{1}{r^3} \right\rangle. \tag{31.9}$$

For an order of magnitude estimate we set $Z = 1$, drop the $(2l+1)$ factor, and replace the expectation value by $(a_0)^{-3}$:

$$\Delta E = \frac{e^2\hbar^2}{4m^2c^2} \left(\frac{\hbar}{mc\alpha} \right)^{-3} = \frac{1}{4} \frac{e^2\hbar^2}{m^2c^2} \frac{m^3c^3}{\hbar^3} \alpha^3$$

$$= \frac{1}{4} \frac{e^2}{\hbar c} \alpha^3 mc^2 = \frac{1}{2} \left(\frac{1}{2} mc^2\alpha^2 \right) \alpha^2 \tag{31.10}$$

which, like the relativistic effect discussed in Eq. (29.44) is of order α^2 of the energy. When the precise expression for $1/r^3$ is used, and when the relativistic correction to the kinetic energy is included, the formula

$$\Delta E = -\frac{1}{2} mc^2(Z\alpha)^4 \frac{1}{n^3} \left(\frac{1}{j+1/2} - \frac{3}{4n} \right) \tag{31.11}$$

results. This turns out to be in excellent agreement with experiment.

The anomalous Zeeman effect In discussing the normal or "classical" Zeeman effect, we considered the shift of the energy levels caused by the change in the Hamiltonian

$$H' = \frac{e}{2mc} \, \mathbf{L} \cdot \mathbf{B}. \tag{31.12}$$

When the magnetic dipole moment of the electron is taken into account, this must be augmented by

$$H'' = -\boldsymbol{\mu} \cdot \mathbf{B} \cong \frac{e}{mc} \, \mathbf{S} \cdot \mathbf{B}. \tag{31.13}$$

Thus the total perturbation is

$$H_{\text{mag}} = \frac{e}{2mc} \, (\mathbf{L} + 2\mathbf{S}) \cdot \mathbf{B} = \frac{e}{2mc} \, (\mathbf{J} + \mathbf{S}) \cdot \mathbf{B}$$

$$= \frac{eB}{2mc} \, (J_z + S_z). \tag{31.14}$$

The final form arises when $\mathbf{B}$ is chosen to define the quantization axis $\mathbf{B} = (0, 0, B)$. For an atom in an eigenstate of $\mathbf{J}^2$ and J_z, the expectation value of J_z is just the eigenvalue

$$\langle J_z \rangle = \hbar m_j. \tag{31.15}$$

To calculate $\langle S_z \rangle$ we require the apparatus of angular momentum theory, such as the use of the form of the eigenstates (30.44), (30.45). Actually one can get the correct answer using a semiclassical argument. Since the total angular momentum $\mathbf{J}$ is conserved, we treat it as a "vector" fixed in space. Because of the interaction, both $\mathbf{S}$ and $\mathbf{L}$, although of constant length, precess about $\mathbf{J}$. Thus the components of $\mathbf{S}$

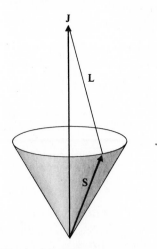

Figure 31.2. Semiclassical representation of precession cone of **S** about the fixed total angular momentum vector **J**.

perpendicular to $\mathbf{J}$ change constantly and average to zero, and all that matters is the component of $\mathbf{S}$ along $\mathbf{J}$. We thus guess, and detailed calculations bear this out, that we may make the replacement (Fig. 31.2)

$$\langle \mathbf{S} \rangle \to \langle \mathbf{J} \rangle \frac{\langle \mathbf{J} \cdot \mathbf{S} \rangle}{\langle \mathbf{J}^2 \rangle} = \langle \mathbf{J} \rangle \frac{\frac{1}{2}\langle \mathbf{J}^2 + \mathbf{S}^2 - (\mathbf{J} - \mathbf{S})^2 \rangle}{\langle \mathbf{J}^2 \rangle}$$

$$= \langle \mathbf{J} \rangle \frac{\langle \mathbf{J}^2 + \mathbf{S}^2 - \mathbf{L}^2 \rangle}{2\langle \mathbf{J}^2 \rangle}$$

$$= \langle \mathbf{J} \rangle \frac{j(j+1) + \frac{3}{4} - l(l+1)}{2j(j+1)} \tag{31.16}$$

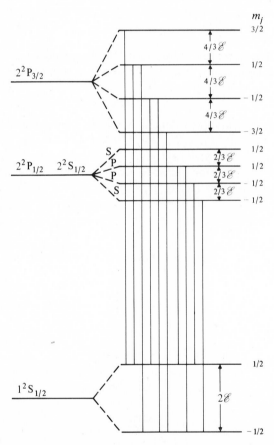

Figure 31.3. Zeeman splittings of the low-lying levels in hydrogen. The magnitudes of the splittings are expressed in units of $\mathscr{E} = e\hbar B/2mc$. The transitions obeying the selection rules $\Delta l = 1$, $\Delta m = 1$, $0, -1$ are drawn in.

where in the last step the expectation value in a state characterized by a definite j, l, and s was substituted.

Thus the expression for the energy shift in an external magnetic field is

$$\frac{eB\hbar}{2mc} m_j + \frac{eB\hbar}{2mc} m_j \frac{j(j+1)-l(l+1)+3/4}{2j(j+1)}$$

$$= \frac{eB\hbar}{2mc} m_j \left(1 + \frac{j(j+1)-l(l+1)+3/4}{2j(j+1)}\right). \qquad (31.17)$$

The factor in the parentheses is the so-called *Landé factor*, discovered by A. Landé empirically, before it was explained by the existence of electron spin. Note that again a given j-multiplet is split into its $(2j+1)$ components, with the splitting of the lines the same for a given (j, l) multiplet. The splitting, in contrast to the normal Zeeman effect, depends on the quantum numbers of the level, and thus in the transition, we no longer need to have just three spectral lines (Fig. 31.3).

Hyperfine structure In atoms there exists a kind of permanent Zeeman effect due to the fact that the nuclei also have spin, and therefore magnetic dipole moments. The interaction of the electronic magnetic moment with the magnetic field due to the magnetic dipole of the nucleus is called the *hyperfine interaction*. Consider, for example, the nucleus of charge Ze mass M_N and spin I. Its magnetic dipole moment will be

$$\mu_N = \frac{eg_N}{2M_N c} \mathbf{I}. \qquad (31.18)$$

The classical expression for the dipole-dipole interaction,

$$\frac{1}{r^5} (3(\mu_N \cdot \mathbf{r})(\mu_e \cdot \mathbf{r}) - r^2 \mu_N \cdot \mu_e) \qquad (31.19)$$

carries over into quantum mechanics. When the expectation value of this is worked out properly for hydrogen, using $\langle r^{-3} \rangle$, and $\mu_e = e\mathbf{S}/mc$, one gets

$$H'_{\text{h.f}} = \frac{4}{3} g_N \left(\frac{m}{M_N}\right) \alpha^4 mc^2 \frac{1}{n^3} \frac{\mathbf{S} \cdot \mathbf{I}}{\hbar^2} \qquad (31.20)$$

for $l = 0$ states. In first approximation there is no hyperfine interaction in the $l \neq 0$ states, because the centrifugal barrier reduces the overlap of the electron wave function with the nucleus; for a point nucleus the electron in P or higher states does not "see" any structure at the origin. To evaluate the expectation value of $\mathbf{S} \cdot \mathbf{I}$, we consider the total spin of the atom

$$\mathbf{F} = \mathbf{S} + \mathbf{I}. \qquad (31.21)$$

This gives us

$$\mathbf{S} \cdot \mathbf{I} = \tfrac{1}{2}(\mathbf{F}^2 - \mathbf{S}^2 - \mathbf{I}^2) = \tfrac{1}{2}(F(F+1)-\tfrac{3}{4}-I(I+1))\hbar^2$$

$$= \begin{cases} \tfrac{1}{2} I\hbar^2 & F = I+\tfrac{1}{2} \\ -\tfrac{1}{2}(I+1)\hbar^2 & F = I-\tfrac{1}{2}. \end{cases} \qquad (31.22)$$

For hydrogen, the measured value of g_N is 5.56, so that the energy difference between the excited state $(F = 1)$ and the ground state $(F = 0)$ is

$$\Delta E = \frac{4}{3}(5.56)\frac{1}{1840}\left(\frac{1}{137}\right)^4 mc^2 \tag{31.23}$$

since the proton has spin $I = 1/2$. The frequency associated with the transition between the two levels is

$$\frac{\Delta E}{h} = \frac{\Delta E}{2\pi\hbar} \cong 1420\,\text{MHz}. \tag{31.24}$$

The wavelength corresponding to this transition is 21 cm. This is the famous 21-cm line that is so important in astrophysics. Its importance lies in that radio waves are not absorbed by interstellar dust. Actually there is very little emission of 21-cm radiation near stars, since other effects dominate. In dust clouds, far from stars, collisional excitations of hydrogen atoms can put atoms in the $F = 1$ state. The radiative decay is very slow (several selection rules have to be overcome), but it does occur. The radiation can be detected, and from the intensity the distribution of atomic hydrogen in the plane of the galaxy can be mapped. Hyperfine transitions have also been useful in the study of the magnetization in solids.

The exchange interaction Spin effects become important in the study of spectra of atoms more complicated than hydrogen. Already in the case of helium, which has two electrons, spin becomes very important because of the role it plays in the anti-symmetry of the two-electron wave function. Let us consider the energy operator for helium, which has a nucleus with charge $2e$ and two electrons. To good approximation it has the form

$$H = \left(\frac{\mathbf{p}_1^2}{2m} - \frac{2e^2}{r_1}\right) + \left(\frac{\mathbf{p}_2^2}{2m} - \frac{2e^2}{r_2}\right) + \frac{e^2}{|\mathbf{r}_1 - \mathbf{r}_2|} \tag{31.25}$$

that we write as

$$H = H_1 + H_2 + \frac{e^2}{|\mathbf{r}_1 - \mathbf{r}_2|}$$

$$\equiv H_0 + \frac{e^2}{|\mathbf{r}_1 - \mathbf{r}_2|}. \tag{31.26}$$

Let us, in first approximation, neglect the electron-electron repulsion. The solution to

$$H_0\psi(\mathbf{r}_1, \mathbf{r}_2) = (H_1 + H_2)\psi(\mathbf{r}_1, \mathbf{r}_2) = E\psi(\mathbf{r}_1, \mathbf{r}_2) \tag{31.27}$$

is just a product of two hydrogen-like eigenfunctions for electrons "1" and "2." Thus in this approximation, the ground-state wave function is

$$\psi_0(\mathbf{r}_1, \mathbf{r}_2) = \psi_{100}(\mathbf{r}_1)\psi_{100}(\mathbf{r}_2), \tag{31.28}$$

where $\psi_{nlm}(\mathbf{r})$ is the hydrogen-like eigenfunction for quantum numbers (n, l, m) with $Z = 2$. The energy for each electron is

$$-13.6Z^2 = -54.4\,\text{eV} \tag{31.29}$$

so that in this approximation

$$E_0 \simeq -108.8\,\text{eV} \tag{31.30}$$

in poor agreement with the experimental value of $-79\,\text{eV}$. The wave function (31.28) is incomplete in that it makes no reference to the spin of the electrons. It is also incorrect as it stands, since it is symmetric under the interchange of the two electrons, that is, under the interchange $\mathbf{r}_1 \leftrightarrow \mathbf{r}_2$. We may maintain the above spatial wave function, provided the spin part of the wave function is antisymmetric under the interchange of the two electrons, that is, under the interchange of the labels "1" and "2." Such an antisymmetric spin wave function is the $S = 0$ (singlet) linear combination of the electron spinors $\xi_{\pm}^{(1)}, \xi_{\pm}^{(2)}$,

$$\chi_{00} = \frac{1}{\sqrt{2}} (\xi_+^{(1)} \xi_-^{(2)} - \xi_-^{(1)} \xi_+^{(2)}). \tag{31.31}$$

The ground-state wave function is therefore, in this approximation,

$$\psi_0 = \frac{1}{\sqrt{2}} (\xi_+^{(1)} \xi_-^{(2)} - \xi_-^{(1)} \xi_+^{(2)}) \psi_{100}(\mathbf{r}_1) \psi_{100}(\mathbf{r}_2). \tag{31.32}$$

Consider next the first excited state. We now take one of the electrons and put it into the $n = 2, l = 0$ state. We now face two choices: either we make the spatial wave function symmetric and the spin wave function antisymmetric, as in

$$\psi_1^s = \frac{1}{\sqrt{2}} (\xi_+^{(1)} \xi_-^{(2)} - \xi_-^{(1)} \xi_+^{(2)}) \cdot \frac{1}{\sqrt{2}} (\psi_{200}(\mathbf{r}_1) \psi_{100}(\mathbf{r}_2) + \psi_{100}(\mathbf{r}_2) \psi_{100}(\mathbf{r}_1)), \tag{31.33}$$

or we take the space wave function antisymmetric, and take the spin wave function symmetric. This corresponds to the total spin 1 state, and the triplet has three possible wave functions corresponding to $S_z = 1, 0, -1$. The total wave function has the form

$$\psi_1^t = \begin{cases} \xi_+^{(1)} \xi_+^{(2)} \\ \frac{1}{\sqrt{2}} (\xi_+^{(1)} \xi_-^{(2)} + \xi_-^{(1)} \xi_+^{(2)}) \\ \xi_-^{(1)} \xi_-^{(2)} \end{cases} \cdot \frac{1}{\sqrt{2}} (\psi_{200}(\mathbf{r}_1) \psi_{100}(\mathbf{r}_2) - \psi_{200}(\mathbf{r}_2) \psi_{100}(\mathbf{r}_1)). \tag{31.34}$$

In this approximation all these states, as well as the singlet and triplet states with the hydrogen-like linear combinations of the form

$$\frac{1}{\sqrt{2}} (\psi_{21m}(\mathbf{r}_1) \psi_{100}(\mathbf{r}_2) \pm \psi_{21m}(\mathbf{r}_2) \psi_{100}(\mathbf{r}_1)), \tag{31.35}$$

have the same energy, namely,

$$-13.6 Z^2 - 13.6 \frac{Z^2}{n^2} = -54.4 - 13.6$$

$$= -68.0\,\text{eV}. \tag{31.36}$$

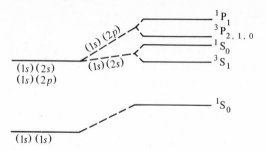

Figure 31.4. Energy level shifts due to inclusion of electron–electron repulsion. The states on the left are the unperturbed states, labeled by (n, l) of the two electrons. The shifts are different for different orbital angular momentum values, and there is a further splitting that depends on the symmetry of the spatial wave function, with triplet states lying lower than singlet states.

If we now take the electron repulsion into account, the degeneracy is lifted. The singlet wave functions, symmetric in the spatial distribution of the electrons, favors the electrons being together, while the triplet wave functions, being spatially antisymmetric, vanish when the electrons are on top of each other, and generally favor the electrons staying far apart. When the electrons are apart, the electron-electron repulsion is less effective in raising the energy. Thus we expect triplet wave functions to have lower energies than the corresponding singlet wave functions. The helium spectrum is shown in Fig. 31.4.

The exclusion principle in effect gives rise to a spin-spin interaction. We call it that because, in an interaction of the form

$$A + B\mathbf{S}_1 \cdot \mathbf{S}_2/\hbar^2, \tag{31.37}$$

the use of

$$\mathbf{S}_1 \cdot \mathbf{S}_2 = \tfrac{1}{2}[(\mathbf{S}_1 + \mathbf{S}_2)^2 - \mathbf{S}_1^2 - \mathbf{S}_2^2] = \tfrac{1}{2}\hbar^2 [S(S+1) - \tfrac{3}{4} - \tfrac{3}{4}]$$

$$= \frac{\hbar^2}{4} \begin{cases} 1 & S = 1 \\ -3 & S = 0 \end{cases} \tag{31.38}$$

leads to different energies for the triplet and singlet states. Note that in helium the splitting between the two spin states is *not* tiny. It is of order of magnitude

$$\left\langle \frac{e^2}{|\mathbf{r}_1 - \mathbf{r}_2|} \right\rangle$$

that is, Coulomb energies, which are measured in electron volts. If the *exchange effect* were not present, the only source of spin dependence would be relativistic, as in a dipole-dipole interaction, and this is usually of order $(v/c)^2 \sim (Z\alpha)^2 \sim 10^{-4}$ times smaller. Heisenberg observed that the phenomenon of ferromagnetism could be understood in this way. In a ferromagnet the spins of the electrons are aligned. If the forces that kept them aligned in the absence of an external magnetic field were weak, of the order of 10^{-4} eV, say, then ferromagnets could keep their magnetization only at very low temperatures. At room temperatures, the kT thermal perturbations are of order 10^{-2} eV and the existence of permanent magnets at these temperatures shows that the effect cannot be relativistic.

Spin and rotational spectra of molecules The lowest excited states of molecules are rotational states in which the molecule as a whole rotates about some axis. If vibrational excitations can be neglected, and this is justifiable at low energies, then for simple *homopolar* molecules such as $H_2, N_2, O_2, \ldots$ the energy operator can be approximated by

$$H = \frac{L^2}{2I} \tag{31.39}$$

where L is the angular momentum of rotation and $I = MR^2/2$ with M the mass of the nucleus and R the internuclear separation. The energy spectrum is given by

$$E(L) = \frac{\hbar^2 L(L+1)}{2I} \tag{31.40}$$

and the frequencies of transitions between adjacent states (transitions with $\Delta L = 1$ are favored by selection rules) are given by

$$\nu(L \rightarrow L-1) = \frac{E(L) - E(L-1)}{2\pi\hbar} = \frac{\hbar^2}{2I} \frac{L(L+1) - (L-1)L}{2\pi\hbar}$$

$$= \frac{\hbar L}{2\pi I} . \tag{31.41}$$

The study of molecular spectra can thus give us information about the internuclear separation. It can also give us information about the nuclear spin. The two nuclei (protons) in the H_2 molecule, for example, each have spin $1/2$, and they can be in a singlet state or in a triplet state. If they are in a singlet state, then, because the nuclei are fermions, the spatial part of the wave function must be symmetric; if they are in a triplet state, the spatial wave function is symmetric. Since in all cases the electronic wave function is completely symmetric about the two nuclei for these low excitation states, the spatial wave function is just the rotational wave function of the nuclei. The eigenfunctions of $L^2/2I$ are proportional to the $Y_{Lm}(\theta, \phi)$ and they have the property that *they are symmetric under the interchange of the nuclei when L is even and antisymmetric under the interchange when L is odd.*

Thus when the two protons are in an $S = 0$ state, only even L states can be excited; when the protons are in an $S = 1$ state, only odd L states can be excited. In a gas of H_2 molecules, there is randomness, so that on the average there will be three times as many H_2 molecules in the $S = 1$ total spin state as there are in the $S = 0$ state. Therefore there will be three times as many molecules excited in L odd rotational states as in even rotational states. This will manifest itself in the intensities of different spectral lines, and from the observations one can deduce the factor 3, that is, $2S + 1$.

More generally, suppose each nucleus has spin I. Then there will be possible spin states $2I, 2I - 2, 2I - 4, \ldots$ that are spin-symmetric, and spin states $2I - 1, 2I - 3$, $2I - 5, \ldots$ that are spin-antisymmetric. If I is an odd half-integer, that is, $I = 1/2$, $3/2, \ldots$, that is, if the nucleus is a fermion, then the total wave function must be antisymmetric, so that the states $2I, 2I - 2, \ldots$ have odd L; if I is an integer, then the nuclei are bosons, and the states $2I, 2I - 2, \ldots$ have even L. The degeneracy, the analog of $3 = 2S + 1$ in the H_2 example, is, for the series $2I, 2I - 2, \ldots$, given by

$$[2(2I) + 1] + [2(2I - 2) + 1] + [2(2I - 4) + 1] + \cdots$$

$$= \sum_{k=0}^{I} [2(2I - 2k) + 1]$$

$$= (4I + 1) \sum_{k=0}^{I} 1 - 4 \sum_{k=0}^{I} k$$

$$= (4I + 1)(I + 1) - 4 \frac{I(I + 1)}{2} = (I + 1)(2I + 1). \tag{31.42}$$

We repeat: This is the number of spin states that are spin-symmetric. The number of states that are spin-antisymmetric is given by

$$[2(2I - 1) + 1] + [2(2I - 3) + 1] + [2(2I - 5) + 1] + \cdots$$

$$= \sum_{k=1}^{I} [2(2I - 2k + 1) + 1]$$

$$= (4I + 3) \sum_{k=1}^{I} 1 - 4 \sum_{k=1}^{I} k$$

$$= (4I + 3)I - 4 \frac{I(I + 1)}{2} = I(2I + 1). \tag{31.43}$$

(Note that these add up to the total number of states, $1 + 3 + 5 + \cdots + (2(2I) + 1) = (2I + 1)^2$.) Hence if I is an integer, the $2I$ series have even L, and thus the fraction of states associated with even L is $(I + 1)(2I + 1)/(\text{total}) = (I + 1)/(2I + 1)$. If I is an odd half-integer, the fraction of states associated with even L is $I(2I + 1)/(\text{total}) = I/(2I + 1)$. Hence for integral I, the intensity ratio (even L)/(odd L) $= (I + 1)/I$, whereas for odd half-integral I the ratio is reversed.

The study of the rotational spectrum of the N_2 molecule in the late 1920s showed that the relative intensity patterns could be understood if I were equal to unity. The finding of $I = 1$ was something of a puzzle: The ^{14}N nucleus was known to have charge 7 and the atomic weight was 14. It was surmised that the nitrogen nucleus consisted of fourteen protons and seven electrons. Such a system would have to have a half-odd integral spin. The discovery of the neutron resolved the difficulty: instead of a nucleus containing electrons to make up the charge, it could consist of protons and neutrons. For ^{14}N the nucleus consists of seven protons and seven neutrons. Since neutrons also have spin 1/2, it is easy for such a nucleus to have spin 1.

PROBLEMS

31.1 Calculate the splitting caused by the spin-orbit coupling and the relativistic effect in the $l = 1 \rightarrow l = 0$ transition for $n = 3$. Express your result in terms of wavelengths.

31.2 Calculate the splittings in the $n = 1$ and $n = 2$ states in hydrogen due to the same relativistic effects.

31.3 What is the effect of a spin-orbit interaction (with unknown coefficient) in a model of a nucleus in which the potential that a single particle moves in is of the form

$$V(r) = \tfrac{1}{2} m \omega_0^2 r^2$$

and the spin-orbit coupling is of the form

$$V_{\text{s.o}}(r) = \tfrac{1}{2} A \mathbf{S} \cdot \mathbf{L} \frac{1}{r} \frac{dV}{dr} \ .$$

Given that the spectrum of the three-dimensional harmonic oscillator is

$$E = (2n_r + l + \tfrac{3}{2}) \hbar \omega,$$

sketch the spectrum of the oscillator as perturbed by the spin-orbit term. Note that the single nucleon has spin $1/2$.

31.4 Sketch the spectrum of the $n = 3, l = 2 \rightarrow n = 2, l = 1$ transition in hydrogen in the presence of a magnetic field. Treat the magnetic field perturbation as small compared with the spin-orbit and relativistic effects, that is, take the unperturbed lines as including the spin-orbit effects as in Eq. (31.11).

31.5 For what value of B will the Zeeman effect be comparable in magnitude to the spin-orbit effects? How does the Zeeman effect look when the spin-orbit terms are completely left out?

31.6 There are clouds of atomic hydrogen in our galaxy. The atoms when in their ground states can be excited to the $F = 1$ state through collisions, that is, through thermal effects. Given that the statistical weight of the $F = 1$ states is three times that of the $F = 0$ ground state, at what temperatures would you expect the $F = 1$ and $F = 0$ states to be equally populated, given the splitting (31.24)? (As a reminder, read the end of Chapter 11).

PART VI
The Structure
of Matter

In this, the last part of the book, a variety of aspects of the structure of matter are discussed. The structure of atoms, molecules, radiative transitions, and radioactivity begin this section of the book. The object here is to give the reader an overview of the phenomena, and to provide a quantitative method of estimating the size of physical effects wherever possible. Because we are now dealing with real, rather than idealized systems, we cannot hope to provide more than rough numerical agreement, while keeping the treatment at a mathematical level accessible to the intended reader, and without making the book impossibly long. Thus in our discussion of atoms we go into some detail in discussing the *Building-up Principle*, which shows us how the shell structure of atoms arises, and at the same time we estimate the binding energies of the last electron. In the discussion of the molecular systems we consider electronic, vibrational, and rotational degrees of freedom, and use some simplified models to discuss the energy-level splittings. The material is supplemented by a brief discussion of the Raman effect, because recent work in spectroscopy has made extensive use of the effect. The chapter on electromagnetic transitions is mathematically more complicated than the other ones, because our objective is to arrive at formulas that show how selection rules emerge. The limitations of a first-order perturbation treatment then naturally lead into a qualitative discussion of the exponential time dependence of the decay rate, and the formulas are generalized to multiple-level systems. Chapter 35 provides a special application of the work on radiative transitions. A detailed discussion of the Einstein derivation of stimulated emission leads to a rather extensive discussion of what makes the laser work, together with a brief description of its many applications.

This introductory section leads us into more specialized topics: solids, nuclei, and elementary-particle physics. In the discussion of solids, I have decided to focus on just a few subjects of major interest: crystal lattices, phonons, and the specific heat of solids, metals, quantum statistics, semiconductors, and superconductors. The treatment is partly qualitative and, whenever possible, quantitative, so that the student can get an idea of why certain quantities have the magnitudes that are observed. In the discussion of nuclear physics, I have had to abandon any thought of discussing scattering and reaction theory. Lack of space has limited me to a discussion of the semi-empirical mass formula, the shell model, the liquid drop model, and some aspects of the collective model. The final three chapters deal with elementary particle physics:

Here we discuss the discovery of antiparticles, neutrinos, and beta decay, the strong interactions, including the Yukawa theory, and above all the symmetries, i-spin, SU(3), and the predictions of that theory. In the final chapter recent developments associated with the notion of permanently confined quarks are briefly discussed.

The amount of material is clearly too extensive to be covered in one academic quarter. The expectation is that the instructor will choose to cover some topics in depth, with the remaining topics left to the student as "cultural" material.

32
The Structure of Atoms

Atoms One of the major achievements of quantum mechanics is that it provides a detailed understanding of the physical and chemical properties of the elements by providing a quantitative description of atomic structure. Atoms consist of a nucleus of charge Ze and Z electrons, each of charge $-e$, bound to it. The nucleus is very massive compared to the electrons: electrons contribute less than 10^{-3} to the total mass of the atom. The nucleus consists of Z protons, each of charge $+e$, and of $N = A - Z$ neutrons, electrically neutral, and both of these particles are approximately two thousand times more massive than the electron. The structure of the nucleus will be discussed in Chapter 39. Because of the large mass ratio, it is possible to treat the nucleus as an infinitely massive point source of positive charge, to which the electrons are attracted by electrostatic forces. If the electrons did not interact with each other, each electron would be moving in a central potential $-Ze^2/r$ and it would be described by hydrogen-like wave functions. The energy levels in that approximation are given by

$$E_n = -\frac{1}{2} m_e c^2 \frac{(Z\alpha)^2}{n^2} \tag{32.1}$$

and there is the familiar degeneracy of levels for all $l \leqslant n - 1$. Even if the electron-electron Coulomb repulsion

$$\sum_{i>j} \sum_j \frac{e^2}{|\mathbf{r}_i - \mathbf{r}_j|} \tag{32.2}$$

is neglected, the fact that there are many electrons has an enormous impact on the structure of atoms because of the requirement of the Pauli Principle. If the exclusion principle were not operative, all of the electrons would be in the lowest ($1s$) orbital state, and in the absence of the Coulomb repulsion, the ground state would have energy

$$E_0 = -\frac{1}{2} m_e c^2 (Z\alpha)^2 Z. \tag{32.3}$$

The *ionization energy*, the energy required to remove one electron from the atom, would then just be

$$\frac{1}{2} m_e c^2 (Z\alpha)^2 = 13.6 Z^2 \text{ eV}. \tag{32.4}$$

The inclusion of the Coulomb repulsion would not change the picture qualitatively. A world made of such atoms would be stark indeed. The least amount of energy needed to get an atom excited would be of the order of $10.2 Z^2$ eV, which corresponds to temperatures of $10^5 Z^2$ degrees K. The chemical reactions that occur in biological systems and that depend on very small energy differences could not occur.

Importance of exclusion principle The exclusion principle states that "no more than two electrons may occupy a given orbital" (that is, a state of given (n, l, m)). This implies that the building up of atoms is much more complicated. For a given Z, there can be no more than two electrons in the $(1s)$ state, no more than two in the $(2s)$ state, and no more than six in the $(2p)$ state, two each for $m_l = 1, 0,$ and -1. The $(3d)$ orbitals, with $l = 2$, can accommodate up to $2(2l + 1) = 2 \times 5 = 10$ electrons, and similarly for the $(4d)$ states. We may use these considerations to build up the structure of atoms. In doing so, we must take into account the electron-electron repulsion which is more important here than it would be for our "bosonic" atom, where it would just shift the ground-state energy somewhat.

The inclusion of the Coulomb repulsion at first sight leads us into consideration of the Schrödinger equation in $3Z$ dimensions, and there is no simple way to handle that. One can, for small Z, solve such an equation on the computer. However, solving

$$\left(-\frac{\hbar^2}{2m} \sum_i \nabla_i^2 - Ze^2 \sum_i \frac{1}{r_i} + \sum_{i>j} \sum_j \frac{e^2}{|\mathbf{r}_i - \mathbf{r}_j|} - E \right) \Psi(\mathbf{r}_1, \mathbf{r}_2, \ldots, \mathbf{r}_2) = 0$$

(32.5)

even for small Z is impossible unless one knows pretty well what to look for (I believe it was Dirac who said that he never solved an equation unless he knew its solution in advance). We must therefore do some approximate thinking about the system, decide what is large and what is small. In the hydrogen-like approximation, the velocities of the electrons are large, of the order of $Z\alpha c/n$, which suggests that each electron "sees" the other electrons as a smear of negative charge distribution. Roughly speaking, the Coulomb repulsion may be approximated by treating it as contributing an average central potential that changes the nuclear potential $-Ze^2/r$. The modification is small for small r, because close to the nucleus the electrons are all likely to be on the outside, canceling each other out. On the other hand, at large r, the electrons will almost screen out the nuclear potential to leave $-e^2/r$ in the large r limit. The mere fact that the potential is no longer purely $1/r$, but more like $-e^2 Z(r)/r$, means that the degeneracy characteristic of the hydrogen-like atoms is lifted. Thus the $(2s)$ and the $(2p)$ states are no longer degenerate. We can also conjecture how the levels split. Electrons in states with large angular momentum spend more of their time at large distances from the nucleus: Classically a large $\langle \mathbf{r} \times \mathbf{p} \rangle$ leads to a large $\langle |\mathbf{r}| \rangle$ for configurations that also minimize the kinetic energy $\mathbf{p}^2/2m$; equivalently, the wave functions for large l behave like r^l near the origin, that is, they are depressed for small r and only grow for large values of r. In the large r region $Z(r)$ tends to be smaller, so that the negative potential energy is smaller. Thus we expect states of larger angular momentum to have a higher energy than those of lower angular momentum for the same n value. Let us now consider a sequence of atoms.

Hydrogen This atom was studied in detail before. There is only one electron, and the ground state configuration is $(1s)$. The ionization potential is 13.6 eV, and the amount of energy needed to excite the first state above the ground state is 10.2 eV. The radius of the atom is 0.5 Å. The spectroscopic description of the ground state is $^2S_{1/2}$.[†]

Helium, $Z = 2$ The lowest two-electron state is one in which both electrons are in the $(1s)$ orbital. We denote the configuration by $(1s)^2$. The binding energy in first approximation is -108.8 eV, but the Coulomb repulsion will increase this. Since the Bohr radius is 0.25 Å, a rough estimate of the average separation between the electrons is 0.35 Å, and thus the Coulomb repulsion is $\sim 1.4e^2/a_0$, where $a_0 = 0.5$ Å. Since $e^2/2a_0 = 13.6$ eV, we get a rough estimate for the total energy of -70 eV. This is an overestimate of the repulsion, and the correct number is -79 eV, a number more consistent with a separation of 0.5 Å. When one electron is removed, the remaining electron is in a $(1s)$ state of a $Z = 2$ hydrogen-like atom. The energy is -54.4 eV and the difference is the ionization potential, 24.6 eV. The two electrons can be in a $S = 0$ state or in an $S = 1$ state, and the former has the lower energy. Thus the spectroscopic description of the state is $^1S_0$. A rough estimate of the energy of the first excited state, with configuration $(1s)(2s)$ is $-13.6Z^2 - 13.6(Z-1)^2/n^2 \cong -58$ eV for $Z = 2$ and $n = 2$ (the second term has a reduced charge to take into account shielding), so that it takes approximately 20 eV to excite an electron to the first excited state. In any reaction with another substance, a substantial amount of energy is thus needed for a rearrangement of the electrons, and it is for this reason that helium is chemically so inactive.

Lithium, $Z = 3$ The exclusion principle forbids a $(1s)^3$ configuration, and the lowest energy electron configuration is $(1s)^2(2s)$. We are thus adding an electron to a *closed shell*. If screening were perfect, the outer electron would see a charge of $Z = 1$, and thus the binding energy of that electron would be -3.4 eV (since $n = 2$). The screening is not perfect, especially since the outer *valence* electron, being in a $2s$ state, has some probability of overlapping the nucleus and feeling the full $Z = 2$ nuclear charge. The experimental value of the ionization potential is 5.4 eV. The spectroscopic description is determined by the single electron outside the closed shell, so that the ground state of lithium may be described as a $^2S_{1/2}$ state. It takes very little energy to excite lithium: The $(2p)$ electronic states lie just a little above the $(2s)$ state, and these $(2p)$ states when occupied make the atom chemically active (see the discussion of carbon). Lithium, like other elements that have one electron outside a closed shell, is a very active element.

Beryllium, $Z = 4$ The natural place for the fourth electron to go is into the empty slot in the $(2s)$ orbital, so that the configuration is $(1s)^2(2s)^2$. As far as the energy is concerned, the situation is very much like that of helium. If screening were perfect, we might expect a binding energy like that of helium, since the inner electrons reduce the effective Z to something like $Z = 2$. The difference is that here $n = 2$, so that we expect the ionization energy to be about 1/4 that of helium, which is to be enlarged since screening is not perfect. A 50% increase, like that encountered for lithium,

---

[†] Recall that the notation is $^{2S+1}L_J$ with $S, P, D, F, G, \ldots$ representing $L = 0, 1, 2, 3, 4, \ldots$.

yields approximately 9 eV for the ionization potential. The experimental value is 9.3 eV. The spectroscopic description is $^1S_0$, since the two electrons in the 2s shell will also pair up with the spin antisymmetric state. We shall see later that atoms in which the outer electrons have their spins "paired up" into singlet states are less reactive. This is the case for beryllium.

Boron, $Z = 5$ The fifth electron can go either into a (3s) state or into one of the (2p) states. Even though the (2p) orbitals are higher lying than the (2s) states, they do have lower energy than the (3s) states, so that it is these states that begin to be filled, beginning with boron. The binding energy might be expected to be somewhat smaller than that of beryllium, since the (2p) state lies higher than the (2s) state. The experimental value of the ionization energy is 8.3 eV. Boron contains one electron outside a closed shell, and is expected to be very reactive.

Carbon, $Z = 6$ The configuration for carbon is $(1s)^2(2s)^2(2p)^2$. The second (2p) electron can spatially stay out of the way of the first electron thus minimizing the repulsive energy, so that the increase in Z is expected to result in something of an increase in the ionization energy. The experimental value is 11.3 eV. The way in which the electrons stay out of each other's way is as follows: the possible $l = 1$ wave functions, $Y_{1,1}$, $Y_{1,0}$, and $Y_{1,-1}$ are proportional to $\sin\theta\, e^{i\phi}$, $\cos\theta$, and $\sin\theta\, e^{-i\phi}$, respectively, and they can be recombined into the wave functions $\cos\theta$, $\sin\theta\cos\phi$, and $\sin\theta\sin\phi$ shown in Fig. 32.1. The probability distributions look like three orthogonal arms, and when two electrons go into different arms, the repulsion is lowered in importance. Note that the electrons do not have to be in a spin singlet system since their spatial wave functions are different. They are therefore not "paired." One might at first sight expect carbon to be divalent, but this is not so because of the subtleties caused by close-lying levels. Its costs relatively little energy to promote one of the (2s) electrons into one of the unoccupied 2p states. The configuration $(1s)^2$ $(2s)(2p)^3$ has four "unpaired" electrons, and the gain in energy from the formation of four bonds with other atoms makes up for the excitation energy. The ionization

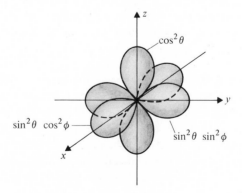

Figure 32.1. Orthogonal distributions of electronic clouds in $L = 1$ state.

energy is expected to be somewhat larger than for boron, in agreement with the experimental result of 11.3 eV.

Nitrogen, $Z = 7$ Here the configuration is $(1s)^2(2s)^2(2p)^3$, or, omitting the closed shells and subshells, $(2p)^3$ for brevity. The three electrons can all be in nonoverlapping spatial wave functions, and the increase in binding energy is expected to be the same as the increase from boron to carbon. This is in agreement with the measured value of 14.5 eV for the ionization potential.

Oxygen, $Z = 8$ Here the configuration is $(2p)^4$ in terms of our abbreviated notation. The fourth electron must go into one of the occupied m-states, and form a singlet state with the electron already in it. This increases the effect of the electron-electron repulsion, and even though Z increases by one, one would expect the ionization energy not to increase as fast as before. In fact, the ionization energy is 13.6 eV, showing that the repulsion overcomes the increase in Z. Because the two electrons in the same orbital are "paired," oxygen is divalent, and the active electrons' wave functions are spatially oriented at right angles to each other.

Fluorine, $Z = 9$ Here the configuration is $(2p)^5$. The monotonic increase in the ionization energy resumes, with the experimental value 17.4 eV. Fluorine is chemically very active, because it can "accept" an electron to form a closed shell $(2p)^6$, which is very stable. One frequently speaks of this as a "hole" in the shell, and this also allows us to determine the spectroscopic description of the atom in the ground state. Since the addition of a single electron with $s = 1/2$ and $l = 1$ yields a $^1S_0$ state, the shell with the hole in it must have $s = 1/2$ and $l = 1$. It is thus a ^2P state. The J value will be discussed below.

Neon, $Z = 10$ With $Z = 10$, the $(2p)$ shell is closed, the configuration is $(2p)^6$ and all electrons are paired off, so that the spectroscopic description is, as already mentioned, $^1S_0$. The ionization energy is 21.6 eV, continuing the monotonic trend. Here, as in helium, the first available state that an electron can be excited into has a higher n value, and thus it takes quite a lot of energy to perturb the atom. Neon shares with helium the property of being an inert gas.

The above, qualitative discussion shows how the exclusion principle brings variety to the elements. As Z increases, the pattern tends to repeat. The next period again has eight elements in it. First the $(3s)$ subshell is filled to give sodium $(Z = 11)$ and magnesium $(Z = 12)$. Then the $(3p)$ shell is filled, giving the elements aluminum $(Z = 13)$, silicon $(Z = 14)$, phosphorus $(Z = 15)$, sulfur $(Z = 16)$, chlorine $(Z = 17)$ and, closing the shell, argon $(Z = 18)$. The properties of the elements are similar to the properties of the $(2p)$ series: Sodium, like lithium, is very reactive; similarly chlorine, like fluorine, consists of a closed shell plus one hole, and it too is very reactive. The elements consisting of a closed shell plus one electron are called alkali metals, and those consisting of a closed shell and a hole are called halogens. For the series $Z = 11$ to $Z = 18$, the ionization energies are somewhat smaller, as befits the $n = 3$ character, and the breaks in their values match those in the $(2p)$ series.

It is at this stage that the departure from the $1/r$ shape of the Coulomb potential manifests itself. The $(3d)$ states are pushed up in energy to bring them up to the same level as the $(4s)$ states. Thus there is a new period, consisting of elements with the configurations $(4s)$, $(4s)^2$, $(4s)^2(3d)$, $(4s)^2(3d)^2$, $(4s)^2(3d)^3$, then, because of the closeness of the levels, an exceptional configuration $(4s)(3d)^5$, then back to $(4s)^2$ $(3d)^5, \ldots, (4s)^2(3d)^9$, $(4s)(3d)^{10}$, and $(4s)^2(3d)^{10}$. After that the $(4p)$ shell gets filled, until the period ends with krypton ($Z = 36$). The chemical properties in this period are similar to those of the earlier periods: Potassium, with the single $(4s)$ electron, is an alkali metal like sodium, and bromine, with the $(4p)^5$ configuration is chemically similar to chlorine and fluorine. The series of elements in which the $(3d)$ shell is being filled, are all rather similar in their chemical properties. The reason is that the "radii" of these electronic distributions are somewhat smaller than those of the $(4s)$ electrons, so that when the $(4s)^2$ shell is filled, these electrons tend to shield the $(3d)$ electrons, no matter how many of them there are, from outside influences. The same effect occurs when the $(4f)$ shell is being filled, just after the $(6s)$ shell has been filled. The elements here are called the *rare earths*.

Spectroscopic description of ground states In discussing the first series of elements, we occasionally drew attention to the spectroscopic description of the ground state. The reason for that is the existence of selection rules in radiative transitions, which makes it important to know the total angular momentum of the ground state. When there are several electrons present, matters get quite complicated. For example, with a $(2p)^2$ configuration outside a closed shell, we can have total spin $S = 0$ or 1, and the orbital angular momenta can add up to 0, 1, or 2. This leads to many possible J values. With a $(3d)^3$ configuration, the spin can be $S = 1/2$ or $3/2$, L can range from 0 to 6, and J can range from $1/2$ to $15/2$. What determines the ground-state quantum numbers is an interplay of spin-orbit coupling and the exchange effect (see Chapter 31). For the lighter atoms, when the motion is nonrelativistic (this turns out to be for $Z \lesssim 40$), the electron-electron interaction effects are more important than the spin-orbit interaction, and the wave functions are more closely approximated by considering the electron orbital angular momenta as added separately from the spins, to form a total L, and a total S. For heavier atoms, a better description is given by the spin and orbital angular momentum for each electron coupling separately, to give a number of J_i, which then couple to yield the total J. In the former case, one speaks of Russell-Saunders coupling; in the latter, one describes the angular momentum states in terms of $j - j$ coupling. For Russell-Saunders coupling, the state that lies lowest can be determined by a set of empirical rules, *Hund's rules*:

1. The state with largest S lies lowest;

2. For a given value of S, the state with maximum L lies lowest;

3. For a given L and S, if the incomplete shell is not more than half-filled, the lowest state has the minimum value of $J = |L - S|$; if the shell is more than half-filled, the state of lowest energy has $J = L + S$.

The rules can be justified by detailed computations within the framework of the effective potential method that was briefly alluded to above.

Figure 32.2.

Care must be taken to apply these rules so as not to violate the Exclusion Principle. This is illustrated by considering the spectroscopic description of carbon $(2p)^2$, oxygen $(2p)^4$ and manganese $(3d)^5$. In the first two cases, we have p states, so that we draw a table, with spaces corresponding to $L_z = 1, 0, -1$. The electrons are, as far as possible, placed in different spaces (to minimize the repulsion). For carbon, we place them in the $L_z = 1$ and 0 states. According to the first Hund rule, the spins will be parallel (Fig. 32.2). Thus we have $S_z = 1$ for the largest possible value, and therefore the value of $S = 1$: we have a triplet state. The largest possible value of L, given the spin assignment, is the one corresponding to $L_z = 1$, that is a P state. The third rule then tells us that it must be a $^3P_0$ state. For oxygen, we fill all three L_z spaces with electrons, and then put the fourth electron in the $L_z = 1$ state, say. The two electrons in the $L_z = 1$ state must form a singlet. Thus only two electrons play a role in determining the spin. Since $S_z = 1, S = 1$ and we again have a triplet state. The total value of $L_z = 2 + 0 + (-1) = 1$, so that $L = 1$ and we have a P state. Since the shell is more than half filled, the state is $^3P_2$.

For manganese we have spaces with $L_z = 2, 1, 0, -1, -2$. One electron goes into each one of these, and the spins are parallel. Thus $S_z = 5/2$ and therefore $S = 5/2$. The value of L_z is zero, and therefore we have an S state: the spectroscopic description is $^6S_{5/2}$.

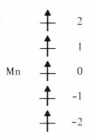

Figure 32.3.

The spectroscopy of elements other than hydrogen is complicated indeed. To get an idea of this, let us consider the ground state configuration of carbon, which consists of two p-state electrons outside a closed shell. The spins can be 0 or 1, the total L can be 2, 1, or 0. The exclusion principle implies that for the spin-antisymmetric singlet states, the spatial wave functions must be symmetric, that is, they must have $L = 0$ or 2; for spin 1, we must have $L = 1$. This holds only for *equivalent electrons*, that is, electrons that have the same n value, so that this part of the wave function is symmetric. The $^1S_0$ state lies above the $^1D_2$ state, and the two singlet states lie above the $^3P_2$, $^3P_1$, and $^3P_0$ states. The simplest excited state has the configuration $(2s)(2p)^3$ and here the states are manifold. The $(2s)(2p)$ electrons can be in a $^3P_{2,1,0}$ or a $^1P_1$ state, and when these are combined with the states already listed for the $(2p)^2$ electrons, we get possible spin values of $2, 1, 0$ and possible L values of $3, 2, 1, 0$. The exclusion principle limits the number of terms, but the structure is surely going to be very rich.

This concludes our brief survey of atomic structure. Perhaps the most important thing that the reader can carry away from this chapter is the recognition that quantum mechanics provides us with a clear, sometimes quantitative understanding of the complexity of atoms and their states of excitation, and an admiration for the way nature manages to achieve the remarkable richness of chemical structure with such an economy of means.

NOTES AND COMMENTS

1. More detailed discussions of the electronic structure of atoms and the chemical consequences are to be found in textbooks on quantum chemistry and physical chemistry. I am quite unfamiliar with the literature in this field. Two books that look like excellent references for this chapter and the next are, in order of difficulty, M. Karplus and R.N. Porter, *Atoms and Molecules*, W.A. Benjamin (Addison-Wesley Advanced Textbook Series), Reading, Mass., 1970, and G.W. King, *Spectroscopy and Molecular Structure*, Holt, Rinehart and Winston, New York, 1964. An older classic is G. Herzberg, *Atomic Spectra and Atomic Structure*, Dover Publications, New York, 1944.

2. The quantitative approach to treating electrons in a complex atom as experiencing a central potential due to the presence of all the other electrons was pioneered by D.R. Hartree as early as 1927. Improvements that took into account antisymmetrization of the electronic wave functions that appeared in the potential terms were introduced by V. Fock.

3. The calculations carried out using the self-consistent Hartree method show that the "size" of the electron distribution grows very slowly with Z (and n), in contrast to the hydrogen-like atom formula $a_0 n^2 / Z$. The largest atomic radii are only about three times as large as the Bohr radius a_0.

TABLE OF ELEMENTS

In the table opposite we list the elements from $Z = 1$ to $Z = 42$, the electronic configuration, the spectroscopic description where known, the masses in atomic mass units defined such that normal oxygen has 16.000 amu, and the ionization energies, where known.

Atom	Z	Configuration	Description	Mass (amu)	I.E. (eV)
H	1	$(1s)$	$^2S_{1/2}$	1.008	13.6
He	2	$(1s)^2$	$^1S_0$	4.003	24.6
Li	3	$He(2s)$	$^2S_{1/2}$	6.939	5.4
Be	4	$He(2s)^2$	$^1S_0$	9.012	9.3
B	5	$He(2s)^2(2p)$	$^2P_{1/2}$	10.811	8.3
C	6	$He(2s)^2(2p)^2$	$^3P_0$	12.011	11.3
N	7	$He(2s)^2(2p)^3$	$^4S_{3/2}$	14.007	14.5
O	8	$He(2s)^2(2p)^4$	$^3P_2$	16.000	13.6
F	9	$He(2s)^2(2p)^5$	$^2P_{3/2}$	18.998	17.4
Ne	10	$He(2s)^2(2p)^6$	$^1S_0$	20.183	21.6
Na	11	$Ne(3s)$	$^2S_{1/2}$	22.991	5.1
Mg	12	$Ne(3s)^2$	$^1S_0$	24.32	7.6
Al	13	$Ne(3s)^2(3p)$	$^2P_{1/2}$	26.98	6.0
Si	14	$Ne(3s)^2(3p)^2$	$^3P_0$	28.09	8.2
P	15	$Ne(3s)^2(3p)^3$	$^4S_{3/2}$	30.97	11.0
S	16	$Ne(3s)^2(3p)^4$	$^3P_2$	32.06	10.4
Cl	17	$Ne(3s)^2(3p)^5$	$^2P_{3/2}$	35.45	13.0
Ar	18	$Ne(3s)^2(3p)^6$	$^1S_0$	39.94	15.8
K	19	$Ar(4s)$	$^2S_{1/2}$	39.10	4.3
Ca	20	$Ar(4s)^2$	$^1S_0$	40.08	6.1
Sc	21	$Ar(4s)^2(3d)$	$^2D_{3/2}$	44.96	6.6
Ti	22	$Ar(4s)^2(3d)^2$	$^3F_2$	47.90	6.8
V	23	$Ar(4s)^2(3d)^3$	$^4F_{3/2}$	50.94	6.7
Cr	24	$Ar(4s)(3d)^5$	$^7S_3$	52.00	6.76
Mn	25	$Ar(4s)^2(3d)^5$	$^6S_{5/2}$	54.94	7.43
Fe	26	$Ar(4s)^2(3d)^6$	$^5D_4$	55.85	7.90
Co	27	$Ar(4s)^2(3d)^7$	$^4F_{9/2}$	58.93	7.86
Ni	28	$Ar(4s)^2(3d)^8$	$^3F_4$	58.71	7.63
Cu	29	$Ar(4s)(3d)^{10}$	$^2S_{1/2}$	63.55	7.72
Zn	30	$Ar(4s)^2(3d)^{10}$	$^1S_0$	65.37	9.39
Ga	31	$Ar(4s)^2(3d)^{10}(4p)$	$^2P_{1/2}$	69.72	6.00
Ge	32	$Ar(4s)^2(3d)^{10}(4p)^2$	$^3P_0$	72.59	8.13
As	33	$Ar(4s)^2(3d)^{10}(4p)^3$	$^4S_{3/2}$	74.92	10.00
Se	34	$Ar(4s)^2(3d)^{10}(4p)^4$	$^3P_2$	78.96	9.75
Br	35	$Ar(4s)^2(3d)^{10}(4p)^5$	$^2P_{3/2}$	79.90	11.84
Kr	36	$Ar(4s)^2(3d)^{10}(4p)^6$	$^1S_0$	83.80	14.00
Rb	37	$Kr(5s)$	$^2S_{1/2}$	85.47	4.18
Sr	38	$Kr(5s)^2$	$^1S_0$	87.62	5.69
Y	39	$Kr(5s)^2(4d)$	$^2D_{3/2}$	88.91	6.60
Zr	40	$Kr(5s)^2(4d)^2$	$^3F_2$	91.22	6.95
Nb	41	$Kr(5s)(4d)^4$	$^6D_{1/2}$	92.91	6.77
Mo	42	$Kr(5s)(4d)^5$	$^7S_3$	95.94	7.18

PROBLEMS

32.1 Use the Hund rules to check the spectroscopic descriptions for the elements $Z = 9, 22, 23, 24$, using the configurations in the table.

32.2 Plot the ionization potential as a function of Z. The shell structure of atoms becomes particularly clear from such a figure.

32.3 Use the ionization potentials given in the table to calculate an effective charge seen by the outermost electrons. The Z_{eff} may conveniently be defined by

$$\text{I.E.} = 13.6 Z_{eff}^2/n^2.$$

32.4 The table ends with $Z = 42$. Assuming that the pattern continues in a manner analogous to the period that ended with Kr, what is the Z value for the element that closes the shell (it is xenon)?

32.5 What would you expect the electronic configuration of the first excited state of sodium $(Z = 11)$ to be? Remember that it takes a lot of energy to break up a closed shell. What are the possible spectroscopic terms of the first excited state? Can you estimate the splitting, using what you know about spin-orbit coupling in the hydrogen atom, and using the Z_{eff} obtained in Problem 32.3?

32.6 What are the allowed spectroscopic terms for the ground state of sulfur $(Z = 16)$ and for nickel $(Z = 28)$? [*Hint*: Use the notion of *holes* in closed shells.]

33
Molecules

A molecule is a stable arrangement of electrons and more than one nucleus. The simplest molecules contain two nuclei and they are evidently more complicated objects than atoms in that, after the center of mass is fixed in space, the nuclei are still free to move. This leads to an increase in the number of degrees of freedom (an increase in the number of coordinates required to specify the motion), and except for the simplest cases, a frontal attack by solving the Schrödinger equation is impossible. In our brief discussion we will content ourselves with a qualitative treatment of why molecules can be formed, and with a discussion of the dynamical consequences of the additional degrees of freedom that result from the presence of more than one nucleus.

An important simplification in the dynamics of molecules arises from the fact that nuclei are very much more massive than electrons. Even for the lightest nucleus, hydrogen, $m_p/m_e = 2 \times 10^3$; a consequence of this is that nuclei move much more slowly than the electrons. The motion of the electrons is, in first approximation, determined by stationary nuclei. The total energy of the system thus consists of the electronic energy, which is a function of the position of the nuclear coordinates $\mathscr{E}$ $(\mathbf{R}_1, \mathbf{R}_2, \ldots)$, together with the Coulomb repulsion between the nuclei, which also depends on the $\mathbf{R}_i$. The set of position coordinates $\mathbf{R}_i$ that minimizes the total energy yield a first approximation to the structure of the molecule. The nuclei, in effect, sit in potential wells created by the rapidly moving electrons and the other nuclei. Given such potential wells, motion about the equilibrium positions becomes possible. Nuclei may have vibrational motion and the whole molecule can have a variety of rotational states. When the electronic structure is not in its ground state, the nuclei sit in a new set of potential wells and the motion is different again. Clearly the energy spectrum is much more complex than it is for atoms.

Ionic bonding The study of the electronic energy function $\mathscr{E}$ $(\mathbf{R}_1, \mathbf{R}_2, \ldots)$ is in general quite complicated. We will limit our discussion to a qualitative treatment of some simple cases. Consider first what happens when an atom consisting of a closed shell + one electron (alkali) is brought near an atom with one "hole" in a closed shell (halogen). Under these conditions a simple electronic rearrangement can take place: The electron outside the shell in the first atom fills the hole in the shell of the second atom. The energy required to remove one electron from an orbit around a closed shell is the ionization potential. For sodium this is 5.1 eV. The capture of the electron into a hole in the shell releases some energy. This is the energy that it would take to knock an electron out of the closed shell and it is called the *electron affinity*. The energy is comparable to the ionization energy; for fluorine, for example, it is 3.5 eV. Thus the electron rearrangement costs about 1.6 eV for NaF. Now we are left with two charged atoms, and at distances that are not too small, there is a sizable Coulomb attraction $- e^2/R_{12}$. This is a very crude estimate, since it is only correct for spherically symmetrical charge distributions, and does not take into account that the presence of charges on one atom polarizes the charge distribution on the other. The potential becomes very strongly repulsive at distances of the order of Ångstrom or so. The reason for this is the *Exclusion Principle*. Consider NaF again. Sodium has $Z = 11$ and an electronic configuration $(1s)^2 (2s)^2 (2p)^6 (3s)$, while fluorine has $Z = 9$ and a configuration $(1s)^2 (2s)^2 (2p)^5$. If the nuclei actually coalesced, the charge would be $Z = 20$, and the lowest electronic configuration would be $(1s)^2 (2s)^2 (2p)^6 (3s)^2 (3p)^6$ $(4s)^2$, so that nine electrons would have to be promoted to higher orbitals. This costs a lot of energy. Thus the potential has the form sketched in Fig. 33.1 and it shows that there is a minimum. For NaF the minimum is at a separation of the order of 2 Å. This is an example of *ionic bonding*.

Valence bonding A more common example is that of two atoms, both of which have unfilled shells. Here one speaks of *valence bonding*. As an example, let us consider the H_2 molecule. The wave function must be antisymmetric under the interchange of the electrons and if the spin of the two electrons is zero, the spin wave function is antisymmetric and therefore the spatial two-electron wave function is symmetric. We expect this to be the ground state, because the total spin 1 state will have a more

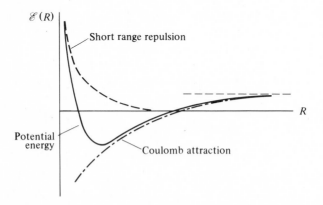

Figure 33.1. Potential energy between Na^+ and F^- ions. At large distances the Coulomb energy vanishes and what is left is just the positive rearrangement energy (the zero is set for uncharged atoms). At close distances there is the repulsion that competes with the Coulomb attraction.

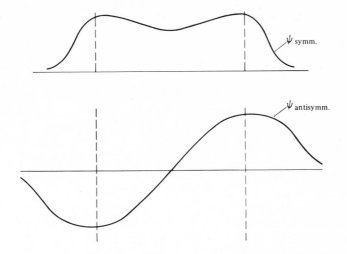

Figure 33.2. Schematic sketch of wave functions that are spatially symmetric and antisymmetric respectively, showing that the latter has more curvature, and thus represents a state with a larger kinetic energy.

curved spatially antisymmetric wave function (Fig. 33.2). The spatially symmetric two-electron wave function has the form

$$\Psi(\mathbf{r}_1, \mathbf{r}_2) = \frac{1}{\sqrt{N}} (\psi_A(\mathbf{r}_1)\psi_B(\mathbf{r}_2) + \psi_A(\mathbf{r}_2)\psi_B(\mathbf{r}_1)) \tag{33.1}$$

for large separation of the nuclei. Here $\mathbf{r}_1$ and $\mathbf{r}_2$ are the electron coordinates, the subscripts A and B denote the protons about which the electrons orbit, and the $\psi(\mathbf{r})$ are hydrogenic ground-state wave functions

$$\psi_A(\mathbf{r}) = (\pi a_0^3)^{-1/2} e^{-|\mathbf{r} - \mathbf{R}_A|/a_0}. \tag{33.2}$$

The normalization constant is evaluated by requiring that

$$N = \int d^3\mathbf{r}_1 \int d^3\mathbf{r}_2 (\psi_A(\mathbf{r}_1)\psi_B(\mathbf{r}_2) + \psi_A(\mathbf{r}_2)\psi_B(\mathbf{r}_1))^2$$

$$= 2 \int d^3\mathbf{r}(\psi_A(\mathbf{r})|^2 \int d^3\mathbf{r}' |\psi_B(\mathbf{r}')|^2$$

$$+ 2 \int d^3\mathbf{r}\,\psi_A(\mathbf{r})\psi_B(\mathbf{r}) \int d^3\mathbf{r}'\,\psi_A(\mathbf{r}')\psi_B(\mathbf{r}')$$

$$\equiv 2(1 + S^2)$$

where

$$S = \int d^3\mathbf{r}\,\psi_A(\mathbf{r})\psi_B(\mathbf{r}). \tag{33.3}$$

The *overlap integral* S depends on the separation of the two nuclei. To get an idea of how the energy depends on the separation of the two protons, we calculate the expectation value of the energy operator in the state (33.1). Although we could not expect this to give a correct description of the system at short distances, one can show quite generally that this expectation value always lies higher than the true energy, so that if our approach predicts binding, then binding will surely occur.

The energy operator contains the kinetic energies of the two electrons, the attraction of each to the two protons, and the electron-electron and proton-proton repulsion terms:

$$\frac{\mathbf{p}_1^2}{2m} - \frac{e^2}{|\mathbf{r}_1 - \mathbf{R}_A|} + \frac{\mathbf{p}_2^2}{2m} - \frac{e^2}{|\mathbf{r}_2 - \mathbf{R}_B|}$$

$$- \frac{e^2}{|\mathbf{r}_1 - \mathbf{R}_B|} - \frac{e^2}{|\mathbf{r}_2 - \mathbf{R}_A|}$$

$$+ \frac{e^2}{|\mathbf{r}_1 - \mathbf{r}_2|} + \frac{e^2}{|\mathbf{R}_A - \mathbf{R}_B|} \ . \tag{33.4}$$

In the above, the nuclei are treated as fixed, so that there is no kinetic energy term to describe their motion. The calculation is tedious. One uses

$$\left(\frac{\mathbf{p}_1^2}{2m} - \frac{e^2}{|\mathbf{r}_1 - \mathbf{R}_A|} \right) \psi_A(\mathbf{r}_1) = E_1 \psi_A(\mathbf{r}_1) \tag{33.5}$$

and symmetry. The results only depend on the separation between the two protons R_{AB}. What emerges from the calculation is a set of contributions, listed below:

i) the Coulomb repulsion between the two protons

$$\frac{e^2}{R_{AB}}, \tag{33.6}$$

ii) the energy of the two hydrogen atoms in the limit of infinite separation

$$2E_1 = -mc^2\alpha^2, \tag{33.7}$$

iii) a term of the form

$$\frac{e^2}{1 + S^2} \int\int d^3\mathbf{r}_1 d^3\mathbf{r}_2 \frac{|\psi_A(\mathbf{r}_1)|^2 |\psi_B(\mathbf{r}_2)|^2}{|\mathbf{r}_1 - \mathbf{r}_2|} \tag{33.8}$$

which represents the repulsion between the two electron charge distributions about the two protons,

iv) an exchange term involving the electron-electron repulsion

$$\frac{e^2}{1 + S^2} \int d^3\mathbf{r}_1 d^3\mathbf{r}_2 \frac{\psi_A(\mathbf{r}_1)\psi_A(\mathbf{r}_2)\psi_B(\mathbf{r}_1)\psi_B(\mathbf{r}_2)}{|\mathbf{r}_1 - \mathbf{r}_2|}, \tag{33.9}$$

v) attractive terms involving the attraction of the electron cloud about one proton to the other

$$- \frac{2e^2}{1 + S^2} \int d^3\mathbf{r} \frac{|\psi_A(\mathbf{r})|^2}{|\mathbf{r} - \mathbf{R}_B|}, \tag{33.10}$$

and an overlap term weighted by the distance of one of the electrons to the proton

$$- \frac{2e^2 S}{1 + S^2} \int d^3\mathbf{r} \frac{\psi_A(\mathbf{r})\psi_B(\mathbf{r})}{|\mathbf{r} - \mathbf{R}_A|}. \tag{33.11}$$

The energy will be large and repulsive at short distances because of (33.6), and it will be attractive provided (33.10) and (33.11) are numerically large. This will be the case if the wave functions $\psi_A(\mathbf{r})$ and $\psi_B(\mathbf{r})$ overlap significantly. This can only take place in the region between the nuclei. The shape of $\mathscr{E}(R)$, the energy as a function of the proton-proton separation can roughly be obtained from the following considerations: In the limit of large R, the system consists of two hydrogen atoms and the energy is $-2 \times 13.6 = -27.2\,\text{eV}$. In the limit that $R \to 0$, we effectively have a helium nucleus (it is irrelevant that it has no neutrons), and the energy is known to be $-78.5\,\text{eV}$. We interpolate between these two values with a smooth curve, and add on the Coulomb repulsion e^2/R to get Fig. 33.3. The experimental value of the separation of the two protons is $0.74\,\text{Å}$ and the binding energy is $-4.75\,\text{eV}$. The result of a calculation with the approximate wave function (33.1) yields a minimum in the $\mathscr{E}(R)$ curve at $0.87\,\text{Å}$, and a binding energy of $-3.14\,\text{eV}$. The discrepancy can be attributed to the poor trial wave function. Although the trial wave function, a linear combination of atomic orbitals (LCAO in molecular literature), is not particularly good, the qualitative features that emerge from our discussion are correct. These are as follows:

a) there is binding only when the electrons are localized between the nuclei and overlap significantly;

b) this can occur only when the electrons are in an antisymmetric $S = 0$ total spin state;

c) a large overlap means that molecules are small, of atomic dimensions. We remark parenthetically that atomic calculations show that orbital radii of the outermost electronic charge distributions cluster around the value of $1\,\text{Å}$, and only occasionally get as large as $2\,\text{Å}$.

W. Heitler and F. London originated the application of quantum mechanics to molecular bonding, and they have generalized the above considerations. In the Heitler-London, or *valence-bond* approach, one associates valence bonds with electrons that overlap in a spin $S = 0$ state. In counting bonds, the following simplifications occur:

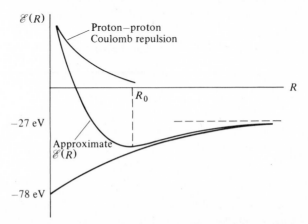

Figure 33.3. Approximate shape of the electronic energy with the proton–proton Coulomb repulsion for the H_2 molecule.

a) only electrons not in closed shells can form bonds, and then only those that are not too tightly bound to the nucleus;

b) if two electrons outside a closed shell are in a spin $S = 0$ state, they are said to be paired, and they do not contribute to the formation of valence bonds. The reason is that when a third electron is introduced into the system, its spin must be parallel to that of one of the paired electrons, and therefore the wave function must be antisymmetric in the spatial coordinates of these two electrons. An antisymmetric spatial wave function always implies a higher energy.

Vibrations in molecules One can also discuss excited electronic states in molecules, but this is outside the range of our brief survey. We shall maintain the nuclei in the lowest electronic state, and consider next the possible motions with a given electronic cloud. As above, our discussion will be limited to diatomic molecules. We discuss vibrations first. In first approximation the nuclei are located at fixed positions, with separation determined by the minimum of the electronic + Coulomb energy

$$V(R) = \mathscr{E}(R) + \frac{Z_1 Z_2 e^2}{R}. \tag{33.12}$$

In the vicinity of the minimum we approximate the smooth function $\mathscr{E}(R)$ by a straight line

$$\mathscr{E}(R) = \mathscr{E}(R_0) + \eta(R - R_0). \tag{33.13}$$

The location of the minimum is given by

$$\frac{d}{dR}\left(\eta R + \frac{Z_1 Z_2 e^2}{R}\right) = 0, \tag{33.14}$$

so that

$$\eta = \frac{Z_1 Z_2 e^2}{R_0^2}. \tag{33.15}$$

If the linear form (33.13) is approximately correct, then the potential energy in the vicinity of the equilibrium point is determined by the Coulomb repulsion alone. In first approximation,

$$\begin{aligned} V(R) &= V(R_0) + \tfrac{1}{2}(R - R_0)^2 (d^2 V/dR^2)_{R=R_0} \\ &= V(R_0) + \tfrac{1}{2}(R - R_0)^2 (2Z_1 Z_2 e^2/R_0^3). \end{aligned} \tag{33.16}$$

We may write this in the form

$$V(R) = V(R_0) + \tfrac{1}{2}(R - R_0)^2 M_{\mathrm{red}}\omega^2$$

where M_{red} is the reduced mass of the two nuclei, so that

$$\omega^2 = \frac{2Z_1 Z_2 e^2}{M_{\mathrm{red}} R_0^3}. \tag{33.17}$$

The potential is harmonic, and the spectrum of the one-dimensional vibrational motion is the familiar simple harmonic oscillator spectrum

$$E_v = (v + \tfrac{1}{2})\hbar\omega \qquad v = 0, 1, 2, \ldots. \tag{33.18}$$

Typically $R_0 \cong 1 \text{ Å} = 2a_0$ (a_0 is the Bohr radius in the lowest state), and we also know that $e^2/2a_0 = m_e c^2 \alpha^2/2$. Hence

$$\omega^2 \simeq 2Z_1 Z_2 \frac{e^2}{2a_0} \frac{1}{(2a_0)^2 M_{\text{red}}}$$

$$= 2Z_1 Z_2 \frac{m_e c^2 \alpha^2}{2} \frac{1}{4 M_{\text{red}} (\hbar/m_e c \alpha)^2}$$

$$= \frac{1}{4} Z_1 Z_2 \left(\frac{m_e}{M_{\text{red}}} \right) \frac{m_e^2 c^2 \alpha^4}{\hbar^2} \tag{33.19}$$

Thus

$$\hbar \omega = \frac{1}{2} m_e c^2 \alpha^2 \left(Z_1 Z_2 \frac{m_e}{M_{\text{red}}} \right)^{1/2}. \tag{33.20}$$

Typically, vibrational energy-level splittings are of the order $(m_e/M)^{1/2} = 10^{-2}$ times smaller than the typical electronic splittings.

Rotations of molecules Molecules can also rotate as rigid bodies. The spectrum for a spherical body is given by the eigenvalues of the operator

$$H = \mathbf{L}^2/2I. \tag{33.21}$$

Strictly speaking, the molecule is more accurately described by the energy operator for an asymmetric top, in which case one has to take into account motion about the symmetry axis of the molecule, as well as rotations of the top as a whole in space. This becomes very complicated, and for our qualitative discussion we discuss the consequences of an energy spectrum of the form

$$E = \frac{J(J+1)\hbar^2}{2M_{\text{red}} R_0^2}. \tag{33.22}$$

Typical splittings are of the order of

$$\Delta E \simeq \frac{\hbar^2}{M_{\text{red}} R_0^2} \simeq \frac{\hbar^2}{M_{\text{red}}} \left(\frac{m_e c \alpha}{\hbar} \right)^2 \simeq \left(\frac{m_e}{M_{\text{red}}} \right) m_e c^2 \alpha^2, \tag{33.23}$$

that is, of the order of 10^{-4} of electronic splittings. The radiation in transitions has wavelength in the millimeter range.

If we ignore electronic excitations, then the energy levels of diatomic molecules can be described in terms of the vibrational quantum number v and the angular momentum J. Figure 33.4 shows the energy-level structure. The existence of *selection rules* in the radiative transitions between energy levels, of the form

$$\Delta v = \pm 1, 0$$

$$\Delta J = \pm 1, \tag{33.24}$$

determines the spectrum. For example, for a rigid rotator,

$$\nu(J \to J-1) = \frac{E(J) - E(J-1)}{2\pi\hbar} = \frac{\hbar^2 \cdot 2J}{2I \cdot 2\pi\hbar} = \frac{\hbar J}{2\pi I}, \tag{33.25}$$

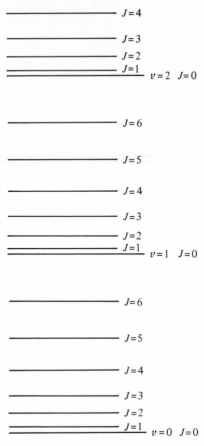

Figure 33.4. Schematic picture of vibrational and rotational spectrum of a diatomic molecule.

leading to a series of equally spaced spectral lines. The spacing may be used to determine the moment of inertia of the molecule

$$\Delta \nu = \hbar/2\pi I. \tag{33.26}$$

For a molecule like H_2, with a nuclear separation of the order of 0.75 Å and with a reduced mass $M_{red} = M_p/2$, the spacing should be of magnitude

$$\Delta \nu = \frac{1.05 \times 10^{-27}}{3.14 \times 1.6 \times 10^{-24}(0.75 \times 10^{-8})^2}$$

$$= 3.7 \times 10^{12} \text{ Hz}$$

and the wavelengths

$$\lambda = \frac{c}{\nu} = \frac{3 \times 10^{10}}{3.7 \times 10^{12} J} = 0.8 \times 10^{-2}/J \text{ cm}.$$

Effects of Exclusion Principle A special situation arises for homonuclear molecules, that is, for molecules in which the two nuclei are identical. The Pauli Principle requires that the nuclear wave function be symmetric (for integer spin nuclei) or antisymmetric (for half odd integer spin nuclei) under the interchange of the nuclei. The eigenfunctions of the rotational energy operator behave like the Y_{LM} functions under the interchange of the nuclei, so that, under interchange of two antipodal points on a sphere ($\theta \rightarrow \pi - \theta$, $\phi \rightarrow \pi + \phi$), the eigenfunctions are multiplied by $(-1)^L$. Since there is another selection rule that requires that the spin state does not change in a radiative transition,

$$\Delta S = 0, \tag{33.27}$$

the L values must either remain even, or they must remain odd. In any case, it is not possible to satisfy the $\Delta L = 1$ rule and thus there are no rotational transitions.

In the special case of homonuclear molecules, the selection rule $\Delta v = \pm 1$ also forbids vibrational transitions to first approximation. Transitions are still observed, albeit very faintly, since departures from simple harmonic motion (anharmonicity effects) allow transitions with $\Delta v \geqslant 2$. In general, beside the rotational transitions and the vibrational transitions, there are also vibration-rotation transitions, that is, transitions from (v, J) to (v', J') with both of these changing by unity. These manifest themselves in fine structure for each of the vibrational transition lines (Fig. 33.5).

Raman effect An important tool in the study of molecules is an effect discovered in 1928 by C.V. Raman and independently by G. Landsberg and L. Mandelstam in the scattering of light by molecules. In addition to light scattered without a change of frequency (Rayleigh scattering, discussed in Chapter 14), it was found that light was also scattered with a frequency shift. Incident light with frequency ν gives rise to lines with frequency ν as well as $\nu \pm \nu_0$. The ν_0 are independent of the incident frequency, and the correct interpretation of the Raman effect is that it represents a situation in which the molecule changes its energy state during the scattering process. If a molecule is in a vibrational state characterized by v, say, it may during the scattering process make a transition to the state characterized by $v + 1$. Thus the molecule absorbs energy $h\nu_0$, where ν_0 is the vibrational frequency, and the emitted light has that much less energy. Its frequency is therefore decreased by ν_0. The molecules in the state $v + 1$ may make a transition to the state v during the scattering process, and the energy is taken up by the scattered photon, whose frequency will therefore be raised to $\nu + \nu_0$. In a gas of molecules, the number of molecules in the vibrational state characterized by the quantum number v falls like $e^{-(h\nu_0/kT)v}$ and therefore there are more molecules with lower v values. Thus there will be more transitions going

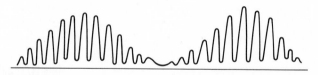

Figure 33.5. Sketch of vibrational-rotational transition lines.

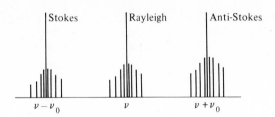

Figure 33.6. Raman spectrum (schematic).

upwards than downwards, and thus the line with the downward shifted frequency, $\nu - \nu_0$, will be more intense than the line with frequency $\nu + \nu_0$.

Surrounding the Rayleigh line, the downward-shifted Stokes line and the upward-shifted anti-Stokes line (the usual terminology) are bands of lines in which the Raman effect proceeds through the excitation or de-excitation of rotational lines during the inelastic photon-scattering process (Fig. 33.6). These even appear for homonuclear molecules, because the scattering process does not obey the $|\Delta J| = 1$ selection rule. Scattering is a two-step process: A photon is absorbed and re-emitted, but it does not proceed through the intermediate step of excitation of a real state of the molecule — that would be resonant scattering or resonant fluorescence; it proceeds through an intermediate step in which any state of the molecule, whatever its energy, can participate. One speaks of the excitation of "virtual" states, and it is then possible to satisfy the $|\Delta J| = 1$ rule in each part of the scattering process. Since there are states in the homonuclear molecule that have odd J values when the rotational series have even J values (and vice versa), such as, for example, "virtual" states in which the electronic cloud is distorted, the overall selection rule for the Raman effect is

$$|\Delta J| = 0, 2. \tag{33.28}$$

Raman spectroscopy is usually done with light of a frequency such that it is below the threshold for the excitation of electronic states, but high enough to excite a reasonably large number of vibrational states. It is of enormous importance in the study of molecules.

Molecular spectroscopy in astrophysics In recent years astrophysicists have become very interested in molecular spectroscopy, because absorption spectra are a good way of studying the presence of molecules in the galaxy and in extragalactic space. As an example, cyanogen is a molecule that has a visible absorption line (that is, it will absorb light with that wavelength) at 3874 Å because the corresponding frequency describes the energy difference between the ground state and the first electronic excitation state. Since the electronic states have rotational states associated with them this line is split. For example, the $J = 0 \to J = 1$ line has wavelength 3874.608 Å, the $J = 1 \to J = 2$ line has wavelength 3873.998 Å, and the $J = 2 \to J = 3$ line has wavelength 3873.369 Å (all of these correspond to absorption, that is, transition from a lower J *up* to a higher J); there is also the absorption line $J = 1 \to J = 0$ at $\lambda = 3875.763$ Å. Thus the molecule CN has a well-defined signature, and when these

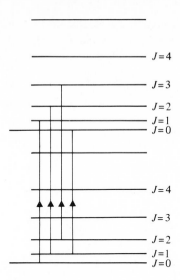

Figure 33.7. Absorption spectrum in CN. Only the lines mentioned in the text are sketched in.

absorption lines were found in the spectrum of a particular star, it was possible to infer the presence of CN in an interstellar cloud between us and the star. Because of the presence of the $J = 1 \rightarrow J = 2$ line (see Fig. 33.7), one could deduce that not all the CN molecules were in the ground state, but that some had somehow been raised to the first rotational excitation state. From the relative strength of the two absorption lines one can estimate the number of $J = 1$ molecules compared with the number of $J = 0$ molecules, and thus the temperature. The temperature turned out to be $2.3°$K, and in the absence of other means of excitation, it was realized, many years after the original discovery by McKellar of the CN radicals, that the temperature was actually the blackbody radiation background discovered in 1965.

As with all of the topics treated in this part of the book, the material discussed barely gives us a glimpse of the vast number of applications and the detailed physical effects that provide fine deviations from the gross structure, and thus important new information. It is nevertheless encouraging how the basic ideas of quantum mechanics allow us to understand what goes on.

NOTES AND COMMENTS

1. Much more material than we were able to cover is discussed in M. Karplus and R.N. Porter, *Atoms and Molecules*, W.A. Benjamin, Menlo Park, Calif., 1970, M.W. Hanna, *Quantum Mechanics in Chemistry*, W.A. Benjamin, 1965, and the many physical chemistry books listed in the reference sections of these books. Some topics not discussed here are treated in U. Fano and L. Fano, *Physics of*

Atoms and Molecules, University of Chicago Press, Chicago, Ill., 1972, and also in many of the more advanced quantum mechanics textbooks.

2.   The absence of rotational spectra in homopolar molecules is a particularly dramatic confirmation of the Pauli spin-symmetry connection when one realizes that whereas there are no such spectra in an O_2 molecule, the molecule in which two different isotopes of oxygen occur, for example, ^{16}O and ^{17}O, which mechanically differs from O_2 only in a few percent change of the moment of inertia, shows the complete rotational spectrum, just because ^{16}O and ^{17}O are *not identical*.

3.   The computation of the potential energy for the H_2 molecule is very crude. If one writes out a more complicated trial wave function that depends on a large number of parameters, calculates the molecular potential energy in terms of those parameters, and then minimizes, one obtains an extremely accurate value for the separation and energy. A discussion of the use of the high-speed computer in chemistry research, with many beautiful diagrams, may be found in A.C. Wahl, "Chemistry by Computer," *Scientific American*, 222 (April 1970): 54.

PROBLEMS

33.1 To get a quantitative feeling for ionic bonding, consider the NaCl molecule. You are given that (a) the electron rearrangement energy, that is, the ionization potential of sodium less the electron affinity of Cl is 1.49 eV; (b) the equilibrium separation of the nuclei is 2.36 Å; (c) the potential energy in the vicinity of the equilibrium separation has the form

$$\tfrac{1}{2} k(R - R_0)^2$$

with $k = 1.09 \times 10^5 \, \mathrm{erg \, cm^{-2}}$; and (d) the shape of the repulsive potential may be approximated by

$$A e^{-\alpha R}.$$

Use these facts to determine the binding energy, and compare your answers with experiment. The experimental value is 4.22 eV.

33.2 Given the above data, what is the energy difference between different vibrational levels in NaCl? What is the wavelength of radiation emitted in a $J = 1 \rightarrow J = 0$ rotational transition?

33.3 The rotational motion of molecules has an effect on the equilibrium position of the nuclei. If R_0 is the separation for zero angular momentum, calculate the new separation by minimizing

$$E(R) = \frac{1}{2} M_{\mathrm{red}} \omega^2 (R - R_0)^2 + \frac{\hbar^2 J(J + 1)}{2 M_{\mathrm{red}} R^2},$$

assuming that the change from R_0 is small. How is the spectrum of the molecule affected? [*Hint*: The moment of inertia changes; calculate how.]

33.4 The wavelength of the $v = 1 \rightarrow v = 0$ transition in carbon monoxide CO is 2.93×10^{-3} cm. At what temperatures would you expect the $v = 1$ state in CO gas to be 1% occupied?

33.5 The $J = 0 \rightarrow J = 1$ absorption line in CO has a wavelength of 2.603 mm. Use this to calculate the moment of inertia of the CO molecule and the equilibrium separation.

33.6 Use the data provided in the text on the CN molecule to calculate the moment of inertia in the lowest electronic state, and thus the internuclear separation. Calculate these same numbers for the first excited electronic state.

33.7 There exists in nature a particle known as the *muon*. It is just a heavy electron with mass $m_\mu = 207\, m_e$. It decays in 10^{-6} seconds, so that experiments with it are difficult. Suppose there existed a molecule that is an analog of H_2^+ (two protons + 1 muon). (a) What would be the size of such a molecule? (b) If a rotational state would be excited, what would be the wavelength of the radiation emitted in the transition to the ground state?

33.8 Consider the H_2 molecule. The two nuclei (protons) have spin 1/2 and can therefore be in a total spin $S = 0$ or an $S = 1$ state. (a) What is the orbital angular momentum of the two-nucleon system in the lowest energy state for the two values of the total spin? (b) What is the wavelength of the radiation emitted in the transition from the lowest rotational excitation in the two cases?

34
Electromagnetic Transitions, Selection Rules, and Radioactivity

The ground states of atoms and molecules are particularly relevant to the chemical properties of these substances. The energy-level structure, the spectrum of these systems, manifests itself in the radiation emitted when the systems make transitions from one energy level to another. The postulate of Bohr that in the transition from an energy level E_1 to an energy level E_2 radiation is emitted or absorbed with frequency $\nu = (E_1 - E_2)/h$ or $(E_2 - E_1)/h$, has been shown to be generally true, and is a consequence of the laws of quantum mechanics. Furthermore, as has frequently been observed, radiation is favored when the angular momenta associated with the energy levels differ by unity. Thus spectroscopy not only yields the energy levels but also their quantum numbers.

Time-dependent perturbation theory To understand the Bohr frequency rule, which, written in the form

$$E_1 = E_2 \pm h\nu, \tag{34.1}$$

just expresses energy conservation, requires a study of the time-dependent Schrödinger equation. When done properly this can become quite lengthy, so that we limit ourselves to a rough outline, which is nevertheless instructive because it shows us how the selection rules emerge from the formalism. Consider, for example, a hydrogen atom, whose energy operator is

$$H_0 = \frac{p^2}{2m} - \frac{e^2}{r} \tag{34.2}$$

and for which the eigenvalues and eigenfunctions E_n and $u_n(\mathbf{r})$ — we leave off other subscripts for brevity — are known. These quantities satisfy the equation

$$H_0 u_n(\mathbf{r}) = E_n u_n(\mathbf{r}). \tag{34.3}$$

Let us now consider an atom that at time $t = 0$ is in a particular excited state, say $u_1(\mathbf{r})$, and place it in an oscillating electric field. If the field is uniform over the dimensions of the atom, that is, if the size of the atom is much smaller than the wavelength associated with the radiation field, then the electron experiences a force that can be described by the potential

$$V(\mathbf{r}, t) = e\mathbf{E} \cdot \mathbf{r} \cos \omega t. \tag{34.4}$$

This is just the classical expression for the interaction of a dipole $\mathbf{d} = (-e)\mathbf{r}$ with the electric field, which has the form $V = -\mathbf{d} \cdot \mathbf{E}$. We check that the expression is correct by calculating the force

$$\mathbf{F} = -\boldsymbol{\nabla} V(\mathbf{r}, t) = -e\mathbf{E}\cos\omega t. \tag{34.5}$$

The system is now described by the time-dependent Schrödinger equation

$$i\hbar\frac{\partial}{\partial t}\psi(\mathbf{r}, t) = (H_0 + V(\mathbf{r}, t))\psi(\mathbf{r}, t) \tag{34.6}$$

with the initial condition

$$\psi(\mathbf{r}, 0) = u_1(\mathbf{r}). \tag{34.7}$$

The expansion theorem At this point we must make use of a general result of the mathematical theory of quantum mechanics, that any function $f(\mathbf{r}, t)$ can be expanded in terms of the eigenfunctions of any hermitian operator, such as H_0. The precise conditions under which the series is well behaved are not important to us at this stage. The expansion theorem is rather obvious for the case of a one-dimensional function defined over a region $(0 \leqslant x \leqslant a)$, if, for H_0 we take the energy operator of an infinite box. In that case the eigenfunctions have the form $\sin n\pi x/a$, and the statement that

$$f(x, t) = \sum_{n=1}^{\infty} a_n(t)\sin\frac{n\pi x}{a} \tag{34.8}$$

is just the statement that $f(x, t)$ can be expanded in a Fourier series. The interpretation of the coefficients $a_n(t)$ is that

$$P_n(t) \equiv |a_n(t)|^2 \tag{34.9}$$

is the probability that a system described by the function $f(x, t)$ will be found to have energy E_n when an energy measurement is made. Recall that an energy measurement can yield only one of the eigenvalues $E_1, E_2, E_3, \ldots$, and the probability of finding a particular one depends on what the state, described by $f(x, t)$, is.

We use the expansion theorem to write

$$\psi(\mathbf{r}, t) = \sum_{n} a_n(t)u_n(\mathbf{r})e^{-iE_n t/\hbar}. \tag{34.10}$$

The initial condition states that

$$a_1(0) = 1; \qquad a_k(0) = 0 \qquad (k \neq 1). \tag{34.11}$$

The coefficients $a_n(t)$ will now be determined by Eq. (34.6). Substituting Eq. (34.10) into Eq. (34.6), we obtain for the left-hand side

$$\sum_{n=0}^{\infty} i\hbar\frac{da_n(t)}{dt}u_n(\mathbf{r})e^{-iE_n t/\hbar} + \sum_{n=0}^{\infty} a_n(t)E_n u_n(\mathbf{r})e^{-iE_n t/\hbar}.$$

On the right-hand side we get, using Eq. (34.3)

$$(H_0 + V)\psi(\mathbf{r}, t) = \sum_{n=0}^{\infty} a_n(t)E_n u_n(\mathbf{r})e^{-iE_n t/\hbar}$$

$$+ \sum_{n=0}^{\infty} a_n(t)V(\mathbf{r}, t)u_n(\mathbf{r})e^{-iE_n t/\hbar}.$$

Thus canceling corresponding terms on the two sides of the equation we get

$$\sum_{n=0}^{\infty} i\hbar \frac{da_n(t)}{dt} u_n(\mathbf{r})e^{-iE_n t/\hbar} = \sum_{n=0}^{\infty} a_n(t)V(\mathbf{r}, t)u_n(\mathbf{r})e^{-iE_n t/\hbar}. \quad (34.12)$$

This equation can be reduced to a simpler looking set of equations by multiplying both sides of the equation by

$$u_m^*(\mathbf{r})e^{iE_m t/\hbar}$$

and integrating over all space. The left side simplifies because of the orthogonality property of the eigenfunctions:

$$\int u_m^*(\mathbf{r})u_n(\mathbf{r})d^3r = \begin{cases} 1 & m = n \\ 0 & m \neq n. \end{cases} \quad (34.13)$$

The equation then becomes

$$i\hbar \frac{da_m(t)}{dt} = \sum_{n=0}^{\infty} a_n(t)e^{-i(E_n - E_m)t/\hbar} \int d^3r\, u_m^*(\mathbf{r})V(\mathbf{r}, t)u_n(\mathbf{r}). \quad (34.14)$$

This is unfortunately an infinite set of equations, all coupled together because of the integral. We can make progress if the potential $V(\mathbf{r}, t)$ is weak, that is, if only small changes in the system occur because of its presence. Let us write out the sum on the right side of Eq. (34.14) in detail. For $m \neq 1$, it is

$$a_1(t)e^{-i(E_1 - E_m)t/\hbar} \int d^3r\, u_m^*(\mathbf{r})V(\mathbf{r}, t)u_1(\mathbf{r})$$

$$+ \sum_{k \neq 1} a_k(t)e^{-i(E_k - E_m)t/\hbar} \int d^3r\, u_m^*(\mathbf{r})V(\mathbf{r}, t)u_k(\mathbf{r}). \quad (34.15)$$

Since initially all the $a_k(t)$ vanish, and they cease to be zero only because of the presence of the *small* perturbation $V(\mathbf{r}, t)$, all terms involving products of $a_k(t)$ and the integral involving $V(\mathbf{r}, t)$ are extra small, of second order in the smallness parameter that characterizes the perturbing potential. Thus only the first term in the series (34.15) is present, and the simplified equation reads

$$i\hbar \frac{da_m(t)}{dt} = e^{-i(E_1 - E_m)t/\hbar} \int d^3r\, u_m^*(\mathbf{r})V(\mathbf{r}, t)u_1(\mathbf{r}). \quad (34.16)$$

We have replaced $a_1(t)$ by $a_1(0) = 1$, since the change in a_1 is also small, so that it should not be taken into account to this order. With the potential energy (34.4) this becomes

$$i\hbar \frac{da_m(t)}{dt} = e^{-i(E_1 - E_m)t/\hbar} \cos \omega t \int d^3r\, u_m^*(\mathbf{r})e\mathbf{E} \cdot \mathbf{r}u_1(\mathbf{r}) \quad (34.17)$$

and it is readily integrated

$$a_m(t) = \frac{e}{i\hbar} \left(\int d^3r\, u_m^*(\mathbf{r})\mathbf{E} \cdot \mathbf{r}u_1(\mathbf{r}) \right) \int_0^t dt'\, \cos \omega t' e^{-i(E_1 - E_m)t'/\hbar}. \quad (34.18)$$

We note that if the right-hand side does not vanish, then after a time $a_m(t)$ will no longer be zero. The probability that the system is in the state "m" after a time t is given by

$$P_m(t) = |a_m(t)|^2. \tag{34.19}$$

Transition rate The rate of change of probability with time is called the transition rate. Thus

$$R_{1 \to m} = \frac{d}{dt} |a_m(t)|^2. \tag{34.20}$$

If we write

$$\omega_0 \equiv \frac{E_1 - E_m}{\hbar}, \tag{34.12}$$

then the time dependence is described by the factor

$$\frac{1}{\hbar^2} \frac{d}{dt} \left| \int_0^t dt' \cos \omega t' e^{-i\omega_0 t'} \right|^2. \tag{34.22}$$

We will return to the implications of this, but first turn to the time-independent factor in Eq. (34.18), the so-called *matrix element*

$$M_{m1} \equiv e \int d^3 r u_m^*(\mathbf{r}) \mathbf{E} \cdot \mathbf{r} u_1(\mathbf{r}), \tag{34.23}$$

from which the selection rules will be derived.

Selection rules To be specific, let us consider the initial state, labeled "1," to be the ground state of the hydrogen atom, and let us consider the state labeled "*m*" to be a particular state with the quantum numbers (n, l, m). Thus

$$u_1(\mathbf{r}) = R_{10}(r) Y_{00}(\theta, \phi)$$
$$u_m(\mathbf{r}) = R_{nl}(r) Y_{lm}(\theta, \phi). \tag{34.24}$$

Let us choose the electric field to be polarized in the $x - y$ plane. This is the plane in which the electric field associated with a wave propagating in the z direction would be polarized. Thus

$$\mathbf{E} \cdot \mathbf{r} = x E_x + y E_y = r \sin \theta (E_x \cos \phi + E_y \sin \phi). \tag{34.25}$$

The integral now becomes

$$\int_0^\infty dr r^2 \int_0^\pi \sin \theta \, d\theta \int_0^{2\pi} d\phi R_{nl}(r) r R_{10}(r)$$

$$N_{lm} P_l^{|m|}(\theta) \sin \theta \, e^{-im\phi}$$

$$(E_x \cos \phi + E_y \sin \phi)(4\pi)^{-1/2}. \tag{34.26}$$

We have replaced Y_{00} by its value $(4\pi)^{-1/2}$, and we wrote for Y_{lm} the function $P_l^m(\theta) e^{im\phi}$, with the normalization factor denoted by N_{lm}. We now observe the following:

a) The radial integration will not vanish because of the factor r that comes from the potential energy, that is, the orthogonality of the radial eigenfunctions does not come into play.

b) Since $\cos\phi$ and $\sin\phi$ can be decomposed into $e^{\pm i\phi}$, the azimuthal integration involves

$$\int_0^{2\pi} d\phi \, e^{-im\phi} e^{\pm i\phi} = \begin{array}{ll} 2\pi & m = \pm 1 \\ 0 & \text{other } m. \end{array} \tag{34.27}$$

We are thus led to a selection rule

$$\Delta m = \pm 1. \tag{34.28}$$

This rule depends on the fact that we chose the wave propagation vector (perpendicular to **E**) to lie in the same direction as the usual z-axis, the quantization axis. Had we chosen **E** to lie in the z-direction, then $\mathbf{E} \cdot \mathbf{r}$ would be $Er \cos\theta$, and we would get

$$\Delta m = 0. \tag{34.29}$$

We will return to this point in a moment. The polar integration over θ involves

$$\int_0^\pi \sin\theta \, d\theta \, P_l^{|m|}(\theta) \sin\theta = \int_0^\pi \sin\theta \, d\theta \, P_l^1(\theta) P_1^1(\theta)$$

$$\propto \delta_{l1} \tag{34.30}$$

The orthogonality of the associated Legendre polynomials tells us that this integral will vanish *unless* $l = 1$. We are thus led to the orbital angular momentum selection rule

$$\Delta l = 1 \tag{34.31}$$

since the initial l-value was zero. More generally, if we consider the transition from a state (n, l, m) to a state (n', l', m'), we get the integral

$$\int_0^\pi \sin\theta \, d\theta \int_0^{2\pi} d\phi \, Y_{lm}^*(\theta, \phi) r \sin\theta \, (E_x \cos\phi + E_y \sin\phi) Y_{l'm'}(\theta, \phi)$$

which involves the integrals

$$\int_0^\pi \sin\theta \, d\theta \int_0^{2\pi} d\phi \, Y_{lm}^*(\theta, \phi) Y_{1, \pm 1}(\theta, \phi) Y_{l'm'}(\theta, \phi). \tag{34.32}$$

One can show quite generally that these imply that

$$m = m' \pm 1 \tag{34.33}$$

and

$$l = \begin{cases} l' + 1 \\ l' \\ l' - 1. \end{cases} \tag{34.34}$$

The transition $l' \to l$ is not possible by *parity rules*: Under reflection $x \to -x, y \to -y, z \to -z$, $\mathbf{E} \cdot \mathbf{r}$ changes sign. Under these reflections $Y_{lm} \to (-1)^l Y_{lm}$, and thus in Eq. (34.32) we cannot have $l' = l$. We may state this in the form of a selection rule

$$\textit{Parity must change.} \qquad \tag{34.35}$$

The selection rules have nothing to do with the radial wave function of the system making the transition: They are *a consequence of angular momentum conserv-*

ation. The electromagnetic field carries angular momentum as well as momentum. In terms of photons, *we associate spin 1 with the photon*. In the limit that the radiation field is described by an electric field that is uniform over the size of the radiation source (the atom), there is no orbital angular momentum associated with the radiation field, and thus angular momentum conservation implies that the atom must change its angular momentum by one unit. The reason for the $\Delta m = \pm 1$ selection rule is that the photon, though having spin 1, does not have its full complement of m-values $1, 0, -1$. Because of the masslessness of the photon, it turns out that the m-values are restricted to ± 1, and this explains the selection rule (34.33). The absence of the $m = 0$ state for the photon translates into the classical observation that the electromagnetic field is polarized transverse to the direction of propagation of the electromagnetic wave. We used this transversality in Eq. (34.25) by choosing the direction of propagation of the electric field wave to be in the z-direction, although in our approximation of a uniform electric field, *the electric dipole approximation*, the direction of propagation never appears.

Selection rules are not absolute. If the electromagnetic field is associated with a photon traveling in the direction described by the vector $\mathbf{k}$, with wavelength λ, then the spatial dependence of the electric field is described by

$$\mathbf{E} = \mathbf{E}_0 \, e^{i\mathbf{k}\cdot\mathbf{r}}, \tag{34.36}$$

where $|\mathbf{k}| = 2\pi/\lambda$, so that instead of $\mathbf{E} \cdot \mathbf{r}$, we should really consider

$$\mathbf{E}_0 \cdot \mathbf{r}(1 + i\mathbf{k}\cdot\mathbf{r} - \tfrac{1}{2}(\mathbf{k}\cdot\mathbf{r})^2 - \ldots) \tag{34.37}$$

in the matrix element (34.23). Thus if $\Delta l \neq 1$ for the atomic states, the leading term yields zero, but the next term will give a contribution of the form

$$ie \int d^3 r u_m^*(\mathbf{r})\mathbf{E}_0 \cdot \mathbf{r}\mathbf{k}\cdot\mathbf{r}u_1(\mathbf{r}) \tag{34.38}$$

which will not vanish when $\Delta l = 2$. We call this an *electric quadrupole* transition. If $\Delta l = 3$, the third term in (34.37) will contribute, and we call it an *electric octupole* transition, and so on.

Size of suppression of forbidden transitions The magnitude of an electric quadrupole matrix element is quite a bit smaller than that for an electric dipole transition. The ratio is of order kR, where R is a measure of the size of the system that is radiating. Since rates are proportional to squares of the transition amplitude,

$$\frac{\text{Rate}\,(\Delta l = 2)}{\text{Rate}\,(\Delta l = 1)} \cong (kR)^2. \tag{34.39}$$

To estimate this for atomic systems, we note that

$$k = \frac{\omega}{c} = \frac{\hbar\omega}{\hbar c} = \frac{\Delta E}{\hbar c} \sim \frac{m_e(c\alpha Z)^2}{\hbar c}$$

and

$$R \simeq \frac{\hbar}{m_e c\alpha Z},$$

so that

$$(kR)^2 \simeq (Z\alpha)^2. \tag{34.40}$$

Thus the intensity of $\Delta l = 2$ transitions is reduced, relative to the intensity of the allowed electric dipole transitions by a factor of 10^{-4} or so. The $\Delta l = 3$ electric octupole transitions are suppressed by another factor of $(Z\alpha)^2$. Incidentally, the parity selection rules change: It is evident from (34.38) that the parities of the initial and final states must be *the same* for electric quadrupole transitions, and opposite for electric octupole transitions.

In the discussion above we made no mention of magnetic transitions. The reason is that the magnetic field interacts with the magnetic dipole moment, so that in the expression

$$V(\mathbf{r}, t) = -\boldsymbol{\mu} \cdot \mathbf{B} \cos \omega t \tag{34.41}$$

it is the matrix element of the form

$$\int d^3 r u_m^*(\mathbf{r}) \boldsymbol{\mu} \cdot \mathbf{B} u_1(\mathbf{r})$$

that must be calculated. In the absence of spin, $\boldsymbol{\mu}$ is proportional to $\mathbf{L}$, and the orthogonality of the radial eigenfunctions for different energies will make this vanish. This will no longer be the case when $\mathbf{B}$ is not uniform over the dimensions of the atom, but that term is suppressed, over and above the characteristic $v/c \simeq Z\alpha$ suppression of magnetic interactions (see Chapter 31 for a discussion of spin-orbit coupling).

Spin-flip transitions The operator that induces the transition, $\mathbf{E} \cdot \mathbf{r}$, does not involve the spin, so that there is an additional selection rule that follows from the orthogonality of the spin wave functions for different spin states:

$$\Delta S = 0 \tag{34.42}$$

the spin state does not change. This selection rule is also not absolute, since, as noted in our discussion of the Zeeman effect, there is an interaction term of the spin (magnetic dipole moment) with the magnetic field of the form

$$V = -\boldsymbol{\mu} \cdot \mathbf{B}$$

$$= \frac{eg}{2mc} \mathbf{S} \cdot \mathbf{B} \tag{34.43}$$

and this does involve the spin. The magnitude of this coupling is of order $(eB)(\hbar/mc)$. If we compare this with the magnitude of the electric dipole matrix element $(eE)(R) \simeq (eE)(\hbar/mcZ\alpha)$, we see that the spin-flip transitions are also suppressed by terms of order $(Z\alpha)^2$.

In some systems the coupling of spin and orbital angular momentum is very strong. In those systems it does not make much sense to speak of Δl selection rules, but one still has

$$\Delta J = 1, 0, -1$$

$$\Delta J_z = \pm 1$$

parity changes (34.44)

for the dominant transitions. There is actually an absolute selection rule connected with angular momentum conservation, and that is the rule that $J = 0 \rightarrow J = 0$ transitions are *absolutely forbidden*.

Bohr frequency condition Now that we have understood the selection rules, let us turn to the time dependence. We have the time factor

$$\frac{1}{i\hbar} \int_0^t dt' \cos \omega t' e^{-i\omega_0 t'} \tag{34.45}$$

where $\omega_0 = \dfrac{E_1 - E_m}{\hbar}$. The integration yields

$$\frac{1}{2i\hbar} \left[\frac{e^{i(\omega - \omega_0)t} - 1}{i(\omega - \omega_0)} + \frac{e^{-i(\omega + \omega_0)t} - 1}{i(\omega + \omega_0)} \right]$$

$$= -\frac{1}{2\hbar} e^{i(\omega - \omega_0)t/2} \frac{e^{i(\omega - \omega_0)t/2} - e^{-i(\omega - \omega_0)t/2}}{\omega - \omega_0} + \cdots$$

$$= \frac{1}{i\hbar} e^{i(\omega - \omega_0)t/2} \frac{\sin(\omega - \omega_0)t/2}{\omega - \omega_0} + \cdots . \tag{34.46}$$

The terms represented by . . . are of the same form, with $\omega - \omega_0$ replaced by $\omega + \omega_0$. The absolute square of this integral (34.22) is involved in the rate, and it will turn out to be large when the denominator vanishes. This will occur when $\omega - \omega_0$ and we concentrate on $\omega \approx \omega_0$. Thus we only square Eq. (34.46) and get

$$\frac{1}{\hbar^2} \frac{\sin^2(\omega - \omega_0)t/2}{(\omega - \omega_0)^2} . \tag{34.47}$$

This curve has the shape shown in Fig. 34.1 and for very large t it is of the form of a factor t multiplied by a very sharply peaked function of unit area under it.[†] This has

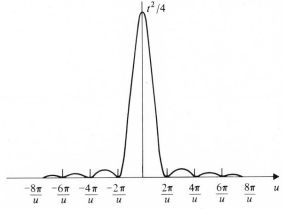

Figure 34.1. Plot of the curve $(\sin^2 ut/2)/u^2$ as a function of u.

[†] The height of the central peak grows like t^2 and the width, as a function of u, decreases as $1/t$, so that the area is proportional to t.

the consequence that the rate, defined in Eq. (34.20) is independent of time, and is proportional to a function that is peaked very sharply only where energy is conserved, that is, where

$$\hbar\omega = E_1 - E_m. \tag{34.48}$$

This equation describes energy conservation when the initial state "1" has higher energy than the state "m", that is, de-excitation of an atom. Such a de-excitation may be induced by an external field **E**, but it could also occur spontaneously, through the emission of the radiation field **E**. The relationship between induced and spontaneous emission will be discussed in the next chapter. The term that we have left out only contributes when

$$\hbar\omega = E_m - E_1, \tag{34.49}$$

that is, when we are describing the excitation of an atom from a lower to a higher energy state by an external electric field. It is in this way that the Bohr condition on emission and absorption frequencies is explained.

The procedure so outlined is oversimplified. We found that the probability for the transition "1" → "m" was of the form

$$P_{1 \to m}(t) \cong tR_{1 \to m} \tag{34.50}$$

for large t, but this cannot be so, because for large enough t this could exceed unity. A more accurate calculation shows that matters can be improved. The situation is the following:

a) There is indeed a constant transition rate, and to good approximation it is of the form obtained above, being proportional to

$$\left| e \int d^3 r u_m^*(\mathbf{r}) \mathbf{E} \cdot \mathbf{r} u_1(\mathbf{r}) \right|^2. \tag{34.51}$$

b) The relation describing the depletion of the initial state obtained using Eq. (34.50),

$$P_1(t) = 1 - t\left(\sum_m R_{1 \to m} \right) \tag{34.52}$$

is an approximation for the correct expression

$$P_1(t) = e^{-\Gamma_1 t} \qquad \Gamma_1 = \sum_m R_{1 \to m}. \tag{34.53}$$

This form is derived in more advanced textbooks.

Exponential decay The formula (34.53) makes a rather simple statement about the decay process: This is that the number of atoms present at a time t is given by

$$N_1(t) = N_1(0)e^{-\Gamma_1 t} \tag{34.54}$$

where $N_1(0)$ is the number present initially. This implies that

$$\frac{dN_1(t)}{dt} = -\Gamma_1 N_1(t) = -\sum_m R_{1 \to m} N_1(t), \tag{34.55}$$

that is, the rate of depletion of atoms from the initial state, is equal to the number present at a given time, multiplied by the sum of all the rates into various decay "channels". The fact that the rate of depletion depends only on the number present and not on the history of the atom is a peculiarity of quantum mechanics. Although expressions like Eq. (34.54) may occur in population statistics, there the expression is an approximation, and subpopulations (for example, smokers or skydivers) may have different decay rates from the average. In the case of quantum systems, the rate is really a characteristic of each atom, but because of the probabilistic nature of the laws of nature, one cannot predict the fate of a single atom. In a way we may say that atoms do not have built-in clocks that tell them when to decay.

Formulas like Eq. (34.55) do not hold only for radiative transitions; they apply to any kind of decay of an excited state. Nuclei can decay by alpha emission or by beta decay: In all cases, similar formulas apply, though the decay rates Γ may be different. The decay rate Γ has the dimensions of $(\text{sec})^{-1}$. We may relate it to the mean life of the atom

$$\tau = \frac{\int_0^\infty dt\, t N(t)}{\int_0^\infty dt\, N(t)} = \frac{\int_0^\infty dt\, t e^{-\Gamma t}}{\int_0^\infty dt\, e^{-\Gamma t}} = \frac{1}{\Gamma}. \tag{34.56}$$

The half-life, a time at which the number of particles initially present is reduced to one half that number, is given by

$$e^{-\Gamma \tau_{1/2}} = \tfrac{1}{2} = e^{-\log 2},$$

that is,

$$\tau_{1/2} \cong 0.69\,\tau. \tag{34.57}$$

Radioactive chains The differential equation (34.55) may be generalized to more complicated situations in which there is a chain of radiactive decays. Consider, for example, a three-level system (Fig. 34.2). The decrease in the occupancy of level A is described by

$$\frac{dN_A}{dt} = -R_{AB} N_A(t) - R_{AC} N_A(t) \equiv -\Gamma_A N_A(t). \tag{34.58}$$

The change in the occupancy of B is given by

$$\frac{dN_B}{dt} = R_{AB} N_A(t) - R_{BC} N_B(t) \tag{34.59}$$

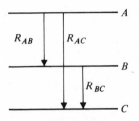

Figure 34.2. Three-level system and associated decay rates.

which shows the accretion due to leakage from A, as well as decay to C. There is no need to write down the equation that describes the number in level C, since

$$N_A + N_B + N_C = \text{const} \equiv N. \tag{34.60}$$

To solve Eq. (34.58) and Eq. (34.59) we note that the first equation has a simple solution:

$$N_A(t) = Ne^{-\Gamma_A t}.$$

This can be inserted into the second equation to read

$$\frac{dN_B}{dt} = R_{AB}Ne^{-\Gamma_A t} - R_{BC}N_B(t).$$

With the substitution

$$N_B(t) = F(t)e^{-\Gamma_A t} \tag{34.61}$$

we get the differential equation for $F(t)$:

$$\frac{dF}{dt} + R_{BC}F(t) - \Gamma_A F(t) = R_{AB}N. \tag{34.62}$$

A further change

$$F(t) = G(t) - \frac{R_{AB}N}{\Gamma_A - R_{BC}} \tag{34.63}$$

yields the differential equation

$$\frac{dG}{dt} = -(R_{BC} - \Gamma_A)G, \tag{34.64}$$

whose solution is

$$G(t) = G(0)e^{-R_{BC}t}e^{\Gamma_A t}.$$

This finally yields

$$N_B(t) = N\frac{R_{AB}}{\Gamma_A - R_{BC}}(e^{-R_{Bc}t} - e^{-\Gamma_A t}) \tag{34.65}$$

when $G(0)$ is determined by the condition that $N_B(0) = 0$.

This equation may be studied in various limits. For example, when $\Gamma_A \gg R_{BC}$, that is, when the initial state decays much faster than the "daughter" state B, then after a time that is short on the lifetime scale of B,

$$N_B(t) \approx N\frac{R_{AB}}{\Gamma_A}e^{-R_{BC}t}. \tag{34.66}$$

What happens is that the state A is depleted, partly into B and partly into C. The fraction that goes into B is then the "initial" number that decays at the decay rate $\Gamma_B = R_{BC}$ from that state. If, on the other hand, $\Gamma_B \gg \Gamma_A$, that is, B decays much faster than the rate at which it is populated, then after a time that is long compared to the lifetime of $B(t \gg 1/\Gamma_B)$,

$$N_B(t) \approx N\frac{R_{AB}}{R_{BC}}e^{-\Gamma_A t} \approx \frac{R_{AB}}{\Gamma_B}N_A(t). \tag{34.67}$$

In general, the state B does not obey the exponential decay law.

The equations just discussed can be generalized to more complicated chains. That such chains can occur can be seen from the example

$$^{235}U \xrightarrow[\alpha]{3.1 \times 10^{-17}\text{sec}^{-1}} {}^{231}Th \xrightarrow[\beta]{7.6 \times 10^{-4}\text{sec}^{-1}} {}^{231}Pa$$

$$\xrightarrow[\alpha]{6.8 \times 10^{-13}\text{sec}^{-1}} {}^{227}A \xrightarrow[\beta]{1.6 \times 10^{-9}\text{sec}^{-1}} {}^{227}Th \xrightarrow[\alpha]{4.2 \times 10^{-7}\text{sec}^{-1}}$$

$$^{223}Ra \xrightarrow[\alpha]{7.4 \times 10^{-7}\text{sec}^{-1}} \ldots$$

The striking aspect of this chain, the enormous variation from 10^9 years to 20 minutes in the lifetimes in this part of the chain, can be understood in terms of the interactions involved and the properties of the nuclei. We will touch on some of these matters later.

Carbon dating A very interesting application of the study of radioactivity outside of nuclear or solid state physics is in geological dating. One approach, applicable to objects made of organic matter 500 to 30,000 years ago, is the ^{14}C *dating method*, developed by Willard Libby. Cosmic rays in the atmosphere cause nuclear reactions which produce radioactive ^{14}C. The present concentration of ^{14}C relative to normal ^{12}C is about one part in 10^{12}. It is believed that this fraction has been constant over a period much longer than the 30,000 years that is set as a practical upper limit for carbon dating. ^{14}C decays by beta emission to ^{14}N with a lifetime of the order of 8000 years. Living matter, plant or animal, exchanges carbon with the atmosphere, so that the composition of carbon in living matter is the same as that in the atmosphere. The carbon in living matter will have the same *activity*, defined by

$$A(t) = N(t)/\tau \tag{34.68}$$

as atmospheric carbon, which turns out to be 15 decays per gm per sec. When the plant or animal dies, exchange with the atmosphere stops, and the ^{14}C is no longer replenished. Thus after approximately $0.69 \times 8000 \cong 5500$ years, there will be only one half as much ^{14}C activity, and this reduction in activity may be used to date objects. There is excellent agreement with the age determined by independent methods, when these are available (for example, Pompeii, thousand-year-old trees ring-dated, and so on).

Similarly, the ratios of transuranic elements in rocks may be used to determine their ages, as are measurements of helium content in minerals. The latter is associated with age because in the conversion of ^{238}U to ^{206}Pb, eight alpha particles (He nuclei) are produced. There are some uncertainties, because some of the helium could have escaped, but these are not large. The methods indicate that the oldest terrestrial minerals are approximately 2.7×10^9 years old, while the oldest meteorites are of the order of 4.5×10^9 years old.

Radiation dosimetry We conclude our discussion of radioactivity with a brief review of some units that are used in radiation dosimetry. This is the language used in discussions of emission of nuclear power plants. Activities, that is, the number of dis-integrations per second, are measured in *curies* (Ci). An activity of 1 curie corresponds to 3.7×10^{10} disintegrations per second (the activity of 1 gm of radium). Another unit that measures the effect of radioactivity is the *Roentgen*, defined in terms of the amount of radiation that produces a certain amount of ionization in 1 cc of air (0.001293 gm). It is the amount of radiation whose associated corpuscular emission per cc of air produces ions carrying one electrostatic unit of charge. This roughly corresponds to the absorption of radiation energy of 84 erg/gm. Actually, for dosimetry, one is more interested in the energy absorbed by soft tissue than by air. A beam delivering one roentgen will deliver 93 ergs/gm of soft tissue. This is called the Rep (roentgen equivalent physical). To specify biological damage further the Rem is used, with

$$1 \text{ Rem } = 1 \text{ Rep} \times 1 \text{ RBE}$$

where RBE (relative biological effectiveness) depends on the kind of radiation involved. RBE varies from 1–20, depending on the amount of ionization caused by the beam. Permissible occupational exposure ranges from 1 Rem/week for skin exposure to 0.3 Rem/week for eyes, blood-forming organs, and so on. Over a long period of time, these should be reduced by a factor of 3. Cosmic rays at sea level deliver 0.005 Rem/week, which is about 1.7% of the tolerance dose.

NOTES AND COMMENTS

1. Time-dependent perturbation theory and selection rules are discussed in all quantum mechanics textbooks. The exponential decay law is more complicated to derive. A relatively simple treatment may be found in a special-topics section in S. Gasiorowicz, *Quantum Physics*, John Wiley, New York 1974. When the transition problem is treated more accurately, then the sharply peaked function (34.47) is replaced by what is called the Lorentz curve of the form

$$\frac{\Gamma/2}{(\omega - \omega_0)^2 + (\Gamma/2)^2} .$$

The quantity Γ is the width of the spectral line, and one can show that there is a relation between the lifetime and the width of the form

$$\tau \Gamma \sim 1.$$

This quantity is called the natural line width.

PROBLEMS

34.1 Consider an atom in its ground state. At $t = -\infty$ it is placed in an electric field that is turned on very slowly, and again turned off very slowly, so that the time dependence in Eq. (34.4) is

$$e^{-t^2/T^2} \qquad T \text{ very large}$$

instead of cos ωt. Obtain an expression for the dependence of the probability of exciting the state "m" (cf. Eq. (34.19)) as a function of T. Show that the probability of making a transition is very small for T large compared with $\hbar/(E_m - E_1)$. It will help to know the integral

$$\int_{-\infty}^{\infty} dx\, e^{-x^2 + iax} = \sqrt{\pi}\, e^{-a^2/4}.$$

34.2 Which of the following transitions are allowed in the electric dipole approximation, and which are suppressed partially?

$$^2S_{1/2} \to {}^2P_{3/2} \qquad\qquad {}^2D_{3/2} \to {}^2D_{5/2}$$

$$^3D_2 \to {}^1P_1 \qquad\qquad {}^4D_{1/2} \to {}^4P_{1/2}$$

$$^2D_{3/2} \to {}^4S_{3/2} \qquad\qquad {}^3P_0 \to {}^1S_0$$

34.3 A nucleus of radius 0.8×10^{-12} cm undergoes a radiative transition by an electric dipole transition, and another one, to a neighboring level, by electric quadrupole transition. Estimate the relative probabilities for making the transitions if the energy of the photon is (a) 10 keV, (b) 100 keV, and (c) 1 MeV.

34.4 Show that the solution of the differential equation

$$\frac{dN}{dt} = -\gamma N + F(t)$$

with the initial condition

$$N(0) = 0$$

is

$$N(t) = \int_0^t dt'\, e^{-\gamma(t-t')} F(t').$$

34.5 Estimate the age of the earth, given that the mean life of ^{238}U is 0.6×10^{10} years, that of ^{235}U is 1.0×10^9 years, and that the present abundance ratio is ^{238}U/^{235}U $= 140$. Assume that the two isotopes were equally abundant at the time of the formation of the earth.

34.6 There exist in nature particles that are in many respects just heavy electrons. These particles, the muons, have a rest mass of 207 electron masses. Given that the transition rate for the $2P \to 1S$ transition in hydrogen is 0.6×10^9 sec^{-1}, can you estimate the rate for the same transition in an atom in which the muon orbits about the proton? [*Hint:* Use dimensional analysis.]

34.7 A nucleus decays into two channels with probabilities 0.62 and 0.38, respectively. Its half-life is 13 hours. What are the partial decay rates into the two channels?

34.8 What is the ^{14}C activity expected in a timber structure that is 800 years old? 1200 years old?

34.9 Consider the function that appears in Fig. 34.2, divided by T,

$$\sin^2 uT/u^2 T$$

in the limit of very large T. Show that for a function $f(u)$ that varies slowly compared with the above function,

$$\lim_{T=\infty} \int_{-\infty}^{\infty} du f(u) \frac{\sin^2 uT}{\pi u^2 T} = f(0).$$

A function for which the above holds is called a Dirac delta function

$$\delta(u) = \lim_{T=\infty} \frac{\sin^2 uT}{\pi u^2 T}.$$

Show that other representations of the Dirac delta function are

$$\delta(u) = \lim_{\alpha \to \infty} \frac{\alpha}{\sqrt{\pi}} e^{-\alpha^2 u^2}$$

and

$$\delta(u) = \lim_{a \to 0} \frac{1}{\pi} \frac{a}{u^2 + a^2},$$

that is, show in all cases that

$$\int_{-\infty}^{\infty} du f(u) \delta(u) = f(0).$$

[*Hint*: In the integral involving a sharply peaked function $P(x)$ and a smoothly varying function $f(x)$,

$$\int P(x) f(x) dx \approx f(x_0) \int P(x) dx,$$

where x_0 is the location of the peak.]

35
Stimulated Emission, Lasers, and their Uses

The Einstein A and B coefficients The conclusion drawn from the last chapter is that an oscillating electric field induces transitions between atomic states with a transition rate proportional to

$$\frac{e^2}{\hbar^2}\left| \mathbf{E} \cdot \int d^3r u_m^*(\mathbf{r})\mathbf{r}u_1(\mathbf{r})\right|^2, \tag{35.1}$$

subject to the condition that the frequency of oscillation ω is related to the energy change in the transition by

$$\hbar\omega = |E_1 - E_m|. \tag{35.2}$$

We wish to concentrate on the aspect of the formula that states that the transition rate is proportional to the square of the electric field, when we average over many atoms in which the directions of the angular momenta are random. Since the energy of the electromagnetic radiation is proportional to $\mathbf{E}^2$, we may rephrase this by stating that the transition rate for exciting an atom from the ground state to an excited state is proportional to the energy density of the electromagnetic radiation of the correct frequency, the frequency that matches the energy of excitation in accordance with the Bohr rule. If we consider an ensemble of atoms, the transition rate is also proportional to the number of atoms present in the state "1." We may thus write down an expression for the rate of transitions from a state "1" *up* to a state "2" in a cavity containing blackbody radiation with energy density $u(\nu, T)$:

$$\Gamma_{1\to2} = N_1 u(\nu, T)B_{12}. \tag{35.3}$$

This could have been written down even without the use of (35.1) and it was indeed done in 1917 by Einstein, who introduced the notion of stimulated and spontaneous absorption and emission. The above formula describes the rate of transitions *upwards*, and such transitions are entirely stimulated by the presence of the radiation field that also supplies the energy necessary for the transition. When considering transitions *downward* from an excited state "2" to the ground state "1," Einstein recognized two contributions: One is the stimulated emission, and the other is spontaneous emission, the transition that, according to the original Bohr model, will occur even if the atom is isolated in a vacuum. Thus Einstein wrote down the rate of transitions from "2" to "1" in the form

$$\Gamma_{2\to1} = N_2(u(\nu, T)B_{21} + A_{21}). \tag{35.4}$$

Here the rate is again proportional to the number of atoms in the state "2," the B-term describes stimulated emission, and the A-term is the spontaneous emission contribution. Only the former involves the energy density of the radiation in the cavity.

Let us now consider radiation in equilibrium in a cavity, and let us assume that the walls are made of atoms that have only two energy levels, the ground state "1" and the excited state "2." Then the energy density at the frequency

$$\nu = \frac{E_2 - E_1}{h} \tag{35.5}$$

is the only one that is relevant. The equilibrium condition requires that the transition rates $\Gamma_{1 \to 2}$ and $\Gamma_{2 \to 1}$ be equal, so that we can write

$$N_1 B_{12} u(\nu, T) = N_2 (B_{21} u(\nu, T) + A_{21}). \tag{35.6}$$

We also have the Boltzmann relation between the population of the levels N_2 and N_1:

$$\frac{N_1}{N_2} = \frac{g_1}{g_2} \frac{e^{-E_1/kT}}{e^{-E_2/kT}} = \frac{g_1}{g_2} e^{h\nu/kT}, \tag{35.7}$$

where g_2 and g_1 are the "degeneracies" of the two levels. We now know that these are just the values $(2J_2 + 1)$ and $(2J_1 + 1)$, respectively, of the degeneracies of states with angular momentum J_2 and J_1. We may thus rewrite Eq. (35.6) in the form

$$\frac{g_1}{g_2} e^{h\nu/kT} u(\nu, T)B_{12} = u(\nu, T)B_{21} + A_{21}. \tag{35.8}$$

Note that A_{21}, B_{12}, and B_{21} do not depend on the temperature. Einstein pointed out that in the limit of high temperatures, the Rayleigh-Jeans law (cf. Eq. (13.9)) shows that $u(\nu, T)$ grows linearly with T, and this means that Eq. (35.8) can only be satisfied in the high T limit if

$$g_1 B_{12} = g_2 B_{21}. \tag{35.9}$$

Thus the Einstein coefficients for stimulated emission and absorption are the same once proper account is taken of the degeneracies of the atomic states. When Eq. (35.9) is inserted into Eq. (35.8), we get

$$u(\nu, T) = \frac{A_{21}/B_{21}}{e^{h\nu/kT} - 1}. \tag{35.10}$$

Comparison with the Planck radiation formula (13.10) then yields

$$\frac{A_{21}}{B_{21}} = 8\pi h\nu^3/c^3. \tag{35.11}$$

The calculation of the A and B coefficients separately requires a computation of the electric dipole matrix element of the transition, which is beyond the scope of this course. We may obtain a better understanding of the stimulated emission process by rewriting the emission transition rate *per atom* in the form

$$\frac{\Gamma_{2 \to 1}}{N_2} = A_{21} + B_{21} u(\nu, T) = A_{21} \left(1 + \frac{1}{e^{h\nu/kT} - 1}\right) \tag{35.12}$$

which can be expressed in terms of the *average number of photons* per unit volume of frequency ν. That number can be obtained from the probability distribution (13.29):

$$P_n = \frac{e^{-nh\nu/kT}}{\sum\limits_{n=0}^{\infty} e^{-nh\nu/kT}} = \frac{e^{-nh\nu/kT}}{(1-e^{-h\nu/kT})^{-1}} \tag{35.13}$$

which yields

$$\bar{n} = \sum_{n=0}^{\infty} ne^{-nh\nu/kT}/(1-e^{-h\nu/kT})^{-1}$$

$$= \frac{\left(-\dfrac{d}{dx}\sum\limits_{n=0}^{\infty} e^{-nx}\right)_{x=h\nu/kT}}{(1-e^{-h\nu/kT})^{-1}} = \frac{e^{-h\nu/kT}}{1-e^{-h\nu/kT}}$$

$$= \frac{1}{e^{h\nu/kT}-1} \tag{35.14}$$

Thus

$$\frac{\Gamma_{2\to1}}{N_2} = A_{21}(1+\bar{n}(\nu)). \tag{35.15}$$

The transition rate is proportional to the spontaneous emission coefficient, and it increases linearly with the number of photons of the right frequency that are present in the enclosure. Before we return to a discussion of the significance of this result, let us note that we can also calculate $\overline{n^2}$. It is given by

$$\overline{n^2} = (1-e^{-h\nu/kT})\left(\frac{d^2}{dx^2}\sum_{n=0}^{\infty} e^{-nx}\right)_{x=h\nu/kT}$$

$$= (1-e^{-h\nu/kT})\left(\frac{d^2}{dx^2}(1-e^{-x})^{-1}\right)_{x=h\nu/kT}$$

$$= \frac{e^{h\nu/kT}+1}{(e^{h\nu/kT}-1)^2}$$

$$= \frac{2}{(e^{h\nu/kT}-1)^2} + \frac{1}{e^{h\nu/kT}-1}$$

$$= 2(\bar{n})^2 + \bar{n} \tag{35.16}$$

after a little algebra. Thus the root mean square deviation

$$\Delta n = (\overline{n^2}-(\bar{n})^2)^{1/2} = (\bar{n}(\bar{n}+1))^{1/2}$$

$$\approx \bar{n}+\tfrac{1}{2} \tag{35.17}$$

is very large for large $\bar{n}$. This implies very large fluctuations in the number of photons present in the blackbody radiation field.

Lasers The remarkable prediction of stimulated emission by Einstein predated quantum mechanics by eight years, and predated the first exploitation of that mechanism by almost forty. The most effective use of that physical principle is in the construction of the laser (light amplification through stimulated emission). The basic components of a laser are the following:

a) a two-level system in which atoms in the upper level "2" make a transition to the lower level "1" that is stimulated by the presence of photons of the right frequency;

b) some sort of cavity to provide the environment of the right kind of photons;

c) some mechanism for repopulating the upper level "2" for repeated operation.

Laser cavity It turns out to be convenient to discuss the nature of the cavity first. In dealing with microwave radiation (as in the ammonia clock discussed in Chapter 23), the wavelength is in the centimeter range, so that a metallic cavity with dimensions in that range will have modes well separated in frequency, allowing for a simple tuning capability. For radiation in the optical (~ 5000 Å) range, a cavity centimeters in size, has very many closely spaced modes. It is rather remarkable that it is nevertheless possible to achieve well-defined frequency separation when one deals with a tube with only slightly transparent mirrors at the ends, and consider light moving parallel to the axis. If we ignore the transverse dimensions of the tube, light will bounce back and forth provided the dimension of the cavity along the main axis, L, is an integral number of half wavelengths, so that

$$L = N \frac{\lambda}{2} \tag{35.18}$$

N is a large integer. For a 1 meter tube, and $\lambda = 5000$ Å, it is of the order of 0.4×10^7. If n is the refractive index of the medium, then the velocity of the light is c/n and the frequency is given by

$$\nu = \frac{c}{n\lambda} = \frac{Nc}{2nL}. \tag{35.19}$$

The traversal time is given by nL/c. If the reflection coefficient of the mirror is r, then the intensity of the light is reduced to r^k of the initial intensity after k traversals. With $r = 1 - \epsilon$, this is

$$I_k/I = (1 - \epsilon)^k = e^{-k\epsilon},$$

so that the intensity is reduced to $1/e$ of its initial value after $k = 1/\epsilon = 1/(1 - r)$ traversals. Thus the "lifetime" of the radiation in the cavity is given by

$$\tau = k \frac{nL}{c} = \frac{nL}{c(1 - r)}. \tag{35.20}$$

This corresponds to a linewidth of

$$\Delta \nu \simeq \frac{1}{\tau} \simeq \frac{c}{nL} (1 - r). \tag{35.21}$$

On the other hand, the frequency separation of two modes that differ in the N value by unity is $c/2nL$, and for a highly reflecting mirror, with $r = 0.98$, for example, this is much larger than the linewidth. It is therefore possible to provide a separation between the axial modes much larger than the size of the cavity might indicate. Put in other words, not all of the modes allowed by the wavelength condition (35.19) can be sustained in the cavity with mirrors, and the ones that are sustained are discrete. Thus a construction of the type outlined above provides an environment in which photons that are highly monochromatic can exist for a long time. In the real world one cannot ignore the sides of the cavity, and the separation is not quite as ideal. On the other hand, one can improve matters by terminating the optical resonator not with mirrors that are perpendicular to the axis, but with Brewster angle windows, beyond which identical spherical mirrors, separated by a distance equal to their common radius of of curvature, are placed (Fig. 35.1). (A Brewster angle window is a window placed at an angle such that the polarization component of the light parallel to the plane of incidence is completely transmitted, so that after many traversals between the mirrors only that component remains, leading to 100% linearly polarized laser light).

Population inversion To see under which conditions continuous operation of a laser is possible, let us consider the rate of change of photons of the right frequency. The number will increase because of stimulated and spontaneous emission, and it will decrease because of stimulated absorption and because of losses. We have

$$\frac{dn(\nu)}{dt} = N_2(u(\nu)B_{21} + A_{21}) - N_1 u(\nu)B_{12} - \frac{n(\nu)}{\tau_0}. \tag{35.22}$$

The last term represents photon losses due to leakages through the mirrors, sideways scattering and so on, which are proportional to the number of photons present. The coefficient τ_0 has the dimensions of time, and it is presumably large compared with the traversal time. Using Eq. (35.9) and Eq. (35.11), and the relation between the energy density and the photon number,

$$u(\nu) = \frac{8\pi\nu^3}{c^3} h\nu n(\nu), \tag{35.23}$$

which holds for blackbody radiation but is more general, since it describes the energy density as the (number of modes per unit frequency interval) × (energy at the

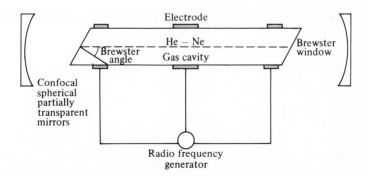

Figure 35.1. Schematic drawing of He-Ne laser.

frequency) × (number of photons with that energy), we get

$$\frac{dn}{dt} = \left(N_2 A_{21} - \frac{g_2}{g_1} N_1 A_{21} - \frac{1}{\tau_0} \right) n + N_2 A_{21}. \qquad (35.24)$$

Thus the number of photons will decrease with time, unless

$$N_2 - \frac{g_2}{g_1} N_1 \geqslant \frac{1}{A_{21} \tau_0}. \qquad (35.25)$$

Note that in thermal equilibrium this cannot be satisfied, since there,

$$\frac{N_2/g_2}{N_1/g_1} = e^{-h\nu/kT} < 1, \qquad (35.26)$$

so that the laser, to operate continuously, must have an excess population in the upper state. One must set up a *population inversion*.

Optical pumping There are several ways of doing this: One involves the use of a three-level system, such as is used in a ruby laser. Consider the system of levels shown in Fig. 35.2. One may set up equations for the rate of change of N_2, N_1, and N_0, with the inclusion of a light beam of some energy density u_p that "pumps" atoms from the ground state "0" to the higher of the two excited states, "2." The rate equation for the highest state, for example, is

$$\frac{dN_2}{dt} = -N_2 A_{21} - N_2 A_{20} + u_p B_{02} N_0 - u_p B_{20} N_2$$

$$-u(\nu)B_{21} N_2 + u(\nu)B_{12} N_1 \qquad (35.27)$$

and in equilibrium, all the time derivatives are set equal to zero. An analysis of the equations shows that if the transition rate Γ_{21} is very fast, and the transition rate Γ_{10} is slow, that is, when the middle level "1" is *metastable*, then this *optical pumping* procedure leads to a population inversion. Another way of getting a population inversion is to populate the upper states by electronic collisions in gas mixtures. In the helium-neon laser, (Fig. 35.3), for example, there are neon levels with approximately the same energy as the metastable (because of the $\Delta S = 0$ selection rule) excited state in which the two electrons are in a ^3S state constructed out of the orbitals $(1s)(2s)$. A

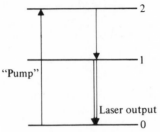

Figure 35.2. Three-level system allowing for optical pumping of level "1".

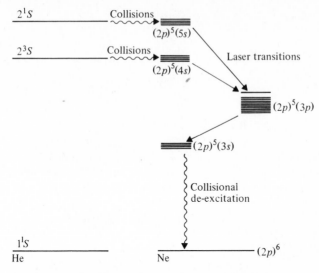

Figure 35.3. Energy levels relevant to He-Ne laser.

discharge will excite helium atoms to the metastable state, and these atoms will transfer their energy to neon atoms by collisions. These neon atoms are then excited to a particular state, in which an excess population is built up. When spontaneous decay occurs, the photons are trapped between the mirrors and provide the environment in which stimulated emission occurs.

Coherence The emission of radiation is stimulated by an electric field with a well-defined phase, and therefore, it turns out, all the atoms undergo stimulated decay in phase, that is *coherently*. The notion of coherence of a light beam is actually quite complicated when quantum effects are taken into account, but one can look at the problem classically, by considering two light sources oscillating with the same frequency but with different phases. The total field (ignoring the vectorial character of the electric field) is

$$E_1 \cos \omega t + E_2 \cos(\omega t + \phi) \tag{35.28}$$

and the intensity is given by the square of that,

$$
\begin{aligned}
I &= (E_1 \cos \omega t + E_2 \cos(\omega t + \phi))^2 \\
&= E_1^2 \cos^2 \omega t + E_2^2 \cos^2(\omega t + \phi) \\
&\quad + E_1 E_2 (\cos(2\omega t + \phi) + \cos \phi).
\end{aligned}
\tag{35.29}
$$

A measurement of the intensity averages over times that are long compared with $1/\omega$, so that

$$\frac{1}{T}\int_0^T dt \cos^2 \omega t = \frac{1}{T}\int_0^T dt \cos^2(\omega t + \phi) = \frac{1}{2}$$

$$\frac{1}{T}\int_0^T dt \cos(2\omega t + \phi) = 0, \tag{35.30}$$

and we are left with

$$I_{Av} = \tfrac{1}{2}E_1^2 + \tfrac{1}{2}E_2^2 + E_1 E_2 \overline{\cos\phi}. \qquad (35.31)$$

If the phase fluctuates rapidly on the time scale of the measurement time T, then

$$\overline{\cos\phi} = \frac{1}{T} \int_0^T dt \cos\phi = 0 \qquad (35.32)$$

and the intensity is just the sum of the intensities of the two sources. If, on the other hand, the fluctuations in the phase difference ϕ are slow on the time scale of the measurement time, then $\cos\phi$ may be viewed as a constant, and interference effects, which are a sign of coherence, manifest themselves. It should be noted that when the frequencies are not perfectly sharp, then ϕ will contain a piece that is just $\Delta\omega t$, where $\Delta\omega$ is the linewidth, and thus the coherence is limited by a time $T < 2\pi/\Delta\omega$. This does not cause any problems in the laser. The laser acts coherently because all of the atoms are stimulated to radiate by a single electric field, and therefore the phases are strongly correlated.

The virtues of the laser, strong monochromaticity, high intensity, and a high degree of coherence, allowing for good beam collimation, make it an important tool in many fields of scientific research. We will discuss just a few applications:

Lunar ranging The high intensity of a laser beam allows one to send signals over a great distance. In 1969 the first accurate earth-moon distance measurement was obtained by sending a laser beam through the 120-inch telescope at Lick Observatory (without a telescope the beam, even though diffraction limited in spread, with $\theta \sim \lambda/$ diameter, would cover an area of diameter 300 miles) at a retroflector package consisting of "corner cubes" arranged in an 18-inch square array and accurately measuring the time of flight of the photons. This can be done to an accuracy of 2×10^{-9} seconds, and thus the distance can be determined to an accuracy ± 1 ft. This may seem puzzling, since the speed of light is not known that well, but the distance measurement is determined to this accuracy relative to all other astronomical measurements, since the speed of light is set at the internationally agreed value of 299, 792.5 km/sec. It is amusing that the first measurements showed a discrepancy which was resolved only when it was discovered that, due to a surveying error, the observatory was about 1800 feet away from the position that was assumed for determining the predicted time of return for the laser beam. A by-product of measurements of the earth-moon distance over the past few years has been the determination of the absence of a variation with a 29-day period in that distance. Such a variation would be predicted on the basis of several theories of gravitation competing with that of Einstein. The absence of the variation to the accuracy measured can be translated into the equality of gravitational and inertial mass to 1.5 parts in 10^{11}.

Separation of isotopes Most elements found in nature are mixtures of isotopes, that is, the nuclei all have the same number of protons (Z), but they may differ in the number of neutrons, and thus in atomic weight. The electronic structure depends primarily on Z, and thus the separation of isotopes by chemical methods is extremely difficult.

Methods like gaseous diffusion rely on the fact that the lighter atoms diffuse slightly more rapidly through a porous barrier. The difference in the mass and other properties of the nucleus leads to slightly different properties of the electronic cloud, and thus to slightly different frequencies at which light is emitted and absorbed. The high degree of monochromaticity, that is, the very narrow linewidth of laser light may be used to separate atoms with the same Z but different values of A. The simplest procedure is to create a beam of atoms by evaporating the substance in an oven and letting the atoms emerge through a small aperture. The beam is irradiated with a laser beam whose frequency is tuned to a transition in one of the isotopes. Thus atoms of that particular isotope are excited, and this gives them quite different electromagnetic properties, allowing for their separation from the rest of the beam by electromagnetic means (Stern-Gerlach setup, for example). There is a difficulty in that if the atoms of the desired isotope are easily excited, then they decay just as easily to the ground state. To by-pass this difficulty, two laser beams are sometimes employed, with the second beam exciting the selected atoms further to a metastable state. Another method takes advantage of the short lifetime of the selected atoms. When a laser photon is absorbed, the atom recoils in the direction of the beam. Emission of a photon with the same frequency, but in an arbitrary direction, leads to a random recoil; a subsequent absorption of a laser photon again imparts momentum in the beam direction. When a large number of laser photons are absorbed and reemitted, there is a net transverse velocity acquired by the radiation pressure of the laser beam, shining at right angles to the atomic beam. In this way separation is again achieved. The latter technique was used in the pioneering work on laser separation in which an atomic beam of barium was enriched in the ^{138}Ba isotope. Somewhat different techniques, based on the same general principle, may be used to separate ^{235}U from the more common and less useful isotope ^{238}U.

High-precision spectroscopy The high intensity of laser beams, with fluxes of the order of 10^{17} photons per second, and the availability of tunable lasers, in which the frequency can be adjusted with high precision to atomic resonances, has made possible a technique for eliminating the broadening of spectral lines that results from the recoil accompanying absorption of a photon as well as from the thermal motion. The idea is to tune *two* lasers so that the frequency of each is *one half* of that corresponding to a transition between levels for which $\Delta l = 2$ and to pass the atomic beam between the two lasers firing in opposite directions. The level is excited by the absorption of two photons — this is analogous to the Raman effect in which a $\Delta l = 2$ transition is obtained by the absorption and reemission of single photons such that the difference is the desired frequency. The Doppler shifts in the two cases cancel, since the photon beams move in opposite directions, so that the atomic motion is $+v$ relative to one beam and $-v$ relative to the other. There is still some broadening due to the fact that some of the atoms can absorb two photons from one of the beams, but if circularly polarized light is used, one can eliminate this Doppler-broadened background. The excited atoms decay and the photons are observed. The opposing-beam technique can be used with two lasers of different frequencies which add up to the one that one wishes to study, and in this way one can sometimes enhance the probability of the transition, because the individual absorptions could be near resonances. The technique can be used to study hyperfine splittings.

Recently laser spectroscopy began to be used in the study of unstable, radioactive isotopes, by measuring shifts and splittings of atomic optical transitions. The combination of high fluxes and large cross sections when one is exciting a resonant level (the cross sections are not determined by the atomic radius in that case, that is, they are not of order πa_0^2, but are of order $\pi(\lambda/2\pi)^2$) means that one does not require a very dense target; even a tenuous beam of radioactive nuclides will provide enough reactions to get accurate data.

Holography In ordinary photography, light is reflected off an object and then allowed to fall on a photographic plate. The exposure rate depends on the intensity, and when light is subsequently passed through the plate, the image reflects that information. Because the light is incoherent, the phase relation between rays reflected from different parts of the object is lost. In 1947 D. Gabor invented a method of wave-front recon-

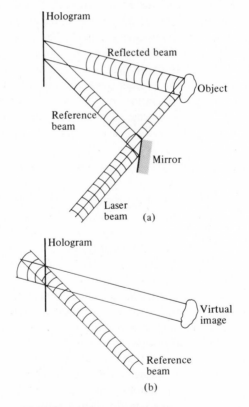

Figure 35.4. Schematic diagram for (a) formation of hologram, and (b) use of hologram for construction of virtual image of original object.

struction, named holography, which requires coherent light, and which therefore became practicable only with the invention of the laser. The idea is the following: Coherent light from a laser beam is split into two: one beam is reflected off an object, and the other is deflected in such a way that the beam reflected off the object, and the "reference beam" combine at the photographic plate (Fig. 35.4). A complex interference pattern will be recorded, and this is the hologram. When such a hologram is illuminated by the reference beam alone, the light rays are absorbed or transmitted in such a way as to create in the outgoing beam a component that reproduces the original beam that reflected off the object when the hologram was made. Thus the observer will "see" the original object, and this provides three-dimensional photographs of the objects.

Laser fusion The hope for tapping nuclear energies available in thermonuclear reactions, such as, for example

$$^{2}H + {}^{3}H \rightarrow {}^{4}He + n + 17.6\,\text{MeV},$$

(cf. Chapter 22), depends on bringing the nuclei close enough together. This requires tremendous compression (equivalently, temperatures of the order of $10^8\,°\text{K}$), and one proposal that is being investigated vigorously is the compression of small frozen deuterium-tritium pellets by a battery of high-power rapidly pulsed lasers. Present laser technology is inadequate to provide sufficient radiation pressure for compression: For example, compressions to densities of $10^3\,\text{gm/cm}^3$ are required to initiate burning, whereas so far compressions only to $1-10\,\text{gm/cm}^3$ have been achieved. Whether a two-order-of-magnitude technological improvement is possible, and whether instability problems which might lead to the break-up of pellets before burning starts can be dealt with, remains to be seen. Suffice it to say that the implications of the 1917 Einstein paper have already revolutionized many fields of research, and there is no reason why the present rate of progress needs to slow down!

NOTES AND COMMENTS

1. The notion of stimulated emission and absorption is currently very closely tied to lasers and masers. An enormous literature has developed in these fields, ranging from esoteric treatments of "quantum optics" to the most practical books on the use of lasers in ocular photocoagulation for the treatment of eye disease connected with diabetes. At the elementary level, occasional articles in *Scientific American* are very interesting. A recent one, with references to older articles and appropriate textbooks, is "White-Light Holograms" by E.N. Leith, *Scientific American*, October 1976. Some technical aspects of laser and maser construction are discussed in J.S. Thorp, *Masers and Lasers*, Macmillan, London, 1967.

2. The use of lasers to study unstable isotopes is in its infancy, and I am indebted to Professor G. Greenlees for telling me about the feasibility of such studies.

3. There are many textbooks on optics, interference phenomena, and the like, in which problems such as coherence are discussed. I particularly like Frank S. Crawford, Jr. *Waves*, Berkeley Physics Course, vol. 3, McGraw-Hill, New York, 1968, for its detailed and very physical coverage, and for the many interesting excursions into nonstandard topics.

PROBLEMS

35.1 Consider N bosons in a one-dimensional box, the eigenstates of which are denoted by $u_n(x)$. Assuming that the bosons are noninteracting, and that they all occupy different states with eigenvalues $n_1, n_2, \ldots, n_N$, the eigenfunction for the N-particle state is given by

$$\Psi = \frac{1}{\sqrt{N!}} \left\{ u_{n_1}(x_1)u_{n_2}(x_2)\ldots u_{n_N}(x_N) + \begin{matrix} \text{all permutations of the } x\text{'s} \\ \text{in the arguments of the } u_{n_1}\ldots u_{n_N} \end{matrix} \right\}$$

a) Check that the normalization constant $(N!)^{-1/2}$ is the proper one.

b) Show that if $N-1$ of the bosons are in the same state, the proper eigenfunction is

$$\Psi = \frac{1}{\sqrt{N}} \{u_{n_1}(x_1)u_{n_1}(x_2)\ldots u_{n_2}(x_N) + u_{n_1}(x_1)\ldots u_{n_2}(x_{N-1})u_{n_1}(x_N) + \cdots\}$$

c) Write down the eigenfunction when all the bosons are in the same state, and show that this event is N times more probable than if $(N-1)$ of the N are in the same state. This is the same effect as appears in Eq. (35.15). [*Hint*: Work this out for $N = 3$ first, to see pattern.]

35.2 The probability of finding n photons in a mode for blackbody radiation is given by Eq. (35.13). Show that this may be written in the form

$$P_n = \frac{(\bar{n})^n}{(1+\bar{n})^{n+1}}.$$

For a coherent beam of photons, that probability is instead given by the Poisson distribution

$$P_n = \frac{(\bar{n})^n}{n!} e^{-\bar{n}}.$$

Show that for a coherent beam, $\Delta n = \sqrt{\bar{n}}$, instead of Eq. (35.17) which holds for an incoherent distribution.

35.3 The angular spread of a laser beam is ideally diffraction-limited, that is, the angular spread is given by $\theta_0 = 1.22 \lambda/D$ where λ is the wavelength of the light and D is the aperture of the source, that is, the diameter of the laser rod. For a ruby laser of diameter 1 cm, with light of wavelength 7000 Å, what is the size of the spot projected by the beam at a distance of 1 km? If the beam has a flux corresponding to 10^{18}

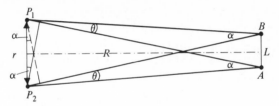

Figure 35.5. Rays of light from point sources P_1 and P_2 to two observers A and B, showing phases for Brown–Twiss experiment (Problem 35.4).

photons per second, what is the energy deposited per cm^2 per second at the target 1 km away?

35.4 Consider two point sources P_1 and P_2 separated by a distance r (Fig. 35.5). Two observers at A and B are separated by a distance L. The field observed at A is

$$E_1 \cos(\omega t - \phi_1) + E_2 \cos(\omega t - \phi_2)$$

where

$$\phi_1 = 2\pi(AP_1)/\lambda + \alpha_1 \qquad \text{and} \qquad \phi_2 = 2\pi(AP_2)/\lambda + \alpha_2$$

with α_1, α_2 representing independently fluctuating phases. The field observed at B is

$$E_1 \cos(\omega t - \psi_1) + E_2 \cos(\omega t - \psi_2)$$

where

$$\psi_1 = 2\pi(BP_1)/\lambda + \alpha_1 \qquad \text{and} \qquad \psi_2 = 2\pi(BP_2)/\lambda + \alpha_2.$$

a) Calculate the intensity $I_A(t)$ and its time average; do the same for the point B. In 1956 R. Hanbury Brown and R.O. Twiss had the interesting idea of measuring the correlation of the intensity at the two points, that is

$$\langle I_A(t)I_B(t)\rangle - \langle I_A(t)\rangle\langle I_B(t)\rangle.$$

b) Calculate this quantity, assuming that the fluctuating parts of the phases from the two sources are independent. Show that the correlation varies with L in such a way that r may be determined if the distance from A, B (two points on earth) to the middle of the star (P_1, P_2 are its edges) is known. This procedure lies behind long baseline interferometry which is used to measure the diameters of stars, and the separation of stars too small to be resolved by ordinary telescopes.

35.5 Consider a laser beam 2 mm in diameter, with $\lambda = 6000$ Å. If the beam delivers 10^{17} photons/sec, what is the power of the laser? What is the radiation pressure that the beam can exert? If we were to consider the laser beam as a "target" made of photons, what is the number of photons/cm^3 in that target?

Suppose a beam of electrons moving with velocity very close to that of light intersects the laser beam at right angles. Assuming that the photon-electron scattering cross-section is 10^{-24} cm^2, how many electrons will be scattered per second out of a beam containing 10^{13} electrons/sec? This is a way of making a monochromatic high-energy photon beam in a high-energy electron accelerator facility, and has been so used at the Stanford Linear Accelerator Center.

36
Crystal Lattices and Phonons

Quantum mechanics has found its widest range of applications in the study of ordinary matter, in the field known as *solid-state physics*. It is a field so vast that books of the size of the present one are needed to provide introductory treatments. Our attempts at discussion of a few facets of solid-state physics are presented only to provide a feeling for the role that quantum physics plays in the properties of solids, beyond its role in explaining the structure of the atoms and molecules that are the constituents of solids.

Crystals Matter, that is, large aggregates of atoms, appears in the form of solids, at least at temperatures low enough such that kT is small compared with the energy required to break or deform the bonds that hold the atoms together. It is something of a mystery that these aggregates of atoms arrange themselves in regular arrays which we call crystals. Crystals have well-defined symmetries: for example, a cubic crystal preserves its form under $90°$ rotations about axes that originate at the center of the cube, and also under a variety of reflections and other rotations. Furthermore, a crystal, an array of atoms arranged on a cubic lattice (Fig. 36.1) say, is invariant in form under displacements by a lattice spacing in the x, y, or z directions, or in fact under a displacement

$$\mathbf{r} \rightarrow \mathbf{r} + \hat{\mathbf{i}}al + \hat{\mathbf{j}}am + \hat{\mathbf{k}}an \tag{36.1}$$

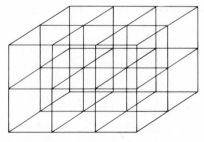

Figure 36.1. Section of a cubic lattice. The symmetry under a variety of reflections, rotations, and displacements is evident.

where l, m, n are integers and a is the lattice spacing for a cubic lattice. This symmetry, and other, more complicated ones for more spectacular crystals, should not distract us from the realization that the arrangement of atoms in lattices in the ground state represents a *loss of symmetry*. A liquid state would have symmetry under an arbitrary rotation and under an arbitrary displacement, and the laws of nature also have this larger symmetry. There is at present no fundamental understanding why the ground state solutions of a complex quantum mechanical system have less symmetry than the equations that they solve, but this phenomenon is a deep characteristic of complicated systems, and appears to be true also of elementary particle physics.

Because of limitations of space, we shall concentrate on crystals with a simple cubic symmetry, that is, crystals in which the equilibrium positions of the nuclei of the atoms are given by

$$\mathbf{R}_{nlm} = a(\hat{\imath}l + \hat{\jmath}m + \hat{k}n) \tag{36.2}$$

where l, m, and n are integers. There are forces between the atoms, and we shall make the approximation of considering the dominant forces to be those between nearest neighbors. As we saw in our discussion of molecular forces, bonds exist in regions of large overlap of electronic wave functions, and since these wave functions fall off exponentially beyond a certain range, this is a reasonable approximation to make in our qualitative considerations. As with molecules, there is a potential energy for the displacement of the ions from their equilibrium positions, and this we approximate by a harmonic potential. The motion of the ions, coupled though it is by the interparticle forces, can be decomposed into normal modes, each of which represents harmonic motion characterized by a definite frequency.

Lattice vibrations To see this in a simple context, let us consider a one-dimensional lattice, that is, a linear array of atoms. If we want to treat a finite array, and at the same time avoid having to treat the atoms at the ends ("surface effects") differently, it is useful to introduce periodic boundary conditions, which treat the array as if it were a long, closed chain (Fig. 36.2). Thus if the displacements of the atoms from equilibrium are denoted by x_i, we shall take

$$x_{N+1} = x_1. \tag{36.3}$$

Let m be the mass of the atom and K the force constant for the harmonic force between nearest neighbors. The equation of motion is then

$$m\frac{d^2 x_n}{dt^2} = -K(x_n - x_{n+1}) - K(x_n - x_{n-1}), \tag{36.4}$$

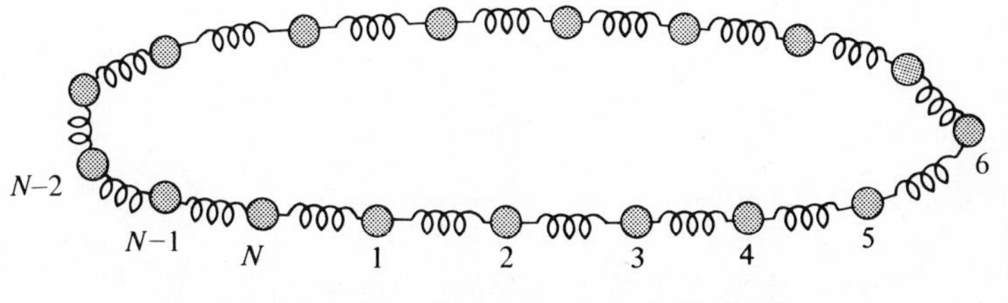

Figure 36.2. One-dimensional periodic lattice.

describing the pulls to the right as well as the pulls to the left. A particular mode of oscillation is characterized by a characteristic frequency ω_r. We expect, with N atoms, that is, with N degrees of freedom, that there will be N modes. For the rth mode, we have

$$x_n = A_{n,r} \cos \omega_r t, \tag{36.5}$$

which implies that

$$-\omega_r^2 m A_{n,r} = -K(2A_{n,r} - A_{n+1,r} - A_{n-1,r}). \tag{36.6}$$

We will guess the answer to this difference equation:

$$A_{n,r} = B_r \cos k_r na, \tag{36.7}$$

where B_r is an amplitude that characterizes the rth mode and k_r is an as-yet undetermined "wave number." It follows from

$$\cos(k_r a(n+1)) + \cos(k_r a(n-1)) = 2 \cos k_r an \cos k_r a$$

that

$$-\omega_r^2 m = -K(2 - 2 \cos k_r a),$$

which we rewrite in the form

$$\omega_r^2 = \frac{4K}{m} \sin^2 \frac{k_r a}{2} \tag{36.8}$$

or, equivalently,

$$\omega_r = 2\omega_0 \left| \sin \frac{k_r a}{2} \right|, \tag{36.9}$$

where

$$\omega_0 = (K/m)^{1/2}. \tag{36.10}$$

Figure 36.3 shows a plot of ω_r/ω_0 as a function of k_r. If we impose the boundary condition (36.3), that is,

$$A_{N+1,r} = A_{1,r}, \tag{36.11}$$

then it follows from Eq. (36.7) that

$$\cos k_r Na = 1,$$

that is,

$$k_r = \frac{2\pi}{Na} r \quad r = 1, 2, \ldots, N. \tag{36.12}$$

The most general motion is a superposition of all modes, with

$$x_n = \sum_{r=1}^{N} B_r \cos\left(na \frac{2\pi r}{Na}\right) \cos \omega_r t. \tag{36.13}$$

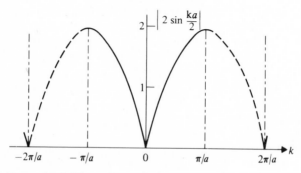

Figure 36.3. Dispersion relation connecting ω and k for a one-dimensional monatomic lattice.

We remark, as an aside, that with

$$q_r \equiv \sqrt{\frac{N}{2}} B_r \cos \omega_r t \qquad (36.14)$$

describing the "coordinate" of a particular mode, the lattice coordinates may be written as

$$x_n = \sum_{r=1}^{N} M_{nr} q_r, \qquad (36.15)$$

with the matrix M_{nr} given by

$$M_{nr} = \sqrt{\frac{2}{N}} \cos \frac{2\pi nr}{N} . \qquad (36.16)$$

It follows from the fact that

$$\sum_{r=1}^{N} \cos \frac{2\pi rl}{N} \cos \frac{2\pi rm}{N} = \frac{N}{2} \delta_{lm}$$

that the matrix has the property that $\sum_r M_{nr} M_{rl} = \delta_{nl}$. Hence the matrix is equal to its inverse, and we may write

$$q_r = \sum_{n=1}^{N} M_{rn} x_n . \qquad (36.17)$$

When N becomes very large, k_r becomes continuous (just like the wave number for a particle in a large one-dimensional box), and thus the frequency ω_r, related to it by Eq. (36.9) also becomes continuous. We denote these variables by k and ω respectively. It is clear from Eq. (36.7) that when $ka = \pi$, adjacent atoms are completely out of phase with each other, that is,

$$A_{n+1} = -A_n . \qquad (36.18)$$

When ka is increased beyond π, the phase relationship is the same as for a negative value of ka above $-\pi$, as is clear from Fig. 36.3. It is therefore sufficient to restrict the values of k to the region

$$-\frac{\pi}{a} \leqslant k \leqslant \frac{\pi}{a}, \qquad (36.19)$$

called the *first Brillouin zone*. Similar restrictions on the values of the vector **k** occur for three-dimensional lattices. With this choice of range, chosen symmetrically about $k = 0$, the discrete case Eq. (36.12) is best rewritten in the form

$$k_r = \frac{\pi}{aN} (2r - N) \quad r = 1 \ldots N. \qquad (36.20)$$

The classical solution in the continuum limit,

$$x_n = B \cos kna \cos \omega t \qquad (36.21)$$

represents a standing wave, which may be viewed as a superposition of waves traveling to the right and to the left:

$$x_n = \tfrac{1}{2} B \left[\cos (kna - \omega t) + \cos (kna + \omega t) \right]. \qquad (36.22)$$

Velocity of sound The group velocity of a disturbance associated with some value of k is given by

$$\left(\frac{d\omega}{dk}\right)_{k_0} = \left(\omega_0 a \cos \frac{ka}{2}\right)_{k_0}. \tag{36.23}$$

Thus a longitudinal wave with small k (long wavelength) propagates with velocity

$$v_s = \lim_{k \to 0} \frac{d\omega}{dk} = \omega_0 a = a\sqrt{\frac{K}{m}}. \tag{36.24}$$

The propagation of a density compression and rarefaction in the direction of propagation is associated with sound, and thus v_s is the *velocity of sound*. For a cubic lattice, if we consider only forces due to nearest neighbors in the longitudinal direction, an identical expression is obtained. Instead of the force constant K, we may use a more measurable quantity, the *compressibility*, or its reciprocal, the *bulk modulus* of the material. This is defined as

$$B = -V\frac{dp}{dV}, \tag{36.25}$$

where p is the pressure, that is, the force per unit applied to the material, and V its volume. For a cubic lattice in which only longitudinal compression is considered, we have $V = a^3$, $p = F/a^2$, and $dF/dV = (1/a^2)dF/dx = -K/a^2$, so that

$$B = \frac{K}{a}. \tag{36.26}$$

Since $m/a^3 = \rho$ is the mass density, we have the relation

$$v_s = \sqrt{\frac{B}{\rho}}. \tag{36.27}$$

Table 36.1 lists some of the relevant parameters for a few metals. The agreement between Eq. (36.27) and the measured values of the speed of sound is fair, although in some cases there are 30% discrepancies. These reflect the unrealistic approximation

Table 36.1 Selected Properties of Some Metals

	Density ρ in gm/cm^3	Nearest neighbor distance in Å	Bulk modulus B in 10^{10} $dynes/cm^2$	$\sqrt{B/\rho}$ in m/sec	Measured velocity of sound in m/sec	$\theta_D = \dfrac{\hbar v_s^{exp}}{k}$ $\times (6\pi^2 n_A)^{1/3}$ in $^\circ K$	Debye temperature θ_D deduced from specific heats $(^\circ K)$
Na	0.97	3.71	6.8	2650	2250	195	158
Cu	8.96	2.55	137.0	3910	3830	497	343
Zn	7.13	2.66	59.8	2900	3700	440	327
Al	2.70	2.86	72.2	5170	5110	591	428
Pb	11.34	3.49	43.0	1950	1320	125	105
Ni	8.90	2.49	186.0	4570	4970	660	450

that the restoring forces in longitudinal displacements are solely due to nearest neighbors in that direction, that is, the treatment of the problem as a one-dimensional one. The ions are actually close enough together that forces coming from ions several lattice spaces away are relevant, and forces in lateral directions are also important. In reality, the speed of sound propagation is a quantity that depends on the direction of propagation in a real solid. Nevertheless our simple considerations do allow us to relate the velocity of sound to atomic properties of solids. Although the discussion has been carried out for a simple cubic lattice, similar relations hold for more complex crystallographic forms, and the general qualitative agreement of Eq. (36.27) with experiment continues to hold.

Diatomic lattice The relation between the frequency and the wave number k is modified significantly when there are different atoms that make up a lattice, as, for example, in sodium chloride. Let us again consider a one-dimensional crystal, with the equilibrium positions of the atoms given by na (n an integer), such that atoms with mass m appear at the even-integer sites and atoms with mass M at the odd-integer sites (Fig. 36.4). The classical equation of motion, assuming a $\cos \omega t$ time dependence for all the displacements from equilibrium, leads to the relations

$$-m\omega^2 A_{2r} = K(A_{2r+1} + A_{2r-1} - 2A_{2r})$$
$$-M\omega^2 A_{2r+1} = K(A_{2r+2} + A_{2r} - 2A_{2r+1}). \tag{36.28}$$

If we choose

$$A_{2r} = Ae^{2ikra}$$
$$A_{2r+1} = Be^{ik(2r+1)a}, \tag{36.29}$$

with the understanding that only the real parts are relevant, we very easily get the relations

$$(-m\omega^2 + 2K)A = 2KB \cos ka$$
$$(-M\omega^2 + 2K)B = 2KA \cos ka. \tag{36.30}$$

The product of the two leads to the dispersion relation

$$Mm\omega^4 - 2K(M + m)\omega^2 + 4K^2 \sin^2 ka = 0, \tag{36.31}$$

whose solution is

$$\omega^2 = K\left(\frac{1}{m} + \frac{1}{M} \pm \sqrt{\left(\frac{1}{m} + \frac{1}{M}\right)^2 - \frac{4 \sin^2 ka}{mM}}\right). \tag{36.32}$$

The solution with the minus sign resembles that for a single atom lattice in the small k region. For small k it reads

$$\omega^2 \cong K\left(\frac{2k^2a^2}{mM}\right)\left(\frac{1}{m} + \frac{1}{M}\right)^{-1} = K\frac{2k^2a^2}{m + M}, \tag{36.33}$$

so that the only difference is that the mass is replaced by the average of the masses of the two atoms. This is the *acoustical branch* of the dispersion relation (see Fig. 36.5).

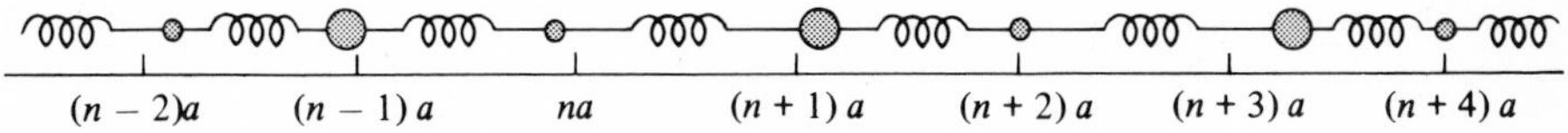

| $(n - 2)a$ | $(n - 1)a$ | na | $(n + 1)a$ | $(n + 2)a$ | $(n + 3)a$ | $(n + 4)a$ |

Figure 36.4. One-dimensional diatomic lattice.

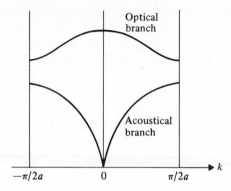

Figure 36.5. Dispersion relation for one-dimensional diatomic lattice. The acoustical and optical branches are shown in the figure. Note the size of the first Brillouin zone.

We note that the first Brillouin zone has its boundary at $k = \pm \pi/2a$. It is not the lattice spacing, but the repeat distance in the lattice, that determines the size of the zone. The solution with the plus sign describes the *optical branch* of the dispersion relation. The reason for this nomenclature is that motion described by that branch can be excited by an electric field of a light wave, when the atoms are oppositely charged, as in an ionic crystal.

Phonons The decomposition of the lattice motion into normal modes makes quantization straightforward. For the one-dimensional lattice, the relation (36.15) shows that the general motion may be viewed as a superposition of that of simple harmonic oscillators described by the coordinates q_r for which the equation of motion (following Eq. (36.14)) is

$$\ddot{q}_r + \omega_r^2 q_r = 0. \tag{36.34}$$

Thus the energy operator is of the form

$$H = \sum_{r=1}^{N} \left(\frac{p_r^2}{2m} + \frac{1}{2} m \omega_r^2 q_r^2 \right) \tag{36.35}$$

and the result of Chapter 24 is that

a) The energy eigenvalues are of the form

$$E = \sum_{r=1}^{N} \hbar \omega_r (n_r + \tfrac{1}{2}) \tag{36.36}$$

where n_r are integers ($n_r = 0, 1, 2, 3, \ldots$.)

b) The energy eigenfunctions are of the form

$$\prod_{r=1}^{N} \psi_{n_r} (q_r, \omega_r) \tag{36.37}$$

where the $\psi_n(q, \omega)$ are eigenfunctions of the harmonic oscillator. In particular, the ground-state eigenfunction for a single oscillator is

$$\psi_0(q, \omega) = \left(\frac{\omega m}{\hbar \pi}\right)^{1/4} e^{-\omega m q^2/2\hbar}, \tag{36.38}$$

so that the ground-state wave function for the lattice is

$$\Psi_0 = \prod_{r=1}^{N} \left(\frac{\omega_r m}{\hbar \pi}\right)^{1/4} e^{-\omega_r q_r^2 m/2\hbar} \tag{36.39}$$

We may describe the general state by saying that it consists of n_1 *phonons* of frequency ω_1, n_2 phonons of frequency ω_2, n_3 phonons of frequency ω_3, and so on. The name phonon is used in analogy with the name *photon* used to describe the state of excitation of the normal modes that describe the electromagnetic field in a cavity. There is a parallelism with electrodynamics: A classical elastic wave in a lattice may be described in terms of phonons traveling through the medium, just as a traveling electromagnetic wave consists of photons. In many ways a phonon of wave number **k** interacts with particles like electrons or neutrons (which are often used to study lattice structure) as if it had momentum **k**. In contrast to the photon, a phonon can exist only within a lattice and thus cannot carry any momentum that it absorbs in a collision out of the crystal: when a finite crystal absorbs a photon, say, the real momentum is, in the final analysis, always absorbed by the crystal as a whole. The fact that a state with a wave number k is indistinguishable from one with wave number $k \pm m\pi/a$, where m is an integer, shows that $\hbar k$ cannot be a real momentum. In a three-dimensional cubic lattice, there is a similar equivalence between a wave vector **k** and **k** ± **g**, where

$$\mathbf{g} = \frac{\pi}{a}(m_1 \hat{\mathbf{i}} + m_2 \hat{\mathbf{j}} + m_3 \hat{\mathbf{k}}) \quad (m_i \text{ integers}). \tag{36.40}$$

The quantity represented by **g** is called the reciprocal lattice vector, and it can be defined for lattices more complicated than a cubic one.

Heat capacity Let us consider a crystal lattice in equilibrium at a temperature T, and study the heat capacity. To do that, we need to calculate the average energy per unit volume U, and evaluate the specific heat at constant volume, defined by

$$C_v = \frac{\partial U}{\partial T}. \tag{36.41}$$

The average energy of a collection of oscillators of fixed frequency ω_0 was calculated in Chapter 13 (Eq. (13.30)) using the expression

$$\langle E \rangle = \frac{\sum\limits_{n=0}^{\infty} P_n E_n}{\sum\limits_{n=0}^{\infty} P_n} = \frac{\sum\limits_{n=0}^{\infty} \hbar\omega_0(n + \tfrac{1}{2}) \exp\left[-\hbar\omega_0(n + \tfrac{1}{2})/kT\right]}{\sum\limits_{n=0}^{\infty} \exp\left[-\hbar\omega_0(n + \tfrac{1}{2})/kT\right]}$$

$$= \frac{\hbar\omega_0}{e^{\hbar\omega_0/kT} - 1} + \frac{1}{2}\hbar\omega_0. \tag{36.42}$$

The simplest model, due to Einstein, assumes that all the oscillators have the same frequency. With N atoms in the lattice, and with each atom having three degrees of freedom, there are $3N$ normal modes. Thus, if the volume of the crystal is V,

$$U = \frac{3N}{V}\left(\frac{\hbar\omega_0}{e^{\hbar\omega_0/kT} - 1} + \frac{1}{2}\hbar\omega_0\right),$$ (36.43)

and the heat capacity is readily calculated. The Einstein model yields results that are much better than the classical result based on equipartition, that is, kT units of energy per mode:

$$U = \frac{3N}{V}kT$$ (36.44)

that obtains in the high-temperature limit.

The Einstein model does not work so well at low temperatures and it was improved on by P. Debye in 1913, before the development of quantum mechanics, but after the basic notion of quantum states was developed. The improvement comes from the realization that it is a poor approximation to assume that all oscillators have the same frequency. We found this not to be the case in Eq. (36.9) and therefore Eq. (36.43) should be replaced by

$$U = \frac{1}{V}\left(\sum_j \frac{\hbar\omega_j}{e^{\hbar\omega_j/kT} - 1} + \sum_j \frac{1}{2}\hbar\omega_j\right).$$ (36.45)

The counting of modes for a crystal containing N^3 atoms is accomplished as follows: Each mode is characterized by a triplet of integers (r_x, r_y, r_z) and, as in our discussion in Chapter 26, the number of modes may be associated with the volume N^3 in the lattice space of the integers. If $\Delta r_x, \ldots$ denote the spacings between the integers ($\Delta r_x = 1$, and so on), then we have

$$\frac{1}{V}\sum_j = \frac{1}{V}\sum_{r_x=1}^{N}\sum_{r_y=1}^{N}\sum_{r_z=1}^{N} 3\,\Delta r_x \Delta r_y \Delta r_z$$

and use Eq. (36.12) to write this as

$$\frac{1}{V}\sum_j = \frac{3}{V}\left(\frac{Na}{2\pi}\right)^3 \sum \Delta k_x \Delta k_y \Delta k_z.$$ (36.46)

The factor of three in front comes from the fact that there are altogether $3N^3$ degrees of freedom for the simple monatomic lattice that we are considering. The sum may be replaced by an integral since the spacing in wave number is so tiny. Thus we have

$$\frac{1}{V}\sum_j = \frac{3}{V}\frac{V}{(2\pi)^3}\int d^3\mathbf{k},$$ (36.47)

since the volume of the crystal is $(Na)^3$. A further approximation is made. The simplifying assumption that

$$|\mathbf{k}| = \frac{\omega}{v_s}$$ (36.48)

is made, and also that the range of $|\mathbf{k}|$, and therefore ω, is bounded by a value such that

$$\frac{3V}{(2\pi)^3} \int d^3\mathbf{k} = \text{total number of modes}$$

$$= 3N^3,$$

so that

$$\frac{3V}{(2\pi)^3} \cdot \frac{4\pi}{3} \left(\frac{\omega_{max}}{v_s}\right)^3 = 3N^3. \tag{36.49}$$

Hence

$$\frac{\omega_{max}}{v_s} = \left(6\pi^2 \frac{N^3}{V}\right)^{1/3}$$

$$= (6\pi^2 n_A)^{1/3}, \tag{36.50}$$

where n_A is the number of atoms per unit volume.
Now

$$U = \frac{3}{8\pi^3} \int_0^{k_{max}} d^3\mathbf{k} \frac{\hbar\omega}{e^{\hbar\omega/kT} - 1}$$

$$= \frac{3}{8\pi^3} \int_0^{\omega_{max}} 4\pi \left(\frac{\omega}{v_s}\right)^2 d\left(\frac{\omega}{v_s}\right) \frac{\hbar\omega}{e^{\hbar\omega/kT} - 1}$$

$$= \frac{3}{2\pi^2} \frac{\hbar}{v_s^3} \int_0^{\omega_{max}} d\omega \frac{\omega^3}{e^{\hbar\omega/kT} - 1}. \tag{36.51}$$

Next we calculate

$$\frac{\partial U}{\partial T} = \frac{3\hbar}{2\pi^2 v_s^3} \int_0^{\omega_{max}} d\omega \frac{\omega^3}{(e^{\hbar\omega/kT} - 1)^2} e^{\hbar\omega/kT} \frac{\hbar\omega}{kT^2},$$

which can be rewritten by making the change in variables

$$y = \frac{\hbar\omega}{kT}$$

and introducing the *Debye temperature* θ_D defined by

$$\theta_D = \frac{\hbar\omega_{max}}{k}. \tag{36.52}$$

A little algebra leads to

$$C_v = \frac{3}{2\pi^2} \frac{k^4 T^3}{\hbar^3 v_s^3} \int_0^{\theta_D/T} dy \frac{y^4 e^y}{(e^y - 1)^2}. \tag{36.53}$$

This may be rewritten using

$$\theta_D^3 = \frac{\hbar^3}{k^3} \omega_{max}^3 = \frac{\hbar^3}{k^3} v_s^3 \cdot 6\pi^2 n_A$$

$$C_v = 9n_A (T/\theta_D)^3 \int_0^{\theta_D/T} dy \frac{y^4 e^y}{(e^y - 1)^2}. \tag{36.54}$$

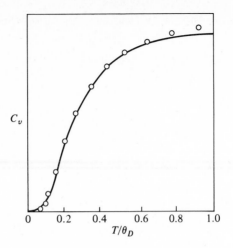

Figure 36.6. Sketch of C_v as a function of T/θ_D in Debye model. The theory agrees very well with experiment, except near $T/\theta_D = 1$.

This formula is in fairly good agreement with experiment. In particular it predicts a T^3 dependence for the specific heat at temperatures $T \ll \theta_D$, where the upper limit of the integral can be set equal to infinity. Figure 36.6 shows the dependence of C_v on T/θ_D. Table 36.1 also gives some values of θ_D as calculated from Eq. (36.50) and Eq. (36.52) and a comparison with θ_D determined from an experimental determination of C_v. The values of θ_D obtained in the two ways agree only qualitatively. The reason is that the formula for the specific heat (36.54) is a very approximate one, and does not take into account the real complexity of the process of mode counting required by the formula (36.45). It is very satisfying that the specific heat formula works as well as it does. In metals, the low temperature data often show the behavior

$$C_v = AT + BT^3 \tag{36.55}$$

with the linear term coming from the specific heat contribution of the free electrons in the lattice.

Anharmonic forces In our discussion of the specific heat of a lattice we considered the phonons to be in thermal equilibrium characterized by a temperature T. With our assumptions, that is, with the description of the energy operator by Eq. (36.35), such an equilibrium state is actually impossible to achieve because the phonons do not interact: the energy operator is a sum of terms, each of which leaves the phonon state unaltered. In our discussion of the ideal gas, we dealt with a similar situation, but noted that in reality the molecules of the gas do collide with each other, and through this process equilibrium is established. For the phonon "gas" the mechanism that leads to equilibrium is the presence of anharmonic terms in the potential experienced by an atom in the lattice. In addition to the terms

$$V_{\text{harm}} = \tfrac{1}{2}K(x_i - x_{i+1})^2 + \tfrac{1}{2}K(x_i - x_{i-1})^2, \tag{36.56}$$

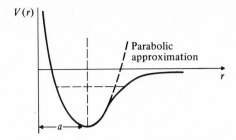

Figure 36.7. A schematic sketch of the potential energy as a function of the separation of two adjacent ions. The deviation from the parabolic shape is due to the anharmonic terms. The horizontal line shows the location of the classical motion.

characterizing the potential energy of a given atom, there are higher order terms, of the form (Fig. 36.7)

$$V_{\text{anharm}} = \tfrac{1}{3} \eta \, (x_{i+1} - x_i)^3 + \tfrac{1}{3} \eta \, (x_{i-1} - x_i)^3$$
$$+ \tfrac{1}{4} \xi \, (x_{i+1} - x_i)^4 + \tfrac{1}{4} \xi \, (x_{i-1} - x_i)^4$$
$$+ \cdots . \tag{36.57}$$

Even though the coefficients $\eta, \xi, \ldots$, are tiny, they modify the energy operator (36.35) by the addition of cubic and quartic terms involving the q's, with couplings between different normal modes. Effectively, a cubic term describes the emission of a phonon by another phonon, or the decay of a phonon into two phonons. Detailed consideration of these processes, and the quartic processes that describe phonon-phonon scattering (see Fig. 36.8) shows that in all cases phonon momentum is conserved, up to the equivalence of a phonon momentum $\mathbf{k}_i$ and $\mathbf{k}_i + \mathbf{g}$, as described

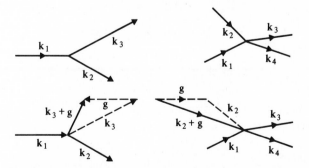

Figure 36.8. Diagrammatic representation of terms trilinear in q and quadrilinear in q, showing phonon-splitting and phonon scattering. Phonon momentum is conserved modulo $\mathbf{g}$, and the second line, in each instance, shows how the momentum vectors look.

by Eq. (36.40). It is these anharmonic terms that change the momenta and energies of individual phonons and allow the system to go through a sequence of states toward the equilibrium state. Anharmonic effects are also responsible for the fact that phonons cannot travel unhindered through a lattice. Because of them, phonons are characterized by a mean free path, that may range from tens to hundreds of Ångstroms, depending on the material and on the temperature. Because of this impedance to unrestricted phonon flow, it is possible for different parts of a solid to be at different temperatures, locally in a seeming state of equilibrium. Thermal energy is transferred from the region of higher temperature to the region of lower temperature by the phonons, and the rate of this transfer, that is, thermal conductivity, depends on the mean free path of the phonons.

Thermal expansion The anharmonic effects also explain the thermal expansion of solids. A qualitative understanding of this can be arrived at by looking at the classical motion of the ions about their equilibrium position (Fig. 36.7). For a parabolic potential, raising the temperature of the ion, that is, raising the energy, lifts the energy of the ion in the potential, but the average displacement a remains fixed; it is just the fluctuations about equilibrium that are larger. With the anharmonicities present, the potential flattens out more for large r, and the mean position of the ion is farther out as the energy increases. If we consider a one-dimensional system, and take the potential energy to be

$$V(x) = A(x-a)^2 - B(x-a)^3 \qquad (36.58)$$

for simplicity, and then calculate the mean value of x using the Boltzmann probability distribution (Eq. 11.51, for example)

$$\langle x \rangle = \frac{\int_{-\infty}^{\infty} dx\, x\, e^{-V(x)/kT}}{\int_{-\infty}^{\infty} dx\, e^{-V(x)/kT}} \qquad (36.59)$$

then, in the region in which Eq. (36.58) is valid,

$$e^{-V(x)/kT} \cong e^{-A(x-a)^2/kT}\left(1 + \frac{B}{kT}(x-a)^3\right). \qquad (36.60)$$

A straightforward calculation shows that $\langle x \rangle > a$, and that the deviation from the mean value a is proportional to T (this follows from a simple dimensional analysis, and is left to be shown by the reader). The dependence is in qualitative agreement with experiment.

Mössbauer effect The coupling of different atom motions through the harmonic (and anharmonic) interactions plays an important role in the *Mössbauer effect*. Consider a particular nucleus of an atom in a lattice, and let it undergo radiative decay. If the energy of the excited nuclear level is E units above the ground state, a first approximation gives

$$\omega = \frac{E}{\hbar}$$

for the photon frequency. This is not quite right, since a photon of frequency ω carries off momentum $\hbar\omega/c$; momentum conservation requires that the nucleus recoils with the same momentum, and therefore the nucleus takes $(\hbar\omega/c)^2/2M$ where M is the nuclear mass, of the total energy. Thus the frequency of the photon is shifted downward. Since its energy is decreased by $E^2/2Mc^2$, approximately, its frequency is shifted by

$$\Delta\omega = \frac{E^2}{2Mc^2\hbar} = \omega\frac{E}{2Mc^2}. \tag{36.61}$$

For the emission of a 100 keV photon by a nucleus for which Mc^2 is $940A$ MeV, approximately, $\Delta\omega/\omega \simeq 0.1/940A$, which is of the order of 10^{-6}. This may seem like a tiny amount, but it is, in fact, quite a bit larger than the natural linewidth associated with the decay. Thus if the photon emitted by an iron nucleus is to be resonantly absorbed by an iron absorber, the shift is disastrous: the photon energy is outside of the frequency range in which absorption can occur. R. Mössbauer discovered that, in fact, the absorption can occur and that happens because there is a finite probability that because of the harmonic forces that bind the emitting nucleus to the whole network of atoms forming the lattice, the crystal *as a whole* absorbs the recoil momentum, and since for the crystal M is 10^{20} times the nuclear mass, there is, effectively, no frequency shift. In phonon language, we might describe the initial state of the lattice by the ground-state wave function (36.39): There are no phonons present in the lattice. When a single nucleus emits a photon, it recoils, and the lattice is locally perturbed. In phonon language we say that the emission of the gamma ray is accompanied by the emission of one or more phonons. In the situation discovered by Mössbauer, the final state of the lattice is again the no-phonon state, that is, no elastic disturbances propagate through the lattice. An analogous situation might be the absence of the recoil of a rifle fired by someone standing on a battleship, with the battleship as a whole absorbing the recoil. This does not happen on battleships, because the probability of recoilless emission turns out to be proportional to e^{-R}, where the number R is given by

$$R = \frac{\text{recoil energy}}{\text{average energy spacing in oscillator}}$$
$$= \frac{\hbar^2 k^2/2M}{\hbar\langle\omega\rangle}. \tag{36.62}$$

Thus if the mean value of the frequency $\langle\omega\rangle$ is large, which corresponds to a large spring constant, that is, a stiff lattice, then the exponent R will be small, and the probability of "recoilless" emission will be sizable. If $\langle\omega\rangle$ is small, corresponding to a small spring constant, the probability falls very rapidly. For the classical emission of a rifle bullet, the density of states for a macroscopic system is enormous, since the average level spacing is incredibly small (recall the box potential), while the recoil energy is large. Thus recoilless emission is impossible for classical systems.

When the emission is recoilless, only thermal effects broaden the spectral line, and cooling reduces that. Thus shifts due to slow motion of the recoiling crystal can be measured. The Mössbauer effect has been used to measure the gravitational red shift of a photon "falling" from the top of a building, by moving the absorber with a velocity appropriate to the shift. A photon falling from a height h in a gravitational field g will experience a shift that, by the equivalence principle, corresponds to the Doppler shift, due to the velocity of free fall through such a height. Thus a receiver, moving with such a velocity "with" the photon, will absorb the photon resonantly. It is in this way that the terrestrial test of the gravitational red shift was first carried out.

NOTES AND COMMENTS

1. The difference equation (36.4) goes over into the wave equation in the continuum limit, that is, when the lattice spacing $a \to 0$. The two equations are discussed in detail in most books on wave motion. In particular, see F.S. Crawford, Jr. *Waves*, Berkeley Physics Course, Vol. 3, McGraw-Hill, New York, 1968.

2. The Debye theory of specific heat is particularly good at low temperatures, because there the simplified dispersion relation, relating ω and k, is a better approximation. The contributions from the free electrons must, of course, be included. These will be discussed in the next chapter. The Debye temperature is not a completely well-defined quantity. The specific heat of a lattice depends, in this model, on that parameter θ_D, and it may be determined from the data in a variety of ways, either by getting the best overall fit of Eq. (36.54) to the data, or by fitting at a variety of temperatures. When the latter is done, one finds a slow variation of θ_D with temperature, indicating that the model is only an approximate one. For further discussion, see C. Kittel, *Introduction to Solid State Physics*, John Wiley, New York, 1976; N.W. Ashcroft and N.D. Mermin, *Solid State Physics*, Holt, Rinehart, and Winston, New York, 1976; or for that matter, any one of the many other books on solid state physics. The data presented in Table 36.1 may be found in the *American Institute of Physics Handbook* (Third Edition) McGraw-Hill, 1971, or in the *Handbook of Chemistry and Physics*, The Chemical Rubber Co., Cleveland, Ohio, any year.

3. Limitations of space have not allowed us to discuss the Mössbauer effect in greater detail. The effect has proved to be an important tool in chemistry and biology. A slightly more complete discussion, together with a number of references, may be found in H. Frauenfelder and E.M. Henley, *Subatomic Physics*, Prentice-Hall, Englewood Cliffs, N.J., 1974.

4. A rough way to see how the probability of recoilless emission is calculated (Eq. (36.62)) is to consider the matrix element for the emission of a photon by the ith ion. This is given by a formula that in a one-dimensional model reduces to

$$\text{const.} \int dq_1 dq_2 \ldots dq_N \Psi_0^*(q_1 \ldots q_N) x_i e^{ikx_i} \Psi_0(q_1 \cdots q_N)$$

where $\Psi_0(q_1 \cdots q_N)$ is the ground-state wave function for the lattice. This involves

$$\frac{1}{i} \frac{\partial}{\partial k} \int dq_1 \ldots dq_N \prod_r e^{-\omega_r M q_r^2 / \hbar} e^{ikx_i}$$

aside from a constant; we have used the explicit form Eq. (36.39). Since

$$x_i = \sum_s M_{is} q_s$$

as in Eq. (36.15), the integral is Gaussian, and can be carried out explicitly. The result, Eq. (36.62), emerges from the calculation.

PROBLEMS

36.1 The Debye model can be improved slightly by recognizing that the velocities of sound, that is, of propagation of elastic waves, are different for longitudinal vibrations (v_l) and for transverse vibrations (v_t). Since there are two transverse directions and one longitudinal, the factor $1/v_s^3$ in Eq. (36.49) should be replaced by

$$\frac{1}{v_s^3} \to \frac{1}{v_l^3} + \frac{2}{v_t^3}.$$

Given that in copper, whose density is $8.96 \, \text{gm/cm}^3$, the values for the velocities are $v_l = 4.56 \times 10^5$ cm/sec and $v_t = 2.25 \times 10^5$ cm/sec. Calculate ω_{max} from this information, and hence θ_D. How does this compare with the experimental value given in the table, and with the value obtained with a single velocity of sound also given in the table?

36.2 Use Eqs. (36.59) and (36.60) to calculate the difference $\langle x \rangle - a$. Use this result to calculate the coefficient of thermal expansion defined by

$$\alpha = \frac{1}{l} \frac{\partial l}{\partial T} \qquad l = \langle x \rangle$$

as a function of the anharmonicity coefficient B. Repeat the calculation, with the term $-C(x-a)^4$ added to $V(x)$, assuming that this term, too, can be treated to first order in C/kT as is the case with the B term in Eq. (36.60).

36.3 Use the Debye method to calculate the specific heat C_v of a two-dimensional periodic lattice consisting of N^2 identical atoms. Show that at low temperatures $C_v \propto T^2$ and obtain the high-temperature behavior.

37
Metals

Classical theory of conductivity Quantum mechanics has proved essential in explaining a large class of phenomena connected with the conduction of electricity. To see this, it is useful to first consider the classical theory of conductivity developed by Drude (1900). After the discovery of the electron, it was realized that the presence of free electrons in metals could explain Ohm's law. The mechanism is the acceleration of the electrons by the electric field imposed from the outside, which is counterbalanced by the resistive forces that arise from the collisions of the electrons with the ions in the metal. The equation of motion of an electron is given by

$$m\frac{dv}{dt} = eE - \frac{m}{\tau}v, \tag{37.1}$$

where E is the electric field, m is the electron mass, and the resistive force is taken proportional to the velocity. The coefficient is parametrized in terms of a quantity with the dimensions of time, τ. The solution of the equation is easily obtained:

$$v(t) = \frac{eE\tau}{m}(1 - e^{-t/\tau}). \tag{37.2}$$

The terminal velocity, also called the *drift velocity*, is given by $eE\tau/m$. The current density for a gas of electrons containing n electrons/cm^3 is

$$j = env_d = en\frac{eE\tau}{m} = \frac{e^2 n\tau}{m}E. \tag{37.3}$$

Thus Ohm's law emerges

$$E = \rho j, \tag{37.4}$$

with the resistivity given by

$$\rho = \frac{m}{e^2 n\tau}. \tag{37.5}$$

It is in the estimate of τ that quantum mechanics plays its crucial role. τ is obviously some sort of relaxation time associated with the frictional forces due to collisions. If the electrons move with some mean velocity $\bar{v}$, and the mean free path of the electrons in the metal is λ, then

$$\tau = \lambda/\bar{v}. \tag{37.6}$$

The mean free path may be related to the effective cross-sectional area for collisions σ. If the density of ions is n_i, then the number of collisions made by an electron moving with velocity $\bar{v}$ is equal to the number of ions encountered in one second, that is

$$n_i \times \text{(volume of cylinder swept out in one second)}$$
$$= n_i \bar{v} \sigma. \tag{37.7}$$

Thus the distance between collisions is

$$\lambda = \frac{\bar{v}}{n_i \bar{v} \sigma} = \frac{1}{n_i \sigma} \tag{37.8}$$

and therefore

$$\rho = \frac{m \bar{v} \sigma}{e^2} \left(\frac{n_i}{n} \right). \tag{37.9}$$

The classical theory takes $\bar{v}$ equal to the mean velocity of a Maxwell–Boltzmann gas at temperature T

$$\bar{v} = \frac{\int_0^\infty 4\pi v^2 \, dv \, v e^{-mv^2/2kT}}{\int_0^\infty 4\pi v^2 \, dv \, e^{-mv^2/2kT}} = \left(\frac{8kT}{\pi m} \right)^{1/2} \tag{37.10}$$

and replaces σ by the effective area of an atom, that is, $\sim 10^{-16} \, \mathrm{cm}^2$. Although at room temperatures the resistivity fortuitously comes out of the right order of magnitude, the time dependence $T^{1/2}$ disagrees with the experimentally measured *linear* T dependence. Actually both the choice of $\bar{v}$ and σ in the classical theory are wrong.

Free electron model[†] The first deviation from classical theory comes from the recognition of the role that the Pauli exclusion principle plays in the description of the "free" electrons in a metal. The electrons are not really free, but the valence electrons are lightly bound to the ions and easily tunnel into the potential well of the adjacent ion, and so on, so that a small number of electrons per ion is free to move through the metal. The number is typically of order 1.0–1.3, though it can be as large as 3.5 (for aluminum). The metal may be viewed as a box containing the free electrons, and the Pauli principle requires that at low temperatures the electrons fill the Fermi sea, that is, fill the energy levels up to the Fermi energy E_F, which for a density of n electrons/cm^3 is

$$E_F = \frac{\hbar^2 \pi^2}{2m} \left(\frac{3n}{\pi} \right)^{2/3}. \tag{37.11}$$

The electrons at the top of the Fermi sea have a velocity

$$v_F = \left(\frac{2E_F}{m} \right)^{1/2} = \frac{\pi \hbar}{m} \left(\frac{3n}{\pi} \right)^{1/3}. \tag{37.12}$$

Let us stop to calculate some numbers: in copper, whose density is $8.96 \, \mathrm{gm/cm}^3$ and atomic weight is approximately 64, the number of atoms per cm^3 is

$$\frac{\rho(\mathrm{gm/cm}^3)}{A m_N(\mathrm{gm})} = \frac{8.96}{64 \times 1.6 \times 10^{-24}} = 8.7 \times 10^{22} \, \mathrm{cm}^{-3},$$

so that

$$E_F = \frac{(3.14 \times 1.05 \times 10^{-27})^2}{2 \times 0.9 \times 10^{-27}} \left(\frac{3 \times 8.7 \times 10^{22}}{3.14} \right)^{2/3} \simeq 1.15 \times 10^{-11} \mathrm{erg}, \tag{37.13}$$

[†] First elaborated by H. Bethe and A. Sommerfeld.

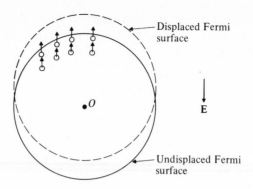

Figure 37.1. Displacement of Fermi sphere
when an electric field is applied.

assuming one free electron per atom. This corresponds to a "Fermi temperature"

$$T_F = E_F/k \cong \frac{1.15 \times 10^{-11}}{1.38 \times 10^{-16}} \simeq 8.3 \times 10^{4}\,^{\circ}\text{K} \qquad (37.14)$$

which is very large compared with room temperature ($\approx 300^{\circ}$K). Thus it does not
make any difference whether we consider the Fermi gas at zero temperature, or at
room temperature.

We will now argue that the relevant velocity in Eq. (37.10) is the Fermi velocity
v_F. When there is no electric field, we may consider the electrons confined to a sphere
in momentum space, with the radius given by the Fermi momentum $p_F = mv_F$. In
reality the Fermi surface is much more complicated, but the argument does not
change. When an electric field is applied in the negative z direction (Fig. 37.1), the
electrons would like to move in the positive z direction. It is, however, only those on
the surface in the upper hemisphere that find empty levels to move into, and *these
electrons have a velocity v_F*. The electrons that move into unoccupied states leave
vacant states into which other electrons move. In a short time the entire Fermi sphere
in momentum space is displaced by a small amount. This displacement cannot go on
forever. Electrons scatter from ions and are deflected. The ions are for all practical
purposes infinitely massive, so that when electrons scatter, their momentum changes
only in direction but not in magnitude. In Fig. 37.2, an electron in the forward part
of the Fermi sphere (momentum represented by OA) can undergo scattering, that is,
go into a state where the momentum OA′ lies in an unoccupied region in the Fermi
momentum space, an electron with a "backward" pointing momentum vector, such
as OB, cannot scatter, since all possible OB′ vectors lie inside the displaced Fermi
sphere in which all states are occupied. Thus there is a preference for the scattering
of electrons that lie in the forward (positive z direction) part of the displaced Fermi
sphere. These electrons get scattered to the rear, leaving empty states which are again
filled by electrons accelerated by the electric field. The net result is that the Fermi
sphere is displaced by a certain small amount (mv_{drift}) and stays there in a state of
dynamical equilibrium as long as the field is on. When the field is turned off, the

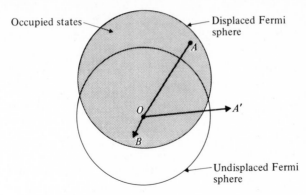

Figure 37.2. Frictional forces in conductivity: the forward momentum vectors (like OA) can be scattered without changing length, while the backward ones cannot, because of occupied states.

backward scattering leaves states empty, and these are no longer filled. The Fermi sphere returns to its original position.

The classical estimate of the mean velocity is of the order of 10^7 cm/sec, according to Eq. (37.10), whereas the Fermi velocity, using Eq. (37.13) is 1.6×10^8 cm/sec for copper.

Electron-lattice scattering A compensating discrepancy in the classical and quantum mechanics values occurs for the scattering cross sections. When we consider the scattering of electrons by ions, we assume that the latter form a crystal lattice. The large difference between the classical estimate for the mean free path, and the quantum mechanical one stems from the result that *for a perfectly rigid lattice, with perfect periodicity, there is never any electron scattering.* A qualitative argument that supports this conclusion may be made by recalling that waves sent through a periodic structure will go through it without attenuation, except for reflections at angles that satisfy the Bragg condition, and that electrons behave like waves. The result actually follows from a mathematical theorem, known as Bloch's theorem (or Floquet's theorem in the mathematical literature), that states:

If we have a potential that is periodic, so that

$$V(x + a) = V(x) \tag{37.15}$$

in one dimension, then solutions of

$$-\frac{\hbar^2}{2m} \frac{d^2 \psi(x)}{dx^2} + V(x)\psi(x) = E\psi(x) \tag{37.16}$$

must be of the form

$$\psi_k(x) = e^{ikx} u_k(x), \tag{37.17}$$

where the $u_k(x)$ are functions that depend on k, and also have the property that

$$u_k(x + a) = u_k(x). \tag{37.18}$$

The value of the parameter k depends on the energy, and there are two values of k for each value of the energy.

The proof of the theorem in one dimension (it is easily generalized to three dimensions) is not particularly difficult, but it is long, and illuminates no physics, so that we do not present it here. We limit ourselves to the observation that it follows from Eq. (37.17) that

$$\begin{aligned}
\psi_k(x + a) &= e^{ik(x+a)} u_k(x + a) \\
&= e^{ik(x+a)} u_k(x) \\
&= e^{ika} \psi_k(x),
\end{aligned} \tag{37.19}$$

or, more generally,

$$\psi_k(x + na) = e^{ikna} \psi_k(x).$$

Thus with a one-dimensional crystal containing N atoms (potential centers), the periodic boundary condition

$$\psi_k(x + Na) = \psi_k(x) \tag{37.20}$$

yields

$$e^{ikNa} = 1,$$

that is,

$$k = \frac{2\pi n}{Na} \qquad n = 0, 1, \ldots N - 1. \tag{37.21}$$

This restriction yields a further restriction on the energy eigenvalues that are connected with k.

As an example of a one-dimensional periodic lattice, R. Kronig and W. Penney studied a series of potential wells of depth V_0, width b, and separation between centers a. Here the Schrödinger equation can be solved exactly and the energy spectrum is easily calculated. The algebra is tedious, so that the reader is referred to other books for the solution; we limit ourselves to quoting the result for the energy in the limit that $b \to 0$, $V_0 \to \infty$ such that

$$bV_0 = \xi \frac{\hbar^2}{2ma}; \quad \xi \text{ fixed.} \tag{37.22}$$

If the energy is written in the form

$$E = \frac{\hbar^2 q^2}{2m}, \tag{37.23}$$

then the eigenvalue condition reads

$$\cos qa + \frac{1}{2} \xi \frac{\sin qa}{qa} = \cos ka, \tag{37.24}$$

giving an example of a relation between k and the energy.

The Bloch solution (37.17) shows quite clearly that there is no attenuation for a wave traveling to the right ($k > 0$), say, which is equivalent to the statement that electrons do not scatter in a perfectly periodic lattice. Does this mean that the mean free path of an electron in a metal is infinite? Not quite, because one can think of a number of ways that real systems deviate from our image of a perfectly periodic lattice. One is a temperature-independent effect that we classify under the general heading of *impurities*. This may include actual foreign atoms in the lattice, missing ions (holes), dislocations, and so on. Such impurities do scatter electrons, and they give rise to a scattering cross section that is of the order of

$$\sigma_{\text{imp}} = f \sigma_{\text{at}} \tag{37.25}$$

where f is the fraction of impurity ions in the lattice, and σ_{at} is a typical atomic cross section of the order of $10^{-16}\,cm^2$. With f typically in the 10^{-5} range, except for purified metals, this gives a contribution of the order of $10^{-21}\,cm^2$.

The much more important temperature-dependent effect arises because at $T \neq 0$ the lattice vibrates, and this destroys perfect periodicity. In terms of the normal modes of the lattice, the phonon states, we may describe this process as *electron-phonon scattering*. A proper treatment of this effect is actually very difficult, but it is possible to give a crude estimate of the cross section, which we take to be

$$\sigma_{latt} = \pi(\langle x^2 \rangle + \langle y^2 \rangle), \tag{37.26}$$

where $\langle x^2 \rangle$ and $\langle y^2 \rangle$ are the mean square deviations of an ion from the equilibrium position. We take z to be the electron propagation direction. For a harmonic oscillator in equilibrium at temperature T we have

$$\tfrac{1}{2}M\omega^2 \langle x^2 \rangle = \tfrac{1}{2}M\omega^2 \langle y^2 \rangle = \tfrac{1}{2}kT \tag{37.27}$$

so that, using $\omega = k\theta_D/\hbar$, where θ_D is the Debye temperature defined in the last chapter, the lattice contribution to the cross section is

$$\sigma_{latt} = \frac{2\pi\hbar^2}{Mk\theta_D}\left(\frac{T}{\theta_D}\right). \tag{37.28}$$

For copper, with $\theta_D \approx 200°K$ and $M = 64 \times 1.6 \times 10^{-24}\,gm$, this contribution is of magnitude $3.7 \times 10^{-18}\,cm^2$, so that for relatively pure metals, the electron-phonon cross section dominates. The cross section is proportional to the temperature T and this behavior has been observed. For very pure metals the linear T behavior persists to very low temperatures, leading to a very small resistivity. The numbers for the conductivity $(1/\rho)$ obtained in this way are typically within a factor of two of the measured values; this must be regarded as a good result, given the crudity of our treatment of the electron-phonon interaction.

Band structure Our discussion so far has not dealt with the question: Why are some materials conductors, while others are insulators or semiconductors? Although the answer to this question for any given material may depend in some complicated way on the structure of that material, the broad classification into conductors, insulators, and semiconductors comes about as a result of the interplay of the periodic structure of the material and the Pauli exclusion principle. The periodicity of the potential has as its consequence the bunching of energy levels of the electrons into *bands*. To see an explicit example of this, consider the Kronig–Penney model in the approximation made in Eq. (37.22), which corresponds to a periodic sequence of very sharp (delta function) potentials. The states may be labeled by the index k which in a finite chain takes on the values indicated in Eq. (37.21) but which in the continuum limit may be chosen to lie in the first Brillouin zone (compare Chapter 36)

$$-\pi < ka < \pi. \tag{37.29}$$

If there are no potentials, then Eq. (37.24), with $\xi = 0$, shows that the energy takes on all possible values allowed by the range of k:

$$E = \frac{\hbar^2 q^2}{2m} = \frac{\hbar^2 k^2}{2m}. \tag{37.30}$$

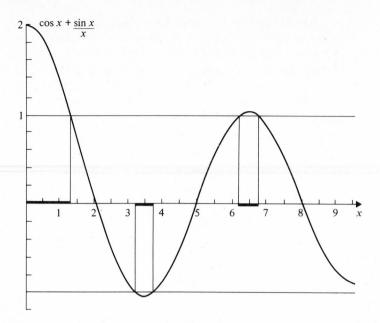

Figure 37.3. Plot of Eq. (37.24) with $qa = x$ and $\xi = 2$. Note that the regions in which $\cos x + (\sin x)/x$ lies outside the $(-1, 1)$ interval are forbidden, so that they correspond to energy gaps. The energy gaps are marked with heavy lines on the x-axis.

When $\xi \neq 0$, then there are constraints on the allowed values of qa that are imposed by the fact that the right side of Eq. (37.24) must lie between -1 and $+1$. These constraints lead to the bunching of energy levels into bands. This is shown in Fig. 37.3; the next Fig. 37.4 shows the energy as a function of k. In both figures we see the existence of an energy gap at the location of the first Brillouin zone.

The existence of the energy gaps can be discussed qualitatively without reference to a particular model. Let us assume that the electrons are almost free. In that case their energy is given by Eq. (37.30) and their wave functions are traveling waves $e^{\pm ikx}$. Even with a weak coupling to the lattice, there will be Bragg reflection when the waves reflected from successive atoms differ in phase by an integral number of 2π's, that is, when

$$k = \pm n\pi/a. \tag{37.31}$$

The back-and-forth pattern of reflections gives rise to standing waves, and these must be the sum or difference of $e^{i\pi x/a}$ and $e^{-i\pi x/a}$:

$$\psi_{\text{even}} = \cos \pi x/a$$
$$\psi_{\text{odd}} = \sin \pi x/a. \tag{37.32}$$

These standing waves are degenerate in energy in the absence of an interaction with the lattice. Once the interaction is taken into account, the energies will differ. The reason is that the ions are located at $x = ma$ where m are integers. The even wave

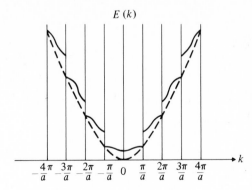

Figure 37.4. Plot of energy as a function of wave number k. The dashed line shows the free-electron relation $E(k) = \hbar^2 k^2/2m$. The energies agree at the onset of the gaps, and they approach each other for increasing k values.

function peaks at these locations, while the odd wave function peaks in between the ion locations. Since the electron-ion interaction is attractive, the energy of the standing wave corresponding to the $\cos^2 \pi x/a$ distribution is lowered relative to the other standing wave. Thus at $k = n\pi/a$ there is a splitting of the energy levels, which shows up as the gaps that we see in Fig. 37.4, for example.

Tunneling and band formation Yet another way of understanding the band structure in a qualitative way is to consider a number of identical atoms placed in a linear array, with the interatomic spacing large, (Fig. 37.5). Each of the atoms will be characterized by a number of energy levels. With N atoms, each energy state will be N-fold degenerate. When the atoms are brought closer together, tunneling between the potential wells becomes possible, and the degeneracy is lifted, as was seen for the case of $N = 2$ at the end of Chapter 23. The lowest lying state now spreads into a band of N very closely spaced states. The spread is small, since for electrons in these states the barrier to tunneling is large. Higher lying states have a bigger spread, and it is even possible for the bands to overlap. This point of view shows that band structure does not require a perfectly periodic lattice for its existence, which is in agreement with experience. For example, sodium, which is an excellent conductor, remains one even when it is melting, even though the periodic structure is no longer intact when this occurs.

Conductors, insulators, and semiconductors Given the band structure, one can understand the conduction properties of materials in a qualitative way. When the bands are partially filled, then there are unoccupied energy levels that can be filled by electrons when they acquire energy from the application of an electric field, and the material behaves like a conductor. When the bands are filled, the Pauli exclusion principle

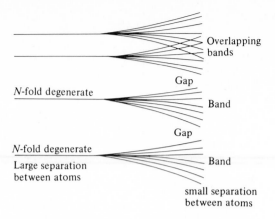

Figure 37.5. Band structure as a consequence of the lifting of degeneracy in N-atom system. Generally the tighter-bound electron orbits overlap less, and therefore the splittings are smaller. This makes for narrower bands and wider gaps.

forbids electronic acceleration by an external field, and we have an insulator. If the external electric field is very large, the lattice will be distorted, and the electron can tunnel through to the new band; the insulator breaks down. For the higher-lying atomic levels, the bands can be separated by quite narrow gaps. Although the material is technically an insulator, it becomes conducting at higher temperatures, because the electrons can be thermally excited into the empty conduction band above the gap. In terms of the picture of the Fermi surface in momentum space, the gap is shown in Fig. 37.6 as surrounding the Fermi sphere, preventing its displacement. For a thin gap, the fuzzing out of the edge of the Fermi surface by thermal effects allows a small shift of the Fermi sphere. We will discuss semiconductors further in the next chapter. At this point we turn to the effect of temperature on the Fermi surface, that is, to the "fuzzing out" of that sharp boundary. This is a matter of interest because of the need to know the contribution of the electrons to the heat capacity of metals. It is known experimentally that the heat capacity is almost completely determined by the properties of the lattice, and the absence of an electronic contribution can be understood to follow from the exclusion principle: most electrons cannot absorb energy because they would have to move to higher energy levels that are already occupied. Only some fraction of electrons near the top of the Fermi sea, roughly of order kT/E_F, can be thermally excited. To put this on a more quantitative basis, we derive the analog of the Boltzmann factor for fermions.

Fermi distribution Let us recall our derivation of the Boltzmann factor. We considered (cf. Chapter 11, including the Notes and Comments) molecules distributed with equal

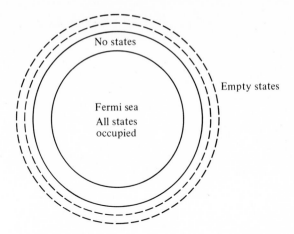

Figure 37.6. Pictorial representation of insulator (and semiconductor) in which lattice structure creates an energy gap above the Fermi surface, so that electrons cannot respond to the electric field at zero temperature.

probability over "cells" characterized by an energy E_i and degeneracy g_i. The number of ways in which N molecules could be so distributed was found to be

$$W(n_1, n_2, \ldots) = \frac{N!}{n_1! \, n_2! \, n_3! \ldots} (g_1)^{n_1} (g_2)^{n_2} \ldots \qquad (37.33)$$

The numbers n_i represent the occupation number for the ith cell, and the most probable distribution was obtained by maximizing $W(n_1, n_2, \ldots)$ subject to the conditions that

$$\sum n_i = N$$

$$\sum n_i E_i = E. \qquad (37.34)$$

The result was found to be

$$n_i = g_i e^{-\lambda} e^{-E_i/kT}. \qquad (37.35)$$

Let us repeat this calculation with fermions. If the ith cell is characterized by a degeneracy g_i, then, since each level can be occupied by at most one fermion, n_i levels will be occupied and $g_i - n_i$ will be unoccupied. The number of ways in which this can happen is

$$\frac{g_i!}{n_i! \, (g_i - n_i)!}. \qquad (37.36)$$

Thus the function $W(n_1, n_2, \ldots)$ is the product of such factors over all cells

$$W(n_1, n_2, \ldots) = \prod_r \frac{g_r!}{n_r! \, (g_r - n_r)!}. \qquad (37.37)$$

We find the extremum of log W as in the Boltzmann case, and include the constraints (37.34) using the method of Lagrange multipliers. This yields

$$d \log W - \alpha \sum_r dn_r - \beta \sum_r E_r dn_r = 0. \tag{37.38}$$

Using Stirling's formula this gives

$$d \sum_r \{ g_r \log g_r - g_r - n_r \log n_r + n_r$$
$$- (g_r - n_r) \log (g_r - n_r) + (g_r - n_r)$$
$$- \alpha n_r - \beta E_r n_r \} = 0. \tag{37.39}$$

A little algebra gives

$$\sum_r dn_r (- \log n_r + \log (g_r - n_r) - \alpha - \beta E_r) = 0. \tag{37.40}$$

Since the constraints have been taken into account, each coefficient of dn_r must vanish. This gives

$$n_r = \frac{g_r}{e^{\alpha + \beta E_r} + 1}. \tag{37.41}$$

Since at high energies this must go over into the Boltzmann (classical) distribution, because for very improbable configurations the constraint imposed by the exclusion principle plays a minor role, we can identify β with $1/kT$. The coefficient α is determined by the requirement that

$$N = \sum_r \frac{g_r}{e^{\alpha} e^{E_r/kT} + 1}. \tag{37.42}$$

It will in general depend on the temperature. At low temperatures, the parameter α should not differ much from its value at $T = 0$. There, however, it must have a form such that for $E_r < E_F$ the occupation number is g_r, while for $E_r > E_F$ it is zero. This is satisfied if we choose $\alpha = -E_F/kT$ so that

$$n_r = \frac{g_r}{e^{(E_r - E_F)/kT} + 1}. \tag{37.43}$$

We can now write down an expression for the expectation value of the energy

$$\langle E \rangle = \sum E_r n_r = \sum_r \frac{g_r E_r}{e^{(E_r - E_F)/kT} + 1}. \tag{37.44}$$

For a free electron gas

$$E = \frac{\hbar^2 k^2}{2m} \tag{37.45}$$

and the sum over cells in the continuum limit is just

$$\sum_r g_r (\quad) \rightarrow \frac{2V}{(2\pi)^3} \int d^3k (\quad). \tag{37.46}$$

This differs from (36.47) calculated for phonon modes only in that there are two electron states (spin up and down) for each level in the box of volume V. Using Eq.

(37.45) we can write

$$d^3k = 4\pi k^2 dk = 2\pi \sqrt{k^2} dk^2$$

$$= (2\pi) \left(\frac{2m}{\hbar^2}\right)^{3/2} \sqrt{E}\, dE. \tag{37.47}$$

Thus we can write, for small T

$$N = \frac{V}{2\pi^2} \left(\frac{2m}{\hbar^2}\right)^{3/2} \int dE \sqrt{E}\, (e^{(E-E_F)/kT} + 1)^{-1}$$

$$\langle E \rangle = \frac{V}{2\pi^2} \left(\frac{2m}{\hbar^2}\right)^{3/2} \int dE \sqrt{E}\, E(e^{(E-E_F)/kT} + 1)^{-1}. \tag{37.48}$$

Strictly speaking, $\partial N/\partial T$ should vanish. It does so, very nearly, for low T. Thus

$$C_v = \frac{\partial}{\partial T} \langle E \rangle$$

$$= \frac{\partial}{\partial T} (\langle E \rangle - N E_F)$$

$$= \frac{\partial}{\partial T} \frac{V}{2\pi^2} \left(\frac{2m}{\hbar^2}\right)^{3/2} \int_0^\infty dE \sqrt{E}\, (E - E_F)(e^{(E-E_F)/kT} + 1)^{-1}$$

$$= \frac{V}{2\pi^2} \left(\frac{2m}{\hbar^2}\right)^{3/2} \frac{1}{kT^2} \int_0^\infty dE \sqrt{E}\, (E - E_F)^2 \frac{e^{(E-E_F)/kT}}{(e^{(E-E_F)/kT} + 1)^2}. \tag{37.49}$$

We now relate V to N, the total number of particles and the Fermi energy E_F,

$$E_F = \frac{\hbar^2}{2m} \left(\frac{3\pi^2 N}{V}\right)^{2/3},$$

so that

$$\frac{V}{2\pi^2} \left(\frac{2m}{\hbar^2}\right)^{3/2} = \frac{3}{2} \frac{N}{E_F^{3/2}}. \tag{37.50}$$

Next, we introduce $u = E - E_F$ as a new variable, and write

$$C_v = \frac{3N}{2E_F} \frac{1}{kT^2} \int_{-E_F}^\infty du \sqrt{1 + \frac{u}{E_F}} \frac{u^2 e^{u/kT}}{(e^{u/kT} + 1)^2}. \tag{37.51}$$

For small T, that is, for $kT \ll E_F$, or $T \ll T_F$, the integrand becomes very tiny for u away from zero, so that we may replace the square root by unity and extend the lower limit in the integral to $-\infty$. We then get

$$C_v = \frac{3N}{2E_F} \frac{1}{kT^2} \int_{-\infty}^\infty du\, u^2 \frac{e^{u/kT}}{(e^{u/kT} + 1)^2}$$

$$= \frac{3N}{2E_F} \frac{1}{kT^2} (kT)^3 \int_{-\infty}^\infty dx\, x^2 \frac{e^x}{(e^x + 1)^2}$$

$$= \frac{\pi^2}{2} Nk \frac{T}{T_F}, \tag{37.52}$$

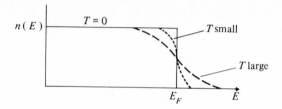

Figure 37.7. A sketch of the number distribution
for $T = 0$ (solid line) and two nonvanishing
temperatures.

where in the last step we used the definition of the Fermi temperature $T_F = E_F/k$ and
the known value of the integral, $\pi^2/3$. Thus the contribution of the electrons to the
heat capacity is very tiny indeed, when it is realized that the value of T_F is of the order
of $10^5\,°K$. Our calculation also shows the degree to which thermal effects smear out
the Fermi surface. This is shown in Eq. (37.43) and qualitatively sketched in Fig. 37.7.

A detailed understanding of metals involves many more complications. What is
heartening, however, is that the salient features of conductivity and of the heat capacity of metals can be understood with the help of some simple concepts that are
basic to quantum mechanics.

NOTES AND COMMENTS

1. Conductivity is discussed in every book on solid state physics. An excellent treat-
    ment may be found in C. Kittel, *Introduction to Solid State Physics*, John Wiley,
    New York, 1976. See also J.P. McKelvey, *Solid State and Semiconductor Physics*,
    Harper and Row, New York 1966, for a fine, leisurely discussion of a number of
    topics contained in this chapter.
2. A simple proof of Bloch's theorem may be found in McKelvey's book.
3. For a solution of the Kronig–Penney model for delta-function potentials, see
    S. Gasiorowicz, *Quantum Physics*, John Wiley, New York, 1974.
4. Our discussion of the energy gaps and their connection with Bragg scattering
    follows that of Kittel (*loc. cit.*) where many more details can be found.
5. The Fermi distribution, and the Bose distribution which we did not discuss,
    receive a very nice treatment in the chapter on Quantum Statistics in R. Eisberg
    and R. Resnick, *Quantum Physics of Atoms, Molecules, Solids, Nuclei, and
    Particles*, John Wiley, New York, 1974.
6. The Bose distribution can be derived by finding the analog of the function
    $W(n_1, n_2, \ldots)$ of Eq. (37.37) and finding its extremum subject to the conditions
    (37.34). For a particular cell with degeneracy g_i and n_i identical particles, we may
    ask for the number of ways that the particles can be distributed into g_i boxes. If
    we arrange particles and box partitions in a linear array, there will be n_i particles
    and $g_i - 1$ partitions, and there are $(n_i + g_i - 1)!$ ways of arranging these. How-

ever, permutations of box partitions on one hand, and of particles on the other do not give a new situation. Thus for a single cell

$$W_i = \frac{(n_i + g_i - 1)!}{n_i!(g_i - 1)!}$$

and

$$W = \prod_i W_i.$$

From here on one follows the procedure leading to the analog of Eq. (37.41). Note that for photons the number N is not fixed, so that a constraint is missing, and the Planck distribution emerges.

PROBLEMS

37.1 The Drude relaxation time τ at $0°C$ for a number of metals is given below:

Metal	Li	Na	Cu	Mg	Fe	Al
In units of 10^{-14} seconds	0.88	3.2	2.7	1.1	0.24	0.80

Assuming the applicability of the classical model, use the Maxwell–Boltzmann value of $\bar{v}$ to calculate the mean free path of electrons in the various metals at $0°C$.

37.2 Given that the Fermi velocities of the above metals, in units of 10^8 cm/sec are given by the numbers 1.29 (Li), 1.07 (Na), 1.57 (Cu), 1.58 (Mg), 1.98 (Fe), and 2.03 (Al), calculate the true mean free path of electrons in the metals.

37.3 A current of 1 amp is flowing in a copper wire 0.2 cm in diameter. Given that 1 amp is equivalent to a flow of 9×10^9 e.s.u. per second, use the density of electrons calculated in the text to estimate the drift velocity.

37.4 Calculate the distribution function for identical bosons, using the approach that was used for the Fermi distribution. See Notes and Comments for the starting point.

37.5 Calculate the heat capacity for an electron gas in *two* dimensions. Note that the change in dimensions manifests itself in (37.46) where

$$\sum_r g_r(\quad) \to \frac{2A}{(2\pi)^2} \int d^2k(\quad)$$

and A is the area of the sample.

38
Semiconductors
and Superconductors

In spite of the common root of the names of the two kinds of materials, semiconductors and superconductors do not lie at opposite ends of a continuum of conducting substances. Whereas the properties of semiconductors can be discussed in terms that applied to metals, superconductors form a new state of matter and will have to be discussed much more qualitatively.

Band structure To begin our discussion of *semiconductors*, let us expand on the formation of energy bands by the interaction of adjacent atoms. When N identical atoms that form a lattice are far apart, each energy level is N-fold degenerate, and can accommodate a total of $2N$ electrons, because of the spin degree of freedom. When the atoms come closer, the degeneracy is lifted, that is, the N levels corresponding to a given state (the $(1s)$ state, for example) spread out into a *band of levels*. The amount of spreading depends on the magnitude of the perturbing potential generated by the neighboring atoms, compared with the binding potential. Thus in an atom with large Z, the low lying $(1s), (2s), (2p), \ldots$ electrons are very tightly bound, and the presence of another atom 1 Å away will not make much difference. Put another way, the barrier to tunneling to another location is very large. Thus the band corresponding to the low-lying levels will be quite narrow. All of the states in the band will be filled, since there are 2 electrons in each $(1s)$ orbital, and thus $2N$ electrons are available to fill N levels. As we get to the outer electrons, the binding is smaller; the electrons are, on the average, farther away from the nuclei, and thus the effect of bringing atoms closer is more dramatic: The bands become wider, and they may overlap. This effect sometimes leads to unexpected behavior, as can be seen by comparing sodium and magnesium. In sodium there is only one valence electron. Thus in the corresponding band there are only N electrons that fill $2N$ states. With a half-filled band, an electric field can readily accelerate electrons, and sodium, like other alkali metals, is a very good conductor. In magnesium, the highest-lying level has two paired electrons in it. One might therefore expect the $2N$ electrons to fill the band, making magnesium an insulator. Because of band overlap, however, (Fig. 38.1), the $2N$ electrons not only have the original $2N$ states that they filled available to them, but also the $2N$ totally empty states from the atomic level above the one that was filled. There is again a partially filled band, and high conductivity.

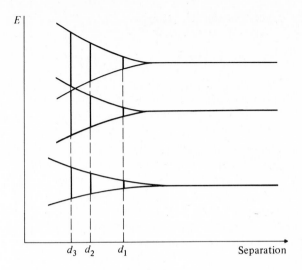

Figure 38.1. Sketch of splitting of energy levels in atoms caused by interaction with neighbors, showing how gaps narrow and disappear when the separation decreases.

Gaps between bands can vary a lot. For example, whereas the gap for carbon in the form of diamond is 5.4 eV wide, the gap width is 1.17 eV for silicon, 0.74 eV for germanium, and only 0.24 eV for InSb (indium antimonide). For small gaps it is possible to induce conductivity by thermally exciting some electrons across the gap. Since the width of the energy gap ΔE is still large compared with kT, the Fermi distribution (37.43) may be replaced by

$$n_r \simeq g_r e^{-(E_r - E_F)/kT}, \qquad (38.1)$$

and a reasonable number of electrons can undergo this excitation. For example, with a gap width of 1.2 eV, at a temperature of $500°K$, the value of the exponential is of the order of 10^{-12}, which is small, but still allows a large number of electrons (since $N \sim 10^{22}$) to be excited. For a gap width of 0.17 eV the fraction at room temperature is of the order of 10^{-3}.

Electrons and holes When an electron is thermally excited across a gap into an empty conduction band, it leaves behind a vacancy, or, to use language that we used in discussing atomic shells, a *hole*. The hole can propagate, because it can be filled by an electron in the same band, which then leaves its state vacant, and so on. It is clear that if the electron that fills the vacancy is moving to the right in an electric field, then the hole is moving to the left. Thus the hole acts as a positive charge. It, too, contributes to the conductivity of the material. As a prelude to the calculation of the conductivity, we calculate the density of electrons (also called *negative carriers* and labeled by the letter n) and the density of holes (also called *positive carriers*, and labeled by the letter p). The formula derived in the last chapter, Eq. (37.48), yields

$$dn = \frac{1}{2\pi^2}\left(\frac{2m}{\hbar^2}\right)^{3/2}\sqrt{E}\,\frac{1}{e^{(E-E_F)/kT}+1}\,dE \qquad (38.2)$$

for the density of negative carriers. This must be modified as follows: The factor $\sqrt{E}$ in Eq. (37.47) came from the $k\ (=p/\hbar)$ and really refers to the kinetic energy of the particle. Thus it should be replaced by $\sqrt{E-E_c}$, where E_c is the location of the bottom of the conduction band. There is the approximation of replacing the Fermi function by the Boltzmann approximation (in which the $+1$ in the Fermi distribution function is left out), and one further modification, which attempts to take into account that beyond the onset of a new energy band the relation between E and k differs somewhat from the free electron relation $E = \hbar^2 k^2/2m$. The way in which this departure from the free electron relation is parametrized is through the definition of an effective mass m^* (Fig. 38.2):

$$m^* = \frac{\hbar^2}{\partial^2 E/\partial k^2}. \qquad (38.3)$$

For the negative carriers we therefore have, for $E > E_c$,

$$dn_n = \frac{1}{2\pi^2}\left(\frac{2m_n^*}{\hbar^2}\right)^{3/2}\sqrt{E-E_c}\,e^{-(E-E_F)/kT}dE. \qquad (38.4)$$

The effective mass should actually also be used in a quantitative treatment of metals using the free electron theory, with m^* determined by fitting the electronic contribution to the specific heat to Eq. (37.52), where T_F is expressed in terms of m^*. The values of m^*/m tend to lie between 1 and 2 for metals, but become very small for semiconductors, with a value of 0.015 for InSb and 0.07 for GaAs, for example.

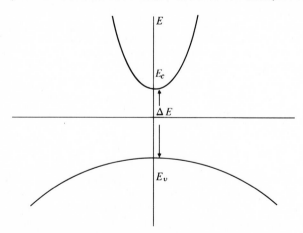

Figure 38.2. Dispersion curve relating energy to wave number for particles and holes. The curvature yields a much smaller m^* for the negative carriers than for the positive ones.

The density of holes, or positive carriers, differs in some respects from Eq. (38.4). Since the shape of the band is different at the top of the valence band from which the electrons make their transition, the effective mass m^* is different, in general, and it is denoted by m_p^*. The factor k in the density now leads to $\sqrt{E_v - E}$, where E_v is the energy at the top of the valence band. Finally, the Fermi distribution

$$f(E) = \frac{1}{e^{(E-E_F)/kT} + 1}$$

is replaced by $1 - f(E)$, since the probability of a hole occurring for a certain energy when added to the probability of the state being occupied, $f(E)$, must give unity. We have

$$1 - f(E) = \frac{e^{(E-E_F)/kT}}{e^{(E-E_F)/kT} + 1} = \frac{1}{1 + e^{-(E-E_F)/kT}}$$

$$= \frac{1}{1 + e^{(E_F-E)/kT}} \approx e^{-(E_F-E)/kT} \tag{38.5}$$

so that

$$dn_p = \frac{1}{2\pi^2} \left(\frac{2m_p^*}{\hbar^2}\right)^{3/2} \sqrt{E_v - E}\ e^{-(E_F-E)/kT} dE. \tag{38.6}$$

The number of negative and positive carriers can now be calculated by a simple integration:

$$n_n = \frac{1}{2\pi^2} \left(\frac{2m_n^*}{\hbar^2}\right)^{3/2} e^{E_F/kT} \int_{E_c}^{\infty} dE \sqrt{E - E_c}\ e^{-E/kT}$$

$$= \frac{1}{2\pi^2} \left(\frac{2m_n^*}{\hbar^2}\right)^{3/2} e^{-(E_c-E_F)/kT} (kT)^{3/2} \int_0^{\infty} du\, u^{1/2} e^{-u}$$

$$= \frac{1}{4} \left(\frac{2m_n^* kT}{\pi\hbar^2}\right)^{3/2} e^{-(E_c-E_F)/kT}. \tag{38.7}$$

The approximation of taking the upper limit on the integral to be infinite, instead of the upper end of the conduction band, is justified by the rapidly decreasing integrand. An identical calculation yields

$$n_p = \frac{1}{2\pi^2} \left(\frac{2m_p^*}{\hbar^2}\right)^{3/2} \int_{-\infty}^{E_v} dE \sqrt{E_v - E}\ e^{-(E_F-E)/kT}$$

$$= \frac{1}{4} \left(\frac{2m_p^* kT}{\hbar^2}\right)^{3/2} e^{-(E_F-E_v)/kT}. \tag{38.8}$$

One consequence of these formulas is a form of mass-action law that expressed the product of n_n and n_p in terms of the energy gap $\Delta E = E_c - E_v$:

$$n_p n_n = \frac{1}{2} \left(\frac{kT}{\pi\hbar^2}\right)^3 (m_n^* m_p^*)^{3/2} e^{-\Delta E/kT}. \tag{38.9}$$

Another consequence is restricted to *intrinsic semiconductors*, in which the number of negative carriers is equal to the number of positive carriers, that is, the only holes are those created by the promotion of electrons into the conduction band. With

$$n_n = n_p, \tag{38.10}$$

Eq. (38.9) yields

$$n_n = n_p = \frac{1}{4} \left(\frac{2kT}{\hbar^2}\right)^{3/2} (m_p^* n_p^*)^{3/4} \, e^{-\Delta E/2kT}. \tag{38.11}$$

We note how strongly the carrier density varies with temperature, and the strong variation from material to material, that is, when ΔE changes.

Hall effect The density of carriers, when either the n-carriers or the p-carriers dominate, can be measured using an effect discovered by E.H. Hall in 1879. Suppose a semiconductor, carrying a current density j in the x direction, is in a magnetic field that is pointing in the z direction. The current density is given by

$$j = env_d \tag{38.12}$$

and charged particles, with a drift velocity v_d, will experience a magnetic force

$$F = -ev_d B/c \tag{38.13}$$

in the y direction. Thus an electric field, given by

$$\mathcal{E} = \frac{F}{e} = -\frac{Bj}{c} \frac{1}{en}, \tag{38.14}$$

will be set up between the top and the bottom of the crystal (Fig. 38.3). The potential difference, given by the field multiplied by the height of the crystal, measures the carrier density, and the sign tells us whether it is positive or negative carriers that are responsible for the current.

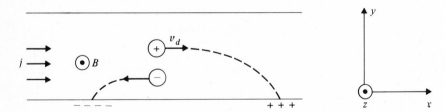

Figure 38.3. Sketch of Hall effect. With current going to the right, both sign carriers get deflected to the bottom, setting up electric fields pointing up and down, respectively. The potential measures the carrier density, and the sign indicates whether it is positive or negative carriers that are responsible for the current.

Electrical conductivity Let us now turn to the electrical conductivity. The formula

$$\frac{1}{\rho} = \frac{e^2 n}{m} \tau = en \left(\frac{e\tau}{m}\right) \equiv en\mu \tag{38.15}$$

may be rewritten in terms of a parameter called the *mobility*. In a metal we have

$$\mu = \frac{e\tau}{m}. \tag{38.16}$$

The utility of writing things in this way is that the density of carriers, which is very temperature dependent in semiconductors, is separated from the other factors. For a semiconductor, with n-type and p-type carriers,

$$\frac{1}{\rho} = e(n_n \mu_n + n_p \mu_p), \tag{38.17}$$

with

$$\mu_n = \frac{e\tau_n}{m_n^*}; \quad \mu_p = \frac{e\tau_p}{m_p^*}. \tag{38.18}$$

Very crudely, the relation

$$\tau = \lambda/\bar{v} \tag{38.19}$$

may again be applied, except that now the use of the classical mean velocity

$$\bar{v} = \left(\frac{8kT}{\pi m^*}\right)^{1/2}, \tag{38.20}$$

instead of v_F, is required, since the exclusion principle plays no role. Since it is m^*, which is smaller than the electron mass m, that appears in the expression for the velocity the value of $\bar{v}$ is larger than for a free particle gas. The calculations, even for a very simplified model, are longer than can be presented here.

Excess carriers In our discussion above, we made reference to intrinsic semiconductors in which the number of negative and positive carriers are equal. Since the holes are created by the excitation of electrons, can it be otherwise? The answer is yes, if there are impurities present. Consider, for example, a crystal of germanium. The addition of atoms of arsenic in small amounts (1 part in 10^5, say) will not change the diamond lattice structure because the arsenic displaces the germanium atom in the lattice. There is a difference in the electronic structure, though. Whereas germanium has four valence electrons, arsenic has five. The extra electron will be very loosely bound to the arsenic ion because of the shielding effect of the electrons in the lattice that are attracted by the ion. Effectively, the Coulomb potential is reduced in strength by a factor ϵ, where ϵ is the dielectric constant of the medium. In germanium $\epsilon = 15.8$ so that the binding energy in the lowest Bohr orbit is changed from

$$\tfrac{1}{2}mc^2 \alpha^2$$

to

$$\tfrac{1}{2}m^*c^2(\alpha/\epsilon)^2.$$

The electron is thus very loosely bound, and it moves in an orbit of radius

$$a = \frac{\hbar\epsilon}{m^*c\alpha}. \tag{38.21}$$

It thus acts as a conduction electron, and the effective number of n-carriers is increased. A material that contains impurities with a larger number of valence electrons, occurring either naturally or through artificial *doping*, is called an *n-type semiconductor*, and the impurity is called a *donor* impurity.

Impurities that have three valence electrons, such as boron, for example, are called *acceptor* impurities, because they can accept electrons from the valence band, giving rise to holes. Semiconductors that have an excess of holes are called *p-type semiconductors*. The relation (38.9) is still correct for semiconductors that have excess of one type carrier or the other. Impurities in small concentrations do not change the mean free path significantly, so that the cross section is still determined by electron-phonon scattering. As the concentration increases, impurity scattering acts to decrease the mean free path.

p-n junction Particular interest in semiconductors has arisen because of their many technical applications. We shall limit ourselves to a brief discussion of the *p-n junction* and its role as a rectifier (Fig. 38.4). A semiconducting crystal is doped with acceptors

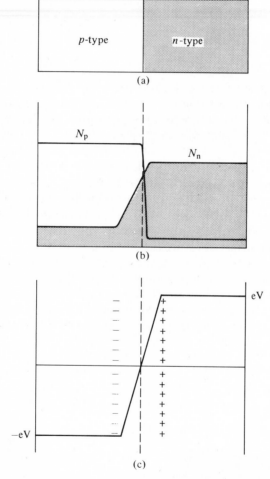

Figure 38.4. (a) *p-n* junction. (b) Densities of holes and electrons in junction. The number of *p*-carriers on the *n*-side and *n*-carriers on the *p*-side is highly exaggerated. (c) When *p*-carriers diffuse to the right, and *n*-carriers to the left, an electric field is set up. The potential is sketched in.

on one side and donors on the other. In the p-type material there will be free holes, and in the n-type material there will be free electrons, with electrical neutrality being maintained by negative acceptor and positive donor sites, respectively. There will be a tendency for the holes to migrate into the n-type region, and for the electrons into the p-type region until the net positive charges remaining behind on the right (cf. Fig. 38.4b) and the net negative charges remaining on the left side of the junction set up an electric field that stops the migration. The figure shows the variation in the density of p and n carriers, with the drawing not to scale in that the low concentrations are highly exaggerated. The ratio of the low levels to the high levels at equilibrium, for a potential difference V, is given by

$$n_p \text{ (right)}/n_p \text{ (left)} = e^{-eV/kT} \tag{38.22}$$

and

$$n_n \text{ (left)}/n_n \text{ (right)} = e^{-eV/kT}. \tag{38.23}$$

Note that this leads to

$$n_n \text{ (right)} \, n_p \text{ (right)} = n_n \text{ (left)} \, n_p \text{ (left)} \tag{38.24}$$

as expected, when the temperature is the same on both sides.

In equilibrium there is no current across the junction. If a wire were to be attached to the two sides, connecting the p- to the n-region, no current would flow, since at the connection point of the wire to the p-end, say, a similar equilibrium state with associated potential difference would be set up. If there are temperature differences in a series of junctions, then currents can flow. If, for example, light is made to shine on a junction, then it can be absorbed to produce an electron-hole pair. The hole will be driven to the left and the electron to the right by the internal field, and if there is an external wire, it will carry a current. This mechanism of converting light energy to electrical energy is at the heart of the workings of solar batteries.

Rectification The detailed way in which a dynamic equilibrium is set up is the following: Large densities of positive carriers approach from the left, but they are turned back by the potential barrier, with only a fraction $e^{-eV/kT}$ getting through. The positive carriers approaching from the right have no barrier to contend with, but their density is smaller by a factor $e^{-eV/kT}$ so that there is no net current. Suppose that the voltage on the right is lowered by U. The current to the left is still the initial equilibrium value which we denote by J_0, but the current to the right is changed because of the lowering of the barrier, so that it is given by

$$n_p \text{ (left)} \, e^{-e(V-U)/kT} = J_0 e^{eU/kT}. \tag{38.25}$$

Thus the net current flowing to the right is

$$J_{\text{net}} = J_0(e^{eU/kT} - 1). \tag{38.26}$$

A plot of J_{net}/J_0 is shown in Fig. 38.5, and it shows how the junction acts as a rectifier, with the net current behaving quite differently for U positive and negative. A discussion of transistors, which involve $p-n-p$ or $n-p-n$ junctions may be found in the references listed at the end of the chapter.

Superconductivity We next turn to the phenomenon of *superconductivity*, which is also becoming very important in a large number of technological applications. Super-

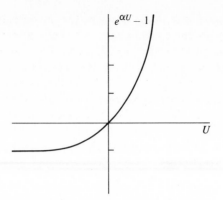

Figure 38.5. Sketch showing net current across a *p-n* junction as a function of the potential *U*. The lack of symmetry under $U \rightarrow -U$ shows how the junction acts as a rectifier.

conductivity is a much more complicated phenomenon than any of the others treated in this book, because it involves quantal collective behavior, which is only now in the process of being understood. Thus we will have few elementary analogies to lean on, and much that will be said will have to be taken as given. The effect was discovered by H. Kamerlingh Onnes in 1911, shortly after he first liquefied helium. This feat enabled him to carry out a whole new class of low temperature experiments, and in one of them he discovered that when mercury is cooled below 4.2°K, the conductivity increases enormously (Fig. 38.6). That this was not the $1/T$ dependence discussed in the last chapter (Onnes did not have the quantum mechanical predictions available to

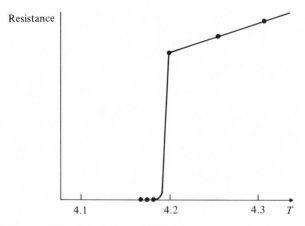

Figure 38.6. Schematic sketch of the resistivity ρ for Hg as a function of temperature in the vicinity of 4°K. The sharp transition to zero resistivity occurs at the critical temperature $T_c = 4.15$°K.

him) was established when it was found that the addition of impurities to mercury did not lead to a residual resistivity. In recent years, persistent currents have been run through a superconducting ring for years, and observed limits on any change in the current show that the "relaxation time" must be larger than 10^5 years! It appears that below a certain critical temperature T_c, the material changes character, or more technically, undergoes a *phase transition*. Superconductivity occurs in many materials. A very abbreviated list of elements and compounds follows, with the critical temperatures.

Material	Tc	Nb	Pb	Hg	Nb_3Ge	Nb_3Ga
Critical temperature in °K	11.2	9.5	7.5	4.2	23.2	20.3

Meissner effect A major discovery, that shed new light on the nature of the superconducting state, was made by W. Meissner and R. Ochsenfeld in 1933, when they discovered that if a superconductor is cooled in a magnetic field below the temperature T_c, then the magnetic flux lines are expelled from the material, leaving $B = 0$ inside except in a very thin surface layer of thickness ranging from 10 Å to 500 Å. A perfect conductor of the normal kind would tend to trap the flux lines inside the material. It was also found that if the external magnetic field was too large, then superconductivity was destroyed. The critical magnetic field $H_c(T)$ has the shape shown in Fig. 38.7. This phenomenon makes it possible to switch a sample back and forth between the normal and the superconducting state.

Figure 38.8 shows the specific heat of a sample of gallium, measured with a strong magnetic field, so that the sample is in a normal state, and also without the field, both at low temperatures. In the superconducting state, at low temperatures, the specific heat can be fitted with the curve

$$C_{es} = Ae^{-bTc/T} \qquad (38.27)$$

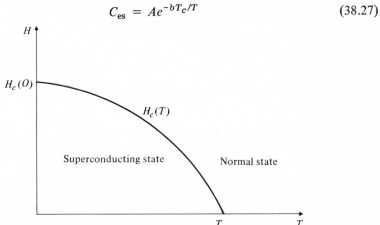

Figure 38.7. Sketch of critical field $H_c(T)$ that destroys the superconducting state.

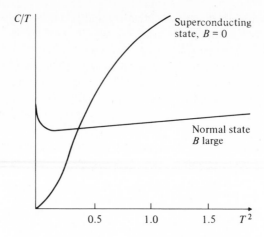

Figure 38.8. Sketch of specific heat (actually C/T) for gallium in normal state (in field $B > B_c$) and in superconducting state ($B = 0$). The behavior of C/T near $T = 0$ shows $\exp(-A/T)$ behavior.

("es" refers to the electronic part of the specific heat, in the superconducting state). This kind of behavior is reminiscent of the behavior of the carrier density in our discussion of semiconductors, and it suggests that there is an *energy gap* that separates the lowest state of the system from the levels whose excitation allows the system to store energy. This is surely not the gap that characterizes insulators, since we are dealing with superconductors. If superconductors have such a gap as one of their characteristics, then there should be other ways of seeing evidence for the gap. A way that involves tunneling was already discussed in Chapter 22. Another way involves transmission of microwave radiation by very thin (approximately 20 Å) films of metal. The reason why such films are not perfectly transparent is that some of the radiation is absorbed. The absorption mechanism is the excitation of electrons from the ground state to excited states. When the metal is in a superconducting state, such electrons, to be excited, must acquire at least 2Δ units of energy, where 2Δ is the region around the top of the Fermi sea out of which the energy levels have been squeezed — Δ is what we call the gap width. For radiation with frequency such that $\hbar\omega < 2\Delta$, absorption is decreased, and therefore microwave transmission is increased.

One more piece of information bears on the nature of superconductors: that is the observation that in the determination of the critical temperatures T_c and the critical fields H_c, these quantities changed when isotopes of the different elements appeared. The specific dependence found was of the form $M^{-1/2}$ where M was the mass of the isotope. Since chemically isotopes of elements all behave alike, except for the effects due to the difference in nuclear mass (an isotope will have the same Z for its nucleus, but different values of A), this result showed that superconductivity is not a property of the electrons alone, but that the lattice plays a very important part.

London theory of superconducting state Major progress in the theoretical understanding of superconductivity came in several stages. The first important step, taken by F. London and H. London in 1935, was the recognition that a superconductor could be described by a single wave function (at $T = 0$), that is, that the superconducting state was a coherent, collective state that involved a very large number of the electrons. With this idea the Londons were able to explain the Meissner effect. Suppose we write this collective wave function in the form

$$\psi = Re^{iS(x)}, \tag{38.28}$$

then the current associated with this wave function is

$$\mathbf{j} = \frac{\hbar q}{2im}(\psi^*\boldsymbol{\nabla}\psi - \psi\boldsymbol{\nabla}\psi^*)$$

$$= \frac{\hbar q}{m}R^2\boldsymbol{\nabla}S \tag{38.29}$$

where q is the charge associated with whatever it is that is flowing. One learns in slightly more advanced treatments of quantum mechanics that the interaction of charged matter with an electromagnetic field is taken into account correctly by the substitutions

$$\boldsymbol{\nabla}\psi \rightarrow \boldsymbol{\nabla}\psi - (iq/\hbar c)\psi\mathbf{A}$$

$$\boldsymbol{\nabla}\psi^* \rightarrow \boldsymbol{\nabla}\psi^* + (iq/\hbar c)\psi^*\mathbf{A} \tag{38.30}$$

where $\mathbf{A}$ is the *vector potential*, which is required to satisfy

$$\boldsymbol{\nabla} \times \mathbf{A} = \mathbf{B}. \tag{38.31}$$

Thus

$$\mathbf{j} = \frac{\hbar q}{m}|\psi|^2\boldsymbol{\nabla}S - \frac{q^2}{mc}|\psi|^2\mathbf{A}. \tag{38.32}$$

The equation

$$\boldsymbol{\nabla} \times \mathbf{j} = -\frac{q^2}{mc}|\psi|^2\boldsymbol{\nabla} \times \mathbf{A} = -\frac{q^2}{mc}|\psi|^2\mathbf{B} \tag{38.33}$$

follows from the preceding equation only if it is assumed that

$$|\psi|^2 \equiv \rho = \text{constant.} \tag{38.34}$$

This is another assumption that is made about the nature of the collective state, namely, that the density of whatever it is that makes up the collective state is uniform throughout the sample. Equation (38.33) immediately leads to the Meissner effect, with the help of Maxwell's equations

$$\boldsymbol{\nabla} \cdot \mathbf{B} = 0; \quad \boldsymbol{\nabla} \times \mathbf{B} = \frac{4\pi}{c}\mathbf{j}. \tag{38.35}$$

Using

$$\boldsymbol{\nabla} \times (\boldsymbol{\nabla} \times \mathbf{B}) = \boldsymbol{\nabla}(\boldsymbol{\nabla} \cdot \mathbf{B}) - (\boldsymbol{\nabla} \cdot \boldsymbol{\nabla})\mathbf{B}$$

$$= -\boldsymbol{\nabla}^2\mathbf{B},$$

we get

$$-\boldsymbol{\nabla}^2\mathbf{B} = \frac{4\pi}{c}\boldsymbol{\nabla} \times \mathbf{j} \quad \text{(Maxwell)}$$

$$= -\frac{4\pi q^2}{mc^2}\rho\mathbf{B} \quad \text{(London).} \tag{38.36}$$

If we consider a slab of superconductor in the $x - y$ plane, then

$$\frac{d^2B}{dz^2} - \frac{4\pi q^2}{mc^2} \rho B = 0 \tag{38.37}$$

describes the variation of the B field. The solution of this,

$$B = B_0 e^{-(4\pi q^2 \rho/mc^2)^{1/2}z} \tag{38.38}$$

shows that the magnetic field penetrates into the superconductor to a penetration depth

$$\lambda_p = \left(\frac{mc^2}{4\pi q^2 \rho}\right)^{1/2} = \left(\frac{m_e c^2}{4\pi e^2 n}\right)^{1/2} \cdot \left(\frac{m}{m_e} \frac{e^2}{q^2} \frac{n}{\rho}\right)^{1/2}. \tag{38.39}$$

The order of magnitude of the first factor, which involves the density of free electrons, is typically 200 Å. The second factor could not be estimated until it was understood what the condensed state consisted of.

Cooper pairs The next step in the understanding of the nature of the collective state came through the work of L.N. Cooper in 1956. The basic idea was that the electron-lattice interaction should give rise to an indirect electron-electron interaction. This interaction is not simply derivable from the known electron-lattice interaction, because the coupling is by no means weak, and its form was guessed by Cooper, who proposed that after taking into account all of the interactions that make up the potential, whose levels up to the Fermi energy are filled by electrons, there should be left over a weak attractive interaction between electrons in the neighborhood of the Fermi sea. This attraction should be strongest between electrons of opposite momentum and spin states; it is now called a *pairing force*. Although the interaction is weak, there is always a bound state of such paired electrons (in three dimensions we know that there is a minimum potential needed to bind particles; the situation here is more like a two-dimensional potential). It should be noted that one expects this to be generally true, and in addition to superconductivity, the pairing force has been detected through its peculiar effects in nuclei (see Chapter 39) and also in liquid ^{3}He at very low temperatures. A collection of such *Cooper pairs* of electrons could give rise to a state lying below the expected normal lowest energy state, that is, a stable ground state with properties different from the usual filled Fermi sea. If it is the Cooper pairs that form the substance described by the collective state, then $m/m_e = 2$, $q = 2e$ in Eq. (38.39) and ρ/n is the fraction of electrons that are in the paired state.

BCS theory In 1957 a monumental paper by J. Bardeen, L.N. Cooper, and R. Schrieffer completely solved the problem of superconductivity, and raised the subject to a new level. The idea was based on the discovery that a coherent state of Cooper pairs gave lower energy than the Fermi sea, and that it was this state that described the super-conductor. One consequence of the condensation of Cooper pairs is the distortion of the Fermi surface. The levels below the Fermi energy are depressed by an amount Δ, and the empty states above that level are pushed up by an amount Δ, giving rise to an overall gap of 2Δ. This gap, however, is not a property of the lattice, as is the case for insulators, but rather of the electrons. Since the gap, so to speak, is attached to the Fermi sphere, currents can get started. What cannot happen is the dissipative slowing down of electrons in one-by-one-collisions. Put another way, the electrons near the Fermi surface form a single, macroscopic quantum mechanical state, and a single electron cannot just scatter, as such an effect would imply the destruction of the state and cost too much energy. The gap parameter Δ depends on the temperature and the BCS theory predicts

$$\frac{\Delta(T)}{\Delta(0)} \cong 1.74 \left(1 - \frac{T}{T_c}\right)^{1/2}. \tag{38.40}$$

A simplified model yields

$$\Delta(0) \cong 1.76 \, kT_c \tag{38.41}$$

and also predicts the critical field to have the behavior

$$\frac{H_c(T)}{H_c(0)} \approx 1 - \left(\frac{T}{T_c}\right)^2. \tag{38.42}$$

The specific heat is also predicted to have the temperature dependence

$$C_{es} \propto e^{-\Delta(0)/kT} \tag{38.43}$$

all in agreement with experiment. Good quantitative agreement is also obtained for microwave absorption and other effects.

Josephson tunneling The wave function of the superconducting state (38.28), with R having no spatial dependence throughout the superconductor, has been used by Feynman to give a simple description of the important effect of electron-pair tunneling discovered by B.D. Josephson in 1962. We follow that discussion below: Let us take two superconductors (consider different ones, for generality) and separate them by a narrow gap. The gap may consist of metal or insulator. If the gap were wide enough, the wave functions of the two superconducting states would be independent. If we focus on the time-dependence, we would have

$$i\hbar \frac{\partial \psi_1}{\partial t} = E_1 \psi_1$$

$$i\hbar \frac{\partial \psi_2}{\partial t} = E_2 \psi_2. \tag{38.44}$$

If there is a potential difference between the two sides of the junction, then

$$E_1 - E_2 = qV. \tag{38.45}$$

This equation can be written in the form

$$i\hbar \frac{\partial}{\partial t} \begin{pmatrix} \psi_1 \\ \psi_2 \end{pmatrix} = H \begin{pmatrix} \psi_1 \\ \psi_2 \end{pmatrix} = \begin{pmatrix} E_1 & 0 \\ 0 & E_2 \end{pmatrix} \begin{pmatrix} \psi_1 \\ \psi_2 \end{pmatrix}, \tag{38.46}$$

characteristic of a two-level system. It is analogous to a spin $\frac{1}{2}$ system, in which there are no transitions between "up" and "down" states. When the gap is narrow, tunneling across it can take place, and the two states are coupled. In the spin $\frac{1}{2}$ case this corresponds to the possibility of a spin flip, and this is described by changing the 2×2 matrix to a form that has off-diagonal elements, so that transitions between the two states can take place. If we write

$$H = \begin{pmatrix} E_1 & K \\ K^* & E_2 \end{pmatrix} \qquad (38.47)$$

in the presence of tunneling, we have the equations we want. We can simplify the equations somewhat by writing

$$E_1 = \bar{E} + qV/2$$
$$E_2 = \bar{E} - qV/2.$$

Then the two-component wave function $\begin{pmatrix} \phi_1 \\ \phi_2 \end{pmatrix}$, defined by

$$\begin{pmatrix} \psi_1 \\ \psi_2 \end{pmatrix} = e^{-i\bar{E}t/\hbar} \begin{pmatrix} \phi_1 \\ \phi_2 \end{pmatrix} \qquad (38.48)$$

satisfies

$$i\hbar \frac{\partial}{\partial t} \begin{pmatrix} \phi_1 \\ \phi_2 \end{pmatrix} = \begin{pmatrix} qV/2 & K \\ K^* & -qV/2 \end{pmatrix} \begin{pmatrix} \phi_1 \\ \phi_2 \end{pmatrix}. \qquad (38.49)$$

If we write

$$\phi_1(t) = e^{-iqVt/2\hbar} \eta_1(t)$$
$$\phi_2(t) = e^{iqVt/2\hbar} \eta_2(t), \qquad (38.50)$$

we get, after a little algebra,

$$i\hbar \frac{d\eta_1}{dt} = K\eta_2 e^{iqVt/\hbar}$$

$$i\hbar \frac{d\eta_2}{dt} = K^* \eta_1 e^{-iqVt/\hbar}. \qquad (38.51)$$

If we now write

$$\eta_1 = R_1 e^{-i\sigma_1/\hbar}, \quad \eta_2 = R_2 e^{-i\sigma_2/\hbar}; \quad K = K_0 e^{i\alpha}, \qquad (38.52)$$

then, with the notation $(\sigma_1 - \sigma_2)/\hbar \equiv \delta$, we get, after taking real and imaginary parts of the equations,

$$\hbar \frac{dR_1}{dt} = K_0 R_2 \sin\left(\frac{qVt}{\hbar} + \delta + \alpha\right)$$

$$\hbar \frac{dR_2}{dt} = -K_0 R_1 \sin\left(\frac{qVt}{\hbar} + \delta + \alpha\right)$$

$$R_1 \frac{d\sigma_1}{dt} = K_0 R_2 \cos\left(\frac{qVt}{\hbar} + \delta + \alpha\right)$$

$$R_2 \frac{d\sigma_2}{dt} = K_0 R_1 \cos\left(\frac{qVt}{\hbar} + \delta + \alpha\right). \qquad (38.53)$$

The first two equations lead to

$$\frac{1}{2} R_1 \frac{dR_1}{dt} = \frac{d\rho_1}{dt} = \frac{1}{2\hbar} K_0 \sqrt{\rho_1 \rho_2} \sin\left(\frac{qVt}{\hbar} + \delta + \alpha\right) = -\frac{d\rho_2}{dt},$$

that is,

$$\frac{d}{dt}(\rho_1 + \rho_2) = 0. \tag{38.54}$$

This is just a conservation law for whatever is described by the wave function, namely, the density of Cooper pairs. The equation

$$\frac{d\rho_1}{dt} = \frac{K_0}{2\hbar} \sqrt{\rho_1 \rho_2} \sin\left(\frac{qVt}{\hbar} + \delta + \alpha\right) \tag{38.55}$$

shows us that even if the potential difference V vanishes, there will be a current of electron pairs flowing provided that $\sin(\delta + \alpha) \neq 0$. This is the so-called *dc Josephson effect*. The second pair of equations yields

$$\frac{d\delta}{dt} = \frac{1}{\hbar}\left(\frac{d\sigma_1}{dt} - \frac{d\sigma_2}{dt}\right) = \frac{K_0}{\hbar}\left(\frac{R_2}{R_1} - \frac{R_1}{R_2}\right)\cos\left(\frac{qVt}{\hbar} + \delta + \alpha\right). \tag{38.56}$$

Because $R_1 \approx R_2$, this equation yields $\delta = $ constant. Thus

$$\frac{d\rho_1}{dt} = \frac{K_0}{2\hbar} \sqrt{\rho_1 \rho_2} \sin\left(\frac{qVt}{\hbar} + \alpha_0\right) \tag{38.57}$$

where α_0 is a constant phase. Thus the current oscillates with frequency $qV/\hbar$. The *ac Josephson effect*, as this is called, permits incredibly accurate measurements of the quantity $2eV/\hbar$. The reason for the appearance of the factor $2e$ is that the wave function describes Cooper pairs, and for these the charge is $q = 2e$. In view of the approximations made, one might wonder whether it is indeed exactly $2eV/\hbar$ that is being measured; one can show more rigorously that this is really the case.

What emerges from our qualitative discussion is that it is the phase of the collective state of Cooper pairs that is the physically interesting quantity. J. Mercereau has taken advantage of this and constructed a sort of interferometer, using two Josephson junctions in parallel. A fairly simple discussion, for which we have no space, ultimately leads to the result that the current through the circuit is proportional to

$$\cos\frac{2e\Phi}{\hbar c} \tag{38.58}$$

where Φ is the magnetic flux enclosed by the circuit. This effect may be used to measure tiny flux changes and is nowadays used in the construction of magnetometers of previously unattainable precision.

Our discussion of these topics has been brief and oversimplified. Nevertheless, even this account should allow the reader to get a glimpse of the subtlety of quantum effects in nature, and the richness of the effects that can emerge from the simple laws of nonrelativistic quantum mechanics. We have also tried to indicate how the exploration of quantum effects, for a long time thought to be very esoteric, has led to important technological advances.

NOTES AND COMMENTS

1. The material discussed in this chapter appears in all solid state textbooks and in most modern physics textbooks. Among the former we list C. Kittel, *Introduction to Solid State Physics* (5th Edition), John Wiley, New York, 1976; N.W. Ashcroft and N.D. Mermin, *Solid State Physics*, Holt, Rinehart and Winston, New York, 1976; J.P. McKelvey, *Solid State and Semiconductor Physics*, Harper and Row, New York, 1966; J.S. Blakemore, *Solid State Physics* (2nd Edition), W.B. Saunders, Philadelphia, 1974; and A.C. Rose-Innes and E.H. Rhoderick, *Introduction to Superconductivity*, Pergamon Press, London, 1969. At this level, the student who does not wish to immerse himself or herself in a detailed study of these subjects can do no better than study R.P. Feynman, F.B. Leighton, and M. Sands, *The Feynman Lectures on Physics*, vol. 3, Addison-Wesley, Reading, Mass., 1965.

2. A discussion of a variety of devices using semiconductors may be found in R. Eisberg and R. Resnick, *Quantum Physics*, John Wiley, New York, 1974.

3. The coupling of charged particles to electromagnetic fields, as described by Eq. (38.30) is discussed in most books on quantum mechanics. See, for example, S. Gasiorowicz, *Quantum Physics*, John Wiley, New York, 1974.

4. The effective mass can be measured by placing the superconductor in a magnetic field, in which the current carriers will move in helical paths with angular frequency $\omega_c = eB/m^*c$. Such a frequency can be measured with high precision by imposing an oscillating electric field perpendicular to the magnetic field and adjusting the frequency so that resonant absorption of energy occurs. Experiments must be carried out at low carrier densities, with pure crystals and low temperatures, so that the helical motion takes place. The method is called the *cyclotron resonance* method.

5. A somewhat advanced, but still accessible, treatment of Josephson junctions may be found in D. Tilley and J. Tilley, *Superfluidity and Superconductivity*, Van Nostrand and Reinhold, 1974.

PROBLEMS

38.1 Show that if the effective masses for the positive and negative carriers are equal, then the Fermi energy lies in the middle of the energy gap.

38.2 The table below lists some effective masses and gap values, the latter at 300°K.

Crystal	$\Delta E(eV)$	m_n^*/m	m_p^*/m (heavy hole)
InSb	0.18	0.015	0.39
InAs	0.35	0.026	0.41
InP	1.35	0.073	0.4
GaAs	1.43	0.07	0.68

Assuming that the energy gap does not change much with T, calculate the carrier densities assuming $n_n = n_p$ for the above crystals, at temperatures of $100°K$, $200°K$, and $300°K$.

38.3 Consider the equation relating the energy $E = \hbar^2 q^2/2m$ to the wave number for the Kronig–Penney model

$$\cos qa - \frac{1}{2} \xi \frac{\sin qa}{qa} = \cos ka.$$

Use this expression to calculate the effective mass (a) near $k = 0$, and (b) near $k = \pi/a$, the zone boundary.

38.4 Calculate the Hall coefficient, defined to be $\mathscr{E}/Bj$, that is, $R_H = -1/ecn_c$, for copper, assuming that there is one negative carrier per atom; repeat the calculation for aluminum, assuming one positive carrier per atom. Compare with the experimental values of -0.6×10^{-24} and $+1.135 \times 10^{-24}$ in c.g.s. units.

38.5 The table below lists the mobilities for positive and negative carriers for a few crystals, with μ in units of $cm^2/volt\text{-}sec$.

Crystal	μ_n	μ_p
InSb	77,000	750
InAs	33,000	460
InP	4,600	150
GaAs	8,800	400

Use this data, and the data listed in Problem 38.2 to calculate values for the relaxation times for the negative and positive carriers, and using Eq. (38.20), the mean free paths of these carriers in all four of the substances.

38.6 Calculate the London penetration depth for lead and for niobium, assuming one carrier per atom at $0°K$.

39
The Structure of Nuclei

The notion of an atomic center, or nucleus, was introduced by Rutherford in 1911, but it is fair to say that nuclear physics really began with the discovery of the neutron by J. Chadwick in 1932. This allowed a proper explanation of the nature of the constituents of the nucleus: a nucleus of atomic weight A and atomic number Z consisted of Z protons and $N = A - Z$ neutrons. Neutrons were soon found to have spin 1/2, which made it possible to understand nuclear spins. A nucleus consisting of protons and electrons could not be made to have the right fermion/boson characteristics. For example, the deuteron, discovered in 1932, would, if it consisted of two protons and one electron, necessarily be a fermion, whereas experimental evidence for spin 1 became available within a few years. A more timely example was the evidence for the integral spin of ^{14}N mentioned in Chapter 31. Although it appeared that nuclear physics was ready to be explored with the powerful tool of quantum mechanics, a quantitative description of the kind that emerged for atomic structure was not possible. For atoms, the law of interaction between the constituents, nucleus, and electrons was known to be the Coulomb potential, so that a good first approximation could be obtained by ignoring the electron-electron interaction, and treating only the potential due to the point nucleus. In nuclei there is no such dominant interaction; all constituents are on the same footing. Furthermore, the nuclear forces were unknown. An attempt to use the deuteron as the "hydrogen atom" of nuclear physics failed. As the uncertainty relations suggest, the exploration of distances Δx require momentum changes of the order of $\Delta p \sim \hbar/\Delta x$, so that a detailed investigation of the potential requires neutron-proton collisions over a very wide range of energies; the static properties of the deuteron depend only on some very gross average of the potential, just as a measurement of the charge and dipole moment of a charge distribution does not give us detailed information about that distribution. In fact, it is only in recent years that a somewhat quantitative picture of the internucleon potential has emerged, and studies of nuclei as a many-body problem subject to this potential are just developing. The understanding of the structure of nuclei has proceeded in another way, through the study of a variety of patterns in properties of nuclei as a whole. In this way, insights from other fields of physics, including atomic physics, have been very helpful. We shall follow this approach in our brief treatment of nuclear physics.

Nuclear size We begin our discussion of the gross structure of nuclei by setting up a *semi-empirical mass formula*. In the process of understanding the various contributions

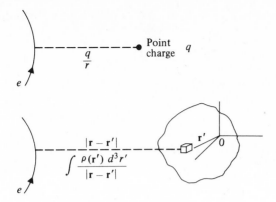

Figure 39.1. Potential energy of electron in field of a point charge and in the field of a distributed charge, with a charge density $\rho\,(\mathbf{r}')$.

to the nuclear mass, we shall learn quite a bit about nuclei and the forces between them. Although some of our discussion will be classical, we shall keep quantum mechanics in mind. It was, in fact, the explanation of alpha decay as a quantum mechanical tunneling phenomenon, which gave the first evidence concerning nuclear sizes. The nuclear radii were found to be approximately

$$R = R_0 A^{1/3}. \tag{39.1}$$

More recently, the charge density of nuclei has been explored by scattering high-energy electrons off them. Here one takes advantage of the fact that one has a good theoretical understanding of the scattering of electrons by point charges, so that for a spread-out charge, one just adds the contributions of the various parts of the charge (Fig. 39.1). Depending on the energy and angle of scattering for the electron, different parts of the charge distribution are probed. A series of experiments, pioneered by R. Hofstadter in the 1950s led to a picture of the nuclear density shown in Fig. 39.2. The

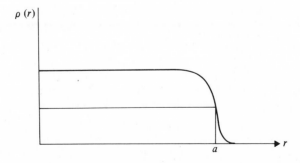

Figure 39.2. Experimentally determined shape of nuclear charge density.

density is fairly constant for a while, and then falls off rather rapidly. The charge density is reasonably well described by the distribution

$$\rho(r) = \frac{\rho_n}{1 + e^{(r-r_0)/a}} \tag{39.2}$$

where $\rho_n = 0.17 \text{ nucleon/fm}^3$
$r_0 = 1.18 A^{1/3} - 0.48 \text{ fm}$
$a = 0.55 \text{ fm}.$

Note that the typical distance in nuclear physics is measured in *fermis*, with $1 \text{ fm} = 10^{-13} \text{ cm}$, just as the atomic scale is described by the Ångstrom. A rougher, but still serviceable description is that of a constant density of $0.17 \text{ nucleons/fm}^3$ up to a radius of

$$R = 1.2 A^{1/3} \text{ fm.} \tag{39.3}$$

This differs by about 20% from the early values obtained from alpha decay (discussed in Chapter 22), but the latter determination uses a much more model-dependent notion of *nuclear radius*, so that Eq. (39.3) is to be preferred. We assume that because the nuclear forces are so much stronger than electrostatic forces, the charge density and the nuclear density are very similar, that is, the protons are not distributed differently from the neutrons.

Short-range forces The density, which incidentally corresponds to about 2.7×10^{14} gm/cm^3, does not depend on the number of nucleons (neutrons or protons) present. This tells us something about the nature of the nuclear forces. It implies that the addition of a nucleon (a change from A to $A + 1$, say) does not increase the density, as might be the case if all of the nucleons initially present attracted the added nucleon. If the forces were similar to electrical forces, the attraction of the extra nucleon would be A times the two-nucleon force. Nuclei would then share the property of atoms that after a certain point they do not increase in size (atoms with Z up to 100 still have radii of the order of a few Ångstroms). Thus the constant nuclear density suggests that *nuclear forces have a short range*, of the order of the size of a nucleon, say $1.0 - 1.2$ fm. They also are *very strong* compared with Coulomb forces. It takes 13.6 eV to remove an electron from hydrogen; it takes approximately 8 MeV to take a nucleon out of a nucleus. This energy is approximately the same for protons and for neutrons. In fact, when the Coulomb repulsion between the protons is properly taken into account, the energy is the same. Nuclear forces are thus *charge independent*. We shall return to this remarkable symmetry later and discuss it in more detail.

Semi-empirical mass formula Many nuclei are found in nature, many stable, many more unstable. The semi-empirical mass formula, which we will discuss below, gives a very good, though not perfect, fit to the masses. The terms that contribute to the mass of a nucleus are the following:

1. The nucleus consists of Z protons and $N = A - Z$ neutrons. The *rest mass* contributions are dominant, and are

$$Zm_p c^2 + (A - Z)m_n c^2, \tag{39.4}$$

with $m_p c^2 = 938.28 \text{ MeV}, m_n c^2 = 939.57 \text{ MeV}.$

2. The nuclear forces are short range, and thus each nucleon feels a force that is independent of the number of nucleons. Since this acts for each nucleon, we expect a negative (binding) contribution of the form

$$-b_{\text{vol}}A. \tag{39.5}$$

Actually the nucleons on the surface feel less attraction, since they have neighbours on one side only, so that the above contribution should be reduced by a term proportional to the surface of the nucleus. Since the radius is proportional to $A^{1/3}$, the surface is proportional to $A^{2/3}$, and thus the contribution is of the form

$$+b_{\text{surf}}A^{2/3}. \tag{39.6}$$

This term can also be written in the form $S(4\pi R^2)$, where S has the interpretation of a surface tension.

3. The protons in the nucleus repel each other, and this increases the energy. If one assumes that the positive charge is uniformly distributed through a volume of radius R_c, the radius determined by the charge distribution, then one obtains the form $(3/5)Z^2 e^2/R_c$, which becomes

$$0.7Z^2/A^{1/3} \text{ MeV}. \tag{39.7}$$

4. If this were all, the energy would be lowered by a reduction of Z. In fact, nuclei are particularly stable when $N \approx Z$, and there are no nuclei with huge neutron (or proton) excesses. The departure from symmetry increases the energy, and the term that describes this is taken to be of the form

$$b_{\text{symm}}(N-Z)^2/2A. \tag{39.8}$$

This effect is purely quantum mechanical, and is due to the effect of the Pauli principle. If there is an excess of neutrons, for example, then a larger number of energy levels get filled up (two neutrons per level) and the energy of the system increases. This, by itself, does not reduce the number of neutrons, of course. However, since it is possible for a neutron to convert into a proton with the emission of an electron and a neutrino by the beta decay mechanism (to be discussed in Chapter 41), systems in which the neutron excess has converted into protons are more stable. Similarly the inverse beta decay of protons into neutrons and positrons (antielectrons) prevents the occurrence of nuclei with large proton excess. This is also inhibited by the Coulomb term (39.7). To see the dependence (39.8) as emerging from the exclusion principle, let us ignore the internucleon interaction and consider the protons and neutrons in a box of volume

$$V = \frac{4\pi}{3}R^3 = \frac{4\pi}{3}R_0^3 A. \tag{39.9}$$

We saw that a collection of noninteracting fermions, with number density n in such a box have energy (cf. Eq. (26.35))

$$E = V\frac{\hbar^2 \pi^3}{10m}\left(\frac{3n}{\pi}\right)^{5/3}. \tag{39.10}$$

Thus for protons, where $n = 3Z/4\pi R_0^3 A$ cm^{-3}, we have

$$E^{(P)} = \left(\frac{9}{4\pi^2}\right)^{5/3}\frac{2\pi^4}{15}\frac{A\hbar^2}{mR_0^2}\left(\frac{Z}{A}\right)^{5/3}. \tag{39.11}$$

Similarly, we have for neutrons

$$E^{(n)} = \left(\frac{9}{4\pi^2}\right)^{5/3} \frac{2\pi^4}{15} \frac{A\hbar^2}{mR_0^2} \left(\frac{N}{A}\right)^{5/3},$$ (39.12)

if we neglect the tiny neutron-proton mass difference. The total energy is the sum of the two terms. Let us now write

$$N = \tfrac{1}{2}A + \tfrac{1}{2}t$$
$$Z = \tfrac{1}{2}A - \tfrac{1}{2}t$$ (39.13)

and write the total energy as a function of $t = N - Z$, for $t \ll A$. We have

$$\left(\frac{N}{A}\right)^{5/3} = \left(\frac{1}{2} + \frac{t}{2A}\right)^{5/3} = \left(\frac{1}{2}\right)^{5/3} \left(1 + \frac{t}{A}\right)^{5/3}$$

$$= \left(\frac{1}{2}\right)^{5/3} \left(1 + \frac{5t}{3A} + \frac{5}{9}\left(\frac{t}{A}\right)^2 + \cdots\right)$$

$$\left(\frac{Z}{A}\right)^{5/3} = \left(\frac{1}{2} - \frac{t}{2A}\right)^{5/3} = \left(\frac{1}{2}\right)^{5/3} \left(1 - \frac{5t}{3A} + \frac{5}{9}\left(\frac{t}{A}\right)^2 + \cdots\right)$$

so that

$$E^{(P)} + E^{(n)} = 2\left(\frac{9}{8\pi^2}\right)^{5/3} \frac{2\pi^4}{15} \frac{A\hbar^2}{mR_0^2} \left(1 + \frac{5}{9}\left(\frac{t}{A}\right)^2\right)$$

$$= \frac{9\pi}{40} \left(\frac{3}{\pi}\right)^{1/3} \left[\frac{A\hbar^2}{mR_0^2} + \frac{10}{9} \frac{\hbar^2}{mR_0^2} \frac{t^2}{2A} + \cdots\right].$$ (39.14)

This is smallest when $t = 0$, that is, when $N = Z$, and the symmetry term is

$$\frac{\pi}{4} \left(\frac{3}{\pi}\right)^{1/3} \frac{\hbar^2}{mR_0^2} \frac{(N-Z)^2}{2A}.$$ (39.15)

The numerical value of the coefficient of $(N - Z)^2/2A$ is 28 MeV in this very crude model, and it is within a factor of two of the best fit to the data. The term proportional to A is positive, with the coefficient of order 20 MeV in this approximation. The attraction between the nucleons overcomes this positive effect. A best fit to the data is

$$b_{vol} = 16\,\text{MeV}$$
$$b_{surf} = 17\,\text{MeV}$$
$$b_{symm} = 50\,\text{MeV}$$ (39.16)

corresponding to a charge distribution radius of $R_c = 1.24\,A^{1/3}$ fm.

5. There is an additional contribution that needs to be included to give a better fit to the masses. That is the so-called pairing term which increases the binding of even-even nuclei, and decreases that of odd-odd nuclei. This is of the form

$$\delta Mc^2 = \begin{cases} -\Delta & N \text{ even}, Z \text{ even} \\ 0 & N \text{ or } Z \text{ odd} \\ \Delta & N \text{ odd}, Z \text{ odd} \end{cases}$$ (39.17)

with

$$\Delta \simeq \frac{12}{A^{1/2}} \, \text{MeV}. \tag{39.18}$$

This term takes into account the fact that the nucleon-nucleon interaction does not merely give rise to an overall average potential for the nucleons, but has some special characteristics reminiscent of the pairing interaction between electrons that gives rise to superconductivity.

The formula so obtained

$$\begin{aligned} E = \; & m_p c^2 Z + m_n c^2 (A - Z) - b_{\text{vol}} A + b_{\text{surf}} A^{2/3} \\ & + 0.7 Z^2 / A^{1/3} + b_{\text{symm}} (N - Z)^2 / 2A + \delta M c^2 \end{aligned} \tag{39.19}$$

fits the masses quite well, though it does not take into account small local variations that arise from the nuclear shell structure that will be discussed in the next chapter. We may use it to calculate the value of Z that minimizes the energy for fixed A. The terms that depend on Z are

$$\begin{aligned} E = \; & Z(m_p c^2 - m_n c^2) + 0.7 Z^2 / A^{1/3} + b_{\text{symm}} (A - 2Z)^2 / 2A \\ = \; & -1.3 Z + 0.7 Z^2 / A^{1/3} + 50 (A - 2Z)^2 / 2A \, \text{MeV}. \end{aligned} \tag{39.20}$$

Hence,

$$\begin{aligned} \partial E / \partial Z = \; & -1.3 + 1.4 Z / A^{1/3} - 100 (A - 2Z) / A \\ = \; & -101.3 + (1.4 / A^{1/3} + 200 / A) Z. \end{aligned} \tag{39.21}$$

This vanishes when

$$Z = 101.3 / (1.4 / A^{1/3} + 200 / A)$$

that is, for

$$Z/A = \frac{101.3}{200 + 1.4 A^{2/3}}. \tag{39.22}$$

The shape of the "stability curve" shows that $Z/A \cong 0.5$ except for heavy nuclei where the fraction decreases somewhat. Near $A = 216 \; (= 6^3)$ this decreases to approximately 0.4 (Fig. 39.3).

Stability of nuclei The semi-empirical mass formula may be used to test for stability of nuclides for a given (A, Z). Thus if $E(A, Z)$ is the total energy of a nucleus, the energy needed to remove a neutron, leading to a nucleus $(A - 1, Z)$ is given by

$$\Delta_n \equiv E(A - 1, Z) + m_n c^2 - E(A, Z) \tag{39.23}$$

which is positive; if it were not, the nucleus would spontaneously emit a neutron with positive kinetic energy.

The energy available for an electron in beta decay is given by

$$Q = E(A, Z) - (E(A, Z + 1) + m_e c^2). \tag{39.24}$$

This process is one in which a neutron turns into a proton and an electron, with the emission of a massless neutrino

$$N \rightarrow P + e^- + \bar{\nu}_e.$$

There is also the potential of the reverse reaction

$$P \rightarrow N + e^+ + \nu_e$$

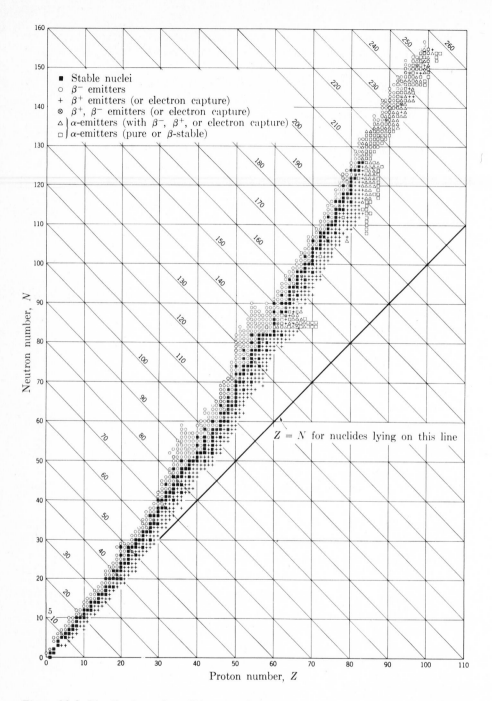

Figure 39.3. Distribution of nuclides as a function of N and Z. (Source: A.P. Arya, *Elementary Modern Physics*, Addison-Wesley, Reading, Mass., 1974, p. 362. Reprinted by permission.)

involving antiparticles, to be discussed together with beta decay in Chapter 41, and the Q value of the reaction, that is, the maximum energy available for kinetic energy of the decay products, is

$$Q = E(A, Z) - (E(A, Z - 1) + m_e c^2).$$ (39.25)

Nuclei are beta stable if $Q < 0$. The energy available for kinetic energy of an alpha particle in alpha decay is given by

$$Q = E(A, Z) - (E(A - 4, Z - 2) + m_\alpha c^2).$$ (39.26)

More refined versions of the semi-empirical mass formula are needed to study potential regions of stability extending beyond the currently known transuranic region extending to $Z \lesssim 110$. In that region fission plays an important role, and in general

$$Q = E(A, Z) - E(A_1, Z_1) - E(A - A_1, Z - Z_1)$$ (39.27)

is positive, because of spontaneous fission

$$^A Z \rightarrow {}^{A_1} Z_1 + {}^{A - A_1}(Z - Z_1).$$

Our first look at the structure of nuclei has not taken into account any of the aspects of a very rich structure of a many-body quantum mechanical system. The various kinds of excitations and some rudimentary discussion of their origin will be taken up in the next chapter.

NOTES AND COMMENTS

1. The semi-empirical mass formula is discussed in every textbook on nuclear physics. Books that discuss these matters on the level appropriate for the reader include

   W.E. Burcham, *Nuclear Physics, An Introduction*, McGraw-Hill, New York, 1963.

   B.L. Cohen, *Concepts of Nuclear Physics*, McGraw-Hill, New York, 1971.

   H.A. Enge, *Introduction to Nuclear Physics*, Addison-Wesley, Reading, Mass., 1966.

   I. Kaplan, *Nuclear Physics*, Addison-Wesley, Reading, Mass., 1955.

   E. Segre, *Nuclei and Particles*, (Second Edition) W.A. Benjamin, Menlo Park, Calif., 1977.

   I particularly liked the second of the books listed. The last one is particularly good in the discussion of the experimental techniques invented for the study of nuclei and particles. A book that is in many ways obsolete (it was written in 1949), but nevertheless contains many gems, is E. Fermi, *Nuclear Physics* (Lecture notes developed by J. Orear, A.H. Rosenfeld, and R.A. Schluter), University of Chicago Press, 1950.

2. The strength of the nuclear forces is very much larger than that of electrical forces, but the short range of the former often leads to relatively weak binding. Thus the bound state of a neutron and a proton, the deuteron, can only exist in an orbital angular momentum zero (S) state; the centrifugal repulsion in higher angular momentum states overcomes the attraction. In contrast, many molecules can exist in rotational excited states with J values up to 20 or 30.

3. The size of the surface tension calculated from the empirically determined value of b_{surf} is $S = 1.5 \times 10^{20}$ erg/cm^2. Compare this with the surface tension of water, which is approximately 70 ergs/cm^2.

4. The calculation of the Coulomb energy, leading to (39.7), is identical with the calculation of the gravitational potential energy for a uniform mass, leading up to Eq. (26.29). The reason is that in both cases the potential energy is of the $1/r$ form. The only difference is that the product GM^2 is to be replaced by $(Ze)^2$.

5. A perusal of nuclear physics textbooks will show that the general form of the semi-empirical mass formula is the same in all of them, but that they differ in the value of the parameters and the A dependence of the pairing term (39.17). All of them give more or less the same fit to the measured nuclear masses.

PROBLEMS

39.1 Consider the semi-empirical mass formula for Sr with $Z = 38$. There are isotopes that have values for $N = 57$ down to $N = 42$.

a) Which is the stablest of these isotopes?

b) What is the smallest value of N for which beta decay
$$(Z, N) \to (Z + 1, N - 1) + e^- + \bar{\nu}_e$$
becomes possible, given that $m_e c^2 = 0.51$ MeV.

c) A possible process in a nucleus is electron capture,
$$e^- + (Z, N) \to (Z - 1, N + 1) + \nu_e.$$
Ignoring the binding energy of electrons, for what range of N's will such a reaction be possible in Sr?

39.2 For the element lead, $Z = 82$ and N ranges from 112 to 132. For what range of N values is alpha decay possible?

39.3 Consider a hypothetical nucleus with $Z = 114$ and N with a range of values around 184. Such nuclei do not exist naturally, but attempts are made to create them by heavy ion collisions. Examine the stability under α decay, e^+ or e^- beta-decay, and under the spontaneous emission of a single nucleon.

39.4 Calculate the numerical value of the kinetic energy per nucleon in the Fermi gas model (cf. Eq. (39.14)), and use this, and the value of b_{vol} to calculate the mean potential energy for the nucleon.

39.5 Use the semi-empirical mass formula to calculate the condition that symmetrical fission, that is, the reaction
$$(A, Z) \to (A/2, Z/2) + (A/2, Z/2)$$
not be energetically possible.

39.6 Consider the nuclei ^{13}N and ^{13}C. The mass difference is measured to be 1.19 MeV/c^2. Use this in conjunction with the expression for the Coulomb energy $3Z(Z-1)e^2/5R_0A^{1/3}$ to calculate R_0.

39.7 Consider the Fermi gas model of a nucleus, for example, ^{63}Cu $(Z = 29)$. What is the largest momentum that a nucleon can have in such a nucleus? Use this information to compare the threshold energy for the production of an antiproton using the reaction

$$p + p \to p + p + p + \bar{p}$$

with that for the reaction

$$p + {}^{63}Cu \to \bar{p} + \cdots$$

with the understanding that the basic reaction is still the same, but the so-called Fermi motion of the proton in the target helps to increase the energy.

40
Nuclear Models

Nucleons move in nuclei with nonrelativistic velocities, and one would expect that, in principle, a reasonable nucleon-nucleon potential, inserted into a many-body Schrödinger equation, would allow for the computation of all relevant properties of nuclei. Such a program has indeed been undertaken for light nuclei, and the results of massive computer calculations, initiated by Y.C. Tang and K. Wildermuth show that such a program makes sense. As in our discussion of atoms and molecules, we are more interested in physical insights than in results of complicated calculations, so that we say no more about the direct assault on the problem. Rather, we discuss a series of nuclear models, each of which deals with a different aspect of the complicated many-body system that is the nucleus. The totality of these descriptions forms our picture of the nucleus.

The liquid drop model The semi-empirical mass formula, originally due to C.F. von Weizsäcker, contains dominant terms that could be associated with the energy of a classical charged liquid drop. Such a picture of the nucleus does not seem unreasonable. The strength of the nuclear interaction, the short range, and the absence of a "central potential" for the nucleons suggest that the independent particle features that characterize electrons in atoms should be absent in nuclei. A classical treatment of a charged liquid drop suggests the dominant excitations of such an object: they are vibrational excitations that have to do with the change in the surface when the nucleus is disturbed from its state of equilibrium, for which we assume a spherical shape of radius R_0. We shall take $R_0 = 1.2A^{1/3}$ fm, and we shall assume that nuclear matter is incompressible. This is a good approximation: the volume energy $-b_0A$ may be written in the form

$$E_{\text{vol}} = -\xi\left(\frac{4\pi}{3}R^3\right).$$

(40.1)

and thus

$$\frac{\partial E_{\text{vol}}}{\partial R}\Delta R = -3\xi\left(\frac{4\pi}{3}R^3\right)\frac{\Delta R}{R}$$

$$= -3b_0A\,\frac{\Delta R}{R}.$$

(40.2)

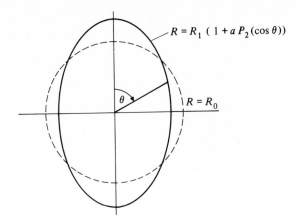

Figure 40.1. Distortion from spherical shape.

Taking $b_0 = 16\,\text{MeV}$, we see that a 1% change in the radius requires approximately $0.5A\,\text{MeV}$ energy of compression, which is a large amount for A in excess of 100. An incompressible droplet may be deformed in many ways. The simplest is an elongation along the z axis, with azimuthal symmetry maintained. A shape of the form

$$\frac{x^2 + y^2}{1 + \epsilon} + \frac{z^2}{1 - 2\epsilon} = \text{const} \quad \epsilon \ll 1 \tag{40.3}$$

yields an ellipsoid of revolution. If we use the fact that in spherical coordinates

$$P_2(\cos\theta) = \frac{3\cos^2\theta - 1}{2} = \frac{3z^2 - (x^2 + y^2 + z^2)}{2(x^2 + y^2 + z^2)}, \tag{40.4}$$

it is not hard to see that the shape (40.3) may be described by the formula (Fig. 40.1)

$$R = R_1(1 + aP_2(\cos\theta)) \quad a \ll 1. \tag{40.5}$$

R_1 will be determined by the requirement that the volume of the drop remains unaltered. We have

$$V = \frac{4\pi}{3} R_0^3 = \int d^3\mathbf{r} = \int_0^{2\pi} d\phi \int_{-1}^1 d(\cos\theta) \int_0^R r^2\, dr$$

$$= 2\pi \int_{-1}^1 d(\cos\theta)\, \frac{1}{3} R^3 = \frac{2\pi}{3} \int_{-1}^1 d(\cos\theta) R_1^3 (1 + aP_2(\cos\theta))^3$$

$$\simeq \frac{2\pi}{3} R_1^3 \int_{-1}^1 d\mu (1 + 3aP_2(\mu) + 3a^2 (P_2(\mu))^2)$$

$$= \frac{2\pi}{3} R_1^3 \left(2 + 3a^2\left(\frac{2}{5}\right)\right) = \frac{4\pi}{3} R_1^3 \left(1 + \frac{3}{5} a^2\right). \tag{40.6}$$

In the integration we used

$$\int_{-1}^1 d\mu\, P_l(\mu) P_{l'}(\mu) = \frac{2}{2l + 1} \delta_{ll'}. \tag{40.7}$$

Incidentally, it is the orthogonality of the spherical harmonics that makes them particularly useful for the parametrization of surfaces that depart only slightly from spherical ones. A more general formula that allows for azimuthally dependent deformations is

$$R = R_1 \left(1 + \sum_{l=2}^{\infty} \sum_{m=-l}^{l} a_{lm} Y_{lm}(\theta, \phi)\right). \tag{40.8}$$

The $l = 1$ term is absent, since it merely represents a displacement of the center of the sphere. The term $P_3(\cos\theta)$ would enter into the description of a pear-shaped droplet.

From Eq. (40.6) we find, using the fact that $a \ll 1$,

$$R_1 = R_0(1 - \tfrac{1}{5}a^2). \tag{40.9}$$

Nuclear vibrations A distortion of the nuclear shape will change the surface energy, because the area of the nucleus changes. A lengthy, though straightforward calculation (cf. Fig. 40.2), shows that the surface area is changed by

$$\Delta S = \tfrac{2}{5}a^2(4\pi R_0^2), \tag{40.10}$$

so that the surface energy is changed to

$$b_{\text{surf}}A^{2/3}(1 + \tfrac{2}{5}a^2) \tag{40.11}$$

to leading order in a^2. The Coulomb energy is also changed, because some of the charge moves further away, while some of the charge is closer in. The result of another long computation is that the Coulomb energy is changed to

$$0.7Z^2(1 - \tfrac{1}{5}a^2)/A^{1/3}, \tag{40.12}$$

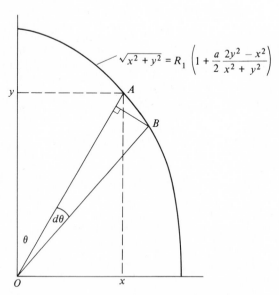

Figure 40.2. Geometry for calculating area of surface given by Eq. (40.5). What one wants is $2\pi x$ (AB) as a function of θ (it is proportional to $d\theta$); then integrate over θ from 0 to π.

so that the energy of deformation is

$$V(a) = \tfrac{1}{2} Ca^2$$

$$C = \frac{2}{5} \left(34A^{2/3} - 0.7 \frac{Z^2}{A^{1/3}} \right). \tag{40.13}$$

If a nucleus is perturbed, the deformation parameter a acquires a time dependence, and there will be a kinetic energy associated with the motion of the fluid. This is a problem in hydrodynamics, and we quote the result only:

$$K = \frac{1}{2} \left(\frac{3MR_0^2}{10} \right) \left(\frac{da}{dt} \right)^2. \tag{40.14}$$

Here M is the mass of the nucleus. Since Eq. (40.13) is expressed in MeV, we must do the same here, and write

$$M = \frac{1}{c^2} Mc^2 = \frac{1}{c^2} 940A. \tag{40.15}$$

The energy of the vibrating system may thus be written in the form

$$H_{\text{vib}} = \frac{1}{2} B \left(\frac{da}{dt} \right)^2 + \frac{1}{2} Ca^2$$

$$B = \frac{3 \times 940A}{10c^2} (1.2 \times 10^{-13} A^{1/3})^2 = 1.50 \times 10^{-44} A^{5/3}$$

$$C = \frac{2}{5} A^{2/3} (34 - 0.7Z^2/A). \tag{40.16}$$

Thus the motion is harmonic, with the angular frequency given by

$$\omega = \sqrt{\frac{C}{B}} = 6.73 \times 10^{22} \sqrt{\frac{1}{A} \left(1 - \frac{Z^2}{48A} \right)}. \tag{40.17}$$

This yields

$$\hbar\omega = 44.2 \sqrt{\frac{1}{A} \left(1 - \frac{Z^2}{48A} \right)} \text{ MeV}. \tag{40.18}$$

We observe that

1. If

$$Z^2 > 48A, \tag{40.19}$$

the frequency becomes imaginary. The solution of the equation of motion that follows from Eqs. (40.16) grows exponentially, so that the system is unstable. Since the equation only describes small deformations a, (40.19) signals trouble, but the details of the unstable motion are not contained in our treatment. What actually happens is that the Coulomb repulsion for large Z exceeds the surface tension and the charged liquid drop *fissions*. The condition (40.19) indicates that spontaneous fission will occur for large Z and A. For example, with $A = 240$, the equation indicates that $Z > 108$. Actually, fission will occur for smaller values of Z. For such values of Z there will be a potential barrier to prevent fission, but quantum mechanical tunneling will allow the process to occur. The liquid drop model yields a good qualitative description (albeit oversimplified) of the fission process.

2. The expression (40.18) shows that $\hbar\omega$ is quite large, even for heavy nuclei: for example, for lead, with $Z = 82$ and $A = 208$, $\hbar\omega = 1.74$ MeV; for aluminum, with $Z = 13$ and $A = 27$, $\hbar\omega = 7.9$ MeV. The reason for our interest in $\hbar\omega$ is that when the vibrational motion is treated quantum mechanically, we have a harmonic oscillator with discrete energy levels given by $\hbar\omega(n + \frac{1}{2})$. Thus the first excited vibrational state will lie $\hbar\omega$ above the ground state. Because the shape deformation is described by $P_2(\cos\theta)$, it turns out that the first excited state has angular momentum $J = 2$. The next excited state has energy $2\hbar\omega$ above the ground state, and the angular momentum, consisting of two "phonons" each with angular momentum 2, can be $4, 2, 0$. Three "surface phonons" will have energy $3\hbar\omega$ and angular momentum $6, 4, 2, 0$. (Odd angular momentum states do not appear because of parity). The states are not quite degenerate, because the deformations can be more complicated than described by Eq. (40.5). At any rate, the liquid drop model predicts a well-defined vibrational spectrum.

Such types of excitations have, in fact, been observed, with the difference that the frequencies are generally quite a bit lower than predicted by Eq. (40.18). For example, in ^{64}Zn, the predicted value of $\hbar\omega$ is 4.62 MeV. The observed levels and their angular momenta are

$J = 2$	0.99 MeV	1 phonon
$J = 0$	1.90 MeV ⎞	
$J = 2$	1.80 MeV ⎬	2 phonons
$J = 4$	2.32 MeV ⎠	
$J = 0$	2.61 MeV ⎞	
$\ldots$	$\ldots$ ⎠	3 phonons

The large discrepancy between the predictions and the observed level energies casts doubt on the quantitative aspects of the liquid drop model. For some reason the nuclei are much more easily deformed (they are less "stiff") than the ground-state energies indicate through the semi-empirical mass formula. This does not mean that the liquid drop model should be totally rejected. It is very useful for understanding some aspects of nuclear structure, and for certain types of nuclear reactions.

Compound nucleus reaction mechanism   Associated with the liquid drop model is a certain picture of how nuclear reactions proceed. The *compound nucleus* mechanism was proposed in 1936 by Niels Bohr, and appeared less explicitly in earlier work of G. Breit and E. Wigner. Bohr proposed that when a particle such as a neutron interacts with a nucleus, it gets captured and a *compound nucleus* is formed. The capture is supposed to be permanent, in that the energy of the incoming particle is immediately shared with all the other nucleons, and there is no energy given to any one particle to make its departure from the nuclear attractive field possible. The mechanism assumes that the energy is rapidly dissipated through frequent collisions among the nucleons. A consequence is that the compound nucleus is very long-lived on a scale determined by the time it takes a nucleon to cross a nuclear diameter, typically 10^{-21} sec. The compound nucleus formation idea can be tested by considering a variety of ways in which a given state can be excited. Thus, for example, the reactions

a) $\alpha + {}^{60}\text{Ni} \rightarrow n + {}^{63}\text{Zn}$

b) $\alpha + {}^{60}\text{Ni} \rightarrow 2n + {}^{62}\text{Zn}$

c) $\alpha + {}^{60}\text{Ni} \rightarrow n + p + {}^{62}\text{Cu}$

d) $p + {}^{63}\text{Cu} \rightarrow n + {}^{63}\text{Zn}$

e) $p + {}^{63}\text{Cu} \rightarrow 2n + {}^{62}\text{Zn}$

f) $p + {}^{63}\text{Cu} \rightarrow n + p + {}^{62}\text{Cu}$

all go through the compound nucleus ^{64}Zn. If the incident α and p energies are so chosen that the total center-of-mass energy of the compound nucleus is the same in all cases, so that in all cases the same states of ^{64}Zn are excited, then the cross sections for the reactions (a)–(f) are expected to obey a definite relation. The compound nucleus hypothesis, expressed quantitatively, states that the cross section for a given reaction is the product of a term that describes the absorption of the incident particle in the formation of the intermediate state, and a term that describes the decay rate of the intermediate state into a given final state. Thus the cross section for the reaction (a) is of the form

$$\sigma(a) = \sigma(\alpha + {}^{60}\text{Ni} \rightarrow {}^{64}\text{Zn}) \, \gamma({}^{64}\text{Zn} \rightarrow n + {}^{63}\text{Zn})/\gamma_{tot}$$

where $\sigma(\alpha + {}^{60}\text{Ni} \rightarrow {}^{64}\text{Zn})$ is the capture cross section, and $\gamma({}^{64}\text{Zn} \rightarrow n + {}^{63}\text{Zn})/\gamma_{tot}$ is the relative decay rate for the intermediate state. With this special factorization hypothesis, we expect that

$$\frac{\sigma(a)}{\sigma(d)} = \frac{\sigma(b)}{\sigma(e)} = \frac{\sigma(c)}{\sigma(f)} \tag{40.20}$$

This has been explicitly verified by experiments carried out in 1950 by S. Ghoshal.

The long life of the compound nucleus states that participate in the reactions as intermediate states, implies that these states have a very small natural linewidth, that is, they are narrow resonances, like those found in atomic spectroscopy. Such resonances have been found in many reactions, such as elastic scattering, (p, α) and (p, γ) reactions (in the standard notation reaction (a) above is described as $^{60}\text{Ni}(\alpha, n)^{63}\text{Zn}$ for example), and lend support to the model. Breit and Wigner derived a general formula for the cross section for a general (a, b) reaction. The formula is

$$\sigma(a, b) = \frac{\lambda^2}{4\pi} \frac{\Gamma_a \Gamma_b}{(E - E_r)^2 + (\Gamma/2)^2} \tag{40.21}$$

where λ is the de Broglie wavelength of the incident particle, E is its energy, E_r is the energy of the resonant level, and Γ_a, Γ_b are partial widths for the decay of the compound nucleus into the initial and final states a, b, respectively. Γ is the total width, given by

$$\Gamma = \sum_{\text{all } i} \Gamma_i. \tag{40.22}$$

This formula resembles the formulas we saw in our discussion of the absorption of light (Eq. (14.44), for example), because the mechanism of light absorption is also one that often is dominated by the excitation of resonances, and their subsequent radiative decay.

The liquid drop model represents one aspect of nuclear structure, that is, the tendency for certain kinds of collective excitations to occur. Curiously enough, nuclei also manifest independent particle motion aspects, as was discovered through work stimulated by the pioneering efforts of H. Jensen and M. Goeppert-Mayer, who proposed a shell model of the nucleus.

The shell model of the nucleus The liquid drop model of the nucleus was based on the strong nucleon-nucleon interaction, a consequence of which would be a very short mean free path of a nucleon in nuclear matter. One may argue that this view is too classical. A nucleon in a nucleus will have a mean free path that is determined by its scattering cross section. While such a cross section is large (geometrical — corresponding to the nucleon area) for free nucleon-nucleon collisions, it is small in nuclear matter, because a nucleon will only scatter if the struck nucleon can recoil into an unoccupied state. The exclusion principle forbids the scattering into a state already occupied by another nucleon. Thus scattering is inhibited, and nucleons can propagate easily inside nuclear matter (compare this with our discussion of electrical conductivity on p. 386). This argument suggests that nucleons could act as independent particles in some average potential created by the presence of the other nucleons, just as an electron sees an average electronic potential in addition to the Coulomb potential of the nucleus. If such a potential is approximately spherically symmetric, one should expect a shell structure of nuclei.

Harmonic oscillator model Evidence for such shell structure was gathered by nuclear physicists in the 1930s and 1940s; particularly intriguing was the existence of the so-called magic numbers: nuclei for which either Z or $N = A - Z$ was equal to one of the numbers

$$2, 8, 14, 20, 28, 50, 82, 126 \qquad (40.23)$$

were found to be particularly stable. This manifests itself in a number of ways: the abundance in nature of various nuclear species reflects this stability, since high abundance is correlated with a high degree of stability; for heavy nuclei the energies of alpha particles in α-decay dips strongly at the magic numbers, indicating a lowered energy for the parent nucleus. Such "magic numbers" may be expected from an independent particle model. Consider, for example, a single proton (or neutron) in a three-dimensional harmonic oscillator potential. One can show that the spectrum has the form

$$E = \hbar\omega(2n_r + l + \tfrac{3}{2}) \qquad (40.24)$$

where $n_r = 0, 1, 2, 3, \ldots$ is the number of radial nodes in the eigenfunction and $l = 0, 1, 2, 3, \ldots$ is the orbital angular momentum. There is a $(2l + 1) -$ fold degeneracy associated with a given l value. If we now put protons into such a potential and treat them as independent particles, that is, assume that the nucleon-nucleon interaction is entirely taken care of by the very existence of the single particle potential, then the lowest energy state for a collection of Z protons is obtained by filling the successive levels with two (spin $\tfrac{1}{2}$) protons each. The spectrum corresponding to

Eq. (40.24) is shown in Fig. 40.3. For such a potential the "magic numbers" are, $2 + 6 = 8, 8 + 12 = 20, 20 + 20 = 40, 40 + 30 = 70$, and so on. The numbers deviate from those observed in real nuclei, and this implies that the potential needs to be modified. The modification, suggested by Jensen and Goeppert-Mayer was to introduce a strong spin-orbit potential of the form

$$V_{s.o.} = -\beta \mathbf{S} \cdot \mathbf{L} = -\tfrac{1}{2}\beta\hbar^2 (j(j+1) - l(l+1) - \tfrac{3}{4}) \qquad (40.25)$$

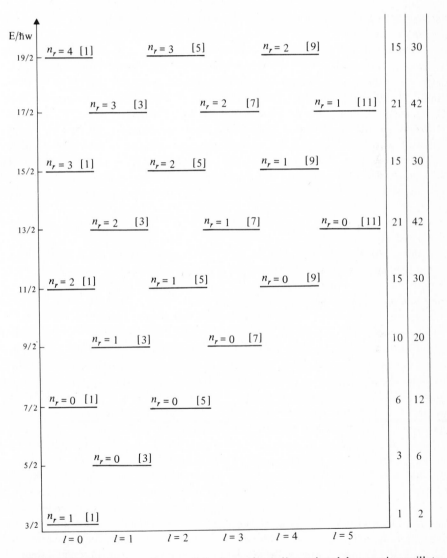

Figure 40.3. Spectrum of a particle in a three-dimensional harmonic oscillator potential. The numbers in [ ] represent the degeneracy and the last column counts the total degeneracy, including the doubling due to spin.

For the simple harmonic oscillator this changes the spectrum to the form

$$
E = \hbar\omega\left[\left(2n_r + l + \frac{3}{2}\right) - \frac{\beta}{2\hbar\omega}\left\{\begin{array}{l} l \\ -l-1 \end{array}\right\}\right]\begin{array}{l} j = l + 1/2 \\ j = l - 1/2. \end{array} \tag{40.26}
$$

Note that the sign of the spin-orbit potential is opposite to that appearing in the hydrogen atom Hamiltonian. The spin-orbit potential introduces the most important modification, but it is also necessary to modify the central potential by flattening the bottom and making the sides steeper. When this is done one reproduces the magic numbers (Fig. 40.4).

In addition to the evidence for shell structure quoted earlier, there is, as expected by analogy with atomic shells and the loose binding of valence electrons, a sharp decrease in the binding energy of nucleons immediately after the magic numbers. Binding energies there are typically 4–5 MeV instead of the average 8 MeV per nucleon. The energies of the first excited states of the magic nuclei lie considerably higher above the ground state than is the case for neighboring nuclei, again with a rough factor of two differences. The increased stability also shows up in a decrease of the nuclear level density, which manifests itself in a reduction of the cross section for the capture of low energy neutrons by nuclei with magic numbers. The nuclear shell model also makes some definite predictions about the spins and electromagnetic properties of nuclei.

i) **Spins of the ground states of nuclei** The spin and parity of a nucleus consisting of a closed shell and a single particle outside of that shell, are just the spin and parity of the extra particle, since a closed shell has $J = 0$ and positive parity. The same is true if we have one *hole* in a closed shell. For example, both ^{13}C and ^{13}N have a single particle outside a closed shell, the $p_{3/2}$ shell. The additional particle, as can be seen from Fig. 40.4 must be in a $p_{1/2}$ shell, and thus must have angular momentum 1/2 and negative parity, since the parity is given by $(-1)^l$. This prediction, and many others like it, are borne out by experiment. When there are two or more particles outside a closed shell, there are many possibilities. For example, two particles in the $f_{7/2}$ shell, even when they are identical, can have many angular momenta, $J = 0, 2, 4, 6$ ($J = 1, 3, 5, 7$ are excluded by the Pauli principle), and in first approximation, all of these states are degenerate. In reality things are simpler. It appears that the effect of all the nucleon-nucleon forces is not only to create the average single particle potential that each nucleon feels, but that there is a residual short-range interaction between the nucleons that has the effect of lowering the energy of states of two nucleons for which the total angular momentum is zero. The effect of this *pairing force* is (a) An even number of particles will have $J = 0$, so that the ground state spins of all even-even nuclei is zero, and (b) The ground state spin of an odd-even nucleus is the J of the shell containing the odd nucleon. For example, ^{93}Nb has $Z = 41$ and $N = 52$. Thus the angular momentum is that of the 41st proton, which, according to Fig. 40.4 lies in the $g_{9/2}$ shell. We thus predict the spin and parity to be $(9/2)^+$, which agrees with experiment.

ii) Magnetic moments of odd A nuclei In our discussion of the Zeeman effect in Chapter 31 (Eq. 31.10) we saw that the magnetic moment of a single electron moving with orbital angular momentum L is given by

$$\mathbf{\mu}_e = -\frac{e\hbar}{2m_e c}(\mathbf{L} + 2\mathbf{S}). \tag{40.27}$$

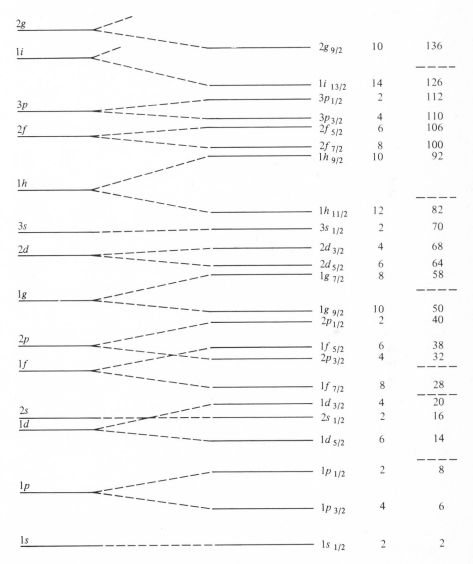

Figure 40.4. Spectrum of particle in a potential with spin orbit coupling adjusted to give the correct level ordering in nuclei. Note that the scale is not correct, but the dashed lines on the right do reflect the large gaps that lead to the magic numbers.

The factor 2 in front of the spin **S** is an approximation. The correct factor can be calculated using quantum electrodynamics and is

$$g_e = 2\left(1 + 0{\cdot}5\left(\frac{\alpha}{\pi}\right) - 0.328\left(\frac{\alpha}{\pi}\right)^2 + \cdots\right) \tag{40.28}$$

with the corrections coming from the structure of the electron that arises from the interaction of the electron with the electromagnetic field that it is a source of. Protons have an even more complicated structure, so that the corrections to the value 2 are quite large. Thus for a proton,

$$\boldsymbol{\mu}_P = \frac{e\hbar}{2M_P c}(\mathbf{L} + g_P \mathbf{S}), \tag{40.29}$$

with $g_P = 5.58$. Note that the sign is opposite to that of the electron, because the charge has the opposite sign. For the neutron, there is no contribution from the orbital motion, since the neutron is electrically neutral. There are, however, structure contributions, and the magnetic moment is

$$\boldsymbol{\mu}_N = \frac{e\hbar}{2M_N c} g_N \mathbf{S} \tag{40.30}$$

with $g_N = -3.82$. In calculating the expectation values of these we do what we did in Eq. (31.12). We write

$$\langle L_z \rangle_{m_j = j} = \langle J_z \rangle_{m_j = j} - \langle S_z \rangle_{m_j = j} \tag{40.31}$$

for the expectation value in the j-eigenstate with $m_j = j$, and

$$\langle S_z \rangle_{m_j = j} = \left\langle J_z \frac{\mathbf{S} \cdot \mathbf{J}}{j(j+1)} \right\rangle_{m_j = j}$$

$$= \frac{j}{j(j+1)} \cdot \frac{1}{2}\left[j(j+1) - l(l+1) + \frac{3}{4}\right] \tag{40.32}$$

where we have used

$$\mathbf{L}^2 = (\mathbf{J} - \mathbf{S})^2 = \mathbf{J}^2 + \mathbf{S}^2 - 2\mathbf{J} \cdot \mathbf{S}.$$

Thus we get, with $M_P = M_N = M$,

$$\hat{\mathbf{i}}_z \cdot \langle \boldsymbol{\mu}_P \rangle_{m_j = j} = \frac{e\hbar}{2Mc}(j - \langle S_z \rangle + g_P \langle S_z \rangle)$$

$$= \frac{e\hbar}{2Mc}\left[j + (g_P - 1)\frac{j(j+1) - l(l+1) + 3/4}{2(j+1)}\right]$$

$$= \frac{e\hbar}{2Mc}\begin{cases} (j - \frac{1}{2} + \frac{1}{2}g_P) & l = j - 1/2 \\[2mm] \dfrac{j}{j+1}(j + \frac{3}{2} - \frac{1}{2}g_P) & l = j + 1/2. \end{cases} \tag{40.33}$$

Similarly, for the neutron magnetic moment we get

$$\hat{\mathbf{i}}_z \cdot \langle \boldsymbol{\mu}_N \rangle_{m_j = j} = \frac{e\hbar}{2Mc}\begin{cases} \frac{1}{2}g_N & l = j - \frac{1}{2} \\[2mm] -\dfrac{j}{j+1}\dfrac{g_N}{2} & l = j + \frac{1}{2}. \end{cases} \tag{40.34}$$

The magnetic moments can be plotted against j. The lines for the values calculated in (40.33) and (40.34) are shown in Fig. 40.5. They are the so-called Schmidt lines, and they represent the values of the magnetic moments based on a calculation which completely ignores any effects due to the closed shells. For light nuclei the formulas work quite well, to within 10–15%. For heavier nuclei the agreement becomes markedly poorer. The accepted explanation is that the nucleons in the core are not totally inert. Because of the *pairing force* the single nucleon outside the core will attract another nucleon with opposite spin, and in effect polarize the core. This acts to reduce the magnitude of the factors g_P and g_N. The experimental evidence shows a reduction of up to 50%. For particularly stable cores, corresponding to magic numbers, one might expect the polarization effect to be small and thus agreement with the results (40.33) and (40.34). This is borne out by the data. The qualitative behavior near closed shells is in agreement with expectations. When a shell is crossed, we expect a sharp change in the magnetic moment from that corresponding to $l - 1/2$ to that corresponding to $l + 1/2$. For example, the formulas predict for ^{39}K, with $j = 3/2$ and $l = 2$ the value 0.12 nuclear magnetons, and for ^{45}Sc with $j = 7/2$ and $l = 3$ the value of 5.79 magnetons. The experimental values are 0.22 and 4.76, respectively. Incidentally, for even-even nuclei, the pairing force leads us to expect the magnetic moments to vanish, and this, too, agrees with experiment.

iii) **Quadrupole moments** The quadrupole moment is a measure of the deviation of the nuclear charge distribution from spherical symmetry. It is defined by

$$Q = \int d^3\mathbf{r}\, \rho(\mathbf{r}) r^2 (3\cos^2\theta - 1). \tag{40.35}$$

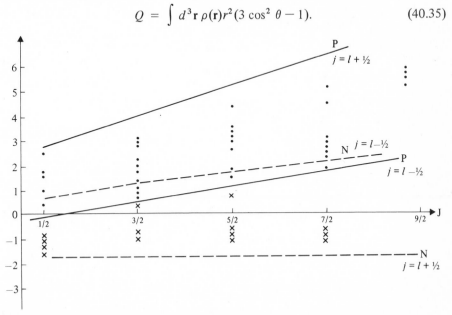

Figure 40.5. The Schmidt lines for the magnetic moments of odd-even nuclei. The solid lines describe the odd proton nuclei, and the dots suggest the distribution of magnetic moments actually measured; the dashed lines and the crosses give the same information for the odd neutron nuclei.

The appearance of the angle θ in the integral involving the charge density implies a choice of z axis relative to which the angle is measured. This axis is defined by taking *the charge density for the nucleus of angular momentum j to be in the state $m_j = j$*. In the classical limit this would correspond to taking the z axis in the direction of the angular momentum.

For a spherically symmetric charge distribution the quadrupole moment vanishes, since

$$Q = \int_0^\infty r^2 dr(r^2 \rho(r)) \int_{-\pi}^{\pi} d\phi \int_0^{\pi} \sin\theta \, d\theta (3\cos^2\theta - 1)$$

$$= 2\pi \int_0^\infty dr \, r^4 \rho(r) \int_{-1}^{1} d\mu (3\mu^2 - 1) = 0. \tag{40.36}$$

Since all even-even nuclei have $j = 0$, there is no directionality associated with such nuclei, and they therefore have vanishing quadrupole moments. This is a quantum mechanical argument which can be proved with more advanced techniques: it is not a natural one, since one could imagine a charge distribution with a bulge in some direction, and a particle with spin circulating in such a way as to cancel the L and yet leaving a distorted charge distribution and an associated quadrupole moment. The argument makes use of the fact that to produce a total $j = 0$ state, one must superimpose, with equal weight all L_z components, and then one makes use of the relation

$$\sum_{m=-l}^{l} |Y_{lm}(\theta, \phi)|^2 = \frac{2l + 1}{4\pi} \tag{40.37}$$

independent of θ or ϕ.

The shell model makes definite predictions for even-odd nuclei. Let us take one proton outside of a closed shell that has $j = 0$ and therefore no contribution to the quadrupole moment. The proton will be in a state described by the single-particle wave function

$$\psi(\mathbf{r}) = R_{nlj}(r) Y_{j,m_j}(\theta, \phi) \tag{40.38}$$

where $R_{nlj}(r)$ is the radial wave function whose form depends on the potential chosen. In general we want

$$Q = \int_0^\infty dr \, r^2 \int_0^{2\pi} d\phi \int_0^{\pi} \sin\theta \, d\theta \, |\psi(\mathbf{r})|^2 (3\cos^2\theta - 1) r^2$$

$$= \int_0^\infty dr \, r^4 R_{nlj}^2(r) \int_0^{2\pi} d\phi \int_0^{\pi} \sin\theta \, d\theta \, |Y_{j,m_j}(\theta, \phi)|^2 (3\cos^2\theta - 1). \tag{40.39}$$

We introduce the notation

$$\langle r^2 \rangle = \int_0^\infty r^2 dr \, R_{nlj}^2(r) r^2 \tag{40.40}$$

and state without proof that

$$\int_0^{2\pi} d\phi \int_0^{\pi} \sin\theta \, d\theta \, |Y_{j,m_j}(\theta, \phi)|^2 (3\cos^2\theta - 1) = \frac{j(j+1) - 3m_j^2}{2j(j+1)}. \tag{40.41}$$

When $m_j = j$, as required by our definition of Q, we obtain

$$Q = -\langle r^2 \rangle \frac{2j - 1}{2(j+1)}. \tag{40.42}$$

The minus sign reflects the fact that the shape of the distribution is oblate, that is, the bulge is in the equatorial plane. That is what we would expect in the classical limit

$j \gg 1$, since then the distribution is concentrated in the equatorial plane for $m_j = j$. We state without proof the result that if there are k protons outside the closed shell, then

$$Q = -\langle r^2 \rangle \frac{2j + 1 - 2k}{2(j + 1)} . \tag{40.43}$$

From this we deduce that

i) if there is a hole in a closed shell, that is, when $k = (2j + 1) - 1 = 2j$, then

$$Q = +\langle r^2 \rangle \frac{2j - 1}{2(j + 1)} \tag{40.44}$$

which is the same as for a single particle outside the shell, except for a reversed sign, indicating a prolate shape.

ii) In general, Q changes sign as the number of protons is increased

$$\begin{array}{ll} Q < 0 & 1 \leqslant k \leqslant j - \frac{1}{2} \\ Q = 0 & k = j + \frac{1}{2} \\ Q > 0 & j + \frac{3}{2} \leqslant k \leqslant 2j. \end{array} \tag{40.45}$$

The neutron contribution to the quadrupole moment might at first sight be expected to vanish, since neutrons are electrically neutral, and therefore do not contribute to the shape of the charge distribution. This is not strictly true: Consider, for example, a single neutron outside a spherically symmetric core. The motion of the neutron implies that the core is also moving, since the center of mass of the whole nucleus is at rest. If the radius of the neutron orbit is some number, the radius of the motion of the core is $1/A$ of that number. Thus we estimate

$$\langle r^2 \rangle_{\text{neutron}} \cong \frac{1}{A^2} \langle r^2 \rangle_{\text{proton}}. \tag{40.46}$$

Since the core contains Z protons, we expect

$$Q_{\text{neutron}} \cong \frac{Z}{A^2} Q_{\text{proton}} \tag{40.47}$$

which is very small.

The unknown parameter $\langle r^2 \rangle$ should, on dimensional grounds, be proportional to the square of the nuclear radius, so that we expect to find that

$$\frac{Q}{A^{2/3}} \cong -(\text{const.}) \frac{2j + 1 - 2k}{2(j + 1)} \tag{40.48}$$

with the constant of the order of magnitude $(10^{-13} \text{ cm})^2$. In Table 40.1 we list some nuclei, giving Z, N, k, j, and the ratios $-(2j + 1 - 2k)/(2j + 2)$, and $Q/A^{2/3}$, the latter in units of 10^{-28} cm^2.

The pattern of sign is in agreement with theory. Numerically the ratio of the fifth to the sixth column should be approximately constant, which it certainly is not. A detailed examination of the data shows that the predictions of the shell model are good for nuclei close to the magic numbers, but tend to be in disagreement in the regions $58 \leqslant Z, N \leqslant 80$, and $91 \leqslant Z, N \leqslant 115$, between the shells. There the experimentally determined quadrupole numbers are a great deal larger than predicted by the theory. This fact, and (a) the fact that there are many more positive Q values than negative ones, and (b) that the neutron contributions to the quadrupole moments are not of order Z/A^2 but rather of order Z/A suggest that the *cores* do, after all, contribute to the quadrupole moments. As with the nuclear magnetic moments, the residual pairing force acts to distort the core. The idea of a permanently distorted core was first suggested by J. Rainwater, and made the basis of the Collective Model of A. Bohr and B. Mottelson, who used this idea as a starting point of an exhaustive exploration of all aspects of collective quantum mechanical behavior in nuclei.

Table 40.1 Quadrupole Moments

Z	N	k	j	$-\dfrac{2j+1-2k}{2j+2}$	$QA^{-2/3}$ in 10^{-28} cm^2
3	4	1	3/2	$-1/3$	-55
5	6	3	3/2	$1/3$	73
11	12	3	5/2	0	136
13	14	5	5/2	$4/7$	167
17	20	1	3/2	$-1/3$	-57
27	32	7	7/2	$2/3$	330
35	44	3	3/2	$1/3$	179
41	52	1	9/2	$-8/11$	-97
43	56	3	9/2	$-4/11$	159
51	72	1	7/2	$-2/3$	-275
53	74	3	7/2	$-2/9$	-308
55	78	5	7/2	$2/9$	-1
57	82	7	7/2	$2/3$	186

The collective model We limit our discussion of the collective model to a few consequences of a permanent distortion of the nuclear shape. Let us assume, as in our discussion of the liquid drop model, that the shape of the nucleus is given by

$$R = R_1(1 + aP_2(\cos \theta))$$
$$R_1 = R_0(1 - a^2/5). \tag{40.49}$$

If we assume a uniform charge distribution, with

$$\frac{4\pi}{3} R_0^3 \rho_0 = Z, \tag{40.50}$$

then

$$Q = \rho_0 \int_0^{2\pi} d\phi \int_0^{\pi} \sin \theta \, d\theta \, (3 \cos^2 \theta - 1) \int_0^{R} r^4 \, dr$$

$$= 2\pi\rho_0 \int_0^{\pi} \sin \theta \, d\theta \, (3 \cos^2 \theta - 1) \tfrac{1}{5} R_1^5 (1 + aP_2(\cos \theta))^5$$

$$= 4\pi\rho_0 \int_0^{\pi} \sin \theta \, d\theta \, P_2(\cos \theta) \tfrac{1}{5} R_1^5 (1 + 5aP_2(\cos \theta) + \cdots)$$

$$\cong \frac{4\pi\rho_0 R_1^5}{5} \int_{-1}^{1} d(\cos \theta) P_2(\cos \theta) (1 + 5aP_2(\cos \theta))$$

$$\cong \frac{8\pi}{5} \rho_0 R_0^5 a \tag{40.51}$$

to leading order in a, which is assumed small. Using Eq. (40.50) we get

$$Q \cong \tfrac{6}{5}aZR_0^2.\tag{40.52}$$

The important result is that even for small a the quadrupole moment can be large, because it is *proportional to Z*, the total number of protons. This *collective effect* implies that the core, even if only slightly deformed ($a \simeq 0.1$) can contribute as much as a factor of ten compared with the single particle value.

Rotational spectra The recognition that nuclei away from the magic numbers can have permanent deformations has had an enormous impact on the understanding of nuclear spectra and the transition rates for gamma and beta decay in nuclei. We briefly discuss the simplest effect, which is that deformed nuclei in some way resemble diatomic molecules and thus will have rotational spectra. The same oversimplified discussion that we used for molecules in Chapter 33 leads to the rotational spectrum

$$E_J = \frac{J(J+1)}{2I},\tag{40.53}$$

with $J = 0, 2, 4, \ldots$, in analogy with homonuclear molecules. If nuclei rotated like rigid bodies, then the moment of inertia of an approximately spherical body is

$$I = \tfrac{2}{5}AMR_0^2\tag{40.54}$$

where M is the nucleon mass, and R_0 the nuclear radius. Experiments show that such rotational spectra exist, and that the moment of inertia deduced from the spacing between the various levels described by E_J is quite a bit smaller than Eq. (40.54) would suggest. This can be understood if one recognizes that the nucleus is not a rigid body, but that the kind of surface oscillations that we discussed in connection with the liquid drop can yield surface waves on the nucleus that make it look as if the nucleus rotated like a rigid body. Since only a small fraction of the nuclear matter moves, the effective moment of inertia is smaller. When the nuclei have intrinsic spin, the motion is much more complicated, since there is interplay of two angular momenta, that of the deformed core, and that of the nucleons outside the core. These matters are certainly beyond the scope of this book. Suffice it to say that the nucleus presents a very complex quantum mechanical system, and the study of nuclear physics has contributed to, as well as drawn from, many other fields of quantum physics.

NOTES AND COMMENTS

1. The reader who wishes to go through the detailed calculations of the various parameters for the liquid drop model will find them in G. Eder, *Nuclear Forces*, M.I.T. Press, 1968. The mathematical level of this book is quite a bit above what is assumed in this book.
2. The parameter that multiplies $(da/dt)^2$ in Eq. (40.14) must be proportional to MR_0^2 on purely dimensional grounds. The numerical factor of 3/10 must be calculated, and cannot be guessed on qualitative grounds alone.

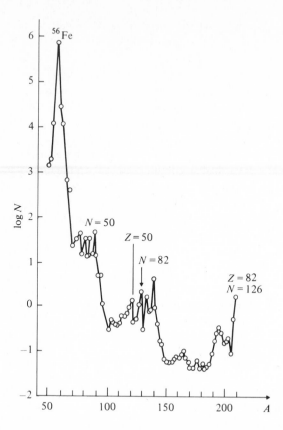

Figure 40.6. Relative abundances of different nuclear species as a function of A. Note the peaks corresponding to the magic numbers. (Source: H. Frauenfelder and E. Henley, *Subatomic Physics*, Prentice-Hall, Englewood Cliffs, N.J., 1974, based on data in *Origins and Distribution of the Elements*, L. H. Arens (Ed.), Pergamon Press, N.Y., 1968. Reprinted by permission.)

3. An excellent description of the compound nucleus picture of nuclear reactions, and the points of contact of that description with the independent particle model, may be found in the essay on Problems of Nuclear Structure, in V.F. Weisskopf, *Physics in the Twentieth Century*, The M.I.T. Press, 1972.

4. The topic of fission and of nuclear reactors as the prime field of application of that topic is hardly mentioned at all in this book. A brief discussion with references for further reading may be found in H. Frauenfelder and E. Henley, *Subatomic Physics*, Prentice-Hall, Englewood Cliffs, New Jersey, 1974.

5. Both the shell model and the collective model are concisely discussed in Frauenfelder and Henley (*loc. cit.*).

6. The data on the abundance of even-even nuclides in the solar system clearly show the stability of the magic number nuclei, as seen in Fig. 40.6.

PROBLEMS

40.1 Given that the successive shells in a nucleus are $s_{1/2}, p_{3/2}, p_{1/2}, d_{5/2}, s_{1/2}, d_{3/2},$ $f_{7/2}, p_{3/2}, f_{5/2}, p_{1/2}, g_{9/2}, g_{7/2}, d_{5/2}, d_{3/2}, s_{1/2},$

a) List the possible angular momenta of nuclei with $(A, Z) = (40, 18), (39, 18),$ $(124, 51), (130, 57), (130, 58), (129, 57), (129, 53), (92, 39), (92, 41)$. Note that for the odd-odd nuclei there are many possible angular momenta. For the others, use the pairing rule.

b) Take the even-odd nuclei in this collection, and calculate the magnetic moments predicted for them by the shell model.

40.2 Sketch the energy levels for a particle in a harmonic oscillator potential with spin-orbit coupling (cf. Eq. (40.26)) as a function of the parameter $x = \beta/2\hbar\omega$, for a range $0 \leqslant x \leqslant 1$.

40.3 Show that an ellipsoid, with azimuthal symmetry about the y axis, described by the surface
$$x^2 + y^2 + (1 - \gamma)z^2 = \text{constant} = R_0^2$$
may have its surface described by
$$r = R_1(1 + aP_2(\cos\theta))$$
if γ and a are small. Express R_1 and a in terms of R_0 and γ.

40.4 Calculate the surface area of the figure described in Problem 40.3, with γ small (do your calculation to lowest order in γ).

40.5 Prove the result given in Eq. (40.43). The steps are the following:

a) If $k \neq 1$, then one proton will be in the $m_j = j$ state and the pairing force will make the even number $(k-1)$ protons to go into states such that if one of them has a given m_j, then the paired one will have $-m_j$, so that they can pair up to give $j = 0$. This pair cannot therefore occupy $m_j = j$ since the single proton already occupies that state. Using Eq. (40.41), calculate the contribution of the paired protons for a given m_j.

b) If the pairs are distributed with equal probability in any one of the $(2j-1)$ states $m_j = j-1, j-2, \ldots -(j-1)$, calculate the average pair contribution
$$\frac{1}{2j-1} \sum_{m_j = -(j-1)}^{(j-1)} (\text{your answer to part (a)})$$

c) Multiply this by the number of pairs, $(k-1)/2$ and add to the single proton contribution from the $m_j = j$ state.

41
Antiparticles, Neutrinos, and Beta Decay

In discussing the tendency of nuclei to have $N \approx Z$ it was pointed out that the beta decay processes

$$N \to P + e^- + \bar{\nu}_e$$
$$P \to N + e^+ + \nu_e$$

made the equalization process possible. This chapter will deal with the obvious questions: what is the e^+? What are the ν_e and $\bar{\nu}_e$? What is beta decay? To answer the first question we turn to the topic of antiparticles, and the remarkable story of the Dirac equation.

Antiparticles The question of a relativistic version of the Schrödinger equation occupied physicists soon after the development of quantum mechanics. The non-relativistic Schrödinger equation may formally be viewed as a translation of the expression

$$E = \frac{\mathbf{p}^2}{2m} + V(\mathbf{r}) \tag{41.1}$$

into

$$E\psi = \left(\frac{\mathbf{p}^2}{2m} + V(\mathbf{r}) \right)\psi \tag{41.2}$$

with the formal identification

$$E \to i\hbar \frac{\partial}{\partial t}; \qquad \mathbf{p} \to \frac{\hbar}{i}\mathbf{\nabla}. \tag{41.3}$$

For a free particle, the relativistic version of $E = \mathbf{p}^2/2m$ is

$$E = (\mathbf{p}^2 c^2 + m^2 c^4)^{1/2} \tag{41.4}$$

but the formal extension of it to

$$i\hbar \frac{\partial \psi}{\partial t} = (-\hbar^2 c^2 \mathbf{\nabla}^2 + m^2 c^4)^{1/2}\,\psi \tag{41.5}$$

is useless, since the square root of the differential operator is not defined, other than as an expansion with an infinite number of terms. Furthermore, we do not really have a relativistic version of the classical equation (41.1), so that even the first step in writing a relativistic theory for interacting particles is absent. To return to free particles, one could, of course, write

$$E^2 = \mathbf{p}^2 c^2 + m^2 c^4 \tag{41.6}$$

and this gives rise to a well-defined equation

$$-\hbar^2 \frac{\partial^2 \psi}{\partial t^2} = -\hbar^2 c^2 \mathbf{\nabla}^2 \psi + m^2 c^4 \psi. \tag{41.7}$$

This equation was actually studied by Schrödinger. In a Coulomb potential, we can write

$$\left(E + \frac{Ze^2}{r}\right)^2 = \mathbf{p}^2 c^2 + m^2 c^4 \tag{41.8}$$

and the corresponding wave equation

$$\left(i\hbar \frac{\partial}{\partial t} + \frac{Ze^2}{r}\right)^2 \psi(\mathbf{r}, t)$$

$$= -\hbar^2 c^2 \mathbf{\nabla}^2 \psi(\mathbf{r}, t) + m^2 c^4 \psi(\mathbf{r}, t) \tag{41.9}$$

is mathematically not so different from the Schrödinger equation. It does, however, have a deep flaw, in that it does not allow for a simple probability interpretation. Recall that in the nonrelativistic Schrödinger equation it is possible to define the probability density as

$$P(\mathbf{r}, t) = |\psi(\mathbf{r}, t)|^2 \tag{41.10}$$

because one can show that

$$\frac{\partial}{\partial t} \int d^3 r |\psi(\mathbf{r}, t)|^2 = 0. \tag{41.11}$$

For the equation (41.7), nowadays known as the Klein-Gordon equation, the density which satisfies

$$\frac{\partial}{\partial t} \int d^3 r P(\mathbf{r}, t) = 0 \tag{41.12}$$

is

$$P(\mathbf{r}, t) = \frac{1}{2i} \left[\psi^* \frac{\partial \psi}{\partial t} - \frac{\partial \psi^*}{\partial t} \psi\right], \tag{41.13}$$

which is not necessarily positive. The positivity requirement must be satisfied for a probability density, and to remedy this, P.A.M. Dirac returned to the task of writing an equation that was linear in $\partial/\partial t$, that is, linear in E. Since E and $\mathbf{p}$ are so closely woven together through the Lorentz transformation, he also insisted on linearity in $\mathbf{p}$. Dirac wrote down

$$E = \alpha_x p_x c + \alpha_y p_y c + \alpha_z p_z c + \beta m c^2 \tag{41.14}$$

and constrained the quantities α_x, α_y, α_z, and β by the requirement that Eq. (41.6) be satisfied. The square of Eq. (41.14) is

$$E^2 = \alpha_x^2 p_x^2 c^2 + \alpha_y^2 p_y^2 c^2 + \alpha_z^2 p_z^2 c^2 + \beta^2 m^2 c^4$$
$$+ mc^3 p_x(\alpha_x \beta + \beta \alpha_x) + mc^3 p_y(\alpha_y \beta + \beta \alpha_y)$$
$$+ mc^3 p_z(\alpha_z \beta + \beta \alpha_z) + p_x p_y c^2(\alpha_x \alpha_y + \alpha_y \alpha_x)$$
$$+ p_x p_z c^2(\alpha_x \alpha_z + \alpha_z \alpha_x) + p_y p_z c^2(\alpha_y \alpha_z + \alpha_z \alpha_y) \tag{41.15}$$

and this implies that

$$\alpha_x^2 = \alpha_y^2 = \alpha_z^2 = \beta^2 = 1$$
$$\alpha_x \beta + \beta \alpha_x = \alpha_y \beta + \beta \alpha_y = \alpha_z \beta + \beta \alpha_z = 0$$
$$\alpha_x \alpha_y + \alpha_y \alpha_x = \alpha_x \alpha_z + \alpha_z \alpha_x + \alpha_y \alpha_z + \alpha_z \alpha_y = 0. \tag{41.16}$$

It is clear from this that the αs and β cannot be numbers. It is, however, possible to satisfy these relations with matrices. Let us try to satisfy them with 2×2 matrices, for example, the Pauli matrices $\sigma_x, \sigma_y, \sigma_z$, which satisfy

$$\sigma_x^2 = \sigma_y^2 = \sigma_z^2 = 1$$

$$\sigma_x \sigma_y + \sigma_y \sigma_x = \sigma_x \sigma_z + \sigma_z \sigma_x = \sigma_y \sigma_z + \sigma_z \sigma_y = 0. \tag{41.17}$$

We can choose

$$\alpha_x = \sigma_x, \quad \alpha_y = \sigma_y, \quad \alpha_z = \sigma_z \tag{41.18}$$

but this leaves us no matrix for β, since if we choose β to be an appropriate linear combination of 2×2 matrices $1, \sigma_x, \sigma_y, \sigma_z$ to satisfy the requirement $\beta^2 = 1$, we cannot satisfy all of the conditions on the second line of Eq. (41.16), Dirac was therefore forced to go to 4×4 matrices. It is easily checked (though that is not our business here), that the matrices

$$\alpha_x = \left(\begin{array}{c|c} 0 & \sigma_x \\ \hline \sigma_x & 0 \end{array} \right); \quad \alpha_y = \left(\begin{array}{c|c} 0 & \sigma_y \\ \hline \sigma_y & 0 \end{array} \right); \quad \alpha_z = \left(\begin{array}{c|c} 0 & \sigma_z \\ \hline \sigma_z & 0 \end{array} \right); \quad \beta = \left(\begin{array}{c|c} 1 & 0 \\ \hline 0 & -1 \end{array} \right)$$

$$\tag{41.19}$$

where each box is a 2×2 matrix, satisfy all the conditions. The Dirac equation, in which E and p are replaced by the differential operators as in Eq. (41.3), reads

$$i\hbar \frac{\partial \psi}{\partial t} = \frac{\hbar c}{i} \left(\alpha_x \frac{\partial \psi}{\partial x} + \alpha_y \frac{\partial \psi}{\partial y} + \alpha_z \frac{\partial \psi}{\partial z} \right) + mc^2 \beta \psi \tag{41.20}$$

and it is an equation that is satisfied by a four-component object

$$\psi = \begin{bmatrix} \psi_1 \\ \psi_2 \\ \psi_3 \\ \psi_4 \end{bmatrix},$$

called a spinor. It must have four components, because 4×4 matrices act on it. Dirac showed that

$$\frac{\partial}{\partial t} \int d^3r \, [\,|\psi_1(\mathbf{r}, t)|^2 + \cdots + |\psi_4(\mathbf{r}, t)|^2\,] = 0 \tag{41.21}$$

so that the usual probability interpretation can be used, with

$$P(\mathbf{r}, t) = \sum_{i=1}^{4} |\psi_i(\mathbf{r}, t)|^2 \tag{41.22}$$

What is the significance of the four components? Dirac showed that, in effect, two of the components describe normal electron states — two components are necessary for a spin 1/2 object. In fact, electron spin emerged quite naturally out of the Dirac equation, as did the fact that $g_e = 2$. When a Coulomb potential was inserted into the Dirac equation by replacing $i\hbar \partial/\partial t$ with

$$i\hbar \frac{\partial}{\partial t} + \frac{Ze^2}{r},$$

excellent agreement with experiment was obtained. The Dirac equation naturally contained the relativistic corrections to the nonrelativistic kinetic energy $\mathbf{p}^2/2m$ but it also naturally incorporated the spin-orbit coupling with the correct Thomas factor of

1/2 that we discussed in Chapter 31. The tremendous success of the Dirac equation made it imperative to clear up the question of the other two components. Dirac showed that the other two components could be associated with electrons that had negative energy, that is,

$$E = -(\mathbf{p}^2 c^2 + m^2 c^4)^{1/2}. \tag{41.23}$$

The existence of negative energy states presents a real difficulty. Electrons could, when slightly perturbed, radiate, and lose energy by going to a lower energy state. Since there is, so to speak, no bottom to the energy spectrum in the Dirac equation, electrons could continually radiate, and accelerate, falling to larger and larger negative energy states. There is no way to avoid this difficulty by specifying initial conditions. Dirac avoided this difficulty in a brilliant way. He postulated that under normal circumstances, *all the negative energy states are occupied.* Because electrons obey the Pauli exclusion principle, all transitions to negative energy states are forbidden. The spectrum of possible states is shown in Fig. 41.1. The region between $-mc^2$ and mc^2 is empty of energy levels for free electrons. If a proton is present, say, then there will be Coulomb bound states just below mc^2. The spectrum of the Schrödinger equation with the correct relativistic and spin-orbit corrections emerges naturally, lending support to the correctness of the Dirac equation. Dirac observed that with an energy input of $2mc^2$ ($= 1.02\,\text{MeV}$), an electron can be lifted from a negative energy state to a positive energy state. Such a transition marks the appearance of an electron out of nothing, but also a *hole* in the negative energy sea. What are the manifestations of such a hole? There is a loss of one unit of negative charge – this shows up as the gain of one unit of positive charge. There is an absence of 1/2 a unit of angular momentum, which shows up as the appearance of 1/2 a unit of angular momentum; there is the absence of negative energy which shows up as the appearance of positive energy, and so on. What happens is that the electron is accompanied by the appearance of a particle that can be called an antielectron, in that it has opposite charge and opposite electromagnetic properties (e.g., magnetic moment), but is otherwise identical to the electron. Such a particle is called the *positron* and is an antiparticle to the electron. An electron

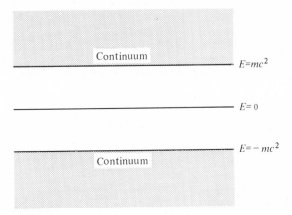

Figure 41.1. Spectrum of states predicted by Dirac equation.

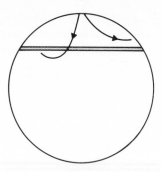

Figure 41.2. Schematic picture of electron and positron tracks in cloud chamber with lead plate.

and a positron can annihilate – this is equivalent to an electron falling into a vacant hole. Left-over energy and momentum emerge in the form of radiation.

This incredible prediction of Dirac was soon confirmed by the discovery of the positron by C. Anderson (1931) in the course of his investigation of charged particle tracks in a cloud chamber. The chamber was placed in a strong magnetic field, so that charged particle tracks were bent (Fig. 41.2). Anderson had placed a lead sheet in the chamber, and he found tracks that had gone through the sheet. Comparing the curvature on the two sides of the sheet he could tell in which direction the particle was going, and thus could conclude that tracks bent a certain way were positive particles going downwards rather than negative particles going upwards. From the ionization along the path and from the curvature, both energy and momentum of the particle could be determined, and thus the rest mass.

Pair annihilation The Dirac theory of relativistic electrons has been reformulated so that one does not have to deal with the infinite sea of negative energy particles. Instead, the modern theory of quantum electrodynamics deals with the interaction of photons, electrons, and positrons, in a manner in which the last two particles appear symmetrically. The possibility of pair creation and pair annihilation emerges naturally out of the theory. When an e^+ and an e^- annihilate, the energy must emerge in the form of radiation. If we consider the collision in the center of momentum frame, with one particle having momentum $\mathbf{p}$ and the other momentum $-\mathbf{p}$, then the energy in the final state is $2((\mathbf{p}c)^2 + (mc^2)^2)^{1/2}$, and the momentum is zero. The total charge is zero, and the angular momentum is an integer, since $\frac{1}{2} + \frac{1}{2} +$ integral orbital angular momentum must add up to an integer. The final state cannot be a single photon, since energy and momentum cannot be conserved. There must be at least two photons, but there can also be annihilation into three photons. A priori this is less likely, since the transition rate in electrodynamics contains one power of α ($= e^2/\hbar c = 1/137$) for each photon, but it can occur when certain selection rules for particular initial states forbid the two photon decay.

Pair creation A single photon cannot convert into an electron-positron pair and conserve energy and momentum. If, however, another body, for example a nucleus, is present, the recoil of that body can take care of momentum and energy conservation (Fig. 41.3). Thus, if the photon direction is defined by the unit vector $\hat{\mathbf{i}}$, then energy and momentum conservation imply that

$$\hbar\omega = [(\mathbf{p}_+ c)^2 + m^2 c^4]^{1/2} + [(\mathbf{p}_- c)^2 + m^2 c^4]^{1/2} + q^2/2M$$

$$\frac{\hbar\omega}{c}\hat{\mathbf{i}} = \mathbf{p}_+ + \mathbf{p}_- + \mathbf{q} \tag{41.24}$$

where $\mathbf{p}_+$, $\mathbf{p}_-$, and $\mathbf{q}$ are the momenta of the positron, electron, and nucleus, respectively (we take the nucleus to be at rest initially), and where M is the large nuclear mass. It turns out that in such a materialization process, the electron and the positron approximately divide the photon energy equally (the nucleus, being very massive, takes very little energy). Thus a very high-energy photon will produce a pair, when it passes through matter. Each member of the pair, when continuing its passage through matter, has a high probability of undergoing a deflection in the Coulomb field of a nucleus. Such a deflection implies an acceleration, and whenever a charge is accelerated, it radiates. This process, known in this context as *Bremsstrahlung* (from the German: "braking radiation"), again leads to an approximately equal division of energy between the photon and the charged particle. The photon makes another pair, the charged particle undergoes Bremsstrahlung, and so on. When the initial energy is large enough, a gigantic shower of electrons and positrons can be created by the cascading process (Fig. 41.4). Large showers initiated by super-high energy cosmic rays at the top of the atmosphere, and covering several square miles, have been detected in large arrays of counters. On a smaller scale, such showers, initiated in high Z material, are used in the construction of photon and electron (positron) detectors. The amount of material needed to make a shower can be estimated from a formula for the mean free path for Bremsstrahlung, called the *radiation length*. The radiation length is approximately given by

$$\lambda_R = \left(\frac{mc}{\hbar}\right)^2 \frac{1}{4nZ^2\alpha^3} \frac{1}{\log 183/Z^{1/3}} \tag{41.25}$$

where m is the electron mass and n is the number of nuclei per cm^3. The mean free path for a photon to produce a pair at high energies ($\hbar\omega \gg mc^2$) is

$$\lambda_{\text{pair}} \cong 1.3\lambda_R. \tag{41.26}$$

The radiation length in air is 330 m, and in lead it is about 0.5 cm.

Positronium When a positron at low energies interacts with matter, it slows down, giving up kinetic energy to ionize some of the material. It frequently slows down enough to be captured by an electron to form an atomic state. This process, in which the atom known as *positronium* is formed, is more probable than annihilation in flight. In first approximation positronium is just a hydrogen-like atom, the only difference being that the reduced mass that appears in the Balmer formula

$$E_n = -\frac{1}{2}\mu c^2 \frac{\alpha^2}{n^2} \tag{41.27}$$

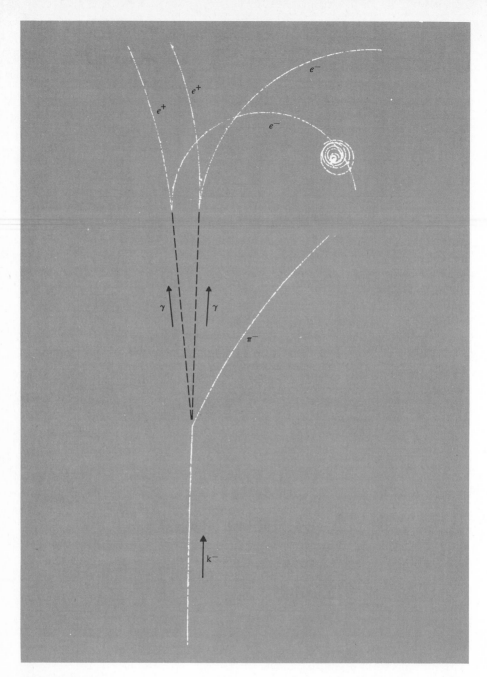

Figure 41.3. A bubble chamber picture showing the materialization of two γ rays into electron–positron pairs. The incident particle is a K^- meson that decays into a π^- (the charged, visible track bending to the right) and a π^0, that decays immediately into two γ rays. These pass through a lead plate positioned in the chamber perpendicular to the γ-ray tracks (invisible, since they are neutral) in which they convert into the visible electron–positron pairs. The tight little loop is an electron ionized by the fast particle. The loop is so tight because of the strong magnetic field and the low energy of the particle. (Courtesy of the Lawrence Berkeley Laboratory, University of California.)

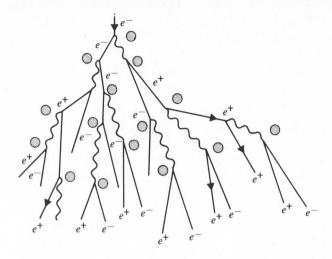

Figure 41.4. Generation of a cascade of charged particles by a very energetic cosmic ray.

is

$$\mu = \frac{m}{2}. \tag{41.28}$$

Thus the binding energy is half that of hydrogen, that is, 6.8 eV, and so on. In detail, the structure is a little different. Thus, for example, the hyperfine structure is different, since for hydrogen, the magnetic moment of the positive particle is of the order of $e\hbar/2Mc$, that is, approximately 2000 times smaller than that of the negative particle. In positronium the magnetic moments of the two particles have the same magnitude. The fact that positronium can annihilate also changes the interaction to a small extent, so that the very fine details of the positronium spectrum are different. So far only the $n = 1$ states have been studied in detail, though the formation of $n = 2$ states has recently been observed.

The electron and the positron are approximately 1 Å apart in the ground state, but the wave functions do overlap, and thus annihilation is possible. When angular momentum considerations do not forbid it, annihilation results in the creation of two photons. The rate is thus proportional to α^2; it will also be proportional to the likelihood of the two particles finding each other, which is inversely proportional to the volume of the atom, that is, it is proportional to

$$\frac{1}{a_0^3} = \left(\frac{\mu c \alpha}{\hbar}\right)^3. \tag{41.29}$$

Thus the rate will be given by a factor of α^5, times the proper combination of $\hbar$, c, and m, to give $(\sec)^{-1}$, with some numerical coefficient such as $1/2$ or π^2 in front. This coefficient can only be obtained by doing a proper quantum-electrodynamic calculation. Our estimate is

$$\text{Rate} = \alpha^5 \frac{mc^2}{\hbar}$$
$$= 0.16 \times 10^{10} \text{ sec}^{-1}$$
$$= \frac{1}{\tau}. \tag{41.30}$$

The experimental value is 1.25×10^{-10} sec for the lifetime, which is just a factor of 5 smaller than the estimate gives. This explains why it is difficult to do experiments with positronium! In spite of this difficulty, the difference in energy between the $n = 1$, $l = 0$ spin 1 and spin 0 states has been determined with extreme precision. The frequency of the radiation corresponding to this energy difference is

$$\nu = 203,396 \pm 5 \text{ MHz} \tag{41.31}$$

one of the most precisely measured numbers in physics. What is astonishing is that it agrees completely with the calculated value, showing that the theory of quantum electrodynamics is extraordinarily successful.

Particle-antiparticle symmetry We observed that the laws of quantum electrodynamics are symmetric under electron-positron exchange. Let us now turn to the broader notion of symmetry under particle-antiparticle conjugation. It turns out that relativistic quantum mechanics predicts antiparticles for all particles. In particular it predicts the existence of antiprotons and antineutrons. Thus there is the potential for a more complete symmetry, namely, that *all* the laws of physics are invariant under the complete change

$$\text{all particles} \longleftrightarrow \text{antiparticles}$$

We would thus expect that there should exist antihydrogen, made up of an e^+ bound by the Coulomb interaction to a negatively charged antiproton, with a spectrum identical in all details with that of hydrogen, and that there should exist an antideuteron, made up of an antiproton and an antineutron, and so on. It is, of course, very difficult to test this conjectured law of nature directly, since protons and antiprotons annihilate, and antihydrogen could not exist for any length of time in an environment in which there are ordinary atoms. Elementary particle physics does, however, afford the opportunity to test this symmetry, and it turns out that the symmetry is almost totally correct.

Before we turn to beta decay, we need to clarify what differentiates a particle from an antiparticle. For the electron and the positron, the answer is obvious: it is the electric charge and all the properties that flow from its existence, such as the magnetic moment, and so on. For the proton we could say the same. But then, what differentiates a neutron from an antineutron? To answer this, we must take note of an obvious, but fundamental fact of nature – *matter is stable!* Thus the reaction

$$P \rightarrow e^+ + \gamma$$

allowed by angular momentum conservation, and by energy and momentum conservation, is not observed. At the very least, protons have a lifetime of 4.5×10^9 years, the age of the earth. Actually the reaction has been looked for in a large volume of scintillator surrounded by light-gathering devices and by material to shield it from cosmic rays. The decay would give a photon ~ 470 MeV energy, which would cause a shower; similarly the positron, with ~ 470 MeV would give rise to a shower. The absence of such showers in a given volume of scintillator for a period of a year or two has been translated into a limit on the lifetime of the proton

$$\tau_{proton} \gtrsim 10^{30} \text{ years.}$$

We thus say that the *proton is stable*. On the other hand, in nuclei, the process

$$P \rightarrow N + e^+ + \nu_e$$

does take place. If the neutron could "disappear," matter would still be unstable. Thus there is something that the proton has, that is passed on to the neutron, that forbids the decay of a neutron into a neutrino, for example. We call the property *baryon number*, and we postulate that *baryon number is conserved*. It is baryon number that distinguishes the neutron from the antineutron. We assign baryon number $+1$ to the proton, to the neutron, and to some other particles that we will discuss in the next chapter. Particles of baryon number $+1$, say, can transform into each other, provided the byproducts have baryon number 0. Electrons, neutrinos, photons, and mesons have baryon number zero, while the antiparticles have the opposite sign baryon number. The number is additive, so that the deuteron has baryon number 2. We extend the law of particle-antiparticle conjugation so that it implies not only that charge $\rightarrow (-$ charge$)$, but also that baryon number $\rightarrow (-$ baryon number$)$. Particles that do not carry any distinguishing quantum number will be their own charge-conjugates, that is, their own antiparticles. For example, the photon is its own antiparticle, and other examples will turn up later. If one introduces an operator C (analogous to parity) which changes particle to antiparticle, then we can write

$$C\psi_{e^-} = \psi_{e^+}$$
$$C\psi_{e^+} = \psi_{e^-}$$
$$\ldots \tag{41.32}$$

and

$$C\psi_\gamma = \pm \psi_\gamma. \tag{41.33}$$

In the last case it turns out that the appropriate sign is $-$, because a change in the sign of the electric charge changes the sign of the electromagnetic field.

Neutrinos and beta decay The beta rays discovered in early experiments on radioactivity were found to be electrons (using an e/m measurement), and it soon became evident that the process involved a nuclear transmutation. What was observed was the process

$$(A, Z) \rightarrow (A, Z + 1) + e^-.$$

The energy spectrum of the electrons was found to have the shape shown in Fig. 41.5. If the electrons were the product of a two-body decay, then electrons of only one energy would emerge. To see this, consider the decay of a particle of rest mass M into two particles of rest mass m_1 and m_2, respectively. In the rest frame of M, the momenta of m_1 and m_2 are p and $-p$, respectively, to conserve momentum. Energy conservation implies that

$$Mc^2 = ((pc)^2 + m_1^2 c^4)^{1/2} + ((pc)^2 + m_2^2 c^4)^{1/2} \tag{41.34}$$

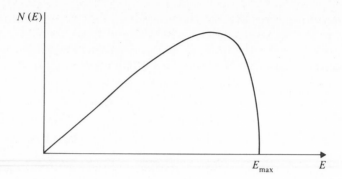

Figure 41.5. Typical electron distribution as a function of energy in beta decay. E_{max}, the endpoint of the spectrum, is equal to the energy difference between the parent and daughter nucleus.

which has a unique solution for the magnitude of p and therefore for the energies of m_1 and m_2. The continuous electron spectrum thus was very mysterious, and it even led Bohr to question the law of energy conservation other than on the average. W. Pauli in 1931 made the (then) daring suggestion that in the process of beta decay, a new, unobserved neutral particle, the *neutrino* be emitted. If the fundamental process is

$$N \rightarrow P + e^- + \bar{\nu}_e \tag{41.35}$$

(we label the particle emitted here as the antineutrino for technical reasons), then energy and momentum conservation can be satisfied for a whole range of electron momenta. The neutrino had to have the following properties:

a) it is clearly neutral, since it leaves no tracks;

b) from the observation of beta decays of even spin nuclei to even spin daughter nuclei (for example $0^+ \rightarrow 0^+$, where the parity also does not change) and the fact that the electron has spin 1/2, it is clear that the neutrino must have odd half-integer spin. The assignment of 1/2 is the simplest one, and it is indeed the correct one.

c) The fact that the maximum electron energy corresponds to $M(A,Z)c^2 - M(A,Z+1)c^2$ implies that the neutrino can, under the most favorable circumstances, carry away zero energy, within the experimental error limits, and this suggests that the neutrino has a very tiny mass. The present experimental limits, and existing theoretical ideas favor *zero rest mass*, so that the neutrino, just like the photon, always travels with the speed of light, and cannot be brought to rest, since the energy and momentum are related by

$$E = pc. \tag{41.36}$$

d) Attempts to stop the neutrino in the walls of a vessel surrounding a radioactive source have shown that the neutrino deposits no energy in the calorimeter. This can only be so if the neutrino is an extremely penetrating particle, and we shall see that this is indeed so. Its mean free path in matter is to be measured in light years!

The laws of relativistic quantum mechanics allow the transfer of a particle from one side of a reaction to the other, provided that the particle is changed to an antiparticle in the process. Thus (41.35) carries with it the implication of other reactions, whose probability amplitudes are all simply related to that for (41.35). These are:

i)
$$P \rightarrow N + e^+ + \nu_e \tag{41.37}$$

mentioned before. Since the neutron is more massive than the proton, this reaction can only occur inside nuclei; the beta process

$$(A, Z) \rightarrow (A, Z - 1) + e^+$$

is frequently seen to occur, when energetically permitted.

ii) The process
$$e^- + P \rightarrow N + \nu_e \tag{41.38}$$

has also been studied. It occurs when an electron from the lowest Bohr orbit around a nucleus gets captured (the historical terminology is K-capture, since the $(1s)$ shell used to be called the K-shell), with a change in the nuclear structure.

iii) In recent years the reactions
$$\nu_e + N \rightarrow P + e^-$$
$$\bar{\nu}_e + P \rightarrow N + e^+ \tag{41.39}$$

have been studied with high-energy neutrino beams, by studying the absorption of neutrinos and antineutrinos in matter. This is possible, since the mean free path decreases at high energies, as we will see below.

Beta decay is just one manifestation of the *weak interactions*. This aspect of elementary particle physics is rather well understood, and in recent years S. Weinberg and A. Salam have taken giant steps in explaining a rather subtle relationship between the electromagnetic interactions and the weak interactions. These matters are clearly beyond the scope of this book, so that we shall limit ourselves to a very rough phenomenological description of the beta decay interaction.

Fermi theory In 1934 E. Fermi constructed a preliminary theory of beta decay. He proposed that the term in the energy operator H responsible for beta decay have the form

$$H' = \frac{\hbar^3 G}{M_P^2 c} \left(\int d^3 r \psi_{e^-}^*(\mathbf{r}) \, \psi_{\nu_e}(\mathbf{r}) \, \psi_P^*(\mathbf{r}) \, \psi_N(\mathbf{r}) \right.$$
$$\left. + \int d^3 r \psi_{\bar{\nu}_e}^*(\mathbf{r}) \, \psi_{e^-}(\mathbf{r}) \, \psi_N^*(\mathbf{r}) \, \psi_P(\mathbf{r}) \right) \tag{41.40}$$

where the ψs represent the wave functions of incoming particles (or outgoing antiparticles, according to the rules of relativistic quantum mechanics), and the ψ^*s represent the wave functions (complex conjugated) of outgoing particles or incoming antiparticles. The combination of $\hbar$ and the factor involving the proton mass (this is a purely conventional choice) are there to make G dimensionless. Given this "perturbation," Fermi was able to calculate the lifetime of the neutron as a function of G. The formula that he derived gave

$$\frac{1}{\tau} = \frac{G^2}{2\pi^3} \left(\frac{m_e}{m_P} \right)^4 \frac{m_e c^2}{\hbar} F(E_{\max}/m_e c^2) \tag{41.41}$$

where

$$F(x) = \int_1^x dy\, y\, \sqrt{y^2 - 1}\, (x - y)^2. \tag{41.42}$$

E_{max} is the maximum energy available to the electron, that is, the difference in energy between the initial and final nuclei. The integrand describes the spectrum of the electron. The number of electrons emitted in beta decay in a given range of energies, say $(E, E + dE)$, is proportional to

$$dN(E) \propto E\, \sqrt{E^2 - m_e^2 c^4}\, (E_{\text{max}} - E)^2 dE. \tag{41.43}$$

The lifetime of the neutron is measured to be 11.7 minutes. This is a very long lifetime, caused in part by the smallness of G, which can be calculated from the above using $E = M_N c^2 - M_P c^2$,

$$G = 1.01 \times 10^{-5},$$

and in part by the smallness of E, or rather the ratio $E/M_P c^2$. There are other weak decays characterized by the same G with lifetimes of 10^{-10} seconds. (The lifetime scales very roughly like the fifth power of the ratio $E/M_P c^2$.)

Neutrino cross sections in matter   The calculation of the neutrino cross section in matter is beyond the scope of the book. We can try a rough estimate, using the fact that (a) the cross section must have the dimension of an area, and (b) the cross section involves the square of a probability amplitude, so that it is quadratic in the coupling strength G, when that coupling is weak. For example, for Thomson scattering, the cross section is given by

$$\sigma = \frac{8\pi}{3}\left(\frac{e^2}{mc^2}\right)^2 = \frac{8\pi}{3}\left(\frac{e^2}{\hbar c}\right)^2\left(\frac{\hbar}{mc}\right)^2 = \frac{8\pi}{3}\alpha^2\left(\frac{\hbar}{mc}\right)^2. \tag{41.44}$$

Written in this form, it involves $(\hbar/mc)^2$, an area written in terms of the only dimension in the problem, and α^2; for the absorption of a photon there is an e, and for the emission there is another, so that the amplitude is proportional to e^2, which appears in the dimensionless guise as α. For neutrino scattering, there is only the absorption of a neutrino, so that G appears in the amplitude, and G^2 appears in the cross section. This must come from the square of the coefficient in Eq. (41.40), that is,

$$\left(\frac{\hbar^3 G}{M_P^2 c}\right)^2 = G^2\left(\frac{\hbar}{M_P c}\right)^2 \frac{\hbar^4}{M_P^2}. \tag{41.45}$$

The details of the relativistic wave functions provide factors to cancel the $\hbar^4$ and to introduce a (mass)^2 factor, to make the dimensions correct. Unfortunately, there are several possibilities of writing natural (mass)^2 factors: one is the square of the mass of the target, the other is the relativistic variable s, which is the square of the center-of mass energy, and in the laboratory frame is equal to $(M_t c^2 + E_\nu)^2 - p_\nu^2 c^2 \approx 2M_t c^2 E_\nu$ for $E_\nu \gg M_t c^2$. It turns out that it is the latter that enters, so that

$$\sigma_\nu = (\text{const})G^2\left(\frac{\hbar}{M_P c}\right)^2 \frac{E_\nu}{M_P c^2}. \tag{41.46}$$

The constant is such that at high energies

$$\sigma_\nu \cong 10^{-38} E_\nu(\text{GeV}) \quad \text{cm}^2. \tag{41.47}$$

The high-energy calculation actually includes the effect of processes other than $\bar{\nu}_e + P$ $\rightarrow e^+ + N$. In fact, one tries to include all processes of the type

$$\nu + \text{nucleon} \begin{array}{l} \nearrow e + \text{anything} \\ \searrow \mu + \text{anything.} \end{array}$$

At low energies, the formula gets quite complicated, and for low-energy neutrinos, the kind that come out of nuclear reactors and out of ordinary beta processes, the cross section is of the order of 10^{-43} cm^2. Recall that the mean free path is given by

$$\lambda = \frac{1}{n\sigma} \tag{41.48}$$

where n is the density of nucleons (nucleons/cm^3). Thus for iron, with $\rho = 7.9$ gm/cm^3

$$n = \frac{\text{nucleons}}{\text{cm}^3} = \left(\frac{\text{nucleons}}{\text{gm}}\right)\left(\frac{\text{gm}}{\text{cm}^3}\right)$$

$$= \frac{1}{1.6 \times 10^{-24}} \rho \cong 5 \times 10^{24}.$$

Thus

$$\lambda \cong \frac{1}{5 \times 10^{24} \times 10^{-43}} = 0.2 \times 10^{19} \text{ cm}$$

$$\cong 2 \cdot 1 \text{ light years.} \tag{41.49}$$

Thus low-energy neutrinos are indeed very penetrating, and it is a tribute to the ingenuity and care with which F. Reines and C. Cowan did their experiment, that the neutrino was finally identified in 1954. At high energies, such as are available at the Fermi National Accelerator Laboratory (FNAL) and at the European Center for Nuclear Research (CERN), with E in excess of 200 GeV, the cross sections are quite measurable, and the reactions

$$\begin{pmatrix} \nu \\ \bar{\nu} \end{pmatrix} + \text{matter} \rightarrow \text{anything}$$

are used to study the structure of the proton and the neutron.

In writing about the neutrino we denoted it by ν_e, and the reason for keeping the subscript must be explained.

The muon A particle of mass 105.7 MeV/c^2 was discovered in cosmic rays in 1938, and its properties have been studied since then. The particle, named the *muon*, has spin 1/2 and interacts both weakly and electromagnetically, since it comes in two charge states, μ^-, with the antiparticle μ^+. At this stage the muon can most succinctly be described as a heavy electron. It decays into an electron and two neutrinos

$$\mu^\pm \rightarrow e^\pm + \nu + \bar{\nu},$$

and this reaction can be described by the same type of H' as appears in (41.40), with N and P replaced by μ^- and ν_μ respectively, but *with the same G*. The question of whether the electron and the muon each have their own neutrino was settled in 1962 when the first neutrino beam experiments were carried out. The neutrino beams are created by making an intense beam of particles called *pions* (these will be discussed in the next chapter). These particles decay according to

$$\pi^+ \begin{cases} \nearrow \mu^+ + \nu_\mu \\ \searrow e^+ + \nu_e \end{cases}$$

$$\pi^- \begin{cases} \nearrow \mu^- + \bar{\nu}_\mu \\ \searrow e^- + \bar{\nu}_e \end{cases}$$

with the *decay into muons about 10^4 more probable than into electrons*. Thus only one neutrino in 10^4 will be of the *e*-type. If the pion beam is not specifically "purified," it will contain a certain amount of K mesons (see next chapter), which decay into muons and electrons with comparable rates, so that they produce roughly equal numbers of muon and electron type neutrinos. What was found in the 1962 experiments was that in the reactions

$$\nu + \text{matter} \rightarrow \begin{pmatrix} \mu \\ e \end{pmatrix} + \text{matter}$$

no electrons were seen when the neutrinos came from a pure pion-originated beam, and some electrons, compatible with the number of ν_e in the incident, K-contaminated beam, were seen when the pion beam was not pure. The ν_μ is very similar to the ν_e, in that it has spin 1/2 and is most probably massless. There is evidence at this point for *e*-type and μ-type conservation laws: the number of e^- + the number of ν_e − the number of e^+ − the number of $\bar{\nu}_e$ is conserved, and similarly for the muons and their neutrinos.

Parity nonconservation   We conclude our discussion of the weak interactions with a fascinating subject, the discovery of *parity nonconservation* in the weak interactions. In 1956, during the investigation of the decay modes of some newly discovered particles, the K mesons, it appeared that the parity of the final state was sometimes even and sometimes odd. For a while one could argue that perhaps there were two particles involved, one of which had even parity and decayed into the even parity final state, and the other one of odd parity. The reasons for clinging to this belief was that the parity conservation law follows from what appears to be a very deep source: the laws of nature are invariant under reflection in a mirror. Two physicists, T.D. Lee and C.N. Yang, had the courage to raise the question: How do we know that parity is conserved in the weak interactions? They pointed out that the soundness of parity conservation in the electromagnetic interactions was checked by the success of a number of selection rules, but the quantitative check that was available (for example, limits on the intensity of parity-forbidden transitions) was not good enough to say anything about the much weaker beta interactions. After numerous calculations, Lee and Yang saw the essence of tests for parity nonconservation.

The conservation law states that physical observation should not enable us to distinguish between our world and "a world reflected in a mirror." We must be subtle in interpreting this: Suppose we have a decay and a muon moves to the left. Under reflection, the muon moves to the right. Does this distinguish between the two worlds? It does not, if the muon could just as well have been moving to the right in our world, that is, if the amplitude for the two directions is the same. Consider an analogy: The law of gravitation is spherically symmetric, since it depends on $1/r$ alone. Suppose we see an orbit that is elliptical. Does this imply a violation? It does not, since different initial conditions could give us all possible families of ellipses, so that our original ellipse upon rotation turns into another permissible ellipse. An experiment that can distinguish between the two worlds is the following: Suppose we have a nucleus with spin that beta decays, such as

$$^{60}\text{Co} \rightarrow {}^{60}\text{Ni} + e^- + \nu_e.$$

If the nucleus is polarized, that is, the initial state is not a mixture of all the possible m_j values, but is in a state such that $m_j = j$, say, then the expectation value of the spin is not zero, that is, $\langle \mathbf{J} \rangle \neq 0$. Suppose the electrons are measured to come out with a distribution such that more electrons come out in the direction of $\langle \mathbf{J} \rangle$ than against it. If the distribution is given by

$$W = A + B \langle \mathbf{J} \rangle \cdot \mathbf{p}_e, \tag{41.50}$$

then this is evidence for parity nonconservation. In a mirror world the momentum changes sign $\mathbf{p}_e \rightarrow -\mathbf{p}_e$. On the other hand, since $\mathbf{r} \rightarrow -\mathbf{r}$, the angular momentum $\mathbf{r} \times \mathbf{p}$ does not change sign, and since spins must behave the same way as orbital angular momentum under parity (otherwise they would add in our world and subtract in the mirror world), $\langle \mathbf{J} \rangle$ does not change sign (Fig. 41.6). Thus in the mirror world, the distribution would be

$$W = A - B \langle \mathbf{J} \rangle \cdot \mathbf{p}_e \tag{41.51}$$

and this is different from Eq. (41.50)! This experiment, and many others suggested by Lee and Yang, were carried out in 1957 and it was clearly established that *parity is not conserved in the weak interactions*. The weak interactions appear to be invariant under mirror inversion combined with particle-antiparticle conjugation, that is, under the combined *CP* operation, but that too is violated on a very tiny scale. The reason for this symmetry breaking will undoubtedly form a fascinating chapter in fundamental particle research. One lesson to be learned from this discovery is that no matter how attractive and compelling a physical theory is, it must be tested by experiment, and that dogmatism has no place in physics.

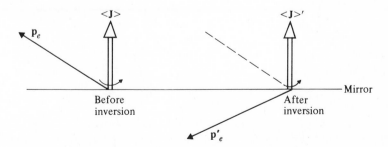

Figure 41.6. Schematic drawing illustrating the behavior of $\langle \mathbf{J} \rangle \mathbf{p}_e$ under inversion in a plane.

NOTES AND COMMENTS

1. The field of elementary particle physics is vast, and growing at a very fast rate. There are two books that can be read to supplement, in much more detail, the material that is covered in this and the next two chapters. These are H. Frauenfelder and E. Henley, *Subatomic Physics*, Prentice-Hall, Englewood Cliffs, N.J., 1974; D.H. Perkins, *Introduction to High-Energy Physics*, Addison-Wesley, Reading, Mass., 1972. Both books devote some space to experimental techniques and make enjoyable reading. There are also occasional articles in *Scientific American* that deal with these matters. Some of the articles relevant to the material of this chapter are:
    "The Overthrow of Parity," P. Morrison, April, 1957.
    "Experiments in Time Reversal," O.E. Overseth, October, 1969.
    "Experiments with Neutrino Beams," B.C. Barish, August, 1973.
    "Unified Theories of Elementary Particle Interactions," S. Weinberg, July, 1974.
    "The Detection of Neutral Weak Currents," D.B. Cline, A.K. Mann, and C. Rubbia, December, 1974.
    "The Search for New Families of Elementary Particles," D.B. Cline, A.K. Mann, and C. Rubbia, January, 1976.

2. The way in which Dirac arrived at a consistent relativistic quantum theory of the electron, by filling all the negative energy states with electrons, and insisting on the exclusion principle to avoid collapse, opened up the whole field of relativistic quantum mechanics. An immediate consequence was that one cannot deal with one particle at a time (as one does in the Schrödinger equation), because at high energies, any number of the electrons can be knocked out of the "sea," or equivalently, any number of electron-positron pairs can make their appearance. Since these particles are also described by a Dirac equation, one effectively ends up with an infinite number of coupled equations. For weak coupling, as is the case for electromagnetic interactions, one can make sensible perturbation approximations. One interesting effect that emerges from the notion of the "sea" is the so-called *vacuum polarization*. Consider an isolated positive charge, a proton, say. Such a proton will attract electrons that are in the negative energy sea, and will thus partially shield its own charge. The shielding depends on the distance from the proton, and it is reduced by a factor of e^2 compared with the dominant Coulomb potential. The effect is to change $-e^2/r$ into

    $$-\frac{e^2}{r}(1 - \alpha f(r)e^{-2mcr/\hbar}),$$

    with $f(r)$ a complicated function of r. The effect is tiny but measurable, and has been identified in high-precision atomic spectroscopy as well as in high-precision measurements of proton-proton scattering.

3. Our description of the beta interaction in Eq. (41.40) is vastly oversimplified in that it does not take into account the possibility that some of the particles can flip their spin in the transition. Thus terms like

    $$\int (\psi_e^* \, \boldsymbol{\sigma} \, \psi_{\nu_e}) \cdot (\psi_P^* \boldsymbol{\sigma} \psi_N) d^3 r$$

should be included. The whole theory should really be formulated relativistically because the electrons are often relativistic, and the neutrino, being massless, always moves with the speed of light. Eq. (41.40) is thus only a symbolic representation of the kind of term that one writes. It does contain the information that particles can convert from one form to another, so that one need not assume that the electron "emerged" from the nucleus, where it was before the process took place. The latter picture has difficulties because the electron has charge and spin, and having an electron staying in the nucleus does not work. One could, of course, say that both the electron and the neutrino were in the nucleus, but then one must explain how their spins always adjusted to give the right angular momentum for the parent nucleus. The notion of creation of particles is intrinsic to relativistic quantum mechanics. Fermi, for all of his enormous accomplishments in physics (he was considered the last "universal physicist"), always viewed his theory of beta decay as his most important work.

4. The function $F(E_{max}/m_e c^2)$ describes all of the energy dependence of beta decay. The function has nothing to do with the actual dynamics of the weak interaction. It is the analogue of the factor $8\pi \nu^2 / c^3$ in the Planck distribution, in that it counts the number of modes into which the decay can go. The decay

$$N \to P + e^- + \bar{\nu}_e$$

will have a rate that is proportional to $d^3 p_e d^3 p_\nu$, and when these are integrated over, subject to energy conservation, the function F emerges. There is no other energy dependence in the nonrelativistic limit since the wave functions are quite smooth, and they all appear at the same point. An amplitude that comes from such a perturbation is necessarily quite structureless.

5. The electron, the muon, and their neutrinos are called *leptons*. Nobody understands why there should be four of them. In fact, there is good evidence now for the existence of a fifth lepton, called the τ, whose mass is around $1800 \, \text{MeV}/c^2$. If previous experience is to guide us, we should assume that the τ, too, has its own neutrino and its own conservation law. For a nontechnical description, see M.L. Perl and W.T. Kirk, *Heavy Leptons*, Scientific American, March, 1978.

PROBLEMS

41.1 Calculate the spectrum of an electron in a Coulomb field using the relativistic Schrödinger equation (41.9). (a) Write

$$\psi(\mathbf{r}, t) = u(\mathbf{r})e^{-iEt/\hbar}.$$

Next separate the angular motion using $u(\mathbf{r}) = R(r)Y_{lm}(\theta, \phi)$ as for the hydrogen atom. Then (b) write the coefficient of $1/r^2$ in the form $l^*(l^* + 1)$ as for the hydrogen atom, and finally write the energy in the form $(const)/(n_r + l^* + 1)^2$ again as for the hydrogen atom.

41.2 Use the formula (41.47) to estimate through how many meters of iron must a 100 GeV neutrino travel if one neutrino out of 10^{10} is to interact.

41.3 When negative muons slow down in matter, they get captured in Bohr orbits about nuclei. In a very short time, compared to their lifetime, they cascade to the lowest ($1s$) Bohr orbit. (a) Show that the capture rate, that is, the rate for the process

$$\mu^- + (A, Z) \rightarrow (A, Z - 1) + \nu_\mu$$

is proportional to Z^4. [*Hint*: Compare the discussion of positronium annihilation, and its dependence on the volume in which the particle is confined.] (b) If the lifetime of the muon is 2.2×10^{-6} sec and for a nucleus of $Z = 10$ half the muons decay and half are captured, what is the capture rate in hydrogen?

41.4 What is the wavelength of the radiation emitted when a muon cascades from a ($2p$) to a ($1s$) orbit in a nucleus with charge Z? What do you expect to happen to the electronic structure when the muon is in the ($1s$) orbit?

41.5 A neutron star whose mass is 2×10^{33} gm has a radius of 10 km. Given that the neutrino-neutron cross section is 10^{-43} cm^2, what is the mean free path of the neutrino inside the neutron star?

41.6 A neutral particle, the π^0, decays by the process

$$\pi^0 \rightarrow \gamma + \gamma.$$

With the π^0 at rest, each γ has an energy of 70 MeV. Suppose one wishes to study this decay by letting it decay in a liquid bubble chamber, and observing the materialization of the γ in the liquid. Use the expression for the mean free path for pair production (41.26) to estimate how large a chamber one would need if the liquid were hydrogen (density $0.07 \, \text{g/cm}^3$). If the liquid is Freon (CF_3Br), with $Z = 6, 9, 35$ for the three elements, respectively, and the density of the liquid is $1.5 \, \text{g/cm}^3$, what size chamber would be needed?

41.7 An antiproton moving with momentum $200 \, \text{GeV}/c$ strikes a proton, and they annihilate. How much energy is available for the creation of particles such as pions? What is the largest possible number of pions that can be produced in such a collision, given that the pion mass is $140 \, \text{MeV}/c^2$?

41.8 Suppose 1 gm of antimatter were to hit the earth someplace and annihilate. A large unit of energy is the TNT equivalent kiloton, used by weapon designers. One TNT equivalent kiloton is 2.61×10^{31} eV. Express the energy released during annihilation in kilotons. How much matter could you lift to a height of 1 km with this amount of energy?

42
The Strong Interactions and Their Symmetries

Soon after the discovery of the neutron in 1932, some basic properties of the nuclear forces were identified. The forces were found to be (a) strong (compared with electromagnetic forces), (b) short range, with a range of the order of $1-1.5$ fermis, and (c) they were charge independent, so that, aside from electromagnetic effects due to the charge on the proton, one could idealize the situation to the assertion

$$V_{PP} = V_{NN} = V_{PN}. \tag{42.1}$$

It was clear that one was dealing with a new force in nature, and the field of exploration of these forces is known as the study of the *strong interactions*.

Yukawa theory One of the deepest contributions to the understanding of the strong interactions was provided by H. Yukawa in 1935. Yukawa noted that the well-understood electromagnetic interactions could be interpreted as follows: Electrons and positrons can emit and absorb photons. The mutual interaction arises from the process of emission of a photon by one particle, and its absorption by the other. This may be represented diagramatically as in Fig. 42.1. The amplitude for emission is proportional to e for electrons and $-e$ for positrons, so that the potential strength is proportional to e^2 and has opposite sign for like charges compared to that for unlike charges. The graphs, incidentally, are called *Feynman graphs*, and they are interpreted as describing a temporal evolution of the system (the bottom of the graph represents the past) only

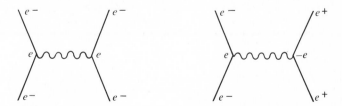

Figure 42.1. Pictorial representation of electron–electron and electron–positron interaction through the exchange of a virtual photon. The virtual photon line has been drawn horizontally, but there is no significance to that. For this diagram, which represents the order e^2 interaction, a single photon exchange, whatever the tilt of the line, represents the whole amplitude.

as far as the "vertices" on a given electron line are concerned. One cannot say that the absorption follows the emission. This is associated with the fact that the emitted photon cannot be real. If an electron emits a photon of momentum q, and remains an electron, it must end up having momentum $-q$ if it started out at rest. Thus energy conservation now reads

$$mc^2 = qc + (m^2c^4 + q^2c^2)^{1/2}, \tag{42.2}$$

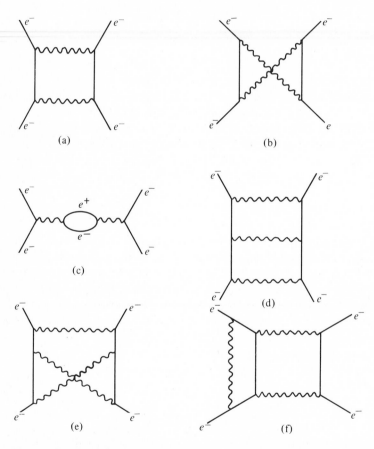

Figure 42.2. Pictorial representation of some higher-order contributions to the electron–electron interaction generated by multiple photon emission. Graphs (a) and (b) represent two-photon exchange contributions, proportional to e^4; (c) represents the exchange of a virtual photon which "materializes" into a virtual electron–positron pair, which then turns back into a virtual photon; this contribution is the *vacuum polarization* referred to in Chapter 41; graphs (d), (e), (f) represent e^6 contributions to the electron–electron interaction. In the case (f) we see a virtual photon being reabsorbed by the emitting electron. These graphs represent just a few of the many that contribute to the e^4 and e^6 terms in the potential.

which cannot be satisfied for real values of q. There are various ways of looking at this, but we shall take the point of view that the momenta are real, and that this implies that in the emission process, energy is not conserved, by an amount

$$\Delta E = ((qc)^2 + m^2 c^4)^{1/2} + qc - mc^2 \approx 2qc \qquad (42.3)$$

for $q \ll mc$. Let us (following an argument due to G.C. Wick) admit that energy need not be conserved for a time Δt allowed by the uncertainty relation,

$$\Delta t \sim \frac{\hbar}{\Delta E}. \qquad (42.4)$$

Thus the "virtual" photon can roam for a time $\Delta t \sim \hbar/2qc$. The distance that it can travel is at most $c\Delta t$, so that here the range is

$$a \simeq \frac{\hbar}{2q}. \qquad (42.5)$$

Since the value of q can be as small as we like, we interpret this to say that the range of the force due to photon exchange is infinite. This is indeed the case for the Coulomb potential. Incidentally, there is no reason why there should not exist multiple photon exchanges as shown in Fig. 42.2. Each exchange contributes an additional factor of e^2 to the potential, so that multiple photon exchanges provide corrections to the Coulomb potential that are of order $\alpha/\pi = 0.2\%$ of the dominant force, given by the potential

$$V(r) = \pm \frac{e^2}{r} \qquad (42.6)$$

This potential can be derived from the one-photon exchange mechanism by well-defined rules of quantum field theory. In applying this approach to the nuclear interactions, Yukawa argued that the observed finiteness of the range had to imply that ΔE could not become arbitrarily small. This can be achieved as follows: Suppose the nucleons interact by the emission of a new kind of quantum (later called *mesons*) which had a rest mass μ. In emission by a nucleon at rest, the energy imbalance is now

$$\Delta E = (\mu^2 c^4 + q^2 c^2)^{1/2} + (M^2 c^4 + q^2 c^2)^{1/2} - Mc^2 \qquad (42.7)$$

which, in the limit $q \to 0$, gives

$$\Delta E \simeq \mu c^2, \qquad (42.8)$$

leading to the range

$$a \sim c \frac{\hbar}{\Delta E} \sim \frac{\hbar}{\mu c}. \qquad (42.9)$$

The range is approximately 1.3×10^{-13} cm, and this tells us that

$$\mu c^2 \cong \frac{\hbar c}{a} \cong \frac{1.05 \times 10^{-27} \times 3 \times 10^{10}}{1.3 \times 10^{-13}}$$

$$\cong 2.4 \times 10^{-4} \text{ ergs}$$

$$\cong 150 \text{ MeV}. \qquad (42.10)$$

The pion The short range of the interaction thus led Yukawa to the prediction of the existence of a new kind of particle, the meson, of mass $\sim 1/6$ that of the nucleon. What else can we say about the meson?

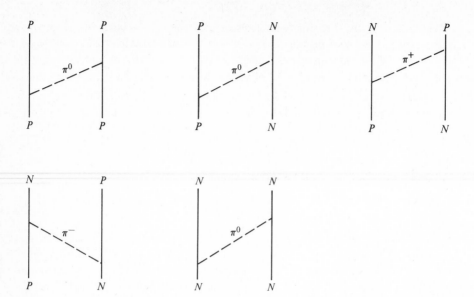

Figure 42.3. Exchanges of neutral and charged mesons in nucleon–nucleon inter-action. In the third and fourth process, the proton turns into a neutron, and vice versa. Such an interaction is called an exchange interaction.

1. It cannot be a fermion, since a fermion cannot emit a fermion and maintain half-integral spin. Thus the meson is expected to have spin $0, 1, 2, \ldots$ Simplicity favors spin 0, while analogy with the photon would favor a "vector meson" with spin 1.

2. The meson could either be neutral, or could come in charge varieties $+$ and $-$, as shown in Fig. 42.3.

3. The strong interactions are strong because the coupling constant at each vertex g is large. A rough estimate gives $g^2/\hbar c \cong 15$, to be compared with the electromagnetic $e^2/\hbar c = 1/137$!

4. A calculation that is beyond the scope of this book shows that the one-meson exchange gives rise to a potential of the form

$$V(r) = -g^2 \frac{e^{-r/r_0}}{r} \tag{42.11}$$

where $r_0 = \hbar/\mu c$. We should note, however, that this last form is only suggestive, since the multiple exchange of mesons is not suppressed. It appears to be enhanced by a factor of 15 for each additional meson, though multiple exchanges have shorter range. We cannot express the potential in a power series in $g^2/\hbar c$ with Eq. (42.11) as the leading term, so that the Yukawa idea did not really solve the nuclear force problem quantitatively. It had, nevertheless, an enormous influence on the whole development of thinking in particle physics.

According to Yukawa, the new quanta should be produced in high-energy reac-tions, in which there would be enough energy to materialize the mesons. At the time that he made the suggestion, energies in excess of 150 MeV were available only in cosmic rays, and searches were made for such heavy particles in cloud chambers. There was an interesting bit of confusion in that in 1938 a particle of mass $\simeq 100$ MeV/c^2 was found in the cosmic rays. Detailed study of the interactions of that particle showed that it only interacted electromagnetically with matter, and therefore could not be the Yukawa meson. This particle is the *muon* that we discussed in the last chapter, and it is just a heavy electron. The true Yukawa meson was first discovered in 1947, and named the *pion*. It came in three charge states, π^+, π^0, π^-, with masses

$$m_{\pi^+}c^2 = m_{\pi^-}c^2 = 139.5 \text{ MeV}$$

$$m_{\pi^0}c^2 = 135.0 \text{ MeV} \qquad (42.12)$$

in excellent agreement with Yukawa's rough estimate. The π^+ and the π^- have exactly the same mass, since they are antiparticles of each other; the π^0 is its own antiparticle. The pions were found to have spin 0, and negative parity, that is, the wave function of a pion at rest changes sign under mirror inversion (a technical point that will come up later). Pions interact with nuclear matter strongly. At high energies, pions are produced copiously, so that the description of nucleon-nucleon forces by a potential to be inserted into a Schrödinger equation is at best a low-energy approximation. Nuclear forces must be studied by means of relativistic quantum mechanics.

Charge independence of nuclear forces Even before Yukawa's work, attention was attracted by the equality (42.1). Physicists do not generally believe in accidents, especially when dealing with fundamental questions. To the charge independence of the nuclear forces should be added the nearly charge independent values of the nucleon masses

$$m_P c^2 = 938.3 \text{ MeV}; \quad m_N c^2 = 939.6 \text{ MeV}. \qquad (42.13)$$

This led Heisenberg and E.U. Condon, independently, to speculate on what the world would be like if electromagnetism could be turned off, that is, what would the interactions look like if $e = 0$. Disregarding the fact that there would be no atoms (perhaps no electrons?), it would be very hard to tell neutrons and protons apart. It is very tempting to ascribe the neutron-proton mass difference to electromagnetism (it is an order α/π effect), so that in the $e = 0$ limit, the neutron and proton would be degenerate. Consider now the following analogy: Take an external magnetic field and place a hydrogen atom in it. The electron has two spin states, spin "up" and spin "down," and in this environment, the two states have different energies. One might be tempted to describe these as "lighter" and "heavier" electrons if one lived in a world in which there was a permanent uniform magnetic field. We know, however, that this is wrong: There is only one electron, but it does have two states, and the energies become degenerate when the magnetic field is turned off. Using this analogy, Heisenberg and Condon proposed that the proton and neutron be "up" and "down" states of a doublet, of isotopic spin (or iso-spin, or i-spin) 1/2. This is not a real spin that couples to orbital angular momentum, but a spin in some internal space. The nucleon thus is an $I = 1/2$ object. The charge independence of the nuclear force can be expressed by the statement that "rotations" in this i-spin space, which "rotate" a proton into a neutron, do not change the interaction between two nucleons. Any invariance under such internal "rotations" carries with it the implication of a conservation law, *the conservation of i-spin*, just as the invariance of the laws of nature under rotations in real space implies the conservation of angular momentum. To recapitulate:

a) there is associated with the strongly interacting particles an internal property, called *i*-spin, which formally is just like spin or angular momentum;

b) the strong interactions conserve *i*-spin;

c) the proton and neutron form an $I = 1/2$ doublet, with the proton having $I_z = 1/2$ and the neutron $I_z = -1/2$.

At this stage, this is just a description of observations, but there are really much deeper implications that stem from the fact that one really postulates that the physics is unaltered under a transformation: proton $\rightarrow$ 30% proton + 70% neutron, say. For example, a two nucleon state can be in a state of total $I = 1$ or $I = 0$, since adding spin 1/2 to spin 1/2 can give these results. The deuteron has no partners, hence it must be an $I = 0$ state. Any peculiarity in the PP system, which must be an $I = 1$ state (since $I_z = 1/2 + 1/2 = 1$), must also be found in the NN system, and the $I = 1$ NP system. The two-nucleon *i*-spin wave functions are

$$
\begin{array}{ccc}
 & PP & I_z = 1 \\[1em]
I = 1 & \dfrac{1}{\sqrt{2}}(PN + NP) & I_z = 0 \\[1em]
 & NN & I_z = -1 \\[1em]
I = 0 & \dfrac{1}{\sqrt{2}}(PN - NP) & I_z = 0.
\end{array}
\tag{42.14}
$$

Because of the technical difficulties of doing neutron-neutron scattering, this particular test has never been carried out, but in more complicated nuclei, such as ^{18}O ($I_z = -1$), ^{18}F ($I_z = 0$), and ^{18}Ne ($I_z = +1$), which can be in a variety of total *i*-spin states, the "analog states," that is, states belonging to the same *i*-spin triplet, have been identified.

i-spin and pions   Where do the pions fit in? The fact that there are three of them, and that their masses are equal to within a few percent (again $0(\alpha)$ effects), strongly suggests that the π^+, π^0, π^- are just the $I_z = 1, 0, -1$ states of an *i*-spin triplet. Nuclear forces due to pion emission and absorption will conserve *i*-spin provided the couplings $PP\pi^0$, $NN\pi^0$, $PN\pi^+$, and $NP\pi^-$ are so arranged that in the emission shown in Fig. (42.4),

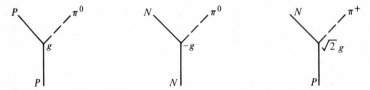

Figure 42.4. Basic couplings leading to *i*-spin conservation. It is only with the couplings so related that the contributions of the various graphs in Fig. 42.3 will lead to Eq. (42.1) when the three pions are taken to have the same mass.

the final state, consisting of a nucleon and a pion again has $I = 1/2$ instead of the possible $I = 3/2$ (the latter possibility would violate i-spin conservation). Many tests for i-spin conservation in pion-nucleon reactions have been proposed, and they all support the notion that i-spin is conserved in the strong interactions between these particles.

The $I = 3/2$ resonance A particular confirmation comes about in the study of pion-nucleon scattering. The pion-nucleon cross section for the π^+P state has the shape schematically sketched in Fig. 42.5. There is a very prominent peak when the center of mass energy of the pion + nucleon (including the rest masses) has the value of $1236 \, \text{MeV}/c^2$. The scattering angular distribution at that energy shows that the scattering takes place in an ordinary angular momentum state $J = 3/2$, with one unit of orbital angular momentum. The parity of this "compound state," or *resonance*, as it is called in particle physics, is $+1$, since $L = 1$, so that the spatial parity is -1, and that due to a single pion is -1 (as noted before). Thus the scattering is interpreted as proceeding through the formation and subsequent decay of a $J^P = 3/2^+$ state. The state has $I_z = 3/2$, and hence it must be pure $I = 3/2$. Such a state has been called Δ^{++}, and the electric charge is $Q = 2$. By i-spin conservation, there should be a corresponding peak in π^-P scattering. Here $I_z = -1/2$, so that there can be $I = 1/2$ contributions to the scattering amplitude in addition to the $I = 3/2$ contribution, but the expectation is that near $1236 \, \text{MeV}/c^2$ the latter should dominate. Thus one expects a peak, and also angular distribution that is predominantly due to a $J^P = 3/2^+, I = 3/2$ intermediate state. Such a state, the Δ^0, has indeed been found, and so have the other two partners, with $I_z = 1/2$ and $I_z = -3/2$. Thus the existence of the i-spin 3/2 quartet Δ^{++}, Δ^+, Δ^0, Δ^-, with mass $1236 \, \text{MeV}/c^2$, lends further support to the notion of i-spin symmetry. Whatever the detailed dynamical theory of pion-nucleon interactions will end up being, the slightly (order α) broken i-spin symmetry will necessarily be part of it.

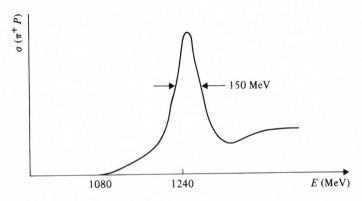

Figure 42.5. Sketch of the energy dependence of the π^+P elastic scattering cross section.

Relation between charge and I_z A final comment related to i-spin is that the charge of a particle can be expressed in terms of I_z. Thus we have, for the nucleons

$$Q = I_z + \tfrac{1}{2} \tag{42.15}$$

and for the pions

$$Q = I_z. \tag{42.16}$$

The two formulas can be put together, using baryon number B, whose value is 1 for the nucleon and 0 for the pion

$$Q = I_z + \tfrac{1}{2}B. \tag{42.17}$$

This also identifies the I_z of the antiproton as $-1/2$, and so on, which should be the case, since the proton and antiproton should be able to annihilate to leave nothing except energy and momentum.

The strange particles In the late 1940s and the early 1950s, cloud chamber experiments and then experiments on the then new Brookhaven National Laboratory 2 GeV Cosmotron showed the existence of some new particles. The first to be discovered were V-shaped tracks in a cloud chamber, exposed to cosmic rays (Fig. 42.6). The curvature of the tracks in a magnetic field could be used to determine the momenta of the particles, and their range (and droplet formation) could yield their energy, with the direction of curvature determining the sign of the charge. It thus appeared that the tracks corresponded to a proton and a negative pion. When the "mass" of the π^-P system was determined by means of

$$(M_{\pi^- P}c^2)^2 = (E_P + E_\pi)^2 - (\mathbf{p}_P c + \mathbf{p}_\pi c)^2 \tag{42.18}$$

the result was not a smooth distribution, as it would be if the two particles were just randomly produced out of a nucleus in the cloud chamber, but a very sharp, unique

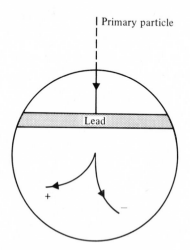

Figure 42.6. Characteristic V-shaped tracks observed in cloud chambers with magnetic fields. These led to the discovery of the Λ^0 and the K^0.

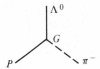

Figure 42.7. Schematic picture of decay of Λ^0 into $P + \pi^-$.

value of $1115\,\text{MeV}/c^2$. By measuring the velocity of this neutral particle that evidently decayed into a P and a π^- after being produced in the lead plate in the chamber, and by measuring the distance of the point of decay from the lead plate where the particle was produced, the lifetime of the particle (named the Λ^0) could be determined. It was found to be $\tau_{\Lambda^0} = 2.6 \times 10^{-10}$ sec. There was no compelling reason for the Λ^0 to have charged partners (like the Δ^0) since the decay was seen to be characterized by a long lifetime compared with the unsuppressed $\hbar/m_\pi c^2$. This implied that the decay proceeded through the weak interactions, and for these, as beta decay shows, there is no i-spin conservation. To make an argument that the decay is really weak, let us describe the decay by the graph shown in Fig. 42.7 and associate the coupling strength G_Λ with the decay amplitude. The decay rate is proportional to the coupling constant squared, G_Λ^2, with something with the dimensions of $(\text{sec})^{-1}$ made out of $\hbar, c$, and the masses, with some unknown, but generally of order 1 coefficient in front. Let us take

$$\frac{1}{\tau_{\Lambda^0}} = G_\Lambda^2 \, \frac{m_\pi c^2}{\hbar} \, \text{sec}^{-1}. \tag{42.19}$$

This, then, tells us that $G_\Lambda \cong 1.4 \times 10^{-7}$, which is a very tiny number, of magnitude comparable to the coupling that appears in beta decay. If the same coupling strength characterized the production mechanism via the reaction shown in Fig. 42.8, for example, then we would expect the production cross section to be of the form $G_\Lambda^2 \times$ (the largest area that can be involved in the process). The area is of the order of magnitude

$$\text{area} \sim \pi \, (\hbar/m_\pi c)^2 \tag{42.20}$$

Figure 42.8. Possible schematic picture of Λ^0 production in NP collision.

so that

$$\sigma \cong \pi G_\Lambda^2 \left(\frac{\hbar}{m_\pi c}\right)^2 \cong 10^{-37} \text{ cm}^2 . \tag{42.21}$$

This number is 10^{10} smaller than the *observed* cross section for production, which is of the order of 10^{-27} cm^2. Clearly this discrepancy is not due to sloppiness in our estimates, but indicates that the new baryon Λ^0 had to be produced via the strong interactions, though it decays weakly (in spite of the absence of a neutrino).

During the same period, other particles were seen to be produced. These, together with their dominant decay modes, masses, and lifetimes are listed in Table 42.1.

Table 42.1. The Experimental Status of the "New" Particles before the Gell-Mann-Nishijima Theory and the Discoveries Stimulated by That Theory

Particle	Baryon number	Mass (GeV/c^2)	Lifetime (sec)	Dominant decay modes	Comment
Λ^0	1	1.115	2.6×10^{-10}	$\Lambda^0 \to P\pi^-$ $\to N\pi^0$	
Σ^+	1	1.189	0.8×10^{-10}	$\Sigma^+ \to N\pi^+$ $\to P\pi^0$	
Σ^-	1	1.197	1.5×10^{-10}	$\Sigma^- \to N\pi^-$	This is *not* the antiparticle of the Σ^+ since both decay into $B = 1$ states.
Ξ^-	1	1.321	1.3×10^{-10}	$\Xi^- \to \Lambda^0 \pi^-$	It is significant that no case of $\Xi^- \to N + \pi^-$ has ever been seen.
$K^\pm$	0	0.494	1.2×10^{-8}	$K^\pm \to \pi^\pm \pi^0$ $\to \pi^\pm \pi^+ \pi^-$ $\to \pi^\pm \pi^0 \pi^0$	The K^+ is the antiparticle of the K^-.
K^0	0	0.498	0.9×10^{-10}	$K^0 \to \pi^+ \pi^-$ $\to \pi^+ \pi^- \pi^0$	It appears as if the K^0, like the π^0, is its own antiparticle. This is wrong!

The spectrum of $B = 1$ states and $B = 0$ states is shown in Fig. 42.9.

The data in all cases contained the same paradox: Production was strong, decay was weak. What would keep the Λ^0 from undergoing the *strong* decay

$$\Lambda^0 \xrightarrow{\text{strong}} P + \pi^-$$

if it was produced strongly? Why is there no G_Λ in the production mechanism? The first clue came from the suggestion of A. Pais that the new particles had to be produced strongly *in pairs*, whereas in the decay, the masses were such that only one of the new particles appeared in the reaction. According to Pais, had the mass of the Λ^0 been larger, a *strong* decay, with lifetime of the order of 10^{-23} sec of the type

$$\underset{\text{new}}{\Lambda^0} \to \underset{\text{old}}{P} + \underset{\text{new}}{K^-}$$

Figure 42.9. $B = 1$ and $B = 0$ spectra before the discovery of strangeness. The states predicted by the scheme and subsequently discovered are sketched in dashed lines.

would have been possible. The notion of *associated production* of the new particles turned out to be correct. In the Λ^0 production, it always turned out that in, say, $\pi^- P$ collisions, one could have

$$\pi^- P \to \underset{\text{new}}{\Lambda^0} \ \underset{\text{new}}{K^0}$$

$$\to \underset{\text{new}}{\Lambda^0} \ \underset{\text{new}}{K^+} \pi^- \ .$$

and the K mesons were always present. In PP collisions, one also always observed pairs of the new particles, such as in

$$PP \to \underset{\text{new}}{\Lambda^0} \ P \ \underset{\text{new}}{K^+}$$

$$\to \underset{\text{new}}{\Sigma^+} \ P \ \underset{\text{new}}{K^0}.$$

Similarly,

$$\pi^- P \to P \ \underset{\text{new}}{K^-} \ \underset{\text{new}}{K^0}$$

$$\to N \ \underset{\text{new}}{K^+} \ \underset{\text{new}}{K^-}$$

was observed, but no single K production, without the production of a Λ^0 or a $\Sigma^\pm$. Still, there remained a problem. Since

$$\Xi^- \xrightarrow{\text{slow}} \Lambda^0 + \pi^-$$
$$\underset{\text{new}}{}$$

was slow, one would expect the Ξ^- to belong to the same class as the nucleon. Yet

$$\Xi^- \to N + \pi^-$$

should then be a strong decay. It has, in fact, never been observed. The problem was solved by M. Gell-Mann and independently by K. Nishijima, who approached it by trying to extend the notion of *i*-spin to the new particles. First of all, this makes sense, since the new particles, at least in pairs, interact strongly with the pion and nucleon, and they should therefore be involved in the strong interactions of the latter. For example, if in the reactions

$$\pi^- + P \to \Lambda^0 + K^0 \to \pi^- + P$$

i-spin were not conserved, one would have difficulty understanding *i*-spin conservation in

$$\pi^- + P \to \pi^- + P.$$

Strangeness Let us now follow Gell-Mann and Nishijima in the i-spin assignment. The Λ^0 really does not have any charged partners that have masses within a few percent of the $1115\,\text{MeV}/c^2$. We must thus assign $I = 0$ to it. Consider now the reaction

$$\pi^- + P \to \Lambda^0 + K^0$$

which is strong. The initial state has $I = 1/2$ or $3/2$. Since the Λ^0 has $I = 0$, the K^0 must have $I = 1/2$ or $3/2$. If it were the latter, it would have three partners, so that a K^{--} or a K^{++} would have to exist. These were not seen, so that we assign $I = 1/2$ to the K^0. Since $I_z = -1/2$ in the initial state, the K^0 must have $I_z = -1/2$. Its i-spin partner with $I_z = +1/2$ must therefore be the K^+, which has a mass very close to that of the K^0. The charge formula

$$Q = I_z + \tfrac{1}{2}B \tag{42.22}$$

does not apply to the Λ^0, and thus a new quantum number *strangeness* S was introduced in

$$Q = I_z + \tfrac{1}{2}(B + S). \tag{42.23}$$

The Λ^0 must have $S = -1$, and using the same formula with $I_z = -1/2$, the K^0 must have $S = +1$; the K^+ must also have $S = +1$. Let us assume that *strangeness is conserved in the strong interactions*. Since the nucleon and pion have $S = 0$, one would not expect to see

$$\pi^- + P \to \Lambda^0 + K^- + \pi^+$$

since the K^-, being the antiparticle of the K^+, presumably has $S = -1$. Thus there are reactions in which the notion of associated production gives the wrong answers.

In the decays

$$\Lambda^0 \to P + \pi^-$$
$$K^0 \to \pi^+ + \pi^-,$$

S changes by one unit, and it was therefore postulated that a reaction with $\Delta S = 1$ must be weak, with a lifetime 10^{13} slower than the strong $\Delta S = 0$ reactions. A reaction with $\Delta S = 2$ should be slower still, and thus reactions like

$$\pi^- + P \to \Lambda^0 + K^- + \pi^+$$

and

$$N + P \to \Lambda^0 + \Lambda^0 + \pi^+$$

should have vanishingly small cross sections. They have been searched for, and never observed.

Let us now go on to the $\Sigma^\pm$. Since there is a Σ^+ and a Σ^-, they must belong to an $I = 1$ state, given that Σ^{++} or Σ^{--} have never been observed. What about the $I_z = 0$ Σ^0 particle predicted by this scheme? Gell-Mann noted that the decay

$$\Sigma^0 \to \Lambda^0 + \gamma$$

could occur electromagnetically, and if electromagnetic interactions conserve strangeness, then such a decay, with a lifetime of the order of 10^{-20} sec (i.e., α/π slower than the strong decays) would have led to the unobservability of the Σ^0. (Note that from the charge formula the Σ triplet has $S = -1$, and were it a bit more massive, it could decay rapidly $\Sigma \to \Lambda^0 + \pi$.) A very careful examination of the Λ^0 decays, in which Λ^0 were produced with high energies, showed that occasionally there was a discrepancy in that the energy and momentum of the Λ^0 and the associated K did not add up (Fig. 42.10). The missing energy and momentum were related by $\Delta E = c\Delta p$, which could

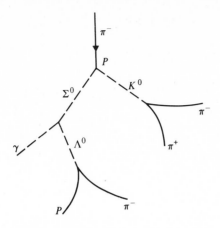

Figure 42.10. Schematic picture of a Σ^0 decay. The visible Λ^0 and K^0 energies and momenta, combined with the known incident energy and momentum, allow a reconstruction of the Σ^0 and hence γ energies and momenta.

be interpreted as an escaping photon. The mass of the invisible Σ^0 was determined in this way, and found to lie at 1192.5 MeV/c^2, right near the charged Σ's.

Let us now consider the Ξ^- decay. If this is a $\Delta S = 1$ decay, and the Ξ^- does not have $S = 0$ (since then $\Xi^- \to N + \pi^-$ would be fast), it must have $S = -2$. The charge formula then tells us that $I_z = -1/2$ for the Ξ^-. The absence of a Ξ^{--} indicates that the Ξ^- must be part of an $I = 1/2$ doublet. The $I_z = 1/2$ partner, Ξ^0, would have to exist, and presumably undergo the hard-to-detect decay

$$\Xi^0 \to \Lambda^0 + \pi^0.$$

This particle was soon identified by looking at some Λ^0s produced in conjunction with K^+s (Fig. 42.11), and measuring missing momenta and energies. Its mass was found to be 1315 MeV/c^2, within a few percent of the 1321 MeV/c^2 of the i-spin partner Ξ^-.

Let us now turn to the K mesons. We saw that (K^+, K^0) form a doublet with $S = +1$. The antiparticles would have $S = -1$ and $I = 1/2$ (S must change sign under particle-antiparticle conjugation if Eq. (42.22) is to hold). They would be

$$K^- \quad I_z = -1/2$$
$$\bar{K}^0 \quad I_z = 1/2,$$

according to the charge formula. Thus there is another prediction: The particle that we called K^0 really has a degenerate antiparticle $\bar{K}^0$. The confirmation of this was achieved in a most interesting way, and we discuss this next.

The neutral K meson decays In the strong interactions, the difference between the K^0 and the $\bar{K}^0$ are manifest, since they have different values of S. Thus they are absorbed

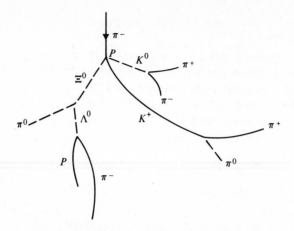

Figure 42.11. Schematic picture of a Ξ^0 production and decay. Here, as in the previous figure, the dotted lines represent neutral particles that leave no traces in the bubble chamber.

differently in matter, since for the K^0 the reactions at low energies are

$$K^0 + P \rightarrow K^0 + P$$
$$\rightarrow K^+ + N$$
$$K^0 + N \rightarrow K^0 + N$$

while for the $\bar{K}^0$, the $S = -1$ value allows many more reactions

$$\bar{K}^0 + P \rightarrow \bar{K}^0 + P$$
$$\rightarrow \Lambda^0 + \pi^+$$
$$\rightarrow \Sigma^+ + \pi^0$$
$$\rightarrow \Sigma^- + \pi^+ + \pi^-$$
$$\cdots$$

(and similarly for a neutron target). Thus their absorption in matter will be quite different. When these particles decay, strangeness is not conserved, so that it is not clear how to tell them apart by their decay modes. Gell-Mann and Pais made the following interesting observation: by the time a decay occurs, strangeness is no longer relevant for the decaying state, but there is a conserved quantum number, CP in the weak interactions. Thus if the final state is a $\pi^+\pi^-$ state (necessarily in an $L = 0$ state since the K's are spinless), then its CP value of $+1$ must also be the CP value of the initial state. What about CP for the K^0 and $\bar{K}^0$? Since the parity of the K was measured to be negative just like for the pion,

$$CP \, \psi_{K^0} = -\psi_{\bar{K}^0}$$
$$CP \, \psi_{\bar{K}^0} = -\psi_{K^0} \qquad (42.24)$$

so that neither the K^0 nor the $\bar{K}^0$ have a definite CP. However, the linear combination

$$K_S = \frac{1}{\sqrt{2}}(K^0 - \bar{K}^0) \qquad (42.25)$$

is an eigenstate of CP, with

$$CP\,\psi_{KS} = \psi_{KS} \tag{42.26}$$

Thus, Gell-Mann and Pais noted, the final state $\pi^+\pi^-$ is the decay product of the coherent (in the sense of a definite phase relationship) mixture of K^0 and $\overline{K}^0$. It is the K_S ("S" stands for *short*, referring to the lifetime) that decays into CP-even states, and it is the K_S that has the lifetime measured from the V's in the cloud chamber that consist of $\pi^+\pi^-$. That lifetime is $\tau_S = 0.9 \times 10^{-10}$ sec.

The orthogonal combination

$$K_L = \frac{1}{\sqrt{2}}(K^0 + \overline{K}^0) \tag{42.27}$$

is also an eigenstate of CP, with

$$CP\,\psi_{KL} = -\,\psi_{KL} \tag{42.28}$$

and it *cannot* decay into $\pi^+\pi^-$. It can, however, decay into $\pi^+\pi^-\pi^0$, $\pi^\pm e^\mp \nu$, and so on. The rates into these channels which involve more particles, are smaller, so that Gell-Mann and Pais expected that the K_L^0 would have a longer lifetime. They suggested looking for this unanticipated particle farther downbeam, with three particle decay modes. The K_L^0 was found, with a lifetime of 5.2×10^{-8} sec! The sequence of what happens is the following: In a reaction like

$$\pi^- + P \to \Lambda^0 + K^0$$

it is a K^0 that is produced, with $S = +1$. This may be viewed as a coherent mixture of K_S and K_L. Using Eqs. (42.25) and (42.27), we see that

$$K^0 = \frac{1}{\sqrt{2}}(K_S + K_L). \tag{42.29}$$

After about 10^{-10} sec the K_S component dies out, with decay

$$K_S \to \begin{cases} \pi^+\pi^- \\ \pi^0\pi^0 \end{cases}.$$

The K_L component travels on, and later decays into $\pi^+\pi^-\pi^0$, for example.

Regeneration A. Pais and O. Piccioni made the following interesting suggestion: Suppose matter is interposed between the region where the K_S decay occurs, and where the K_L decay occurs (Fig. 42.12). What is traveling along the beam in that region is the piece

$$\frac{1}{\sqrt{2}}K_L,$$

which, using Eq. (42.27), may be written as

$$\frac{1}{\sqrt{2}}K_L = \tfrac{1}{2}(K^0 + \overline{K}^0). \tag{42.30}$$

Now in matter, $\overline{K}^0$ is absorbed more strongly than K^0. Suppose we idealize the situation and assume that $\overline{K}^0$ is totally absorbed. What then emerges from the slab of

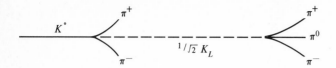

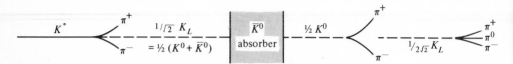

Figure 42.12. (a) Picture of K^0 decay without absorber. (b) Picture of K^0 decay and associated regeneration in the presence of absorber.

matter is

$$\tfrac{1}{2}K^0 = \frac{1}{2\sqrt{2}}(K_S + K_L) \tag{42.31}$$

and in a distance corresponding to 10^{-10} sec beyond the slab of material, $\pi^+\pi^-$ pairs again appear. The interposed matter in effect gives rise to a *regeneration* of the K^0 beam. This has been observed.

The physics behind these remarkable experiments is just the superposition principle, and is no different from the transport of a beam of light through a sequence of polarizers with different axes of polarization. It does demonstrate the applicability of quantum mechanics in this new domain of physics.

Resonances and octets The notion of strangeness, its conservation in the strong and electromagnetic interactions, and the rule that $\Delta S = 1$ in the weak interactions has passed every test. If we introduce a new quantum number, $Y = (B + S)$, called *hypercharge*, then

$$Q = I_z + Y/2 \tag{42.32}$$

and the known particles fit into the patterns:

		$B = 1$			$B = 0$		
$I = 1/2$	$Y = 1$	P	N		K^+	K^0	
$I = 1$	$Y = 0$	Σ^+	Σ^0	Σ^-	π^+	π^0	π^-
$I = 0$	$Y = 0$	Λ^0					
$I = 1/2$	$Y = -1$	Ξ^0	Ξ^-		$\bar{K}^0$	K^-	

All of the baryons have spin 1/2, and all of the mesons have spin 0 and parity -1. Why these patterns, and where is the missing $Y = 0, I = 0$ meson?

The filling of the gap and the understanding of the familial relationship between the various baryons and mesons did not happen until 1961. An important component

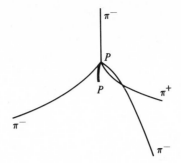

Figure 42.13. Schematic picture
of a bubble chamber photograph
in which $\pi^- + P \to N^* + \pi^-$, with
a rapid decay $N^* \to P + \pi^+ + \pi^-$.

was experimental progress in the construction of high-energy accelerators and the
development of the hydrogen bubble chamber, which behaved somewhat like the
cloud chamber, but gave incomparably better photographs of tracks left by charged
particles. High-energy accelerators made possible the copious production of pions and
K's. The pions and the $K^\pm$ have lifetimes of the order of 10^{-8} sec, so that they can
travel large distances when they have energies much larger than their rest masses. A
particle of mass m and energy $E \gg mc^2$ can travel a distance

$$L \simeq \tau c \, \frac{E}{mc^2} \simeq 300 \, \frac{E}{mc^2} \text{ cm}$$

which for a 10 GeV pion, for example, is of the order of 200 m. It is thus possible to
set up various magnetic focusing devices to bend the beams and produce fairly monochromatic secondary beams of pions and K's. When these particles strike a proton in a
liquid hydrogen target, a reaction takes place, and charged particles leave curved tracks
in the chamber when a magnetic field is present. Pictures of the type shown in Fig.
42.13 are typical. Suppose that the nucleon had an excited state, similar to the Δ mentioned before, that undergoes a decay like

$$N^* \to P + \pi^- + \pi^+.$$

Then the "mass" of such a particle would be given by

$$(M^*c^2)^2 = (E_P + E_{\pi^-} + E_{\pi^+})^2 - c^2(\mathbf{p}_P + \mathbf{p}_{\pi^-} + \mathbf{p}_{\pi^+})^2. \qquad (42.33)$$

What was done in hundreds of thousands of pictures was to look at pairs, triplets, etc.
of tracks, and plot the "mass" obtained above. Occasionally the mass plots (Fig. 42.14)
show rather prominent peaks, and these are tentatively identified with highly unstable
particles or *resonances*. Further study of the angular distributions in the decay in that
mass region can be used to determine the spin and parity of the decaying state N^*. In
this way hundreds of new resonant states were discovered. Among them, the following
decays

$$\eta^0 \to \pi^+ + \pi^- + \pi^0$$

showed a definite narrow peak at 548.8 MeV/c^2. An analysis showed that (a) there
were no charged partners to the η^0, indicating that it had $I = 0$; the decay was not

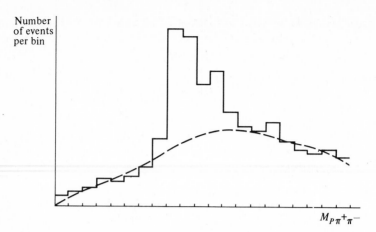

Figure 42.14. Typical mass plot showing resonance in $P\pi^+\pi^-$ system. When the number of events is large, the fluctuations usually present and indicated above get smoothed out. Statistical methods must be used to assess the significance of a given peak. The dashed curve indicates an estimate of the distribution if there were no resonance.

weak, so that $Y = 0$; (b) the angular distribution was characteristic of a spin 0 particle with negative parity. This was then the missing meson.

Thus both baryons and mesons came in octets. Why? The search for a higher symmetry occupied several years, and it was complicated by the fact that the symmetry was badly broken. Whereas it does not take much persuasion to put the Σ^+, Σ^0 and Σ^- together into a near degenerate mass multiplet, the η^0 mass is four times larger than that of the pion!

The Eightfold Way In 1961 M. Gell-Mann and Y. Ne'eman independently discovered a generalization of i-spin with the technical name SU(3). Gell-Mann named it the "Eightfold Way," which we used in the heading of the section. The rotations are no longer in a three-dimensional internal space, but in an eight-dimensional one. The details are beyond the scope of this book, but we can summarize some of the results:

1) Particles come in families that are labeled by i-spin and Y. The simplest families have three members (hence SU(3)), with $I = 1/2$ and $Y = y$, $I = 0$, $Y = y - 1$; the antiparticles have $I = 1/2$, $Y = -y$, $I = 0$, $Y = -y + 1$. These members have not been observed. However, the family with an *octet* of members is just the $I = 1/2$, $Y = 1$; $I = 1$, $Y = 0$; $I = 0$, $Y = 0$; and $I = 1/2$, $Y = -1$ family seen before. There is also a ten-fold family, with members $I = 3/2$, $Y = 1$; $I = 1$, $Y = 0$; $I = 1/2$, $Y = -1$; and $I = 0$, $Y = -2$, and more complicated families with 27 members, and so on.

Particular attention turned to identifying the SU(3) partners of the $I = 3/2$, $Y = 1$, $\Delta(1236)$, with $J^P = 3/2^+$ discussed before. In 1960, a $\Lambda^0\pi$ resonance, called the $\Sigma^*(1385)$ – the mass in MeV/c^2 is given in the parentheses – was discovered, and it was found to have $I = 1$ and $Y = 0$, and also $J^P = 3/2^+$. Gell-Mann and others suggested that it belong to the ten-fold representation. Soon a $\Xi^*(1535)$ was discovered. It had $I = 1/2$ and $Y = -1$, and underwent the strong decay

$$\Xi^* \to \Xi + \pi.$$

According to the SU(3) scheme, there should exist one more particle to fit into the scheme, with $I = 0$ and $Y = -2$. The fact that the mass spacing was uniform for the observed resonances

$$m(\Xi^*) - m(\Sigma^*) \cong m(\Sigma^*) - m(\Delta) \cong 150 \text{ MeV} \qquad (42.34)$$

suggested that the remaining particle would have mass around $1685 \text{ MeV}/c^2$. This was a very interesting result, since this particle, called the Ω^-, could not undergo a *strong* decay such as

$$\underset{Y=-2}{\Omega^-} \to \underset{Y=-1}{\Xi^0} + \underset{Y=-1}{K^-}$$

because its mass is too small. It must therefore decay weakly, into

$$\Omega^- \to \Xi + \pi,$$

for example, and it would be expected to have a long life, with 10^{-10} sec a good guess. It would thus leave a track in a bubble chamber, and it would also have a distinctive production signature, since in production with a K^-P collision, two $Y = +1$ states would have to be produced with it.

In 1964 a perfect picture of an Ω^- was found in a long search carried out at Brookhaven National Laboratory (Fig. 42.15). The production was indeed

$$K^- + P \to \Omega^- + K^+ + K^0$$

and the decay proceeded according to

$$\begin{aligned}
\Omega^- &\to \pi^- + \Xi^0 \\
&\qquad\quad \hookrightarrow \Lambda^0 + \pi^0 \\
&\qquad\qquad\qquad\quad \hookrightarrow \gamma + \gamma \\
&\qquad\qquad\qquad\qquad\qquad \hookrightarrow e^+ e^- \\
&\qquad\qquad\qquad\qquad\quad \hookrightarrow e^+ e^- \\
&\qquad\quad \hookrightarrow P + \pi^-.
\end{aligned}$$

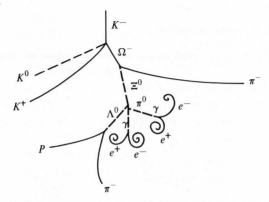

Figure 42.15. Picture of the Ω^- production and decay.

The conversion of both γ rays into e^+-e^- pairs was rather fortuitous, and helped to make the picture more dramatic. The Ω^- was found to have a mass of 1672 MeV/c^2 and a lifetime of 1.3×10^{-10} sec, and its discovery convinced all doubters about the validity of SU(3) as the underlying symmetry of the strong interactions.

Many resonances have been found in the $B = 1$ (baryon) and $B = 0$ (meson) sectors. The properties of some of them are listed in Table 42.2. Interestingly, the only supermultiplets that have been found are *octets* and *singlets* ($I = Y = 0$) for the mesons, and *octets*, *decuplets*, and *singlets* for baryons. Octets of mesons with J^P equal to 1^-, 1^+, 2^+ have definitely been established, and so have accompanying singlets. There is nothing in the symmetry to force this restriction (in an octet-octet collision, the 27-plet could in principle be produced, if it existed), and there is no explanation why multiplets such as the triplet, and the antitriplet do not occur. Recent theoretical speculations, dealing with this and other topics, form the substance of the next, last chapter.

NOTES AND COMMENTS

1. Supplementary material may be found in the books listed in the Bibliography (Appendix D) and at the end of Chapter 41.

2. The question of parity conservation, discussed in the last chapter, arose out of attempts to identify the quantum numbers of the newly discovered strange particles. What we called the K meson was not, at first, seen to be a single particle. The particle that decayed into 2π's was called the θ. Since there was a θ^0 that decayed into $\pi^0 + \pi^0$, two identical bosons, the orbital angular momentum had to be even, and since there are *two* pions, the parity also had to be even. Thus the θ was forced to have $J^P = 0^+$, $2^+, \ldots$ with the first preferred on grounds of simplicity. The particle that decayed into three pions, for example $\pi^+ + \pi^+ + \pi^-$, was called the τ^+. The three pions come off with low kinetic energy, so that it was most likely a state without orbital angular momentum, that is, a state with spin 0, and parity -1, since each of the pions had parity -1. Thus the τ had to be a 0^- particle. R.H. Dalitz devised a method of analyzing the distribution of the decay products into three particles so as to determine J^P, and these qualitative considerations were sustained. By 1956, however, it became clear that the θ and the τ differed very little in mass and in lifetime, and it became more and more attractive to view the 2π and the 3π decays as different decay modes of a single particle. The problem that agitated everybody was, how can a particle have decay products of both positive and negative parity, and it was this problem that led Lee and Yang in the summer of 1956 to raise the question of parity nonconservation in the weak decays.

3. The search for the symmetry among the particles started very early after the discovery of the first few strange particles, and many false leads were followed. The notion of symmetry was not one that was stressed in the teaching of physics, and although much of spectroscopy could be explained very simply with the help of the theory of the rotation group, the use of such higher mathematics was

TABLE 42.2. Low-lying Particles and Resonances

Particle Symbol	J^P	B	SU(3) assignment	Y	J	I_z	Q	Mass in MeV	Width in MeV	Lifetime in sec if stable under strong interaction	Dominant decay modes
P	$1/2^+$	1	8	1	1/2	1/2	1	938.3		$> 2 \times 10^{30}$ yrs	STABLE
N				1	1/2	$-1/2$	0	939.6		918 ± 14 sec	$Pe^-\bar{\nu}$
Σ^+				0	1	1	1	1189.4		0.80×10^{-10} sec	$P\pi^0, N\pi^+$
Σ^0				0	1	0	0	1192.5		$< 10^{-14}$ sec	$\Lambda^0\gamma$
Σ^-				0	1	-1	-1	1197.4		1.48×10^{-10} sec	$N\pi^-$
Λ^0			8	0	0	0	0	1115.6		2.58×10^{-10} sec	$P\pi^-, N\pi^0$
Ξ^0				-1	1/2	1/2	0	1314.9		2.96×10^{-10} sec	$\Lambda\pi^0, \Lambda\gamma$
Ξ^-				-1	1/2	$-1/2$	-1	1321.3		1.65×10^{-10} sec	$\Lambda\pi^-$
K^+	0^-			1	0	1/2	1	493.7		1.24×10^{-8} sec	$\mu^+\nu_\mu; \pi^+\pi^0; \pi^+\pi^+\pi^-$
K^0				1	0	$-1/2$	0	497.7		$\tau_S = 0.89 \times 10^{-10}$ sec	$K_S \to \pi^+\pi^-, \pi^0\pi^0$
π^+				0	0	1	1	139.6		2.60×10^{-8} sec	$\mu^+\nu_\mu$
π^0				0	0	0	0	135.0		$0.83 \pm .06 \times 10^{-16}$ sec	$\gamma\gamma$
π^-				0	0	-1	-1	139.6		2.60×10^{-8} sec	$\mu^-\bar{\nu}_\mu$
η^0				0	0	0	0	548.8		0.8×10^{-18} sec	$3\pi^0, \pi^+\pi^-\pi^0, \pi^+\pi^-\gamma$
$\bar{K}^0$				-1	0	1/2	0	497.7		$\tau_L = 5.18 \times 10^{-8}$ sec	$K_L \to \pi e\nu, \pi\mu\nu, \pi^+\pi^-\pi^0$
K^-				-1	0	$-1/2$	-1	493.7		1.24×10^{-8} sec	$\mu^-\bar{\nu}_\mu, \pi^-\pi^0, \pi^+\pi^-\pi^-$
Δ^{++}	$3/2^+$	1	10	1	3/2	3/2	2	1232	115		$P\pi^+, P\pi^0$
Δ^+				1	3/2	1/2	1				$N\pi^+, P\pi^0$
Δ^0				1	3/2	$-1/2$	0				$N\pi^0, P\pi^-$
Δ^-				1	3/2	$-3/2$	-1				$N\pi^-$
Σ^{*+}				0	1	1	1	1385	35		$\Lambda^0\pi^+, \Sigma\pi$
Σ^{*0}				0	1	0	0				$\Lambda^0\pi^0, \Sigma\pi$
Σ^{*-}				0	1	-1	-1				$\Lambda^0\pi^-: \Sigma\pi$
Ξ^{0*}				-1	1/2	1/2	0	1532	9.1		$\Xi^0\pi^0, \Xi^-, \pi^+$
Ξ^{*-}				-1	1/2	$-1/2$	-1				$\Xi^0\pi^-, \Xi^-, \pi^0$
Ω^-				-2	0	0	-1	1672.2		$1.3 \pm 0.3 \times 10^{-10}$ sec	$\Xi^0\pi^-; \Xi^-\pi^0; \Lambda^0K^-$

restricted to specialists in spectroscopy, with relatively little overlap on particle physics. As mentioned before, the fact that the symmetry was strongly broken did not help.

PROBLEMS

42.1 The spin-orbit potential observed in nuclear shell structure is also observed in proton-proton scattering, for example. It is now believed that this potential is due to an exchange of a spin 1 ("vector") meson, whose rest mass is approximately $760 \, \text{MeV}/c^2$. Use this to estimate the range of the spin-orbit potential.

42.2 ^{3}H and ^{3}He are nuclei that consist of (PNN) and (PPN), respectively. The masses of the two nuclei are nearly equal, so that they are believed to belong to the same i-spin multiplet. What are the possible values of the i-spin? Suggest an experiment to test which of your possible values is the correct one.

42.3 The deuteron is known to have i-spin 0. The reaction
$$D + D \rightarrow \alpha + \pi^0$$
is forbidden. If this forbiddenness is due to i-spin conservation, what is the value of the i-spin of the α (^{4}He nucleus)? Given that there is no stable $A = 4, Z = 0$ state, what is the only possible value of the i-spin for the α?

42.4 Which of the following reactions are allowed, which must go through the weak interactions, and which are absolutely forbidden?

a) $N + P \rightarrow \Lambda^0 + \Lambda^0 + \pi^+$

b) $P + P \rightarrow \Lambda^0 + N + K^+ + \pi^+$

c) $\pi^+ + P \rightarrow \Sigma^+ + K^0 + \pi^+$

d) $\pi^+ + P \rightarrow \Lambda^0 + K^+$

e) $\pi^- + P \rightarrow N + e^-$

f) $\Sigma^0 \rightarrow P + K^-$

g) $K^+ \rightarrow \pi^+ \pi^0 \pi^0$

h) $P + \bar{P} \rightarrow \Lambda^0 + \bar{\Lambda}^0$

i) $P + \bar{P} \rightarrow \Sigma^+ + \Sigma^-$

j) $K^- + P \rightarrow K^+ + \bar{P}$

k) $K^- + P \rightarrow K^0 + N$

l) $\bar{\Lambda} + P \rightarrow \Sigma^+ + \bar{N}$

m) $\pi^+ + P \rightarrow \Sigma^+ + K^+ + K^-$

n) $\Sigma^+ \rightarrow \Lambda^0 + \pi^+$

In reactions that have two particles in the initial state, always assume that the collisions are sufficiently energetic so that threshold effects never forbid a reaction. In decays one cannot make that assumption.

42.5 Derive the coupling relations shown in Fig. 42.4 by the following procedure: Derive the PP, NN, and PN potentials by assuming that they are identical, except for a multiplicative factor equal to the product of the couplings at the two vertices, so that, for example,
$$V_{PP} = g_{PP\pi^0} \, g_{PP\pi^0} \, V(r),$$
and so on. Assuming that only a single charged pion is exchanged (so that the third and fourth of the graphs in Fig. 42.3 are identical, and one is redundant), and assuming that $g_{PN\pi^\pm} = g_{NP\pi^\pm}$, show that
$$g_{NN\pi^0} = -g_{PP\pi^0}, \qquad g_{PN\pi^\pm} = \sqrt{2} \, g_{PP\pi^0}.$$

42.6 The uncertainty in the energy of the Δ (its "width") is measured to be 150 MeV. What is the lifetime of the Δ?

42.7 What is the uncertainty in the mass of the Λ^0 caused by its weak decay?

42.8 A particle is observed to decay rapidly (that is, via the strong interactions) into *two* π^+. What information does this provide about the X^{++} concerning (a) its mass, (b) its total angular momentum, (c) its parity, (d) its i-spin, and (e) its strangeness? Will the X^{++} have partners? Suppose the mass of the X is 1100 MeV. Should one expect to see the decay

$$X^0 \rightarrow K^+ + K^-$$

for the neutral member of the multiplet?

43
Quarks, Color, Charm, and other Flavors

Puzzles of SU(3) The undeniable successes of SU(3) in classifying the known particles, and in explaining (roughly) relationships among lifetimes of the strongly interacting particles within a given SU(3) family, raises a number of questions:

a) Why is SU(3) the correct symmetry?

b) Why are mesons found only in octets and singlets, and baryons only in octets, decuplets, and singlets?

c) Why are the mass patterns, representing symmetry, breaking such as they are? Why, for example, are the various i-spin multiplets within the 10-fold supermultiplet ($\Delta, \Sigma^*, \Xi^*, \Omega$) equally spaced in mass?

d) Finally, if all the particles are to be built up out of a few fundamental building blocks, is eight the smallest possible number?

Quarks In 1964 M. Gell-Mann, and G. Zweig independently proposed that the basic building blocks be the lowest nontrivial representations of SU(3). These states, called *quarks* by Gell-Mann, are the triplet, consisting of an i-spin 1/2 state, with hypercharge $Y = 1/3$, and an i-spin 0 state, with hypercharge $-2/3$, and the antitriplet (antiquark) consisting of an i-spin singlet with hypercharge $Y = 2/3$ and an i-spin doublet with $Y = -1/3$ (Fig. 43.1). The charges are such that the Gell-Mann-Nishijima formula

$$Q = I_z + Y/2 \tag{43.1}$$

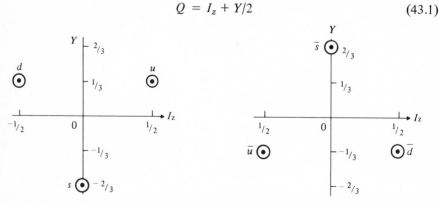

Figure 43.1. Representations of quarks and antiquarks on an I_z-Y plot.

continues to hold. The rules of combination (the analog of the addition of angular momentum) for SU(3) states are

$$3 \times \overline{3} = 8 + 1, \tag{43.2}$$

meaning that the product of a triplet and an antitriplet wave function can be decomposed into an octet and a singlet (just as the product of two spin 1/2 functions can be decomposed into a spin 1 and a spin 0 wave function). For two quarks we have

$$3 \times 3 = 6 + \overline{3}$$

and for three quarks we have

$$3 \times 3 \times 3 = 3 \times 6 + 3 \times \overline{3} = 10 + 8 + 8 + 1. \tag{43.3}$$

This suggested to Gell-Mann and Zweig to assign baryon number $B = 1/3$ to each quark (and $B = -1/3$ to each antiquark) and to postulate that (a) all mesons are of the form $q\overline{q}$; (b) all baryons are of the form qqq. Table 43.1 gives the quantum numbers and names of the quarks. The doublet have the apellation "up" and "down" (u, d) and the singlet is called the "strange" quark (s). *The quarks and antiquarks are assumed to have spin 1/2.*

Table 43.1

Particle	I	I_z	Y	Q	B
u	1/2	1/2	1/3	2/3	1/3
d	1/2	$-1/2$	1/3	$-1/3$	1/3
s	0	0	$-2/3$	$-1/3$	1/3
$\overline{s}$	0	0	2/3	1/3	$-1/3$
$\overline{u}$	1/2	$-1/2$	$-1/3$	$-2/3$	$-1/3$
$\overline{d}$	1/2	1/2	$-1/3$	1/3	$-1/3$

A search for particles with fractional charge has, despite heroic efforts, proved fruitless, so that free quarks have not been observed so far. This does not nullify the theory, as we shall see. If the theory were such that all the quarks had the same mass, and the equations of motion were unchanged under "rotations" of one quark into another, then SU(3) symmetry emerges naturally. Thus the questions (a), (b), and (d) are answered, at the cost of introducing a set of unobserved particles, and by raising some questions such as: how can three quarks bind to make a nucleon, say, but two quarks do not bind, since no six-fold states (which would also have fractional charge) have been observed. There are further complications, which emerge when we look at individual wave functions.

Decuplet states To illustrate what these look like, consider the 10-fold representation. The state of highest charge is the Δ^{++}, and it must be made of (uuu). To get the other Δ's, we convert, in succession $u \rightarrow d$, which lowers the I_z value by unity, but does not change strangeness (cf. Fig. 43.2). Thus

$$\Delta^{++} = uuu$$
$$\Delta^{+} = (duu + udu + uud)/\sqrt{3}$$
$$\Delta^{0} = (ddu + dud + udd)/\sqrt{3}$$
$$\Delta^{-} = ddd. \tag{43.4}$$

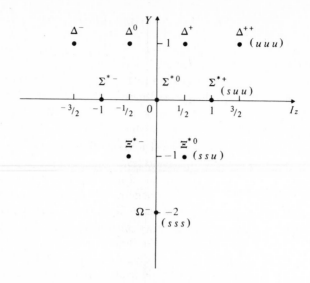

Figure 43.2. Representation of decuplet on an I_z-Y plot.

To get to the highest charge Σ^*, we must start with the Δ^{++}, and convert a u into an s, which lowers the hypercharge by 1. Thus the Σ^{*+} is obtained. The partners are obtained by the conversion $u \to d$ as before. Hence

$$\Sigma^{*+} = (uus + usu + suu)/\sqrt{3}$$
$$\Sigma^{*0} = (dus + uds + dsu + usd + sud + sdu)/\sqrt{6}$$
$$\Sigma^{*-} = (dds + dsd + sdd)/\sqrt{3} . \tag{43.5}$$

To get to the Ξ^*, we proceed from the Σ^* and convert another u into an s, and take it from there

$$\Xi^{*0} = (uss + sus + ssu)/\sqrt{3}$$
$$\Xi^{*-} = (dss + sds + ssd)/\sqrt{3} . \tag{43.6}$$

Finally, the remaining $Y = -2$ state can only be

$$\Omega^- = sss. \tag{43.7}$$

We note that the four multiplets differ from one another by the number of strange quarks. Thus, if we assume that the symmetry of pure SU(3) is broken by the fact that the masses have the pattern

$$m_u = m_d \neq m_s$$
$$m_s - m_u = 150 \, \text{MeV}, \tag{43.8}$$

then we can understand (a) the splitting of the multiplet into equal 150 MeV intervals, and (b) why i-spin symmetry is good, whereas SU(3) is broken.

Octet states The wave functions for the octets (Fig. 43.3) are a bit more complicated. For example, the proton has the same Q and I_z as the Δ^+, so it must be made up of the same quarks, but its wave function must be orthogonal to the Δ^+. This suggests

$$P = u(ud - du)/\sqrt{2} \tag{43.9}$$

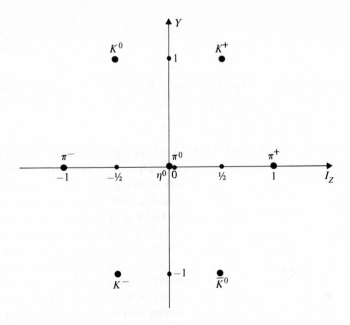

Figure 43.3. Representation of octet on an I_z-Y plot.

and, by the usual $u \to d$ procedure,

$$N = (dud + udd - ddu - udd)/\sqrt{2}$$
$$= d(ud - du)/\sqrt{2}. \tag{43.10}$$

The Σ^+ must have two u's and an s, orthogonal to the Σ^{*+}. It can be obtained from the proton by the change $d \to s$:

$$\Sigma^+ = u(us - su)/\sqrt{2} \tag{43.11}$$

and so on.

The mesonic states are a little simpler. Thus the K^+ must have $Q = 1$ and $Y = -1$, so that

$$K^+ = u\bar{s}. \tag{43.12}$$

From this it immediately follows that

$$K^0 = d\bar{s}$$
$$\bar{K}^0 = \bar{d}s$$
$$K^- = \bar{u}s. \tag{43.13}$$

The pions are

$$\pi^+ = u\bar{d}$$
$$\pi^0 = (d\bar{d} - u\bar{u})/\sqrt{2}$$
$$\pi^- = -\bar{u}d. \tag{43.14}$$

It is a somewhat technical point (and of no consequence to us) why the sign of the π^- wave function is what it is, and we also state without proof that

$$\eta^0 = (d\bar{d} + u\bar{u} - 2s\bar{s})/\sqrt{6}. \tag{43.15}$$

Gell-Mann-Okubo mass formula To get mass relationships for the mesons, with the s and $\bar{s}$ split off from the others, we have

$$M(K) = m_u + m_s - B, \qquad (43.16)$$

where B is some binding contribution common to all members of the octet. We then also get

$$M(\pi) = 2m_u - B. \qquad (43.17)$$

To get an expression for the η^0 mass, we observe in Eq. (43.15) that the probability of finding a u or d quark is $1/3$, and that of finding an s-quark is $2/3$. Thus the mass is given by

$$M(\eta) = \tfrac{1}{3}(2m_u) + \tfrac{2}{3}(2m_s) - B. \qquad (43.18)$$

This gives rise to the so-called Gell-Mann-Okubo mass formula

$$3M(\eta) + M(\pi) = 4M(K), \qquad (43.19)$$

which works even better when written in terms of the squares of the masses. In

$$3M^2(\eta) + M^2(\pi) = 4M^2(K) \qquad (43.20)$$

the left-hand side takes on the value 9.22×10^5 MeV^2, and the right-hand side 9.83×10^5 MeV^2. It is not understood why one should use the squares. For the baryon octet, the analogous formula is

$$3M(\Lambda^0) + M(\Sigma) = 2(M(N) + M(\Xi)) \qquad (43.21)$$

which also works to a few percent accuracy. This formula, and the equal spacing formula for the decuplets has proved very useful in identifying resonances with different Y values that belong together: The $q\bar{q}$ system can have a variety of bound states. For example, there is the $^1S_0$ state, which we identify with the octet that the pion belongs to. The spin is 0, and the parity is *odd*: the reason is that the S state has even parity (since L is even), but the antiquark has parity opposite to that of the quark. The $^3S_1$ state will have spin and parity 1^-: such an octet of spin 1 mesons have been found as resonances in the $K\pi$ system (the $K^*(894)$), in the 2π system (the isotriplet $\rho(765)$ meson) and in the 3π system (the i-spin 0, $\omega(784)$). There is also good evidence for a $^3P_2$ state, an octet of 2^+ resonances, and most of the $^3P_1$ and $^3P_0$ states have been identified. The quark model has also been successful in classifying baryonic resonances (qqq), and in explaining the magnetic moments of the baryons, assuming that the (u, d) quarks have masses in the neighborhood of 300 MeV.

Two new questions arise:

a) If the quarks are so light, how is it that none have ever been seen?

b) How does the spin-statistics connection work? The way the second question arises is the following: consider the Δ^{++} state, which is a (uuu) state, with $J = 3/2$, and consider the $J_z = 3/2$ state. Since this is the lowest decuplet state, it is a good assumption (and there is experimental support) that the quarks are in zero orbital angular momentum states. That, however, means that the wave function of the Δ^{++} is $u_\uparrow u_\uparrow u_\uparrow$, with all the spin states the same. Such a wave function is symmetric under the interchange of any two particles, and this seems to violate the theorem according to which spin $1/2$ states have antisymmetric wave functions.

Color and flavor A very ingenious solution was proposed by O.W. Greenberg, M.Y. Han, Y. Nambu and by M. Gell-Mann, who suggested that the quarks be distinguished by yet another label, which Gell-Mann named *color*. Thus each kind of quark, u, d, and s (Gell-Mann named these labels *flavor*) comes in three varieties, or colors. Thus the quarks really number nine, and are

$$u_R \qquad u_B \qquad u_Y$$
$$d_R \qquad d_B \qquad d_Y$$
$$s_R \qquad s_B \qquad s_Y$$

and the Δ^{++} wave function, though symmetric in spin, space, and "flavor," can be made totally antisymmetric in color. The wave function

$$(u_R u_B u_Y - u_R u_Y u_B - u_B u_R u_Y + u_B u_Y u_R$$
$$+ u_Y u_R u_B - u_Y u_B u_R)/\sqrt{6} \qquad (43.22)$$

(we suppress the spin labels) has this property. Such a wave function is called a *singlet* in "color space," because under the kind of transformations that "rotate" one color into another (just like the SU(3) we spoke about, rotates one flavor into another), the wave function does not change. One can also make a "color singlet" out of quark-antiquark. For example, the pion could be

$$(\bar{u}_R d_R + \bar{u}_B d_B + \bar{u}_Y d_Y)/\sqrt{3} \; . \qquad (43.23)$$

It is possible to write all the known particle wave functions either as totally antisymmetric in color, when they are baryons, or totally symmetric in color, when they are mesons (as in Eq. (43.21) and (43.22)), and it is then possible to solve the nonobservability of quarks, and two-quark systems (and the proliferation of new states that come from the existence of a new quantum number), by postulating that an integral part of the quark model is that *only color singlet states can be physically observable*.

Color confinement How can such a thing come about? Present speculations center around a set of theories called gauge theories. These are generalizations of quantum electrodynamics, except that instead of one photon, there exist an octet of photon-

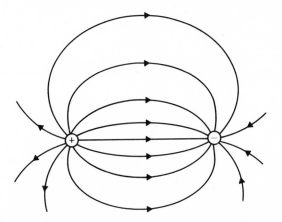

Figure 43.4. Electric field lines connecting charges of opposite sign.

Figure 43.5. Color octet "electric" field lines connecting quark and antiquark.

like objects. The fact that there is an octet is connected with the three-color labels, just as the meson octet is connected with the three flavors u, d, s. Since these objects are color octets, and not singlets, they are unobservable. The speculation is that for reasons not yet understood, the color electric and magnetic field lines connecting a quark and an antiquark, for example, do not fan out all over space (Fig. 43.4) as for a magnet, but rather are confined to a stringlike shape (Fig. 43.5). The field energy is proportional to the volume, and thus to the length of the cylinder. One thus has a potential energy that is proportional to the separation of the quark and antiquark, and it therefore takes an infinite amount of energy to separate a quark from an antiquark. When a lot of energy is transferred to a quark-antiquark system, it is energetically "cheaper" for a new pair to be created out of the vacuum, so that new mesons are produced, as shown in Fig. 43.6.

e^+e^- annihilation at high energies This line of argumentation explains the success of a simple model in the reaction

$$e^+ + e^- \to \text{baryons, mesons}$$

that has been studied in the last five years. The data can be quite well understood in terms of the reaction

$$e^+ + e^- \to \gamma \to q + \bar{q} \tag{43.24}$$

but again, quarks are not seen. Presumably the strong color electric and magnetic fields between the produced quark and antiquark break down, and create a lot of mesons, as

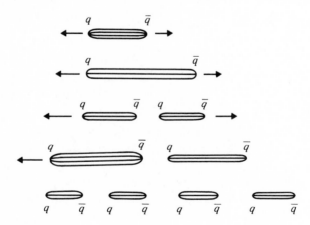

Figure 43.6. Schematic picture for the creation of mesons in quark—antiquark separation at high energies.

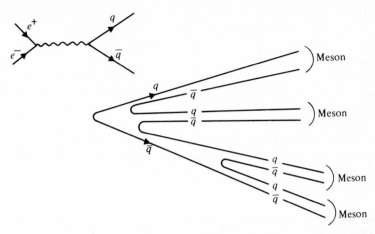

Figure 43.7. Schematic picture of reaction $e^+ + e^- \rightarrow$ mesons. The basic process is the quark creation process. The quarks then "clothe" themselves with pairs out of the vacuum to make physically observable color singlets (mesons here).

shown in Fig. 43.7 and in agreement with experiment. A simple test of the idea is the following: if the quarks are light, and pointlike, then the process

$$e^+ + e^- \rightarrow \gamma \rightarrow q + \bar{q}$$

will be identical with the process

$$e^+ + e^- \rightarrow \gamma \rightarrow \mu^+ + \mu^-$$

at very high energies, the only difference being the difference in the quark charge, which enters at the photon decay vertex only, and which therefore enters into the cross section quadratically. The model predicts that

$$\frac{\sigma(e^+e^- \rightarrow \text{mesons, baryons})}{\sigma(e^+e^- \rightarrow \mu^+\mu^-)} = \sum_i Q_i^2, \tag{43.25}$$

the sum extending over all quark types, with Q_i being the charge of the quark in units of e. If we had only u, d, s quarks,

$$\sum_i Q_i^2 = \left(\frac{2}{3}\right)^2 + \left(\frac{1}{3}\right)^2 + \left(\frac{1}{3}\right)^2 = \frac{2}{3}. \tag{43.26}$$

With each quark coming in three colors, this is tripled, so that

$$\sum_i Q_i^2 = 3\left(\left(\frac{2}{3}\right)^2 + \left(\frac{1}{3}\right)^2 + \left(\frac{1}{3}\right)^2\right) = 2. \tag{43.27}$$

The data up to about 3–4 GeV total "mass" for the final state is consistent with the second number, indicating that the notion of *three colors* is correct. In the 3–4 GeV region, the ratio rises, and suggests the appearance of something new.

Charm In the years following the discovery of SU(3), a number of people speculated on the possibility that perhaps there could be a still higher symmetry, SU(4), based on *four quarks*. This was suggested by analogy with the four leptons, e, ν_e, μ, and ν_μ, and the fourth quark was assigned a new quantum number called *charm* by J.D. Bjorken and S.L. Glashow. The latter author found compelling reasons for the prediction of the fourth quark in studies of the weak decays of strange particles, and a number of people anticipated the possible discovery of new particles, carrying charm, with or without strangeness. The new quark, c, was anticipated to have *i*-spin 0 and charge 2/3, just like the u quark. With this, one expects the ratio in Eq. (43.25) to rise to 10/3 when $c\bar{c}$ pairs begin to be produced. In addition one expects the existence of a new class of mesons, such as the 0^- ($^1S_0$) $\bar{c}u$ and $\bar{c}d$ states, called the D^0, D^+, and the 1^- ($^3S_1$) states called the D^{*0}, D^{*+}. There are also $\bar{c}s$ states, that should exist. The D and D^* were discovered in 1976, with masses $m(D) = 1865$ MeV/c^2 and $m(D^*) = 2006$ MeV/c^2.

Charmonium The most dramatic, though not toally direct, confirmation came in late 1974 with the discovery by S. Ting and collaborators at Brookhaven National Laboratory, and B. Richter and collaborators at the Stanford Linear Accelerator Center, of a very large and narrow peak (see Fig. 43.8) in the cross section for the reactions

$$e^+ + e^- \begin{array}{c} \nearrow \mu^+ + \mu^- \\ \searrow \text{mesons, baryons.} \end{array}$$

The interpretation of this peak is that it is due to the creating of a very massive spin 1 meson of mass 3105 MeV/c^2, followed by the decay of that meson into muons, or strongly interacting particles, collectively called *hadrons*. A number of other states were found at both higher and lower energies in that range, and they can be understood as $\bar{c}c$ states. The first one to be discovered, called the J/ψ (each discoverer has stuck to his label) is the $^3S_1$ state of what might be called *charmonium* in analogy with positronium, but there is no reason why $^1S_0$, $^3P_0$, . . . states should not exist. The spectro-

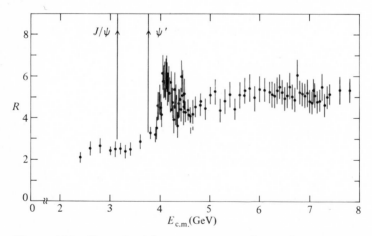

Figure 43.8. Plot of $\sigma(e^+e^- \to \text{hadrons})/\sigma(e^+e^- \to \mu^+\mu^-)$ showing the very sharp charmonium states at 3105 and 3680 MeV, and the increase in the R value to ~ 4, consistent with the color counting (10/3) and the production of $\tau^\pm$ lepton pairs decaying into hadrons. (Courtesy of Standford Linear Accelerator Center.)

scopy of the charmonium states is currently being actively pursued, and the pattern emerging supports the nonrelativistic positronium analogy. The attempts to reproduce the observed spectrum have concentrated on a potential of the form $a/r + br$, in which the first term is associated with the exchange of a massless spin 1 particle, perhaps the color analog of the photon, and the second term is associated with the quark confinement mechanism. It is too early to tell whether such a potential really works and whether this simple form is a consequence of the color-confining quark theory, but there is a whole new area of experimental physics in which much may be learned about the quark-antiquark potential.

The story is hardly finished. In the summer of 1977 two new states, named the Υ ($mc^2 = 9460$ MeV) and Υ' ($mc^2 = 10020$ MeV) and their decay pattern is very reminiscent of the $^3S_1$ $c\bar{c}$ states discovered in 1974. They may be signaling the "quarkonium" structure of yet another flavor, represented by what is called the b-quark whose charge, determined from the Υ decay, appears to be $-1/3$. The lepton-quark symmetry promises to hold up: it appears that there is a still heavier electron, the τ particle, which shares many of the properties of the electron and the muon. If it has its own neutrino, the ν_τ, will there be a sixth quark of charge $2/3$? Whatever the future holds, there is every evidence that the twin pillars of fundamental physics that we have explored in this book, relativity and quantum mechanics, will continue to support the edifice of understanding of nature for a long time to come.

NOTES AND COMMENTS

1. A brief discussion of the quark model may be found in H. Frauenfelder and E. Henley, *Subatomic Physics*, Prentice-Hall, Englewood Cliffs, N.J. Some *Scientific American* articles, in addition to the ones listed at the end of Chapter 41, are "The Structure of the Proton and the Neutron," H.W. Kendall and W.K.H. Panofsky, June, 1971; "Electron-Positron Annihilation and the New Particles," S.D. Drell, June, 1975; "Quarks with Color and Flavor," S.L. Glashow, October, 1975; "The Confinement of Quarks," Y. Nambu, November, 1976.

2. The quark-antiquark SU(3) singlet state is

$$\eta_S = \frac{1}{\sqrt{3}}(u\bar{u} + d\bar{d} + s\bar{s})$$

and its mass cannot be predicted, since the parameter B that appears in (43.16) need not be the same for singlets as it is for octets. Such a state has $I = Y = 0$, and because SU(3) symmetry is not exact, an octet can transform into a singlet provided the octet also has $I = Y = 0$. Effectively the pure singlet above and the η^0 (if we are speaking of 0^- mesons) get "mixed," so that the physical states may be

$$(\eta_0)_{\text{phys}} = \eta^0 \cos \beta + \eta_S \sin \beta$$

$$(\eta_S)_{\text{phys}} = -\eta^0 \sin \beta + \eta_S \cos \beta.$$

Among the vector mesons the mixing is such that the physical states are

$$(\omega^0)_{\text{phys}} = \frac{1}{\sqrt{2}}(u\bar{u} + d\bar{d})$$

$$(\omega_S)_{\text{phys}} \equiv (\phi)_{\text{phys}} = s\bar{s}.$$

3. The idea that there are three colors is confirmed through theoretical calculations of the decay $\pi^0 \rightarrow 2\gamma$ that are beyond the scope of this book. The number of colors enters quadratically here, and neglect of color leads to a factor of 9 discrepancy between experiment and theory.

4. A fascinating account of the discovery of the J/ψ particles may be found in *Adventures in Experimental Physics*, vol. 5, B. Maglich, Ed. That issue also contains an account of the discovery of the ω^0 meson.

5. The alert reader might notice that the baryon octet states, as described in Eqs. (43.9), (43.10), (43.11), all have the property that they are of the form of a quark wave function multiplied by an antisymmetric combination of a two-quark wave function. This is no accident. In a two-quark product, the symmetric part will belong to a **6** representation, and the antisymmetric part to a $\bar{\mathbf{3}}$. It is in the product of $3 \times \bar{3}$ that the baryon octet is placed.

PROBLEMS

43.1 The wave function of the proton in the octet is given in Eq. (43.9) as

$$P = \frac{1}{\sqrt{2}} (uud - udu)$$

Obtain the wave functions of the other members of the octet, using the following procedure:

a) To get the i-spin partners of a state, start with the highest I_z value and then lower it by the change $u \rightarrow d$.

b) To get to the Σ^+ from the P, you need to lower Y by one unit, and raise I_z by $1/2$, which can be done by making the change $d \rightarrow s$. Check that it is orthogonal to Σ^* (43.5).

c) To get the Λ^0, change $u \rightarrow s$ in the proton. This will be a combination of Λ^0 and Σ^0 from (b), the orthogonal combination will give Λ^0.

43.2 Use the wave functions obtained in Problem 43.1 to calculate the probabilities of finding u, d, and s quarks in each i-spin multiplet, and use this to derive the mass relation Eq. (43.21).

43.3 The following $5/2^+$ resonances have been identified:

$I = 1/2, \ Y = 1, \ Mc^2 = 1688 \ \text{MeV}$

$I = 1 \quad Y = 0, \ Mc^2 = 1915 \ \text{MeV}$

$I = 0 \quad Y = 0, \ Mc^2 = 1815 \ \text{MeV}$

What other particles would you expect to find with spin-parity $5/2^+$, and what mass would you expect?

43.4 Suppose the formulas (43.16) and (43.17) are taken seriously, and $m_u = 330$ MeV/c^2. Calculate m_s and B. Next suppose that the charmed quark has mass $m_c =$

$1630\,\text{MeV}/c^2$. Use this, with the assumption that B does not change when charmed quarks appear, to calculate the mass of the D meson.

43.5 Shortly after the discovery of the J/ψ with mass $3105\,\text{MeV}/c^2$ another very sharp resonance with mass $3680\,\text{MeV}/c^2$ was discovered in the same channel. They are conjectured to be the $n = 1$ and $n = 2$ $^3S_1$ states of "charmonium." Suppose charmonium is very much like positronium, that is, the attractive force is Coulomb-like, with an unknown α, however, and unknown m_c. Use the mass information to calculate m_c and α. Note: do not forget the rest masses of the quarks in the total mass formula.

Use this information to predict the mass of the ^3P states and the ^1P states. What sort of effects would make this prediction useless?

Appendix A
Units

I have chosen to use the Gaussian (c.g.s.) system of units rather than the MKSA system of units because physicists, engineers, and chemists who deal with quantum phenomena use these units more frequently than they use the MKS units. Actually this makes very little difference, because each field develops its natural units. Thus in atomic physics, the natural unit of energy is not the erg or the joule, but the electron-volt; the natural unit of distance is the Ångstrom, which is 10^{-8} cm or 10^{-10} m. The masses are frequently expressed in units of energy/c^2, where c is the velocity of light. In elementary particle physics, the unit of energy is the giga-volt (GeV), which is 10^9 eV, and the unit of distance is the fermi, equal to 10^{-13} cm or 10^{-15} m. The common unit for magnetic induction is the gauss; I cannot remember hearing the unit tesla (10^4 gauss) ever used in conversation with scientists dealing with matters that form the subject matter of this book.

A very useful discussion of units and dimensions may be found in J.D. Jackson, *Classical Electrodynamics* (2nd Edition), J. Wiley, New York (1975). We limit ourselves to a brief translation between the MKS system and the c.g.s. system,

Table A.1

Physical quantity	Rationalized MKS	Conversion factor	Gaussian
Length	1 meter (m)	10^2	centimeters (cm)
Mass	1 kilogram (kg)	10^3	grams (gm)
Time	1 second (sec)	1	second (sec)
Force	1 newton	10^5	dynes
Work/Energy	1 joule	10^7	ergs
Power	1 watt	10^7	erg/sec
Charge	1 coulomb	3×10^9	statcoulombs (e.s.u.)
Charge density	1 coulomb/m^3	3×10^3	e.s.u./cm^3
Current	1 ampere (amp)	3×10^9	statamp
Current density	1 amp/m^2	3×10^5	statamp/cm^2
Electric field	1 volt/m	$\frac{1}{3} \times 10^{-4}$	statvolt/cm
Potential	1 volt	1/300	statvolt
Conductivity	1 mho/m	9×10^9	sec^{-1}
Resistance	1 ohm	$\frac{1}{9} \times 10^{-11}$	sec/cm
Magnetic induction	1 tesla	10^4	gauss

Appendix B
Physical Constants

N_0 (Avogadro's number)	$6.0220943(63) \times 10^{23}$ mole^{-1}
c (velocity of light)	$2.99792458(1.2) \times 10^{10}$ cm/sec
e (electron charge)	$4.803242(14) \times 10^{-10}$ e.s.u.
1 eV (electron volt)	$1.6021892(46) \times 10^{-12}$ erg
$\hbar$ (Planck constant/2π)	$1.0545887(57) \times 10^{-27}$ erg sec
	$6.582173(17) \times 10^{-22}$ MeV sec
α $(e^2/\hbar c)$	$1/137.035982(30)$
k (Boltzmann constant)	$1.380662(44) \times 10^{-16}$ erg/^0K
m_e (electron mass)	$9.109534(47) \times 10^{-28}$ gm
	$0.5110034(14)$ MeV/c^2
m_p (proton mass)	$938.2796(27)$ MeV/c^2
	$1836.15152(70)m_e = 1.007276470(11)$ amu
1 amu $(m(^{12}C)/12)$	$931.5016(26)$ MeV/c^2
a_0 (Bohr radius)	$0.52917706(44)$ A
μ (Bohr magneton)	
$\quad = e\hbar/2m_e c$	$0.57883785(95) \times 10^{-14}$ MeV/gauss
R_∞ (Rydberg constant)	
$\quad = m_e c^2 \alpha^2/2$	$13.605804(36)$ eV
G (Gravitational constant)	$6.6732(31) \times 10^{-8}$ cm^3 gm^{-1} sec^{-2}
Calorie	4.184×10^7 erg

Compiled by Stanley J. Brodsky, as presented in *Reviews of Modern Physics*, vol. 48, page S-35, 1976. The figures in parentheses correspond to the statistical uncertainty (one standard deviation) in the last digits of the main number.

Appendix C
Table of the Elements

The table below lists the elements, and their mass number, density, bulk modulus, and Debye temperature, wherever known. Physical properties of elements and many compounds may be found in the *American Institute of Physics Handbook* (Third edition), McGraw-Hill (1971). Detailed references to recent measurements on materials may be found in C. Kittel, *Introduction to Solid State Physics* (Fifth edition), John Wiley and Sons, New York (1976) or in N.W. Ashcroft and N.D. Mermin, *Solid State Physics*, Holt, Rinehart and Winston, New York (1976).

Element	Z	Mass number	Density in g/cm^3	Bulk modulus in 10^{12} $dynes/cm^2$	Debye temperature in $°K$
Hydrogen (H)	1	1.0079	0.089	0.002	110
Helium (He)	2	4.0026	0.179	0.00	26
Lithium (Li)	3	6.941	0.53	0.116	400
Beryllium (Be)	4	9.0122	1.85	1.003	1000
Boron (B)	5	10.81	2.34	1.78	1250
Carbon (C)	6	12.01	2.26	5.45	1860
Nitrogen (N)	7	14.007	1.03	0.012	79
Oxygen (O)	8	15.999	1.43		46
Fluorine (F)	9	18.998	1.97		
Neon (Ne)	10	20.18	1.56	0.010	63
Sodium (Na)	11	22.9898	0.97	0.068	150
Magnesium (Mg)	12	24.305	1.85	0.354	318
Aluminum (Al)	13	26.982	2.70	0.722	394
Silicon (Si)	14	28.086	2.33	0.988	394
Phosphorus (P)	15	30.974	1.82	0.304	
Sulfur (S)	16	32.064	2.07	0.178	
Chlorine (Cl)	17	35.453	2.09		
Argon (Ar)	18	39.948	1.78	0.016	85
Potassium (K)	19	39.09	0.86	0.032	100
Calcium (Ca)	20	40.08	1.54	0.152	230
Scandium (Sc)	21	44.956	2.99	0.435	359
Titanium (Ti)	22	47.90	4.51	1.051	380
Vanadium (V)	23	50.942	6.1	1.619	390

Element	Z	Mass number	Density in g/cm^3	Bulk modulus in 10^{12} $dynes/cm^2$	Debye temperature in $^\circ K$
Chromium (Cr)	24	52.00	7.19	1.901	460
Manganese (Mn)	25	54.938	7.43	0.596	400
Iron (Fe)	26	55.85	7.86	1.683	420
Cobalt (Co)	27	58.93	8.9	1.914	385
Nickel (Ni)	28	58.71	8.9	1.86	375
Copper (Cu)	29	63.55	8.96	1.37	315
Zinc (Zn)	30	65.38	7.14	0.598	234
Gallium (Ga)	31	69.72	5.61	0.569	240
Germanium (Ge)	32	72.59	5.32	0.772	360
Arsenic (As)	33	74.922	5.72	0.394	285
Selenium (Se)	34	78.96	4.79	0.091	150
Bromine (Br)	35	79.91	4.10		
Krypton (Kr)	36	83.80	3.07	0.018	73
Rubidium (Rb)	37	85.47	1.53	0.031	56
Strontium (Sr)	38	87.62	2.60	0.116	147
Yttrium (Y)	39	88.91	4.46	0.366	256
Zirconium (Zr)	40	91.22	6.49	0.833	250
Niobium (Nb)	41	92.91	8.4	1.702	275
Molybdenum (Mo)	42	95.94	10.2	2.725	380
Technetium (Tc)	43	98.91	11.5		
Ruthenium (Ru)	44	101.07	12.2	3.208	382
Rhodium (Rh)	45	102.90	12.4	2.704	350
Palladium (Pd)	46	106.40	12.0	1.808	275
Silver (Ag)	47	107.87	10.5	1.007	215
Cadmium (Cd)	48	112.40	8.65	0.467	120
Indium (In)	49	114.82	7.31	0.411	129
Tin (Sn)	50	118.69	7.30	1.11	170
Antimony (Sb)	51	121.75	6.62	0.383	200
Tellurium (Te)	52	127.60	6.24	0.230	139
Iodine (I)	53	126.90	4.94		
Xenon (Xe)	54	131.30	3.77		55
Cesium (Cs)	55	132.91	1.90	0.020	40
Barium (Ba)	56	137.34	3.5	0.103	110
Lanthanum (La)	57	138.91	6.17	0.243	132
Cerium (Ce)	58	140.12	6.77	0.239	139
Praseodymium (Pr)	59	140.91	6.77	0.306	152
Neodymium (Nd)	60	144.24	7.00	0.327	157
Promethium (Pm)	61	145.0			
Samarium (Sm)	62	150.35	7.54	0.294	166
Europium (Eu)	63	151.96	7.90	0.147	107
Gadolinium (Gd)	64	157.25	8.23	0.383	176
Terbium (Tb)	65	158.92	8.54	0.399	188
Dysprosium (Dy)	66	162.50	8.78	0.384	186
Holmium (Ho)	67	164.93	9.05	0.397	191

Element	Z	Mass number	Density in g/cm^3	Bulk modulus in 10^{12} dynes/cm^2	Debye temperature in $°K$
Erbium (Er)	68	167.26	9.37	0.411	195
Thulium (Tm)	69	168.93	9.31	0.397	200
Ytterbium (Yb)	70	173.04	6.97	0.133	118
Lutetium (Lu)	71	174.97	9.84	0.411	207
Hafnium (Hf)	72	178.49	13.1	1.09	
Tantalum (Ta)	73	180.95	16.6	2.00	225
Tungsten (W)	74	183.85	19.3	3.232	310
Rhenium (Re)	75	186.2	21.0	3.72	416
Osmium (Os)	76	190.20	22.6		400
Iridium (Ir)	77	192.22	22.5	3.55	430
Platinum (Pt)	78	195.09	21.4	2.783	230
Gold (Au)	79	196.97	19.3	1.732	170
Mercury (Hg)	80	200.59	13.6	0.382	100
Thallium (Tl)	81	204.37	11.85	0.359	96
Lead (Pb)	82	207.19	11.4	0.430	88
Bismuth (Bi)	83	208.98	9.8	0.315	120
Polonium (Po)	84	210	9.4		
Astatine (At)	85	210.0			
Radon (Rn)	86	222.0			
Francium (Fr)	87	223.0			
Radium (Ra)	88	226.0			
Actinium (Ac)	89	227.0	10.1		
Thorium (Th)	90	232.04	11.7	0.543	100
Protactinium (Pa)	91	231.0	15.4		
Uranium (U)	92	238.03	19.07	0.987	210
Neptunium (Np)	93	237.05	20.3		188
Plutonium (Pu)	94	244.0	19.8		150
Americium (Am)	95	243.0	11.8		
Curium (Cm)	96	247.0			
Berkelium (Bk)	97	247.0			
Californium (Cf)	98	251.0			
Einsteinium (Es)	99	254.0			
Fermium (Fm)	100	257.0			
Mendelevium (Md)	101	256.0			
Nobelium (NO)	102	254.0			
Lawrencium (Lw)	103	257.0			

Appendix D
Bibliography

Many books have been written on the topics treated in *The Structure of Matter*. They vary in the level and in the special applications that they choose to concentrate on. I could not properly write a critical, descriptive bibliography that does justice to the authors, and I have therefore chosen to list the books referred to in the text, and a few more. Some of these are at a level above that of this book, but they are listed because the reader may find the treatment of a particular topic enlightening.

ESSAYS, BIOGRAPHIES, AND HISTORY

J. Bernstein, *Einstein*, Viking Press, New York (1973).
M. Fierz and V.F. Weisskopf (Ed.), *Theoretical Physics in the Twentieth Century*, Interscience Publishers, New York (1960).
A. Hermann, *The Genesis of Quantum Theory 1899–1913*, M.I.T. Press, Cambridge, Mass. (1971).
B. Hoffmann, *Albert Einstein, Creator and Rebel*, New American Library, New York (1972).
G. Holton, *Thematic Origins of Scientific Thought, Kepler to Einstein*, Harvard University Press, Cambridge, Mass., (1973).
M. Jammer, *The Conceptual Development of Quantum Mechanics*, McGraw-Hill, New York (1968).
P. Schlipp (Ed.), *Albert Einstein, Philosopher-Scientist*, Library of Living Philosophers, Evanston, Ill., (1949)
V.F. Weisskopf, *Physics in the Twentieth Century*, M.I.T. Press, Cambridge, Mass. (1972).

RELATIVITY AND COSMOLOGY

M. Born, *Einstein's Theory of Relativity* (Revised Edition), Dover Publications, New York (1965).
P.G. Bergmann, *The Riddle of Gravitation*, Charles Scribner and Sons, New York (1968).

R.H. Dicke, *The Theoretical Significance of Experimental Relativity*, Gordon and Breach Science Publishers, New York (1964).

A.P. French, *Special Relativity*, W.W. Norton, New York (1968).

O. Gingerich (Ed.), *Cosmology + 1*, a collection of *Scientific American* reprints, W.H. Freeman, San Francisco (1977).

F. Hoyle, *From Stonehenge to Modern Cosmology*, W.H. Freeman, San Francisco (1969).

N.D. Mermin, *Space and Time in Relativity*, McGraw-Hill, New York (1968).

W. Rindler, *Essential Relativity*, Van Nostrand Reinhold Co., New York (1969).

D.W. Sciama, *Modern Cosmology*, Cambridge University Press (1971).

E.F. Taylor and J.A. Wheeler, *Spacetime Physics*, W.H. Freeman, San Francisco (1963).

S. Weinberg, *The First Three Minutes: A Modern View of the Origin of the Universe*, Basic Books, New York (1977).

The Principles of Relativity, a collection of reprints of the basic papers, Dover Publications, New York, (1952).

MODERN PHYSICS

A. Beiser, *Perspectives of Modern Physics*, McGraw-Hill, New York (1969).

R.M. Eisberg, *Fundamentals of Modern Physics*, John Wiley, New York, (1961).

R. Eisberg and R. Resnick, *Quantum Physics of Atoms, Molecules, Solids, Nuclei, and Particles*, John Wiley, New York (1974).

G. Feinberg, *What Is the World Made Of?* Anchor-Doubleday, Garden City, New York (1977).

R.P. Feynman, R.B. Leighton, and M. Sands, *The Feynman Lectures on Physics*, Addison-Wesley, Reading, Mass., (1963).

J.D. McGervey, *Introduction to Modern Physics*, Academic Press, New York (1971).

F.K. Richtmyer, E.H. Kennard, and J.N. Cooper, *Introduction to Modern Physics*, McGraw-Hill, New York (1969).

P.A. Tipler, *Foundations of Modern Physics*, Worth Publishers, New York (1969).

KINETIC THEORY AND STATISTICAL PHYSICS

F.L. Friedman and L. Sartori, *The Classical Atom*, Addison-Wesley, Reading, Mass. (1965).

R. Reif, *Statistical Physics* (Berkeley Physics Course, vol. 5) McGraw-Hill, New York (1964).

E. Schrödinger, *Statistical Thermodynamics*, Cambridge University Press (1948).

A. Sommerfeld, *Thermodynamics and Statistical Mechanics*, Academic Press, New York (1956).

W. Weaver, *Lady Luck, The Theory of Probability*, Anchor Books, Doubleday, Garden City, New York (1963).

QUANTUM MECHANICS

D. Bohm, *Quantum Theory*, Prentice-Hall, Englewood Cliffs, N.J. (1951).

S. Borowitz, *Fundamentals of Quantum Mechanics*, W.A. Benjamin, New York (1967).

S. Gasiorowicz,[†] *Quantum Physics*, John Wiley, New York (1974).

W. Heisenberg, *The Physical Principles of the Quantum Theory*, Dover Publications, (1930).

E. Wichmann, *Quantum Physics* (Berkeley Physics Course, vol. 4) McGraw-Hill, New York (1968).

ATOMS, MOLECULES, SPECTRA, AND OPTICS

F.S. Crawford, Jr., *Waves* (Berkeley Physics Course, Vol. 3) McGraw-Hill, New York (1968).

U. Fano and L. Fano, *Physics of Atoms and Molecules*, University of Chicago Press, Chicago, Ill. (1972).

M.W. Hanna, *Quantum Mechanics in Chemistry*, W.A. Benjamin, Menlo Park, Calif. (1965).

G. Herzberg, *Atomic Spectra and Atomic Structure*, Dover Publications, New York (1944).

M. Karplus and R.N. Porter, *Atoms and Molecules*, W.A. Benjamin, (Addison-Wesley Advanced Textbook Series) Reading, Mass. (1970).

G.W. King, *Spectroscopy and Molecular Structure*, Holt, Rinehart and Winston, New York (1964).

SOLID STATE PHYSICS

N.W. Ashcroft and N.D. Mermin, *Solid State Physics*, Holt, Rinehart and Winston, New York (1976).

J.S. Blakemore, *Solid State Physics* (2nd Edition), W.B. Saunders, Philadelphia (1974).

C. Kittel, *Introduction to Solid State Physics* (5th Edition), John Wiley, New York (1976).

J.P. McKelvey, *Solid State and Semiconductor Physics*, Harper and Row, New York (1966).

A.C. Rose-Innes and E.H. Rhoderick, *Introduction to Superconductivity*, Pergamon Press, London (1969).

NUCLEAR PHYSICS

W.E. Burcham, *Nuclear Physics, An Introduction*, McGraw-Hill, New York (1963).

B.L. Cohen, *Concepts of Nuclear Physics*, McGraw-Hill, New York (1971).

[†] This book contains a fairly extensive bibliography of textbooks on quantum mechanics.

G. Eder, *Nuclear Forces*, M.I.T. Press, Cambridge, Mass. (1968).

H.A. Enge, *Introduction to Nuclear Physics*, Addison-Wesley, Reading Mass. (1966).

E. Fermi, *Nuclear Physics*, University of Chicago Press, Chicago, Ill. (1950).

I. Kaplan, *Nuclear Physics*, Addison-Wesley, Reading, Mass. (1955).

M.A. Preston, *Physics of the Nucleus*, Addison-Wesley, Reading, Mass. (1962).

ELEMENTARY PARTICLE PHYSICS

H. Frauenfelder and E.M. Henley, *Subatomic Physics*, Prentice-Hall, Englewood Cliffs, N.J. (1974).

D.H. Perkins, *Introduction to High Energy Physics*, Addison-Wesley, Reading, Mass. (1972).

E. Segre, *Nuclei and Particles*, (2nd Edition) W.A. Benjamin, Menlo Park, Calif. (1977).

Recent developments in this field, as in others, are covered by articles in *Scientific American.*

Answers to
Selected Problems

Answers to
Selected Problems

1.2 3.3 MeV; 6.6 MeV.

1.3 $T = \dfrac{2L}{c} \left(\dfrac{1 - (v \sin \theta /c)^2}{1 - (v/c)^2} \right)^{\frac{1}{2}}$

1.6 $\theta = \theta * / 2$

2.1 28 cm; 9.7×10^{-10} sec; 34.7×10^{-10} sec

2.3 $\gamma = 105$

2.5 53 turns

2.7 $v_{\text{rel}}/c = 35/37$

2.10 0.8×10^{15} Hz

3.2 $a'_x = \dfrac{a_x}{\gamma^3 (1 - v u_x/c^2)^3}$

3.4 $u' = c/n + w(1 - n^2)/n^2$

3.5 In frame of clock $x = 0$, $t = n/v$.
In frame of clock intercepted signals occur at $x' = 0, t' = n/\gamma v$.

5.2 0.53×10^{-5} dynes/cm^2

5.4 $m_\gamma c^2 < 2.46 \times 10^{-5}$ eV

5.6 5.0058×10^{-24} gm

5.8 0.175 gm

5.10 930 gm

5.12 With $\mathbf{L} = \mathbf{r} \times \mathbf{p}$ and $\mathbf{K} = \mathbf{p}t - \mathbf{r}E/c^2$ we have
$\mathbf{L}' = \gamma(\mathbf{L} - \mathbf{v} \times \mathbf{K})$
$\mathbf{K}' = \gamma(\mathbf{K} + \mathbf{v} \times \mathbf{L}/c^2)$.

6.1 1.8×10^7 tons

6.3 3760 MeV

6.5 (i) 898 MeV (ii) 1500 MeV

6.7   930 GeV

6.9   For $E \gg Mc^2$ and $\theta < 0.3$ $W = 2E(1 - \theta^2/8)$.

7.1   If the distance of nearest approach of the observer to the charge is b, then

$$E'_x = + \frac{\gamma Q v t}{(\gamma^2 v^2 t^2 + b^2)^{3/2}} \; ; E'_y = - \frac{\gamma \bar{y} Q}{(\gamma^2 v^2 t^2 + b^2)^{3/2}} \; ; E'_z = - \frac{\gamma \bar{z} Q}{(\gamma^2 v^2 t^2 + b^2)^{3/2}}$$

when the charge Q is located at $(0, \bar{y}, \bar{z}\,)$.

7.3 $j'_x = \gamma(j_x - v\rho); \; j'_y = j_y; \; j'_z = j_z; \; \rho' = \gamma(\rho - vj_x/c^2)$

8.2 $(\Delta v/v)_{\text{grav}} = 0.296; \; (\Delta v/v)_{\text{rot}} = 0.33 \times 10^{-2}$

9.2 3.85×10^{-7}

9.4 $P = 0.57$

9.5 $P(5000) = 0.8 \times 10^{-2}; \; P(5400) = 1.0 \times 10^{-16}; \; P(6000) = 10^{-89}$

9.7 $P = 0.52$

9.11 $P_0 = 0.79; \; P_1 = 0.19; \; P_2 = 0.02$

10.1   22.7 liters; 1.27×10^{-3} gm/cm^3

10.3 1.7×10^5 cm/sec for H_2; 0.45×10^5 cm/sec for N_2

10.5 $U = 0.46 \times 10^{-4}$ erg/cm^3; 1.53×10^{-5} dynes/cm^2; 0.77×10^{-5} dynes/cm^2

10.7   0.49 cm/sec

11.2 $P = 5.21 \times 10^{-5}$

11.4 $x = 2.75; \; y = -0.27$

11.6 $\langle E^2 \rangle = \dfrac{1}{Z} \dfrac{\partial^2 Z}{\partial \beta^2}; \quad \langle E^2 \rangle - \langle E \rangle^2 = \dfrac{1}{Z} \dfrac{\partial^2 Z}{\partial \beta^2} - \left(\dfrac{1}{Z} \dfrac{\partial Z}{\partial \beta} \right)^2$

$$= \frac{\partial}{\partial \beta} \left(\frac{1}{Z} \frac{\partial Z}{\partial \beta} \right)$$

11.8 $\langle E \rangle = \left(\dfrac{3}{2} + \dfrac{3}{n} \right) kT$

12.1 1.06×10^{44} molecules

12.4 $\langle L \rangle = Na \tanh (mga/2kT)$

12.6 $\Delta v = 10^{11}$ Hz

12.9 $\sqrt{\langle r^2 \rangle} = 1.1 \times 10^{-4} (T/273)^{1/2} \sqrt{t(\text{sec})}$

12.10 $k = 1.373 \times 10^{-16}$ erg/deg

13.2 $T = 3000 \, (\theta/2)^{1/2}$ deg

13.4   Flux $= 1.88 \times 10^8$ erg/cm^2 sec

13.8 $W = 1.8$ eV

13.9 1.4×10^8 atoms/photoelectron

13.12 $\theta = 88.3°$

14.1 (i) 0.2×10^{-8} cm; (ii) 0.2 cm; (iii) 200 cm; (iv) 2×10^{14} km

14.4 $N = 5.3 \times 10^{20}$; $\lambda_0 = 78$ Å

14.5 $\omega_0 = 3.71 \times 10^{16}$; $\lambda_0 = 510$ Å

15.1 $a_0 = \hbar^2(m_1 + m_2)/Gm_1^2 m_2^2$

15.3 $E_1 = -12.5$ KeV; $E_2 = -3.13$ KeV; $a_{op} = 4.9 \times 10^{-4}$ Å

15.5 $E = \dfrac{k+z}{zk} \left(\dfrac{n^2\hbar^2}{ma^2}\right)\left(\dfrac{n^2\hbar^2}{k\,m\,V_0 a^2}\right)^{-2/(k+z)}$

The potential approaches the shape of an infinite box as $k \to \infty$.

15.7 $Z = 137\sqrt{2}$.

16.1 (a) $\lambda = 12$ Å; (b) $\lambda = 0.39$ Å; (c) $\lambda = 390$ f; (d) $\lambda = 1.8$ Å
(e) $\lambda = 8.94$ f; (f) $\lambda = 1800$ f $= 1.8 \times 10^{-2}$ Å

16.2 $E = 6.36\ mc^2$

16.5 $\psi(x) = \pi e^{-\alpha|x|}/2\alpha$
$\Delta x \approx 2/\alpha$

16.7 $v_g/c = (v_0^2\lambda^2/c^2 + 1)^{-\frac{1}{2}}$

16.9 $v_s = \dfrac{1}{2}(g\lambda/2\pi)^{\frac{1}{2}}$; $v_p = 2v_g$

17.1 The equations are
$R\partial S/\partial t = (\hbar/2m)(\partial^2 R/\partial x^2 - R(\partial S/\partial x)^2)$
$\partial R/\partial t = -(\hbar/2m)(2(\partial R/\partial x)(\partial S/\partial x) + R\partial^2 S/\partial x^2)$
with the last equivalent to the conservation equation
$\partial R^2/\partial t + \partial/\partial x\,((\hbar/m)R^2\partial S/\partial x) = 0$.

17.3 $\int_{-\infty}^{\infty} dx\,W(x)|\psi(x)|^2 < 0$

17.5 $P(x,t) = -\dfrac{i}{c^2}\left(\psi^*\dfrac{\partial\psi}{\partial t} - \psi\dfrac{\partial\psi^*}{\partial t}\right)$

18.1 $E_{min} = \dfrac{3}{2}\dfrac{\hbar^{2/3}}{m^{1/3}}\left(\dfrac{V_0}{a}\right)^{2/3}$

18.3 $W_{min} = 2(2\hbar L/p)^{1/2}$

18.5 $\hbar/mc$; 200 MeV

18.7 (a) 0.25×10^{-5} eV; (b) 0.3×10^{-9} eV; (c) 9.3×10^{-19} eV.

18.9 $\Delta\lambda = 1.2 \times 10^{-2}$ Å.

19.2 (a) $<x> = b/2$; $<x^2> = (1 - 3/2\,n^2\pi^2)b^2/3$; $(\Delta x)^2 = b^2\left(1 - \dfrac{6}{n^2\pi^2}\right)/12$
(b) $<p> = 0$; $<p^2> = n^2\pi^2\hbar^2/b^2$; $(\Delta p)^2 = n^2\pi^2\hbar^2/b^2$

(c) $\Delta p \, \Delta x = n\hbar\pi(1 - 6/n^2\pi^2)^{1/2}/\sqrt{12}$

19.4 $\psi(x) = Ae^{x^2/2\lambda}; \ \lambda < 0.$

19.6   (b), (c), (d), (f) are linear operators.

19.8   (a) $<p> = 0; \ <p^2> = \hbar^2\alpha$

(b) $<p> = \hbar b; \ <p^2> = \hbar^2(a + b^2)$

(c) $<p> = \hbar \int dx \, R^2 \, dS/dx; \ <p^2> = \hbar^2 \int dx \left(\left(\frac{dR}{dx} \right)^2 + R^2 \left(\frac{dS}{dx} \right)^2 \right)$

20.3 $<H> = \sum_n E_n |a_n|^2$

20.6 $\lambda(n + 1 \to n) = \dfrac{325}{n+1} \text{Å}$

21.1   (a) $r = (k-q)^2/(k+q)^2$ and $t = 1 - r$, where

$k^2 = 2mE/\hbar^2 , q^2 = 2m(E - V_0)/\hbar^2$

(c) $r' = (q - k)^2/(q + k)^2$

21.5   Flux conservation leads to $|A|^2 - |B|^2 = |C|^2 - |D|^2$ from which the results follow.

21.8 $C = (S_{11}S_{22} - S_{12}S_{21})A/S_{22} + S_{12}B/S_{22}$

$D = -S_{21}A/S_{22} + B/S_{22}$

22.1   (a) $|T|^2 = \exp(-G), G = \dfrac{8}{3}\left(\dfrac{2mVa^2}{\hbar^2} \right)^{1/2} (1 - E/V)^{3/2}$

(b) $|T|^2 = \exp\left[-\dfrac{4}{3}\sqrt{\dfrac{mVa^2}{\hbar^2}} \right]$

(c) $|T|^2 = 0.47$

22.3   Eq. (22.10) gives 0.76, while the $\pi/2$ approximation yields 1.57. The calculated value is $\log_{10}\tau = 12.14$, compared with the experimental value 10.7.

22.6 $t = \left[1 + \left(\dfrac{m\lambda}{k}\dfrac{1}{\hbar^2} \right)^2 \right]^{-1}$

23.2 $E = m\lambda^2/2\hbar^2$; the wave function is symmetric about the $x = 0$ line, with an exponential fall off on each side, and a discontinuity in the derivative at $x = 0$.

23.4   The argument fails because the assumption that the wave function is localized inside the potential is false.

23.6 $E = W(mWa^2/2\hbar^2)$

23.9 $E_B = \hbar^2 f_0^2/2m; \ R = e^{-2ika}\dfrac{f_0 + ik}{f_0 - ik}$

24.1 $E_n = \hbar\omega(n + \frac{1}{2}) - e^2E^2/2m\omega^2$ where $k = m\omega^2$

$u_0(x) = C \exp - [m\omega(x - eE/k)^2/2\hbar]$

24.3   Taking the lowest power coefficient to be unity, we get

$1, y, 1 - 2y^2 , y - 2y^3/3.$

24.6 $\beta/2$, $3\beta/2$, $7\beta/2$, where $\beta = (\hbar/m\omega)^2$

Using $p^2 = 2m(H - m\omega^2 x^2/2)$, we get what we need.

24.7 $\sqrt{(M+1)/2}$

25.3 $(2/L) \sin \dfrac{\pi x_1}{L} \sin \dfrac{\pi x_2}{L}$

$\dfrac{\sqrt{2}}{L} \left(\sin \dfrac{\pi x_1}{L} \sin \dfrac{2\pi x_2}{L} - \sin \dfrac{2\pi x_1}{L} \sin \dfrac{\pi x_2}{L} \right)$

25.5 $2E_1 = 2(\hbar^2 \pi^2/2mL^2) = 2\mathscr{E}$

$2E_1 + E_2 = 6\mathscr{E}$

$2E_1 + 2E_2 + 2E_3 + E_4 = 44\mathscr{E}$

$\dfrac{1}{4} N^3 \mathscr{E}$ $(N \gg 1)$

25.7 $\psi(x_1 \cdots x_N) = \dfrac{1}{\sqrt{N}} (u_1(x_1) \cdots u_1(x_{N-1}) u_2(x_N) + u_1(x_1) \cdots$

$+ u_2(x_{N-1}) u_1(x_N) + \cdots + u_2(x_1) u_1(x_2) \cdots u_1(x_N))$

25.8 (a) $m_r = m_e (1 + m_e/M)^{-1} \cong m_e (1 - 5 \times 10^{-4})$

(b) $m_r = m_e/2$

26.1 The energy eigenvalues are

$$E = \frac{\hbar^2 \pi^2}{2m} \left(\frac{n_1^2}{L_1^2} + \frac{n_2^2}{L_2^2} + \frac{n_3^2}{L_3^2} \right) .$$

The energy surface is

$$\frac{n_1^2}{L_1^2} + \frac{n_2^2}{L_2^2} + \frac{n_3^2}{L_3^2} = \frac{2mE}{\hbar^2 \pi^2} .$$

The number of states is

$$\frac{1}{3} \pi \left(\frac{2mE}{\hbar^2 \pi^2} \right)^{3/2} L_1 L_2 L_3 .$$

26.3 The result follows from $P \propto V^{-5/3}$. The numbers in units of 10^{10} dynes/cm^2 are 5.5, 1.9, 38.3, and 135.8.

26.5 $E_F^p = 55.1 (Z/A)^{2/3}$ MeV; $E_F^n = 55.1 (1 - Z/A)^{2/3}$

$E_F^p = 29.6$ MeV; $E_F^n = 39.4$ MeV

26.8 2.8×10^4 km

27.3 $[L_x, p_x] = 0$, etc.

$[L_x, p_y] = i\hbar p_z$ (cycl.)

$[L_y, p_x] = -i\hbar p_z$ (cycl.)

28.2 The differential equation is

$$(1-u^2)\frac{d^2P}{du^2} - 2(m+1)\frac{dP}{du} + (\lambda - m^2 - m)P = 0.$$

The recursion relation is

$$a_{r+2} = \frac{r(r-1) + 2r(m+1) + m^2 + m - \lambda}{(r+1)(r+2)} a_r .$$

28.4 The eigenvalues are $E = \hbar^2 m^2/2I \quad m = 0, \pm 1, \pm 2, \ldots.$

The eigenfunctions are $e^{im\phi}/\sqrt{2\pi}$.

The degeneracy is associated with the fact that the eigenfunctions are also eigenfunctions of L_z, and it is two-fold since the rotation can be clockwise or counterclockwise in the x-y plane.

28.6 The eigenfunctions are $e^{iNm\phi}/\sqrt{2\pi}$.

The eigenvalues are $E = \hbar^2 N^2 m^2/2I.$

With a gap, there is no symmetry, and the solution is the same as for Problem 28.4.

29.1 (a) $<E> = -\frac{1}{2}mc^2\alpha^2 \ (0.35)$

(b) $<L^2> = 0.8\,\hbar^2$

(c) $<L_z> = 0.3\,\hbar$

29.2 $\Delta p_x\,\Delta x = \frac{1}{3}\,\hbar \left(\frac{5n^2 + 1 - 3l(l+1)}{2}\right)^{1/2}$

29.4 $\dfrac{(\nu\alpha)_H}{(\nu\alpha)_D} \cong 1 - \dfrac{m_e}{2m_H}$

30.1 $\chi_+ = \frac{1}{\sqrt{2}}\begin{pmatrix}1\\1\end{pmatrix}; \ \chi_- = \frac{1}{\sqrt{2}}\begin{pmatrix}1\\-1\end{pmatrix}$

30.5 $<S_x> = \frac{1}{2}\hbar\cos\alpha; \ <S_x> = -\frac{1}{2}\hbar\cos\alpha$

30.8 (i) $J = 5, 4, 3, 2, 1$ multipl. $= 35$
(ii) $J = 5/2, 3/2$ multipl. $= 10$
(iii) $J = 1, 0$ multipl. $= 4$
(iv) $J = 5/2, 3/2, 1/2$ multipl. $= 12$

30.10 (i) $S = 0 \quad {}^1S_0, {}^1P_1, {}^1D_2, {}^1F_3$

$S = 1 \quad {}^3S_1, {}^3P_{2,1,0}, {}^3D_{3,2,1}, {}^3F_{4,3,2}$

(ii) ${}^1S_0, {}^3P_{2,1,0}, {}^1D_2, {}^3F_{4,3,2}$

(iii) ${}^3P_1$

31.1 8.72 cm

31.3 $\quad E = \left(2n_r + l + \dfrac{3}{2}\right)\hbar\omega + \dfrac{1}{4}A\,m\,\hbar^2\omega_0{}^2 \quad \begin{cases} l & j = l + \frac{1}{2} \\ -l-1 & j = l - \frac{1}{2} \end{cases}$

31.5 $\quad 2.5 \times 10^5$ gauss

31.6 $\quad 6.2 \times 10^{-2}\ {}^\circ K$

32.1 $\quad {}^2P_{3/2};\ {}^3F_2;\ {}^4F_{3/2};\ {}^7S_3$

32.4 $\quad Z = 48$

32.5 $\quad (1s)^2(2s)^2(2p)^6(3p);\ {}^2P_{3/2}$ or ${}^2P_{1/2}$; with $Z_{eff} = 1.84$

$\quad\quad \Delta E = \dfrac{1}{2}mc^2\,\alpha^2\,(1.13 \times 10^{-5})$

33.1 $\quad -3.85$ eV

33.2 $\quad 2.2$ cm

33.4 $\quad 110^\circ K$

33.7 $\quad 2.4 \times 10^{-3}\,\text{Å};\ 3 \times 10^2$ cm

34.1 $\quad$ Integral in (34.18) is $\propto e^{(-\omega T)^2/4}$

34.3 $\quad (kR)^2 = 16.5 \times 10^{-4}\,(\Delta E_{MeV})^2$

34.5 $\quad 5.9 \times 10^9$ years

34.8 $\quad 13.6;\ 12.9.$

35.3 $\quad$ Diameter of spot is 8.54 cm; 4.9×10^{-3} J/ cm² sec.

35.5 $\quad 3.3 \times 10^{-2}$ w; 3.5×10^{-4} dynes/cm²; 10^8 photons/cm³; 3.2×10^7 sec⁻¹

36.1 $\quad \omega_{max} = 3.83 \times 10^{13}$ sec⁻¹ ; $\theta_D = 290^\circ K$

36.3 $\quad U = \dfrac{k^2 T^2}{\hbar^2 \pi v_s^2} \displaystyle\int_0^{\theta_D/T} dy\,\dfrac{y}{e^y - 1}$

37.1 $\quad$ In Å: 9.1, 33.0, 27.8, 11.3, 2.5, 8.2.

37.2 $\quad$ In Å: 114, 342, 424, 174, 48, 162.

37.3 $\quad 0.7 \times 10^{-2}$ cm/sec

38.4 $\quad -0.78 \times 10^{-24};\ 1.11 \times 10^{-24}$ in c.g.s. units

38.6 $\quad 285\,\text{Å};\ 221\,\text{Å}$

39.4 $\quad 21$ MeV; -37 MeV

39.5 $\quad Z^2/A > 17.1$

39.7 $\quad 240$ MeV; 6580 MeV; 5080 MeV

40.4 $\quad S = 4\,\pi\,R_0^2\,(1 + 2a^2/5)$

41.2 $\quad 20$ cm

41.4 $6 \times 10^{-8}/Z^2$ cm; Capture becomes highly probable since the radius of the orbit is 2×10^{-11} cm.

41.5 340 m

41.7 19.4 GeV; 138 pions

42.1 0.26 f

42.3 $I = 0$ or 2 ; $I = 0$

42.6 0.4×10^{-23} sec

42.8 $M > 2m_\pi$; J is even; parity is even; $I \geqslant 2$; $S = 0$; at least four partners; process must be electromagnetic.

43.3 An $I = 1/2$, $Y = -1$ state with Mc^2 1990 MeV

43.5 $\alpha = 1.27$; $m_c c^2 = 1950$ MeV. Spin-orbit and spin-spin forces.

Index